I0486478

Steel Structures: Roof Members Design and Detailing

SECOND EDITION

Saad Hasan Tantawi
M.S.CE, B.S.CE, EI, A.M.ASCE

ISBN: 1387872753
ISBN-13: 978-1387-872756

DEDICATION

To my Mom and Dad, Dr. Hasan Tantawi and Maram Tantawi. It is impossible to thank you adequately for everything you have done, from loving me unconditionally to raising me always believing in myself. I could not have asked for better parents or role-models.

And to Anwar, Noor, Diya, Lara, Dr. Sulieman, Khalid, Omar, Abdul Rahman, thanks for all of the wonderful memories of growing up, and for your continued support and encouragement.

A special thank you to Dr. Burdette, and to Dr. Ma who taught me everything I needed to know about structural engineering.

A special thanks goes also to Radhia Ladj for being a great mentor, a life advisor, and a best friend.

CONTENTS	PAGE NO.

PREFACE

The Objective of this series is to guide newly graduate engineers, and engineering students in the process of designing, planning, and calculating loads for steel framed buildings in accordance with the American Institute of Steel Construction (AISC Version 14.0), and the Minimum Design Loads for Buildings and Other Structures in accordance with ASCE 7-10. The building in this book is an imaginary building and it is believed that the material presented is comprehensive enough to serve as a guide for a variety of structural engineering courses as well as for self-study.

This series incorporates both calculations as well as a step by step design check calculations using structural engineering software programs such as RISA, SAP, and REVIT Structures. In addition, graphics using AutoCAD 2017, and REVIT 2017 is implemented throughout this book.

It would not be possible to produce this book without the assistance of many individuals and organizations. I would like to also express my appreciation to my family, colleagues and friends for their helpful comments and suggestions.

Comments from readers regarding errors and any suggestions for further improvements are solicited. Please send to: saad.tantawi@ibcstructures.com

Saad Hasan Tantawi

1 INTRODUCTION

1-1 Steel Frames Components

Steel frames are the most commonly used framing system here in the U.S. for commercial buildings, and industrial buildings. Typically, a steel frame would consist of columns, girders, beams, roof and floor decks, joists, and connections. Columns are used as compression members in buildings; as such they would be used to transmit the weight of the building to the foundations. In addition to gravity loads columns must be designed to resist lateral loads such as wind loads and earthquake loads imposed on a building. Girders are beams used to support other beams; they are the main horizontal component in a structural frame (See Figure 1-1), generally in steel frames girders would have I-beams cross section, in some but few instances HSS (boxed shape) cross section would be used.

Figure 1-1 Steel Frame for A Commercial Pre-Engineered Metal Building

1-2   Steel Classifications According to ASTM

The American Society for Testing and Materials (ASTM) had specified structural steel into four categories:

Carbon Steels
- A36 - structural shapes and plates.
- A53 - structural pipes.
- A500 - structural pipes.
- A501 - structural pipes.
- A529 - structural shapes and plates.

High Strength Low Alloy Steels
- A441 - structural shapes and plates.
- A572 - structural shapes and plates.
- A618 - structural pipe.
- A992 – Possible for W or S I-Beams.
- A913 - Quenched and Self Tempered W shapes.
- A270 - structural shapes and plates.

Corrosion Resistant High Strength Low Alloy Steels
- A242 - structural shapes and plates.
- A588 - structural shapes and plates.

Quenched and tempered alloy steels
- A514 - structural shapes and plates.
- A517 - boilers and pressure vessels.

These steels in the four categories listed above have an alloy identification beginning with A and two or three numbers that signifies the AISI steel grades. Typically angles and stitch plates used in steel construction are made of carbon steels (ASTM A36); while W-shape members are made of high strength low allow steels (ASTM A992).

1-3   Codes And Regulations

Each Country has its own standard codes and specification for steel construction. The one code used for structural steel construction in the United States is the AISC Code of Standard Practice for Steel Buildings and Bridges. Since our building of choice will be designed in the United States, then the AISC code Version 14.0 will be the code that will be used throughout this book. In addition to AISC code, sections from ASCE 7-10 Minimum Design Loads for Building and Other Structures will be used to determine snow, wind, and live loads, and appropriate load combinations.

1-4   References

1.   *Manual of Steel construction,* 8th ed. American Institute of Steel Construction, Inc., Chicago, IL, 1987.

2.   Oppenheimer, Samuel P. *Erecting Structural Steel.* New York: McGraw-Hill, 1960.

2 DESIGN EXAMPLE OF ROOF MEMBERS IN A BUILDING

2-1 Building Geometry and Location

The building is a three-story office building with four bay at 40 ft in the East-West direction and three bays at 30 ft. in the North-South direction (Figure 2-1). The floor-to-floor height for the three floors is 14 ft, and the height from the third floor to the roof is 15 ft.

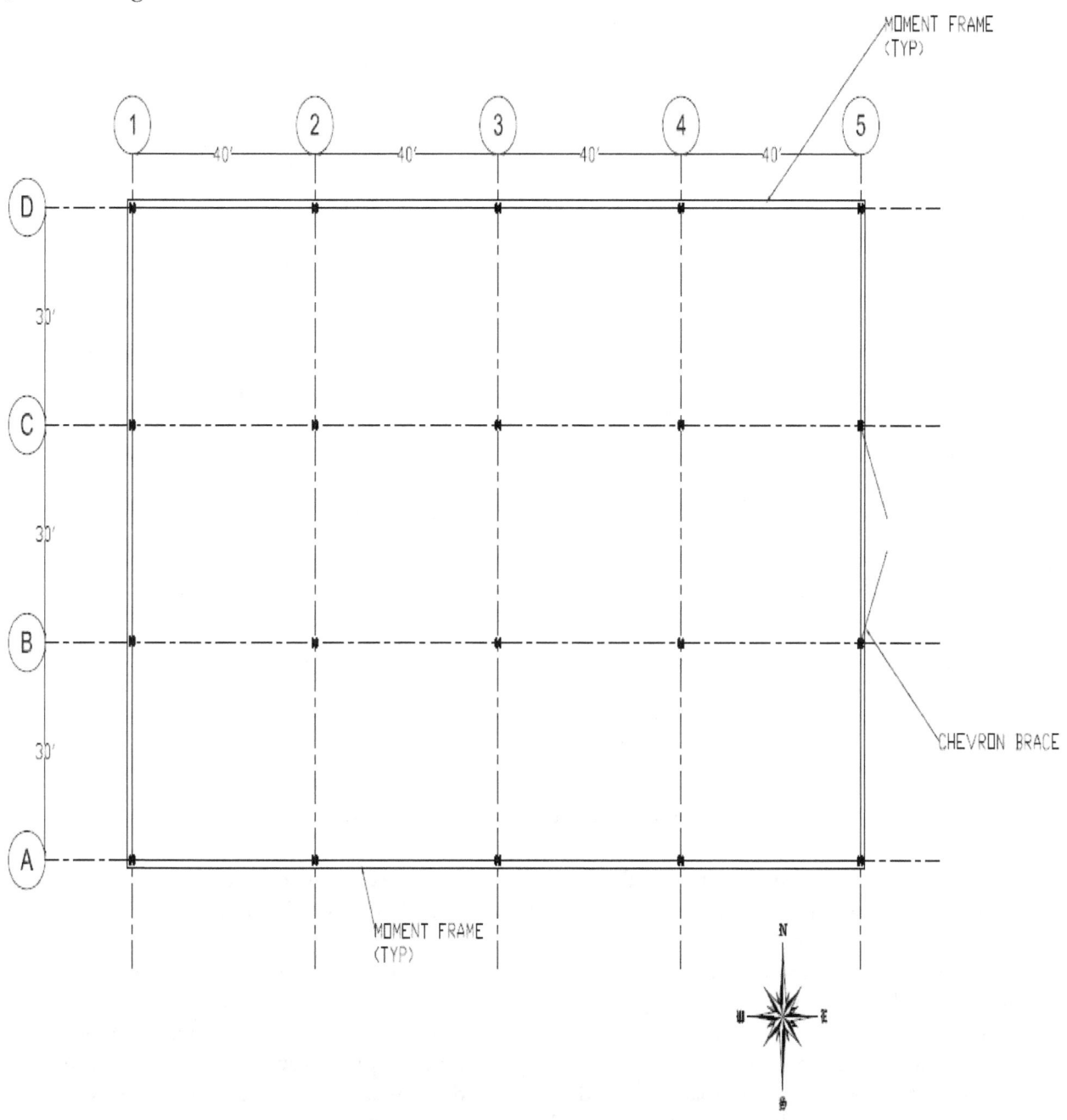

Figure 2-1 Building layout plan

Based on discussions with the architect, the exterior of the building will be a spandrel ribbon wall system, back braced with steel, and in filled with 4 in. metal studs. The wall extends 3 ft. above the elevation of the edge of the roof.

The roof system consists of 1.5 in. wide rib(type B) gauge 22 painted metal deck, placed in a 36/7 pattern (see Figure 2-2) with the deck been continuous over three span, and with joists spaced at 6ft O.C. The roof loads extends 6 in. past the grid centerline. With a span of 6 ft. Center to Center of support, the allowable total (Dead + Live) uniform load for the deck is 89 psf.

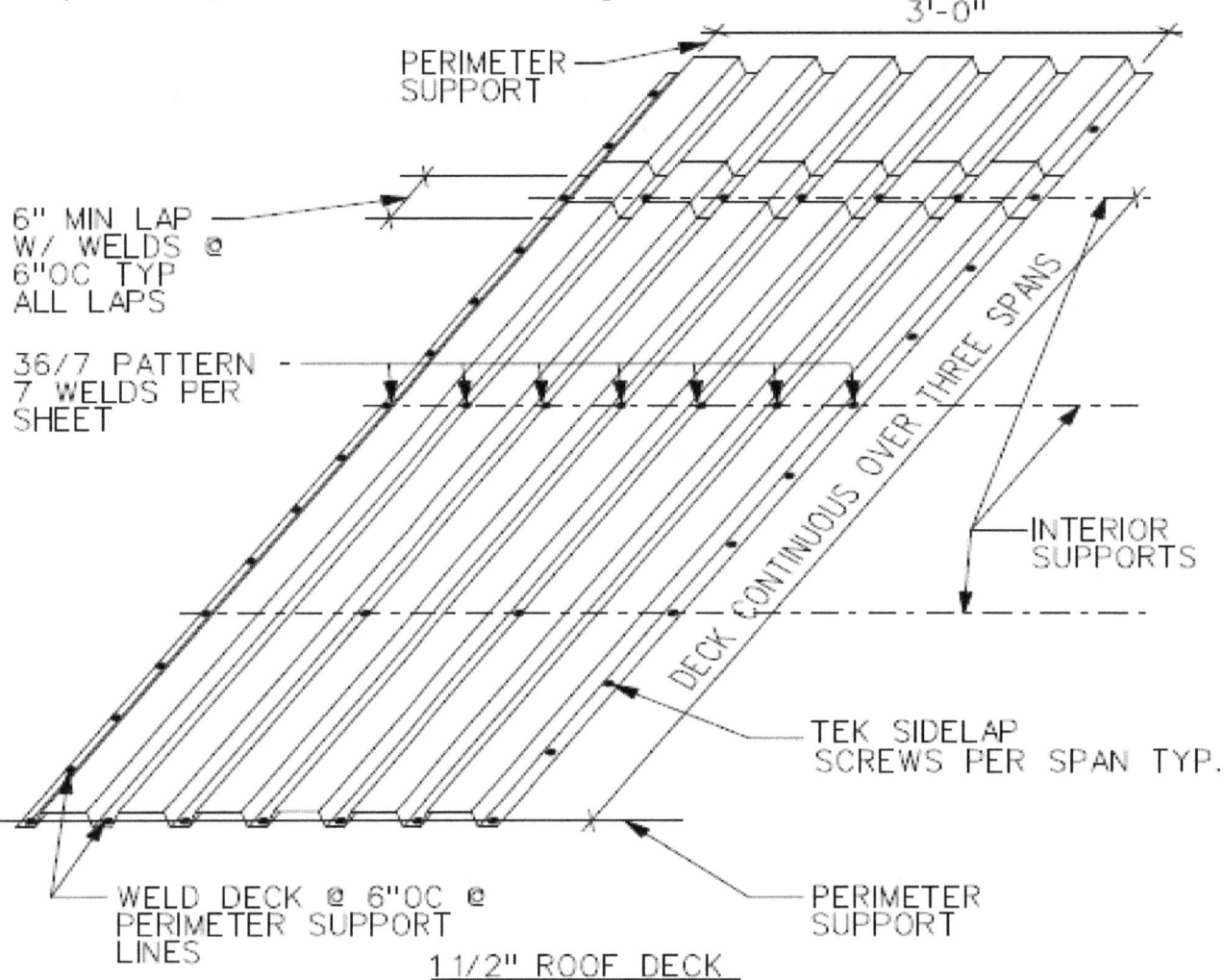

Figure 2-2 Typical 1.5 in. wide rib deck

The middle two bays have a 6 ft tall screen wall around them and house the mechanical equipment and the elevator over run. This will produce extra gravity load, as such steel beams will be placed in the middle two bays instead of steel bar joists.

The two elevated floors have a total slab thickness of 6 in. (3in. of normal weight concrete over 3-in. composite deck). The first floor is a slab on grade, and the foundation consists of spread footings. As shown in figure 2-1 the building will have one chevron brace in the North-South direction, which will be concealed and adjacent to the stairway. Since the building will be equipped throughout with an automatic sprinkler system, and has large open spaces around it (provided the area of the floor opening between stories does not exceed twice the horizontal projected area of the escalator or stairway), consequently fireproofing for the floors is not required.

2-2   Determining Gravity And Lateral Loads

Live Loads:

Table 2.1 Minimum Uniformly Distributed Live Loads, L$_O$

	Minimum Live Load (L$_O$)	ASCE 7-10 Section or table number
Roof	20 psf Reduction in roof life load is permitted	Table 4-1 Section 4.8.1 and 4.8.2
1st Floor	Offices = 50 psf Corridors = 100 psf Stairs ways= 100 psf Partition load = 15 psf (not required where the minimum specified live load exceeds 80 psf) Reduction in uniform live load is permitted (Since the locations of partitions and, consequently, corridors will be subjected to change, the entire floor will be designed for a live load of 80 psf.)	Table 4-1 Section 4.3.2 Section 4.7.1 and 4.7.2
2nd Floor	Offices = 50 psf Corridors = 80 psf Stairs ways= 100 psf Partition load = 15 psf (not required where the minimum specified live load exceeds 80 psf) Reduction in uniform live load is permitted (Since the locations of partitions and, consequently, corridors will be subjected to change, the entire floor will be designed for a live load of 80 psf)	Table 4-1 Section 4.3.2 Section 4.7.1 and 4.7.2

Snow Load Parameters: Building location: Knoxville, TN.
- Ground Snow Load (p_g) = 10 psf (ASCE 7-10 Figure 7-1)
- Snow Importance Factor (I_s) = 1.00 (ASCE 7-10 Table 1.5-2)
- Exposure Factor (C_e) = 1.0 (ASCE 7-10 Table 7-2, for partially exposed roof, note that the roof may be classified as fully exposed roof, but this approach will give a smaller flat roof snow load which is less conservative)
- Thermal Factor (C_t) = 1.0 (ASCE 7-10 Table 7-3)
- Flat Roof Snow Load (p_f) = 7 psf (ASCE 7-10 Eq. 7.3-1: $p_f = 0.7*Ce*Ct*Is*p_g$)
- Roof Slope = 0.4 in./ft (slope angle = 1.9°)
- Roof Slope Factor (C_s) = 1.0 (ASCE 7-10 Figure 7-2a)
- Sloped Roof (Balanced) Snow Load (p_s) = 7 psf (ASCE 7-10 Eq. 7.4-1: $p_s = C_s*p_f$)
- Minimum Snow Load (p_m) = 10 psf (ASCE 7-10 Section 7.3.4: $p_m = I_s * p_g$)
- Rain-on-Snow Surcharge Load: Since the roof slope = 1.9° is less than W/50 = (160/50), then per ASCE 7-10 Section 7.10, a 5 psf rain-on-snow surcharge load should be included
- Ponding Instability: Per ASCE 7-10 Section 7.11, since the roof slope is > ¼ in./ft. ponding instability design calculations are not required
- Snow drifts shall be considered and it will be calculated for the roof in Chapter 3

Wind Load Parameters:

- ▪ Risk Category: II   (ASCE 7-10 Table 1.5-1 and Section C1.5.1)
- ▪ Basic Wind Speed = 115 mph (3 second gust) (ASCE 7-10 Figure 26.5.1-A)
- ▪ Exposure Category = C (ASCE 7-10 Section 26.7.3, there is open terrain with scattered obstructions having heights less than 30 ft. at an upwind distance of 2000 ft. from the building)
- ▪ Wind Directionality Factor (K_d) = 0.85 (ASCE 7-10 Table 26.6-1)
- ▪ Topographic Factor (K_{zt}) = 1.0 (there is no isolated hills, ridges, and escarpments) based on ASCE 7-10 section 26.8.2, the topographic factor K_{zt}=1.00
- ▪ Enclosure Classification: Enclosed Building (ASCE 7-10, section 26.2)
- ▪ Mean Roof Height (height to eaves)= 43 ft
- ▪ Building Classification: Since the mean roof height h is less than 60 ft. then based on ASCE 7-10 Section 26.2 the building is classified as a low-rise building

Seismic Load Parameters: Based on geotechnical report:
- • Average shear wave velocity at small shear strains in top 100 ft = 800 ft/s
- • Average field standard penetration resistance for the top 100 ft = 25
- • Average un-drained shear strength in top 100 ft = 15 psf
- • Site class = D (stiff soil) based on ASCE 7-10 Table 20.3-1
- • Mapped MCE_R, 5 percent damped, spectral response acceleration parameter at short periods (S_s)= 0.415g (ASCE 7-10 Figure 22-1)
- • Mapped MCE_R, 5 percent damped, spectral response acceleration parameter at a period of 1 second (S_1) = 0.125g (ASCE 7-10 Figure 22-2)
- • Seismic Importance Factor (I_e) = 1.00 (ASCE 7-10 Table 1.5-2)
- • Short-Period Site Coefficient (F_a) = 1.468 (ASCE 7-10 Table 11.4-1 using linear interpolation)
- • Long-Period Site Coefficient (F_v) = 2.3 (ASCE 7-10 Table 11.4-2 using linear interpolation)
- • The MCE_R, 5 Percent Damped, Spectral Response Acceleration Parameter at Short Period Adjusted for Site Class Effect (S_{MS}) = 0.609g (ASCE 7-10 Equation 11.4-1: $S_{MS} = F_a * S_s$)
- • The MCE_R, 5 Percent Damped, Spectral Response Acceleration Parameter for Long Period Adjusted for Site Class Effects (S_{M1}) = 0.288g (ASCE 7-10 Equation 11.4-2: $S_{M1} = F_v * S_1$)
- • Design, 5 Percent Damped, Spectral Response Acceleration Parameter at Short Periods (S_{DS}) = 0.406g (ASCE 7-10 Equation 11.4-3: $S_{DS} = (2/3) * S_{MS}$)
- • Design, 5 Percent Damped, Spectral Response Acceleration Parameter at a Period of 1 Second (S_{D1}) = 0.192g (ASCE7-10 Equation 11.4-4: $S_{D1} = (2/3) * S_{M1}$)
- • Seismic Design Category Based on Short Period Response Parameter = C (ASCE 7-10 Table 11.6-1)
- • Seismic Design Category Based on 1-Second Period Response Acceleration Parameter = C (ASCE 7-10 Table 11.6-2)
- • Steel Ordinary Moment Frame is permitted with R (Response Modification Coefficient = 3.5 ASCE 7-10 Table 12.2.1)
- • Over-strength Factor (Ω_0) = 3 (ASCE 7-10 Table 12.2-1)

2-3   Gravity Loads Calculations

The roof members shall be designed so that their design strength equals or exceed the effects of the factored loads in the following combinations:

Table 2-2 Basic Load Combinations

Strength Design Basic Combinations Based on ASCE 7-10 Section 2.3.2	Allowable Stress Design Basic Combinations Based on ASCE 7-10 Section 2.4.1
1.4D	D
1.2D + 1.6L + 0.5(L_r or S or R)	D + L
1.2D + 1.6(L_r or S or R) + (L or 0.5W)	D + (L_r or S or R)
1.2D + 1.0W + L + 0.5(L_r or S or R)	D + 0.75L + 0.75(L_r or S or R)

1.2D + 1.0E + L + 0.2S	D + (0.6W or 0.7E)
0.9D + 1.0W	D + 0.75L + 0.75(0.6W) + 0.75(L$_r$ or S or R)
0.9D + 1.0E	D + 0.75L + 0.75(0.7E) + 0.75(S)
	0.6D + 0.6W
	0.6D + 0.7E

Roof Dead Load:

Roofing	= 5 psf
Insulation	= 2 psf
Deck	= 2 psf
Beams	= 3 psf
Joists	= 3 psf
Misc.	= 5 psf
Total	= 20 psf

Roof Live Load: L_O = 20 psf

Roof Snow Load + Rain on Snow Load = 15 psf. < L_O

From the load combinations in the previous table, either the roof live load or the roof snow load should be chosen with the roof dead load; however, to be conservative, roof members will be designed to resist dead load, roof live load, and the snow drift load.

Check the Snow Drift Load: Slopped roof snow load(p_s) = p_m = 10 psf.

p_g = 10 psf

Snow density(γ) = 0.13p_g + 14 = (0.13*10) + 14 = 15.3 pcf (ASCE 7-10 Equation 7.7-1)

h_b = height of balanced snow load determined by dividing, p_s by γ = 0.654 ft (ASCE 7-10 Section 7.1). See the figure below (ASCE Figure 7-8) that shows h_b, h_d, p_d, h_c in relation to the roof parapet height.

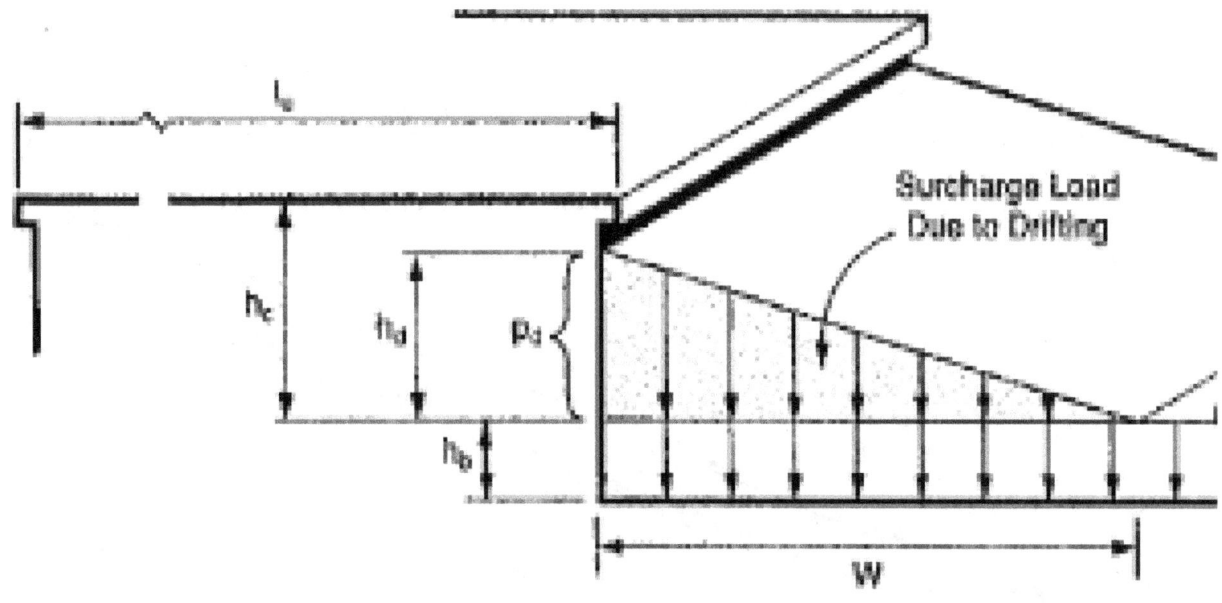

ASCE 7-10 Figure 7-8

Note: $h_d = 0.43 * (l_u^{0.333})*((p_g+10)^{0.25}) - 1.5$ (ASCE 7-10 Figure 7-9)

Table 2-3 Snow Drift Loads

	Upwind Roof	Proj.	Maximum Drift	Maximum Drift Width

	Length (l_u)	Height	Load	(W)
Side Parapet	90 ft + 6 in. + 6 in. = 91 ft. Note: the extra 12 in. is due the roof diaphragm and roof loads extend 6 in. past the centerline of grid in as shown on Sheet ?	2 ft	h_d = 0.43 *($91^{0.33}$)*((10 + 10)$^{0.25}$)-1.5 = 2.59 ft > projected parapet height. Thus, h_d = parapet height - h_b = 2ft – 0.654ft = 1.35 ft p_d = Maximum drift load = 1.35 ft * 15.3 pcf = 20.66 psf	W = smaller of {($4*h_d^2/h_c$), $8*h_c$)}. $4*h_d^2/h_c$ = 4*(2.59^2)/(1.346) = 19.93 ft. $8*h_c$ = 8*1.346 = 10.77 ft. Choose W = 10.77 ft
Side Parapet	160 ft + 6 in. + 6in. = 161 ft.	2 ft	20.66 psf	10.77 ft
Screen Wall (E-W Wind) and with an upwind fetch from the parapet to the centerline of the columns at the penthouse	40 ft + 6 in. = 40.5 ft (grids 2-3)	6 ft	h_d = 1.62 ft; however, based on ASCE 7-10 Section 7.7.1: for windward drifts, the drift height should equal to three-quarters of h_d. h_d = 0.75 * 1.62 ft = 1.22 ft maximum drift load = 1.22 ft * 15.3 pcf = 18.67 psf	W = 4 * 1.22 ft = 4.88 ft

2-4   Roof Joists Selection

Open web steel joist is a light weight steel truss consisting of parallel chords, and triangulated web system. The steel joist institute has a load tables and weight tables for steel joists and joists girder. You can download steel joists standard specification manual by going to www.steeljoist.org. Figure 2-3 shows a typical 12K1 joist.

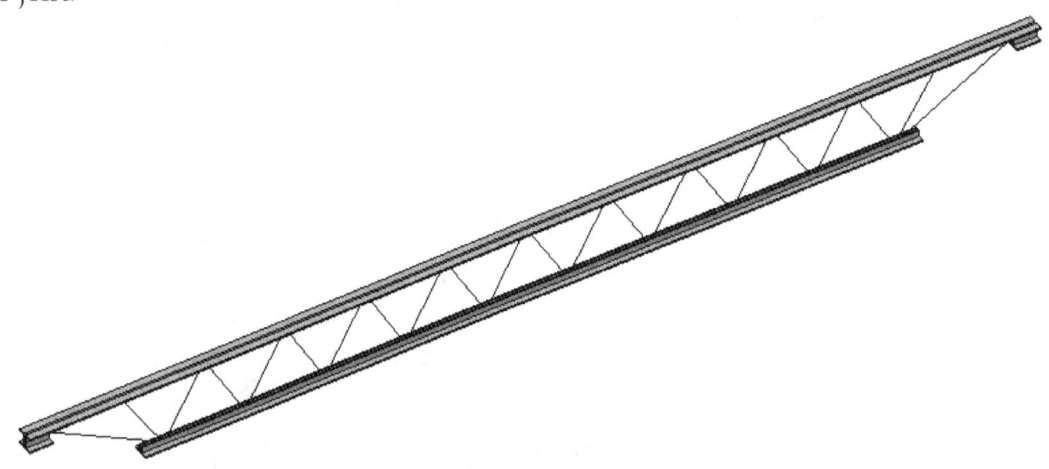

Figure 2-3 12K1 serious joist system

Joists may be designed using ASD or LRFD but are most commonly designed using the ASD method. In this design example we will use ASD method to design roof joists.

The side of the roof with the heaviest loads outside of the screen wall area is shown circulated in Figure 2-4:

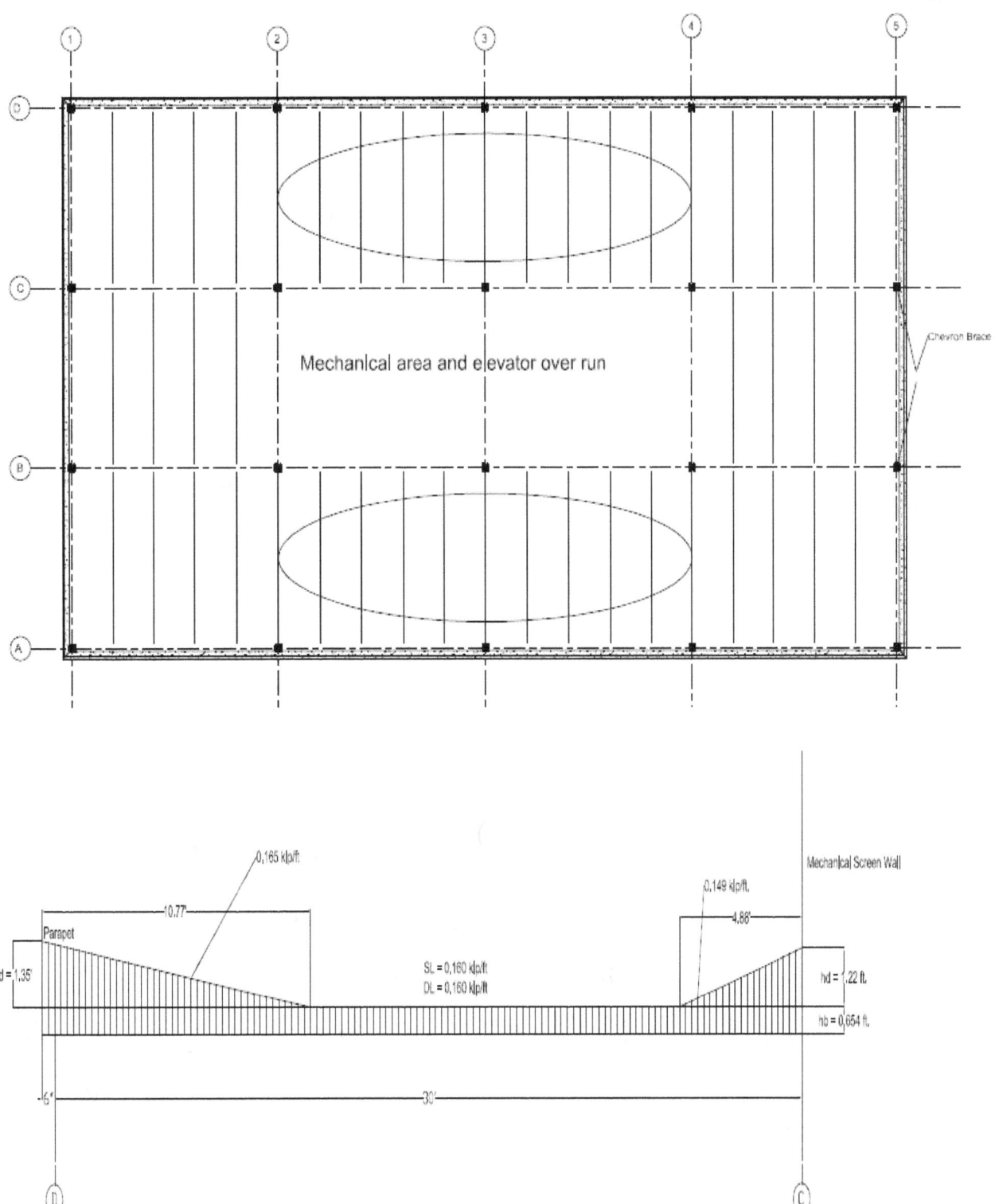

Figure 2-4 Joist loading & bracing diagram (continuous bracing)

With both supports assumed as pin supports, the shear and bending moment diagram for the roof joists is shown in Figure 2-5:

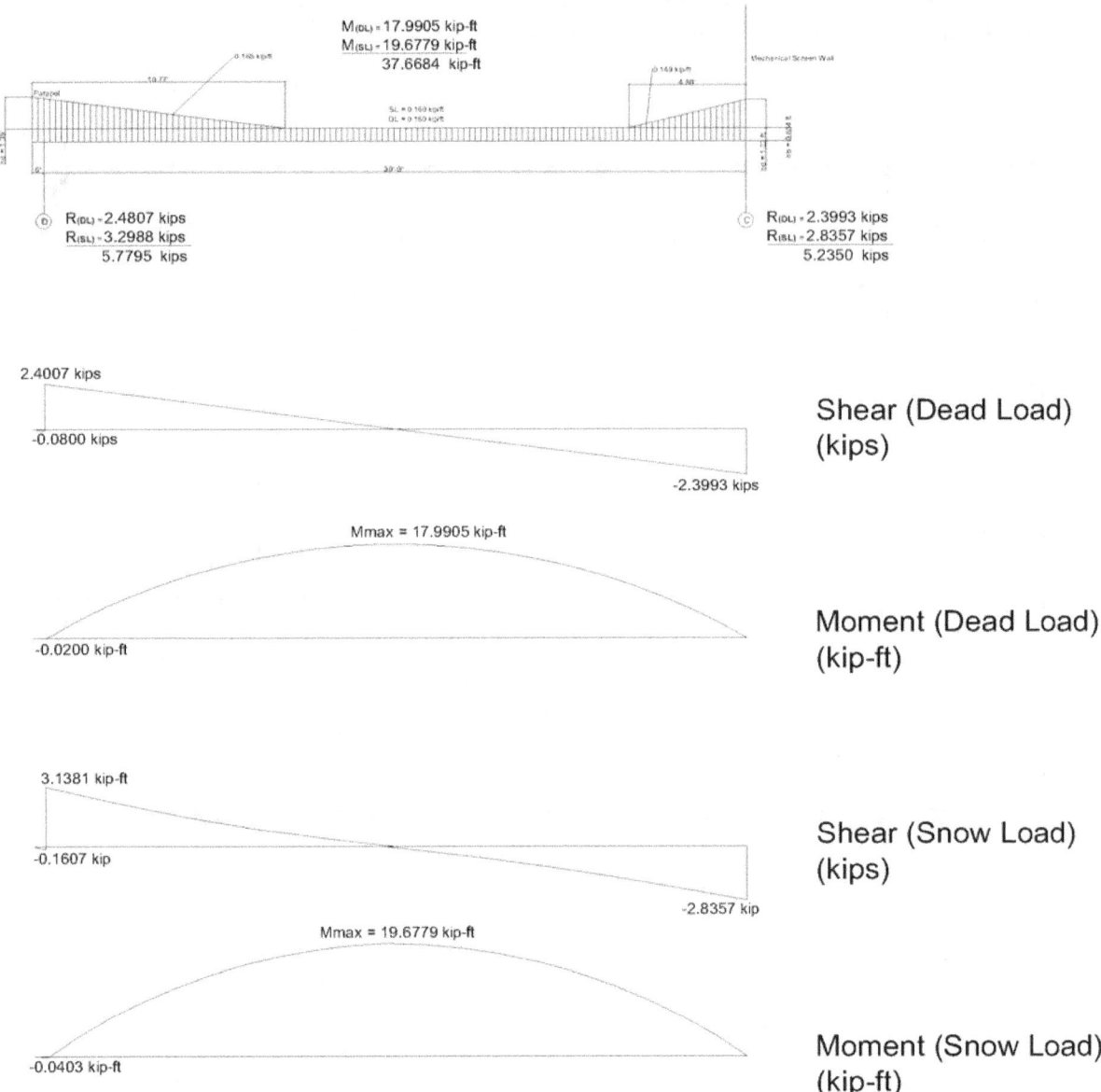

Figure 2-5 Shear and bending moment diagrams

From the Steel Joist Institute load tables (SJI, 2005), select 22KCS2 joist. This joist has an allowable moment capacity of 40.67 kip-ft, and allowable shear of 5.90 kips, in addition this joist has a gross moment of inertia of 194 in.[4], an approximate weight of only 10.0 plf, and since the span is only 30.5 ft, no erection stability bridging is required.

The first joist away from the parapet is loaded with a uniform snow drift load along the entire length of the member as shown in Figure 2-6:

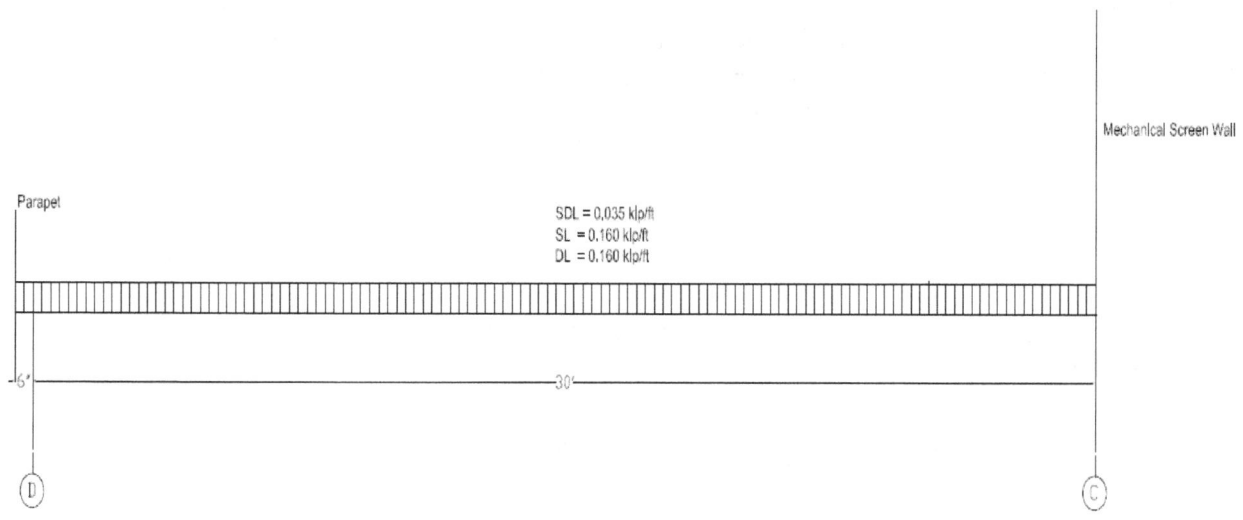

Figure 2-6 Joist with a uniform snow drift load

Based on analysis, a 22KCS2 joist is also acceptable for this uniform snow drift load case (M_{max} = 39.914 kip-ft, $R_{max (DL + SL)}$ = 5.504 kips).

The 30-ft joist in the middle bays as shown circulated in Figure 2-7 will have a uniform load of :

w = (20 psf + 20 psf)(8 ft) = 320 plf

w_{LL} = (20 psft) (8 ft) = 160 plf

From the Steel Joist Institute load tables, select a 22K5 joist which weighs approximately 7.7 plf and satisfies both strength and deflection requirements.

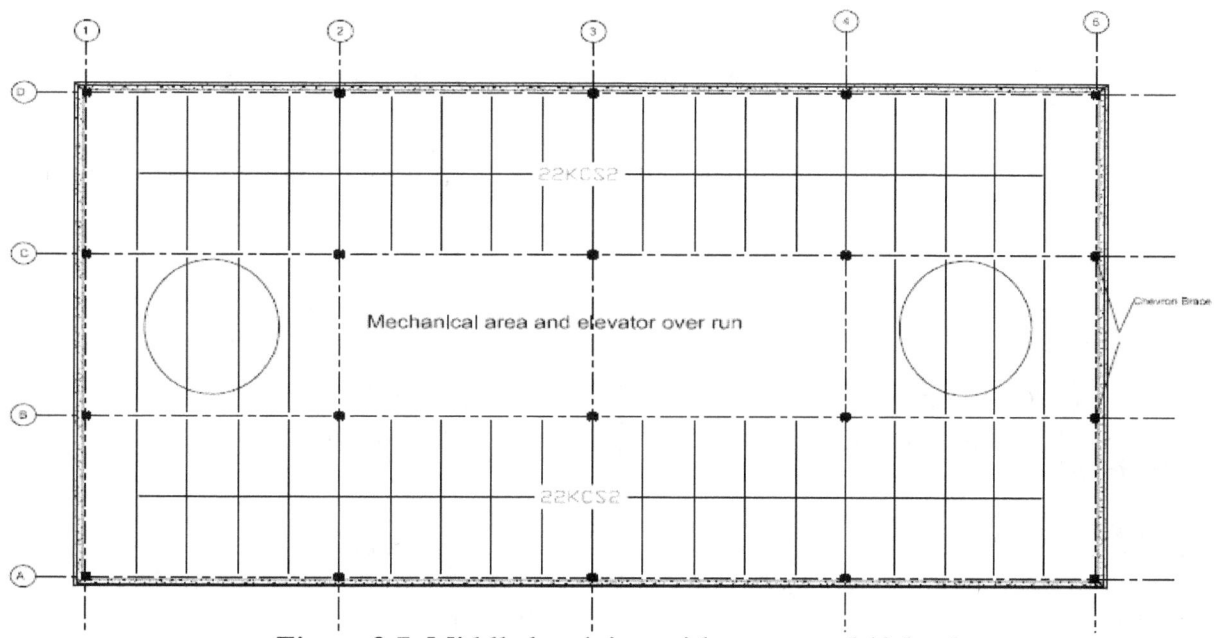

Figure 2-7 Middle bay joists with no snow drift load

The first joist away from the screen wall, and the first joist away from the end of the buildings carry a uniform snow drift load. Use a 22KCS2 in these locations.

2-5 Roof Beams Selection

Calculate loads and select beams in mechanical area.

For the beams in the mechanical area, the mechanical units could weigh as much as 20 psf. Use 40 psf additional dead load, which will account for the mechanical units and the screen wall around the mechanical area. Use 15 psf additional snow load, which will account for any snow drift load that could occur in the mechanical area. In this design example the beams in the mechanical area are spaced at 8 ft o.c.

The deflection should be limited to maximum $l/360 = (30*12)/360 = 1.00$ in. Assuming a plaster ceiling will be used in the main corridors. Use (15 psf + 15 psf = 30 psf) as an estimate of the snow load, including some drifting that could occur in this area, for deflection calculations.

Note: the beams and supporting girders in this area should be rechecked when the final weights and locations for the mechanical units have been determined.

Assuming, the end supports are pinned supports (see Figure 2-8). The required minimum I_x to limit deflection to $l/360$ is:

$$I_{req}(\text{Live Load}) = \frac{5 * (30 \text{ psf} * 8 \text{ ft} * 1\text{kip}/1000 \text{ lb})(30.0 \text{ ft})^4 * (12 \text{ in.}/1\text{ft})^3}{(384)(29,000 \text{ ksi})(1.00 \text{ in.})}$$

$$= 151 \text{ in.}^4$$

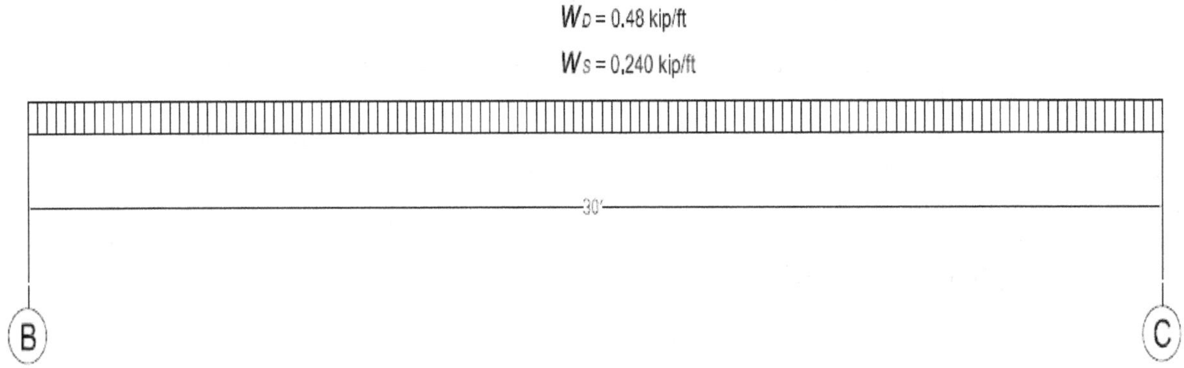

$W_D = 0.48$ kip/ft

$W_S = 0.240$ kip/ft

Figure 2-8 Mechanical area beam loading & bracing diagram (full lateral support)

Calculate the required strengths from ASCE/SEI 7-10 Chapter 2, and select the beams in the mechanical area.

LRFD	ASD
$w_u = 8.00 \text{ ft}[1.2 (0.060 \text{ kip}/\text{ft}^2$ $+1.60(0.030 \text{ kip}/\text{ft}^2)]$ $= 0.96 \text{ kip}/\text{ft}$	$w_a = 8.00 \text{ ft}[0.060 \text{ kip}/\text{ft}^2$ $+0.030 \text{ kip}/\text{ft}^2]$ $= 0.72 \text{ kip}/\text{ft}$
$R_u = \dfrac{30.0 \text{ ft}(0.96 \text{ kip}/\text{ft})}{2}$ $= 14.4 \text{ kips}$	$R_a = \dfrac{30.0 \text{ ft}(0.72 \text{ kip}/\text{ft})}{2}$ $= 10.8 \text{ kips}$
$M_u = \dfrac{0.96 \text{ kip}/\text{ft}(30.0 \text{ ft})^2}{8}$ $= 108 \text{ kip-ft}$	$M_a = \dfrac{0.72 \text{ kip}/\text{ft}(30.0 \text{ ft})^2}{8}$ $= 81 \text{ kip-ft}$
Assuming the beam has full lateral support, from AISC Manual Table 3-2, select an ASTM A992 W14x22 (W12x22 is also acceptable), which has a	Assuming the beam has full lateral support, from AISC Manual Table 3-2, select an ASTM A992 W14x22, which has a design flexural strength of 82.8

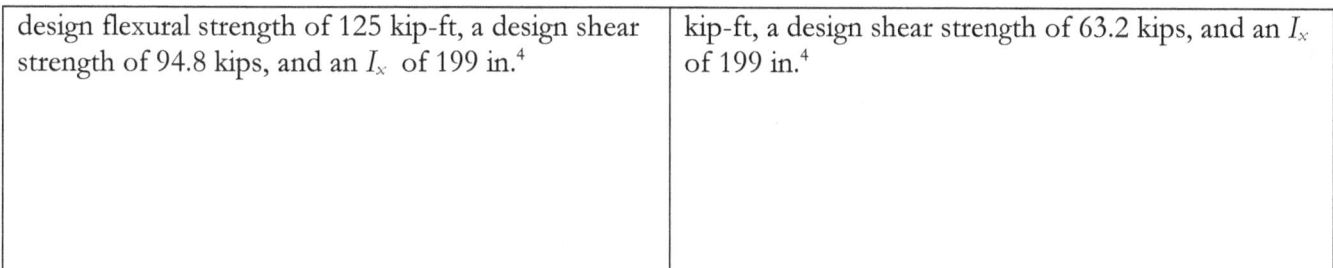

design flexural strength of 125 kip-ft, a design shear strength of 94.8 kips, and an I_x of 199 in.[4]	kip-ft, a design shear strength of 63.2 kips, and an I_x of 199 in.[4]

Calculate loads and select beams at the end (east & west) of the building:

The beams at the ends of the building carry the brick spandrel panel and a small portion of roof load (see Figure 2-9). For these beams, the cladding weight exceeds one quarter of the total dead load on the beam. Therefore, per AISC Design Guide 3, limit the vertical deflection due to cladding and initial dead load to $L/600$ or $3/8$ in. maximum, (whichever is lower). In addition, because these beams are supporting brick above and there is continuous glass below, limit the superimposed dead and live load deflection to $L/600$ or 0.3 in. max to accommodate the brick and $L/360$ or 0.25 in. max to accommodate the glass. Therefore, combining the two limitations, limit the superimposed dead and live load deflection to $L/600$ or 0.25 in. The superimposed dead load includes all of the dead load that is applied after the cladding has been installed. In calculating the wall loads, the spandrel panel weight is taken as 55 psf. The spandrel panel weight is approximately:

$w_D = (6.0 + 2.0)\text{ft} * (0.055 \text{ kip/ft}^2)$

$\quad = 0.440 \text{ kip/ft}$

The dead load from the roof is equal to :

$w_D = 4.50 \text{ ft}(0.020 \text{ kip/ft}^2)$

$\quad = 0.090 \text{ kip/ft}$

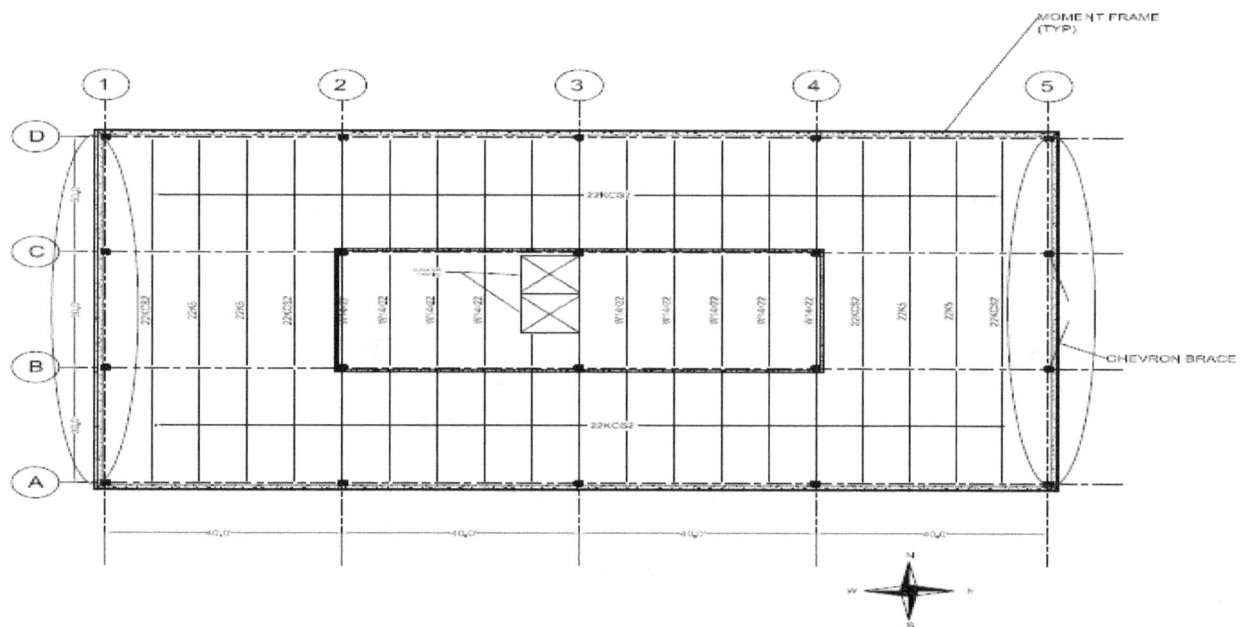

Figure 2-9 Beams at the end (east& west) of the building

Use 8 psf for the initial dead load.

$w_{D(initial)} = 4.50 \text{ ft}(0.008 \text{ kip/ft}^2)$

$\quad\quad = 0.036 \text{ kip/ft}$

Use 12 psf for the superimposed dead load.

$w_{D(super)} = 4.50 \text{ ft}(0.012 \text{ kip/ft}^2)$

= 0.054 kip/ft

The snow load from the roof can be conservatively taken as:

$w_s = 4.50$ ft(0.020 kip/ft² + 0.02066 kip/ft²)

= 0.183 kip/ft

to account for the maximum snow drift as a uniform load.

Assume the beams are simple spans of 30.0 ft.

Calculate minimum I_x to limit the superimposed dead and live load deflection to 0.25 in.

$$I_{req} = \frac{5*(0.054 + 0.183) \text{ kip/ft}(30.0 \text{ ft})^4 * (12 \text{ in./ft})^3}{(384)(29,000 \text{ ksi})(0.25 \text{ in.})}$$

= 596 in.⁴

Calculate minimum I_x to limit the cladding and initial dead load deflection to 0.375 in.

$$I_{req} = \frac{5*(0.036 + 0.440) \text{ kip/ft}(30.0 \text{ ft})^4 * (12 \text{ in./ft})^3}{(384)(29,000 \text{ ksi})(0.375 \text{ in.})}$$

= 798 in.⁴

The beams are fully supported by the deck as shown in detail ?. The loading diagram is shown in Figure 2-10:

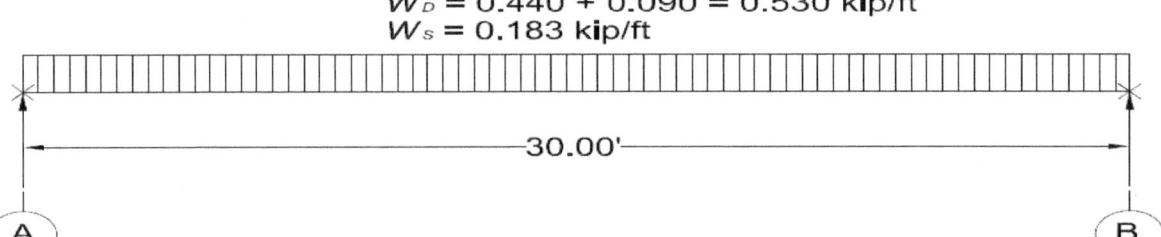

$$W_D = 0.440 + 0.090 = 0.530 \text{ kip/ft}$$
$$W_S = 0.183 \text{ kip/ft}$$

30.00'

A   B

Figure 2-10 Beam loading & bracing diagram (full lateral support)

Calculate the required strengths from Chapter 2 of ASCE/SEI 7 and select the beams for the roof ends.

LRFD	ASD
$w_u = 1.2(0.090 \text{ kip/ft} + 0.440$ kip/ft) + 1.60(0.183 kip/ft) = 0.929 kip/ft	$w_a = (0.090 \text{ kip/ft} + 0.440 \text{ kip/ft})$ +0.183 kip/ft = 0.713 kip/ft
$R_u = \dfrac{30.0 \text{ ft}(0.929 \text{ kip/ft})}{2}$ = 13.93 kips	$R_a = \dfrac{30.0 \text{ ft}(0.713 \text{ kip/ft})}{2}$ = 10.70 kips
$M_u = \dfrac{0.929 \text{ kip/ft}(30.0 \text{ ft})^2}{8}$ = 104.51 kip-ft	$M_a = \dfrac{0.713 \text{ kip/ft}(30.0 \text{ ft})^2}{8}$ = 80.21 kip-ft
Assuming the beam has full lateral support, from AISC Manual Table 3-2, select an ASTM A992 W21x44, which has a design flexural strength of 358 kip-ft, a design shear strength of 217 kips, and an I_x of 843 in.⁴	Assuming the beam has full lateral support, from AISC Manual Table 3-2, select an ASTM A992 W21x44, which has a design flexural strength of 238 kip-ft, a design shear strength of 145 kips, and an I_x of 843 in.⁴

Calculate loads and select beams along the side (north & south) of the building:

The beams along the side of the building (see Figure 2-11) carry the spandrel panel and a substantial roof dead load and live load. For these beams, the cladding weight exceeds one quarter of the total dead load on the beam. Therefore, per AISC Design Guide 3, limit the vertical deflection due to cladding and initial dead load to $L/600$ or 3/8 in. maximum, (whichever is lower). In addition, because these beams are supporting brick above and there is continuous glass below, limit the superimposed dead and live load deflection to $L/600$ or 0.3 in. max to accommodate the brick and $L/360$ or 0.25 in. max to accommodate the glass. Therefore, combining the two limitations, limit the superimposed dead and live load deflection to $L/600$ or 0.25 in. The superimposed dead load includes all of the dead load that is applied after the cladding has been installed. These beams will be part of the moment frames on the side of the building and therefore will be designed as fixed at both ends. The roof dead load and snow load on this edge beam is equal to the joist end dead load and snow load reaction. Treating this is a uniform load, divide this by the joist spacing.

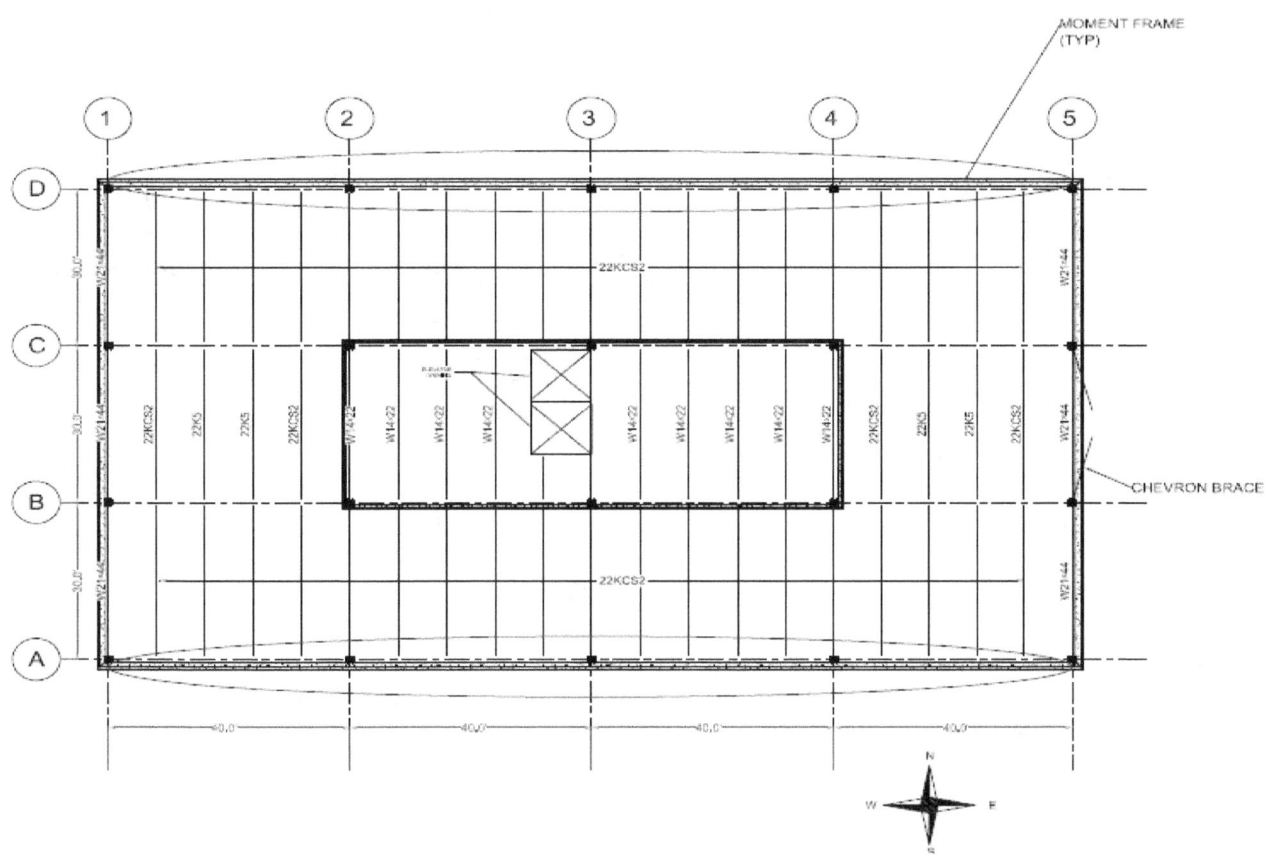

Figure 2-11 Beams along the side of the building (N & S)

w_D = 2.4807 kips/8.00 ft
    = 0.310 kip/ft
w_s = 3.2988 kips/8.00 ft
    = 0.412 kip/ft
$w_{D(initial)}$ = (15.0 + 0.5) ft (0.008 kip/ft²)
        = 0.124 kip/ft
$w_{D(super)}$ = (15.0 + 0.5) ft (0.012 kip/ft²)
        = 0.186 kip/ft

Calculate minimum I_x to limit the superimposed dead and live load deflection to 0.25 in.

$$I_{req} = \frac{(0.412 + 0.186) \text{ kip/ft}(40.0 \text{ ft})^4 * (12 \text{ in./ft})^3}{(384)(29{,}000 \text{ ksi})(0.25 \text{ in.})}$$

    = 950.2 in.⁴

Calculate minimum I_x to limit cladding and initial dead load deflection to 0.375 in.

$$I_{req} = \frac{(0.440 + 0.124) \text{ kip/ft}(40.0 \text{ ft})^4 * (12 \text{ in./ft})^3}{(384)(29{,}000 \text{ ksi})(0.25 \text{ in.})}$$

    = 896.2 in.⁴

Figure 2-12 shows beam loading and bracing diagram for roof beams along the side (north & south) of the building.

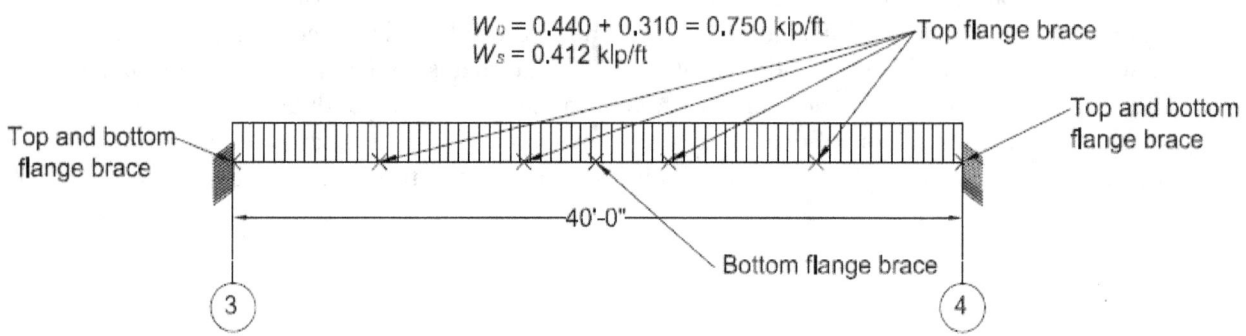

Figure 2-12 Beam loading & bracing diagram (bracing at ends & one fifth points)

Calculate the required strengths from Chapter 2 of ASCE/SEI 7 and select the beams for the roof sides.

LRFD	ASD
$w_u = 1.2(0.440 \text{ kip/ft} + 0.310$ $\text{kip/ft}) + 1.60(0.412 \text{ kip/ft})$ $= 1.559 \text{ kip/ft}$	$w_a = (0.440 \text{ kip/ft} + 0.310 \text{ kip/ft})$ $+0.412 \text{ kip/ft}$ $= 1.162 \text{ kip/ft}$
$R_u = \dfrac{40.0 \text{ ft}(1.559 \text{ kip/ft})}{2}$ $= 31.18 \text{ kips}$	$R_a = \dfrac{40.0 \text{ ft}(1.162 \text{ kip/ft})}{2}$ $= 23.24 \text{ kips}$
Calculate C_b for compression in the bottom flange braced at the midpoint and supports using AISCE Specification Equation F1-1:	Calculate C_b for compression in the bottom flange braced at the midpoint and supports using AISCE Specification Equation F1-1:
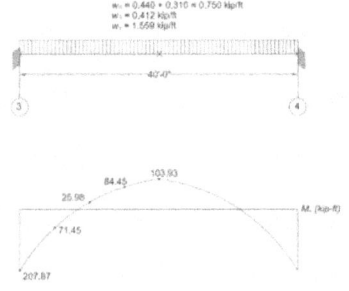	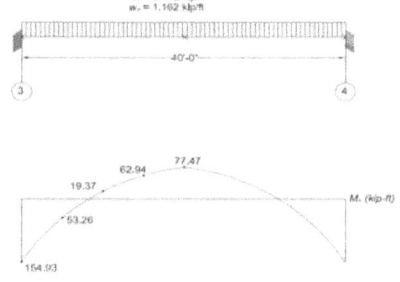
$M_{uMax} = \dfrac{1.559 \text{ kip/ft}(40.0 \text{ ft})^2}{12}$ $= 207.87 \text{ kip-ft}$	$M_{aMax} = \dfrac{1.162 \text{ kip/ft}(40.0 \text{ ft})^2}{12}$ $= 154.93 \text{ kip-ft}$
$M_{uMidpoint} = \dfrac{1.559 \text{ kip/ft}(40.0 \text{ ft})^2}{24}$ $= 103.93 \text{ kip-ft}$	$M_{aMidpoint} = \dfrac{1.559 \text{ kip/ft}(40.0 \text{ ft})^2}{24}$ $= 77.47 \text{ kip-ft}$
At quarter point of unbraced segment, $M_{uA} = 71.45$ kip-ft	At quarter point of unbraced segment, $M_{aA} = 53.26$ kip-ft
At midpoint of unbraced segment, $M_{uB} = 25.98$	At midpoint of unbraced segment, $M_{aB} = 19.37$ kip-ft

kip-ft

At three quarter point of unbraced segment, $M_{uC} =$ 84.45 kip-ft

Using AISC Specification Equation F1-1,

$$C_b = \frac{12.5 M_{max}}{2.5 M_{max} + 3 M_A + 4 M_B + 3 M_C}$$
$$= 2.38$$

Try ASTM A992 W21x50 ($I_x = 984$ in.[4]):
From AISC Manual Table 3-10:

For $L_b = 8$ ft and $C_b = 1.0$
$\mathbf{M}_b M_n = 350$ kip-ft > 103.93
  kip-ft   **o.k.**

For $L_b = 20$ ft and $C_b = 2.38$
$\mathbf{M}_b M_n = (133.7$ kip-ft$)(2.38)$
  $= 318.2$ kip-ft $\leq \mathbf{M}_b M_p$
$\mathbf{M}_b M_n = 318.2$ kip-ft > 207.87
  kip-ft   **o.k.**
From AISC Manual Table 3-2, a W21x50 has a design shear strength of 237 kips and an I_x of 984 in.[4]   **o.k.**

At three quarter point of unbraced segment, $M_{aC} =$ 62.94 kip-ft

Using AISC Specification Equation F1-1,

$$C_b = \frac{12.5 M_{max}}{2.5 M_{max} + 3 M_A + 4 M_B + 3 M_C}$$
$$= 2.38$$

Try ASTM A992 W21x50 ($I_x = 984$ in.[4]):
From AISC Manual Table 3-10:

For $L_b = 8$ ft and $C_b = 1.0$
$M_n / \Omega_b = 233$ kip-ft > 77.45
  kip-ft   **o.k.**

For $L_b = 20$ ft and $C_b = 2.38$
$M_n / \Omega_b = (89$ kip-ft$)(2.38)$
  $= 212$ kip-ft $\leq M_p / \Omega_b$
$M_n / \Omega_b = 212$ kip-ft > 154.93
  kip-ft   **o.k.**
From AISC Manual Table 3-2, a W21x50 has a design shear strength of 158 kips and an I_x of 984 in.[4]   **o.k.**

Note: To improve later frame drift performance during the lateral load analysis, some of the roof beams may need to be upsized to increase their stiffness and strength.

Calculate loads and select beams along the interior lines of the building (north & south):

At the middle bay of the building, there are two individual beam loadings that occur along grids B and C (see Figure 2-13). The beams from 1 to 2 and 4 to 5 have a uniform snow load except for the snow drift at the ends. The snow drift from the far ends of the 30-ft joists is negligible.

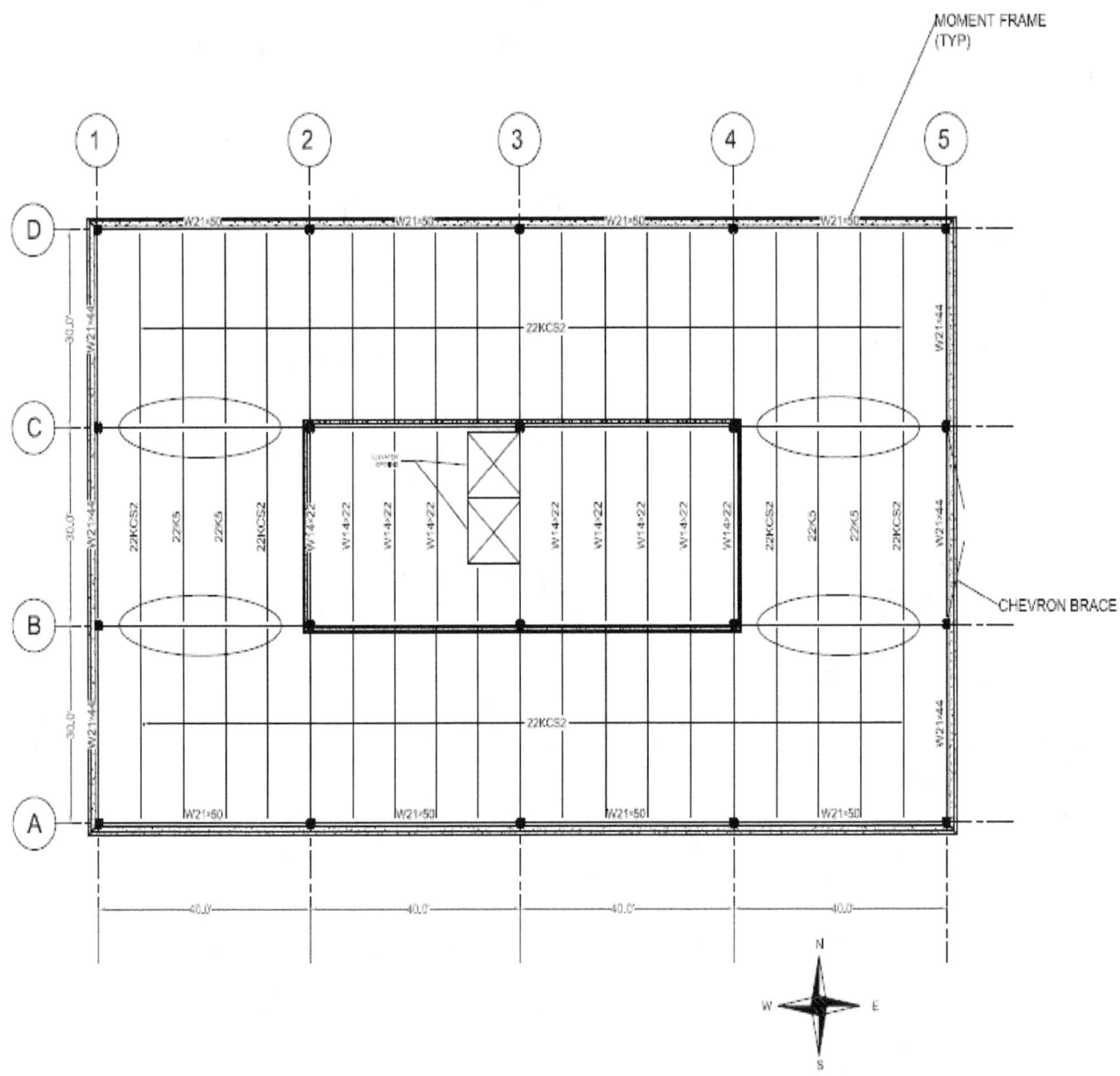

Figure 2-13 Beams along the side of the building (N & S)

The loading diagrams and a summary of the moments, left and right reactions are shown in Figure 2-14.

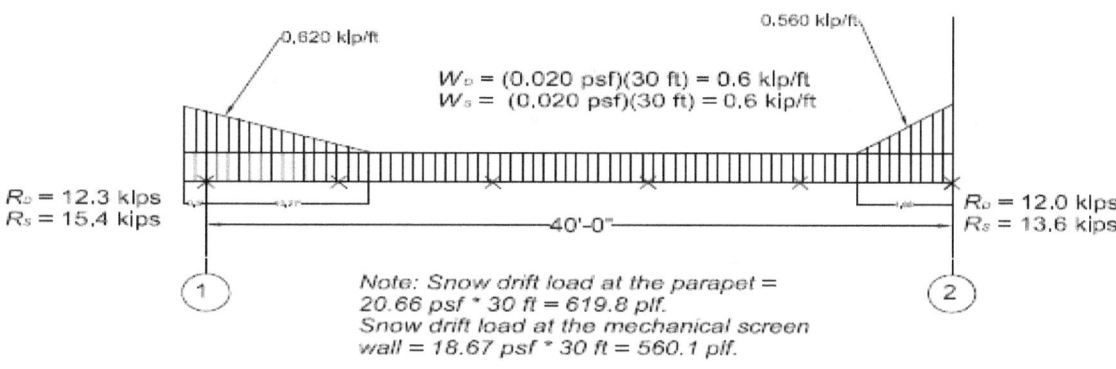

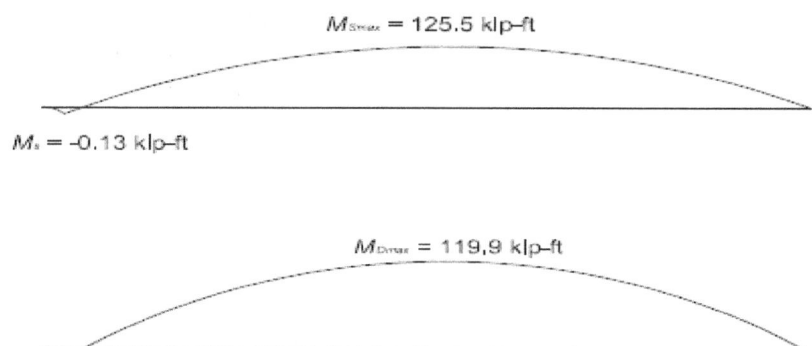

Figure 2-14 Beam loading & bracing diagram (bracing at ends & one fifth points)

Calculate the required strengths from Chapter 2 of ASCE/SEI 7 and select the beams for the roof sides.

LRFD	ASD

R_u = [1.2(12.3) + 1.6(15.4)] kips
   = 39.4 kips
 	
M_u =[1.2(119.9) + 1.6(125.5)] k-ft
   = 344.7 kip-ft
 	
The deflection due to live load should be equal to or less than the span divided by 240 (or 2.0 in.) therefore, the required I_x is:
 	
$I_{x\,req'd} = \dfrac{(0.60 \text{ kip/ft})(40.0 \text{ ft})^4}{1290(2.0 \text{ in.})}$
   = 596 in.^{4}
 	
Try ASTM A992 W18x55 (I_x = 890 in.^{4}):
From AISC Manual Table 3-10:
 	
For L_b = 8 ft and C_b = 1.0
$\mathbf{M}_b M_n$ = 391 kip-ft > 344.7 kip-ft
	R_a = [(12.3) + (15.4)] kips
	   = 27.7 kips
	
	M_a =[(119.9) + (125.5)] k-ft
	   = 245.4 kip-ft
	
	The deflection due to live load should be equal to or less than the span divided by 240 (or 2.0 in.) therefore, the required I_x is:
	
	$I_{x\,req'd} = \dfrac{(0.60 \text{ kip/ft})(40.0 \text{ ft})^2}{1290(2.0 \text{ in.})}$
	   = 596 in.^{4}
	
	Try ASTM A992 W18x55 (I_x = 890 in.^{4}):
	From AISC Manual Table 3-10:
	
	For L_b = 8 ft and C_b = 1.0
	M_n/Ω_b = 261 kip-ft > 245.4 kip-ft

o.k.	**o.k.**
From AISC Manual Table 3-2, a W18x55 has a design shear strength of 212 kips **o.k.**	From AISC Manual Table 3-2, a W18x55 has a design shear strength of 141 kips **o.k.**

The third individual beam loading occurs at the beams from 2 to 3, and 3 to 4. This is the heaviest load.

Calculate loads and select beams along the sides of the mechanical area building:

The beams from 2 to 3, and 3 to 4 have a uniform snow load outside the screen walled area, except for the snow drift at the parapet ends and the screen wall ends of the 30-ft joists. Inside the screen walled area the beams support the mechanical equipment. A summary of the moments, left and right reactions, and required I_x to keep the live load deflection to equal or less than the span divided by 240 (or 2.0 in.) is shown in Figure 2-15.

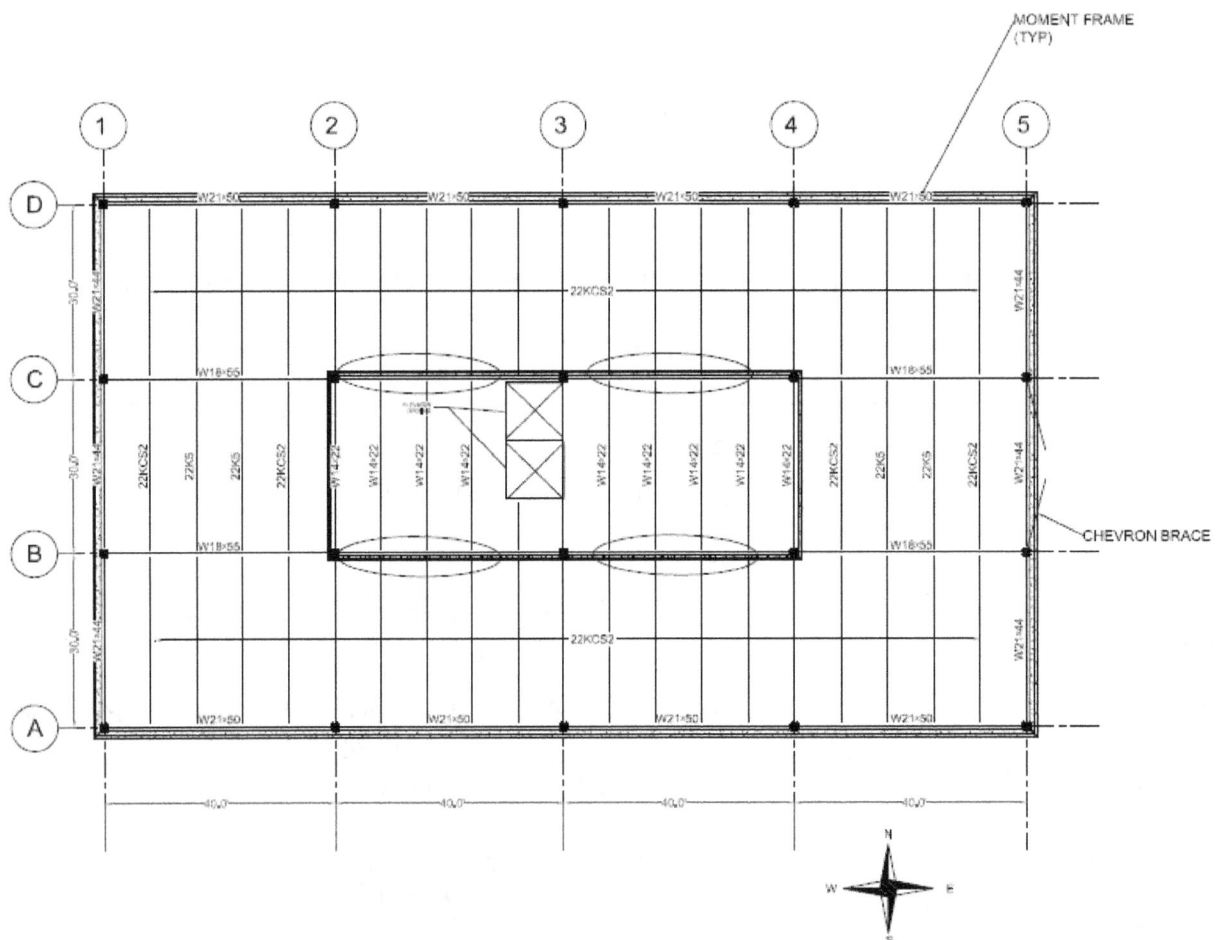

$$W_D = 1.20 \text{ kip/ft}$$
$$W_S = 0.80 \text{ kip/ft}$$

40'-0"

③   ④

Note: Snow load = (18.67 psf * 4.88 ft *0.5) + (30 psf * 15 ft) + (20 psf * 15 ft) = (snow drift load * $\frac{1}{2}$ snow drift width in the mechanical area) + ((original calculated snow load + additional snow load in the mechanical area)* $\frac{1}{2}$ Mechanical area span length (N-S) + (snow load * $\frac{1}{2}$ bays A-B or C-D span length) = 0.0795 kip/ft.
Dead load= (0.060 psf * 15 ft) + (0.020 * 15) = 1.20 kip/ft.

Figure 2-15 Beam loading & bracing diagram (bracing at ends & one fifth points)

LRFD	ASD
$w_u = [1.2(1.20 \text{ kip/ft}) + 1.6(0.80 \text{ kip/ft})] = 2.72$ kip/ft	$w_a = 1.20 \text{ kip/ft} + 0.80 \text{ kip/ft}$ $= 2.00 \text{ kip/ft}$
$M_u = \dfrac{2.72 \text{ kip/ft } (40.0 \text{ ft})^2}{8}$ $= 544 \text{ kip-ft}$	$M_a = \dfrac{2.00 \text{ kip/ft } (40.0 \text{ ft})^2}{8}$ $= 400 \text{ kip-ft}$
$R_u = \dfrac{2.72 \text{ kip/ft } (40.0 \text{ ft})}{2}$ $= 54.4 \text{ kips}$	$R_a = \dfrac{2.00 \text{ kip/ft } (40.0 \text{ ft})}{2}$ $= 40.0 \text{ kips}$
$I_{x \, req'd} = \dfrac{(0.80 \text{ kip/ft})(40.0 \text{ ft})^4}{1290(2.0 \text{ in.})}$ $= 794 \text{ in.}^4$	$I_{x \, req'd} = \dfrac{(0.80 \text{ kip/ft})(40.0 \text{ ft})^4}{1290(2.0 \text{ in.})}$ $= 794 \text{ in.}^4$
Try ASTM A992 W24x68 ($I_x = 1830 \text{ in.}^4$): From AISC Manual Table 3-10:	Try ASTM A992 W24x68 ($I_x = 1830 \text{ in.}^4$): From AISC Manual Table 3-10:
For $L_b = 8$ ft and $C_b = 1.0$ $\mathbf{M}_b M_n = 633 \text{ kip-ft} > 544 \text{ kip-ft}$ **o.k.**	For $L_b = 8$ ft and $C_b = 1.0$ $M_n/\Omega_b = 422 \text{ kip-ft} > 400 \text{ kip-ft}$ **o.k.**
From AISC Manual Table 3-2, a W24x68 has a design shear strength of 295 kips **o.k.**	From AISC Manual Table 3-2, a W24x68 has a design shear strength of 197 kips **o.k.**

Figure 2-16 shows the roof final joists and beams design. (Note the north-south edge beams will be cambered upward 2.5 in. at center of beams to allow rain water flow into roof gutters that are usually present next to columns).

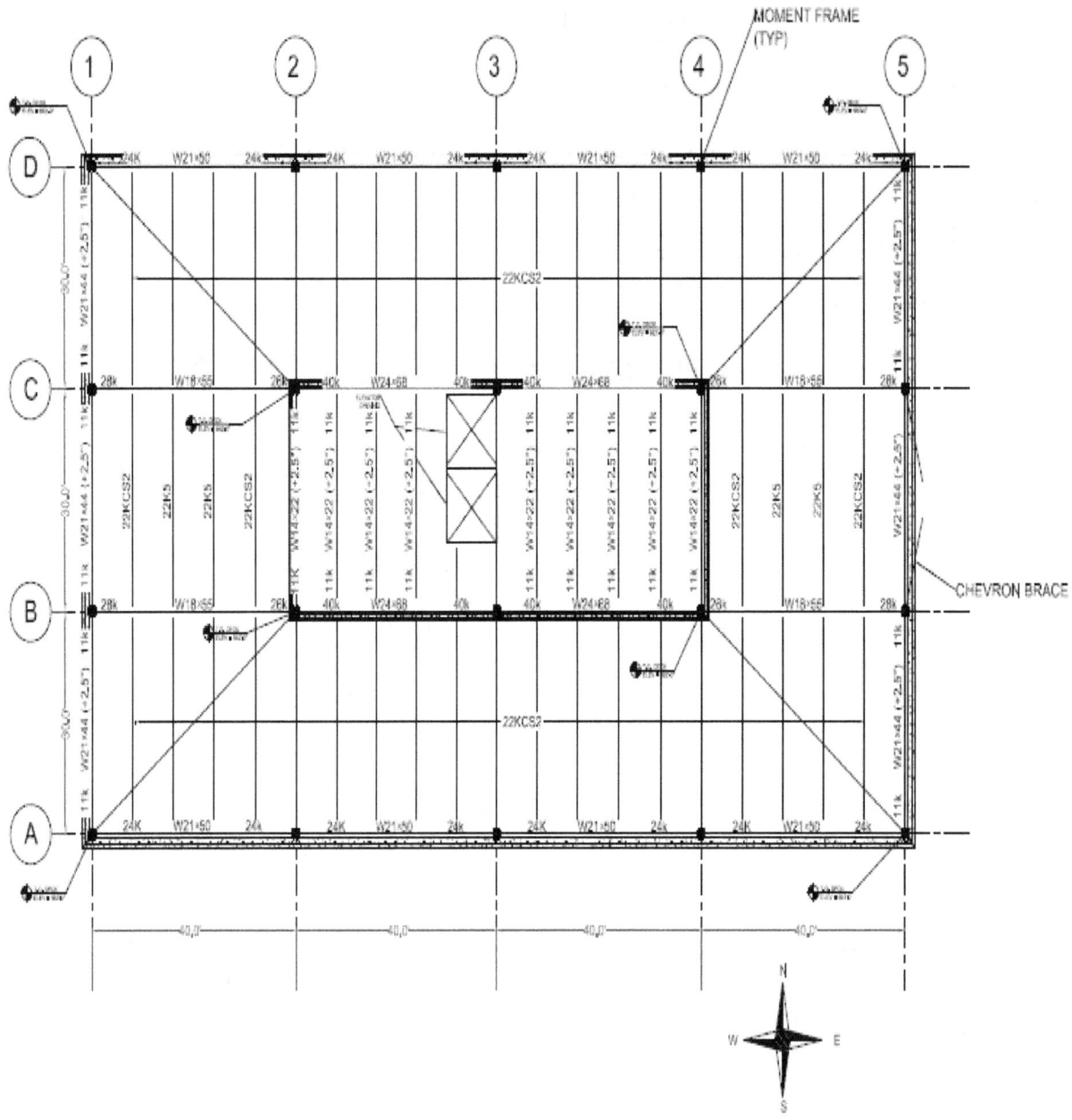

Figure 2-16 Roof beams and joists layout plan

2-6   Sample Hand Calculation Including Lateral Analysis:

The following example is a hand calculations for roof members for a two story building in Ohio.

JOB NAME: *TWO-STORY OFFICE BUILDING DESIGN*	JOB NUMBER: *17-089*	
JOB LOCATION: *LINCOLN WAY WEST, MASSILLON, OHIO*	DATE: *11-10-2017*	
DESIGNED BY: *SAAD HASAN TANTAWI (M.S.CE, B.S.CE, E.I.)*	SHEET 1	OF

△ GEOMETRY: (I) TWO-STORY BUILDING (OFFICE BUILDING) 132'×70'. SEE LAYOUT BELOW

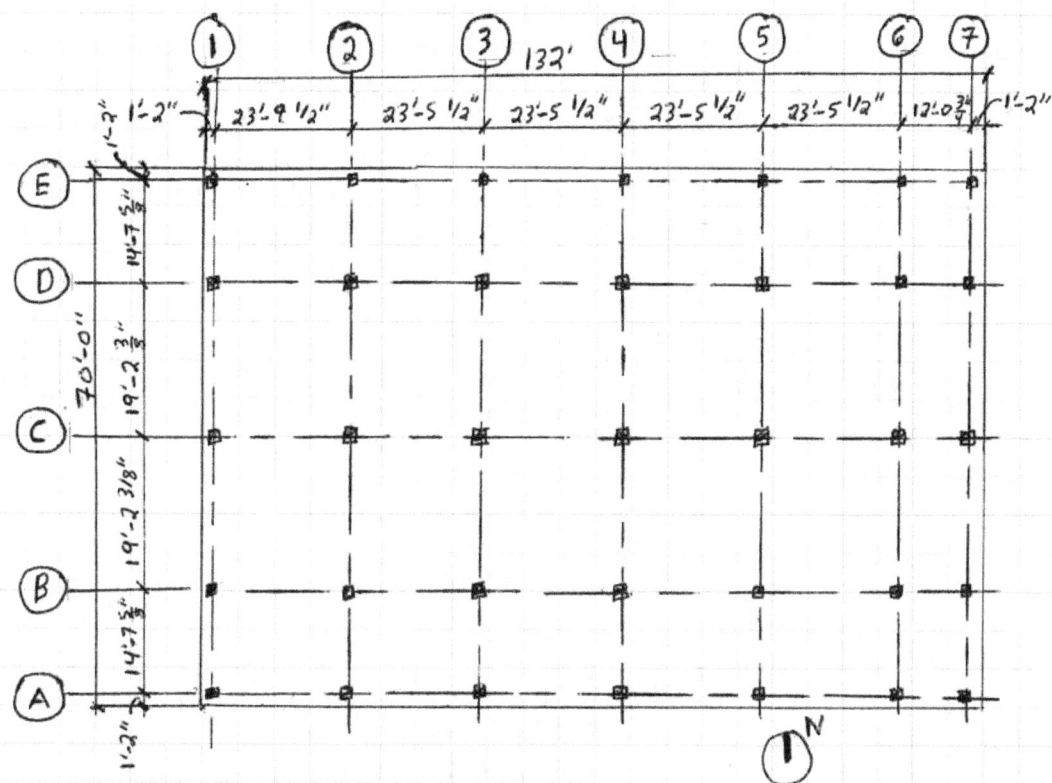

(II) THE FLOOR-TO-FLOOR HEIGHT IS 13 FT. 6 IN. AND THE HEIGHT FROM SECOND FLOOR TO ROOF IS 14FT 6 IN.

(III) THE SAME COLUMN SIZE WILL BE USED FOR THE WHOLE HEIGHT OF THE BUILDING.

(IV) THE EXTERIOR OF THE BUILDING IS A RIBBON WINDOW SYSTEM WITH BRICK SPANDRELS SUPPORTED AND BACK-BRACED WITH STEEL AND INFILLED WITH METAL STUDS.

→

JOB NAME: *TWO-STORY OFFICE BUILDING DESIGN*	JOB NUMBER: *17-089*
JOB LOCATION: *LINCOLN WAY WEST, MASSILLON, OHIO*	DATE: *11-10-2017*
DESIGNED BY: *SAAD HASAN TANTAWI (M.S.CE, B.S.CE, E.I.)*	SHEET 2 OF

Ⓥ THE SPANDREL WALL EXTENDS 3FT 6 IN. ABOVE THE ELEVATION OF THE EDGE OF THE ROOF.

Ⓥ̄Ⅰ THE ROOF SYSTEM IS $1\frac{1}{2}$ IN. METAL DECK ON BAR JOISTS.

Ⓥ̄ⅠⅠ THE ELEVATED FLOORS HAVE $1\frac{1}{2}$ IN. OF NORMAL WEIGHT CONCRETE OVER 3-IN. COMPOSITE DECK FOR A TOTAL SLAB THICKNESS OF $4\frac{1}{2}$ IN.

ⓘⅩ THE BUILDING WILL RESIST LATERAL LOADS THROUGH BRACING IN THE MIDDLE COLUMNS (AND POSSIBLY FRAMES IN THE SOUTH AND WEST WALLS.

Ⓧ THE BUILDING IS SPRINKLERED AND HAS A LARGE OPEN SPACE AROUND IT, AND CONSEQUENTLY DOES NOT REQUIRE FIRE ROOFING FOR THE FLOORS.

ⓧ̄Ⅰ CODES USED: Ⅰ- IBC 2012
                 Ⅱ- AISC V14.1
                 Ⅲ- ASCE 2010
                 Ⅳ- ACI 318-14
                 Ⅴ- SDI (2004)
                 Ⅵ- SJI (2005)
                 Ⅶ- IECC (2012) (FOR INSULATION TYPE AND REQ'MENT)

JOB NAME: *TWO-STORY OFFICE BUILDING DESIGN*	JOB NUMBER: *17-089*	
JOB LOCATION: *LINCOLN WAY WEST, MASSILLON, OHIO*	DATE: *11-10-2017*	
DESIGNED BY: **SAAD HASAN TANTAWI (M.S.CE, B.S.CE, E.I.)**	SHEET 3	OF

△ <u>WIND SPEED</u>: RISK CAT. = II
EXPOSURE CAT. = "<u>C</u>" (SITED IN AN OPEN, RURAL AREA)

$V = 115$ MPH (ASCE 7-10, FIG. 26.5-1, A)
(3-SEC. GUEST)

△ <u>SEISMIC PARAMETERS</u>: $S_S = 0.127 g$ (IBC 2012 FIG. 1613.3.1(1))
$S_1 = 0.055 g$ (IBC 2012 FIG. 1613.3.1(2))
$I_e = 1.0$ (ASCE 7-10 TABLE 1.5-2)

△ <u>ROOF AND FLOOR LOADS</u>: $P_g = $ 20 PSF (ASCE 7-10 FIG. 7-1)

➡ THE SLOPE OF THE ROOF IS $\frac{1}{4}$ IN./FT OR MORE @ ALL LOCATIONS, BUT NOT EXCEEDING $\frac{1}{2}$ IN./FT, THEREFORE, <u>5</u> PSF RAIN-ON-SNOW SURCHARGE IS TO BE CONSIDERED PER ASCE 7-10, SECTION 7.10, BUT PONDING INSTABILITY DESIGN CALCULATIONS ARE NOT REQ'D PER ASCE 7-10, SECTION 7.10.

➡ THIS ROOF CAN BE DESIGNED AS FULLY EXPOSED ROOF, BUT PER ASCE 7-10, SECTION 7.3, CANNOT BE DESIGNED FOR LESS THAN $P_g = (I_s) P_g = $ 20 PSF. UNIFORM SNOW LOAD.

➡ SNOW DRIFT WILL BE APPLIED @ THE EDGES OF THE ROOF BELOW THE PARAPET.

➡ THE ROOF LIVE LOAD FOR THIS BUILDING IS 20 PSF. BUT MAY BE REDUCED PER ASCE 7-10, SECTION 4.8, WHERE APPLICABLE.

JOB NAME: *TWO-STORY OFFICE BUILDING DESIGN*	JOB NUMBER: *17-089*
JOB LOCATION: *LINCOLN WAY WEST, MASSILLON, OHIO*	DATE: *11-10-2017*
DESIGNED BY: *SAAD HASAN TANTAWI (M.S.CE, B.S.CE, E.I.)*	SHEET 4 OF

→ <u>FLOOR LOADS</u> : → THE BASIC LIVE LOAD FOR THE FLOOR IS <u>50 PSF</u> PER ASCE 7-10, TABLE 4-1. AN ADDITIONAL PARTITION LIVE LOAD OF 15 PSF IS SPECIFIED. CORRIDORS LIVE LOAD IS <u>100 PSF</u>. FOR SIMPLICITY THE ENTIRE FLOOR WILL BE DESIGNED FOR A LIVE LOAD OF <u>80 PSF</u> THUS, PER ASCE 7-10, SECTION 4.3-2, PARTITION LIVE LOAD IS NOT REQUIRED WHERE THE MINIMUM SPECIFIED LIVE LOAD EXCEEDS <u>80 PSF</u>.

→ <u>WALL LOADS</u>: A WALL LOAD OF <u>55 PSF</u> WILL BE USED FOR THE BRICK SPANDRELS, SUPPORTING STEEL, AND METAL STUD BACK-UP. A WALL LOAD OF <u>15 PSF</u> WILL BE USED FOR THE RIBBON WINDOW GLAZING SYSTEM.

⚠ <u>GRAVITY LOADS</u>: CALCULATE DEAD AND SNOW LOAD.

→ <u>DEAD LOAD</u>:

$$
\begin{aligned}
\text{ROOFING} &= 5 \text{ PSF} \\
\text{INSULATION} &= 2 \text{ PSF} \\
\text{DECK} &= 2 \text{ PSF} \\
\text{BEAMS} &= 3 \text{ PSF} \\
\text{JOISTS} &= 3 \text{ PSF} \\
\text{MISC.} &= 5 \text{ PSF} \\
\hline
\text{TOTAL} &= 20 \text{ PSF}
\end{aligned}
$$

⚠ SNOW LOAD FROM ASCE 7-10, SECTION 7.3 AND 7.10:

$$
\begin{aligned}
\text{SNOW} &= 20 \text{ PSF} \\
\text{RAIN ON SNOW} &= 5 \text{ PSF} \\
\hline
\text{TOTAL} &= 25 \text{ PSF}
\end{aligned}
$$

JOB NAME: *TWO-STORY OFFICE BUILDING DESIGN*	JOB NUMBER: *17-089*	
JOB LOCATION: *LINCOLN WAY WEST, MASSILLON, OHIO*	DATE: *11-10-2017*	
DESIGNED BY: *SAAD HASAN TANTAWI (M.S.CE, B.S.CE, E.I.)*	SHEET **5**	OF

→ SINCE RAIN AND SNOW LOAD (25 PSF) IS GREATER THAN ROOF LIVE LOAD (20 PSF), PER ASCE 7-10, SECTION 2.4, RAIN AND SNOW LOAD WILL CONTROL

→ THE STEEL DECK IS $1\frac{1}{2}$ IN., WIDE RIB, (VULCRAFT 1.5B), PAINTED ROOF DECK, PLACED IN A PATTERN OF TWO CONTINUOUS SPANS MINIMUM. THE TYPICAL JOIST SPACING IS 6 FT ON CENTER. AT 6 FT ON CENTER, THIS DECK HAS AN ALLOWABLE TOTAL LOAD CAPACITY OF: 70 PSF. NOTE THE ROOF DIAPHRAGM AND ROOF LOADS EXTEND 6 IN. PAST THE CENTERLINE OF GRID.

→ SNOW DRIFT CALCULATIONS:

$$\gamma = 0.13 P_g + 14 \; PCF \quad (ASCE \; 7-10 \; EQ. \; 7.7-1)$$
$$\gamma = 0.13(20) + 14 = 16.6 \; PCF$$

$$P_\ell = 0.7 C_e C_t I_s P_g \; PSF \quad (ASCE \; 7-10 \; EQ. 7.3-1)$$
$$P_\ell = (0.7)(0.9)(1.0)(1)(20) = 12.6 \; PSF.$$
$$WHERE, \; C_e = 0.9 \quad (ASCE \; 7-10 \quad TABLE \; 7-2)$$
$$C_t = 1.0 \quad (ASCE \; 7-10 \quad TABLE \; 7-3)$$

BUT, $P_m = I_s P_g = 20 \; PSF$ (ASCE 7-10 SECTION 7.3.4)

HOWEVER, $P_g + 5 PSF$ (RAIN ON SNOW) = 25 PSF
THUS, $P_m = 25 \; PSF$.

→ FROM ASCE 7-10 FIG. 7-8 : $h_b = \dfrac{P_s}{\gamma}$ (ASCE 7.5 SECTION 7.7.1)

$$P_s = (1.0)(12.6) = C_s P_\ell \geq P_m \quad (ASCE \; 7-10 \; EQ. 7.4-1)$$
$$WHERE, C_s = 1.0 \quad (ASCE \; 7-10 \; FIG. 7-2a)$$

$$P_s = C_s P_\ell = (12.6 \; PSF) < P_m \quad \rightarrow \; CHOOSE \; \underline{P_s = P_m = 25 \; PSF.}$$

$$P_s = \text{~~BALANCED~~} \overset{SLOPED}{} \; SNOW \; LOAD. \; (SLOPED \; ROOF \; BALANCED \; SNOW \; LOAD)$$

→

JOB NAME: *TWO-STORY OFFICE BUILDING DESIGN*	JOB NUMBER: *17-089*
JOB LOCATION: *LINCOLN WAY WEST, MASSILLON, OHIO*	DATE: *11-10-2017*
DESIGNED BY: *SAAD HASAN TANTAWI (M.S.CE, B.S.CE, E.I.)*	SHEET **6** OF

→ h_b = HEIGHT OF BALANCED SNOW LOAD = $\dfrac{P_s}{\gamma}$ = $\dfrac{25\,PSF}{16.6\,PCF}$ = $\underline{\underline{1.50\;FT}}$

→ l_u = LENGTH OF THE ROOF UPWIND OF THE DRIFT
   = 20 FT   FOR LEEWARD   (MIN. PER ASCE 7-10 FIG.7-9)
   = 132 FT   FOR WINDWARD (E-W) DIRECTION
   = 70 FT   FOR WINDWARD (N-S) DIRECTION.

→ h_d = HEIGHT OF SNOW DRIFT = $0.43\sqrt[3]{l_u}\sqrt[4]{P_g+10} - 1.5$ (FIG. 7-9)
   = $[0.43\sqrt[3]{132}\sqrt[4]{20+10} - 1.5] = 3.62\;FT$ > $h_c = 3.5\,FT - h_b = 2\,FT$
   THUS, $\underline{h_d = 2.0\;FT}$ (WINDWARD E-W DIRECTION) $\dfrac{3}{4}h_d = 2.7\,FT\;o.k$

   w = DRIFT WIDTH = $\dfrac{4h_d^2}{h_c}$ (ASCE 7-10 SEC.7.7.1)
   $w = 4(2.70)^2/(2.0)$
   $w = 14.58\;FT$ (WINDWARD E-W DIRECTION)

→ $h_d = [0.43\sqrt[3]{70}\sqrt[4]{20+10} - 1.5] = 1.79\; 2.65\;FT > 2\,FT = h_c$
   $\dfrac{3}{4}h_d = 1.99\,FT < 2.0\,FT = h_c$ (SAY $\dfrac{3}{4}h_d = 2.0\,FT$) FOR WINDWARD
   $w = 4(2.0)^2/(2.0) = 8.0\,FT$ (WINDWARD N-S DIRECTION)

→ (LEEWARD) (E-W): $h_d = 3.62\,FT$ (CALCULATED)
                    $h_d = 2.0\,FT$ (ACTUAL)
                    $w = 4(3.62)^2/2.0 < 8(2.0) = 8(h_c)$
                    $w = 26.21\,FT > 8\,h_c$ ← (CHOOSE $8\,h_c$ PER
                    $\underline{w = 16.0\,FT}$ ASCE 7-10 SEC.7.7.1)

                    $P_d = h_d\gamma = (2.0)(16.6) = \underline{\underline{33.2\;PSF}}$

→ (LEEWARD) (N-S): $h_d = 2.65\,FT$ (CALCULATED)
                   $h_d = 2.0\,FT$ (ACTUAL)
                   $w = 4(2.65)^2/2.0 = \underline{14.045\,FT} < 16\,FT = 8\,h_c.$
                   (CONSERVATIVELY)

USE:
E-W $\begin{cases} h_d = 2.0\,FT \\ w = \dfrac{4(0.75\times 3.62)^2}{2} = 14.74\;\underset{(FT)}{PSF} \\ P_d = 33.2\,PSF \end{cases}$ $\begin{cases} h_d = 0.75\times 2.65 = 1.99\,FT \\ w = 4(1.99)^2/2 = 7.9\,FT \\ P_d = 16.6\times 1.99 = 33\,psf \end{cases}$ →

28

JOB NAME: *TWO-STORY OFFICE BUILDING DESIGN*	JOB NUMBER: *17-089*
JOB LOCATION: *LINCOLN WAY WEST, MASSILLON, OHIO*	DATE: *11-10-2017*
DESIGNED BY: *SAAD HASAN TANTAWI (M.S.CE, B.S.CE, E.I.)*	SHEET 7 OF

△ SNOW DRIFT SUMMARY:

E-W DIRECTION:

PARAPET HT. = 3'-6" ➡ h_c = 2'-0"

P_d = 33.2 PSF P_d = 33.2 PSF
P_s = 25 PSF
h_d = 2'
h_b = 1.5'
14.74'   14.74'
132'-0"

N-S DIRECTION:

h_d = 2'
P_d = 33.0 PSF P_d = 33 PSF
P_s = 25 PSF
h_b = 1.5'
7.9'   7.9'

JOB NAME: *TWO-STORY OFFICE BUILDING DESIGN*	JOB NUMBER: *17-089*
JOB LOCATION: *LINCOLN WAY WEST, MASSILLON, OHIO*	DATE: *11-10-2017*
DESIGNED BY: *SAAD HASAN TANTAWI (M.S.CE, B.S.CE, E.I.)*	SHEET *8* OF

⚠ <u>SELECT ROOF JOISTS</u>: JOISTS SPAN IN THE LONG DIRECTION
(E-W) @ 5'-0" C.C.

➡ JOISTS SPANING FROM COL. LINE ①-② : ⑧-⑩

$$W_{DL} = \frac{20\,PSF * (5\,FT)}{1000} = 0.100\,KLF$$

$$W_{SL} = \frac{25\,PSF * 5\,FT}{1000} = 0.125\,KLF$$

$$W_{DRIFT} = \frac{33.2\,PSF * 5\,FT}{1000} = 0.166\,KLF$$

$W_{DRIFT} = 0.166\,KLF$ $W_{DL} = 0.100\,KLF$ $W_{SL} = 0.125\,KLF$

$8-\frac{1}{2}"$ ├— 14.74 FT —┤ ├———— 23.80 FT ————┤

①   ②

$$R_① = \frac{\left[\begin{array}{l}0.225\,KLF * 24.50' * 24.50' * 0.50 \\ + 0.166\,KLF * 15.45' * 0.50 * (24.5'-5.15')\end{array}\right]}{23.80\,FT}$$

$$\underline{R_① = 3.88\,KIPS}$$

$$R_② = [0.166\,KLF * 15.45' * 0.50] + [0.225\,KLF * 24.50] - [3.88\,KIPS]$$

$$\underline{R_③ = 2.915\,KIPS}$$

∴ M_{MAX} OCCURS WHEN SHEAR = 0 KIPS

➡

JOB NAME: *TWO-STORY OFFICE BUILDING DESIGN*	JOB NUMBER: *17-089*
JOB LOCATION: *LINCOLN WAY WEST, MASSILLON, OHIO*	DATE: *11-10-2017*
DESIGNED BY: *SAAD HASAN TANTAWI (M.S.CE, B.S.CE, E.I.)*	SHEET **9** OF

M_{MAX} OCCURS WHEN $V = 0 = 2.915 K - 0.225 X - Y(x-9.06)(\frac{1}{2})$

$$\frac{0.166 \, KLF}{15.448 \, FT} = \frac{Y \, (KLF)}{(x-9.06) \, FT}$$

$\therefore \quad Y = 0.0107(x-9.06) \quad KLF$

$V = 0 = 2.915 - 0.225x - 0.0107(x-9.06)^2$

SOLVING FOR X: $\quad x = 12.65 \, FT$

THUS, $m_{MAX} = 2.915(12.65) - \dfrac{0.225(12.65)^2}{2} - 0.0384(3.59)^2 (0.5)(1.2)$

$\underline{M_{MAX} = 21.7 \, KIP\text{-}FT}$

→ BECAUSE THE LOAD IS NOT UNIFORM, SELECT A $\underline{16KCS2}$ WITH (2) ROWS OF BRIDGING FROM THE STEEL JOISTS INST. (SJI, 2005).

(16KCS2) $\quad M_{all.} = 29.08 \, KIP\text{-}FT > M_{REQ'D} = 21.7 \, KIP\text{-}FT.$ (O.K)

COL. LINE: $\quad V_{all.} = 4.0 \, KIPS > 3.88 \, KIPS$ (O.K.)
①-②

$I_{xx} = 99 \, IN^4 \Rightarrow \Delta \approx \dfrac{5\left(\frac{21.7*8}{24.5^2}\right)(24.5)^4 (12)^3}{384 \,(29000)(99)} = 0.817 IN. = \dfrac{\ell}{361}$

$L = 23.80'$

$Wt. = 8.5 \, PLF$ O.K

JOB NAME: *TWO-STORY OFFICE BUILDING DESIGN*	JOB NUMBER: *17-089*
JOB LOCATION: *LINCOLN WAY WEST, MASSILLON, OHIO*	DATE: *11-10-2017*
DESIGNED BY: *SAAD HASAN TANTAWI (M.S.CE, B.S.CE, E.I.)*	SHEET **23** OF

△ BRACED FRAME ANALYSIS: (SECOND FLOOR TO ROOF BRACED FRAME))

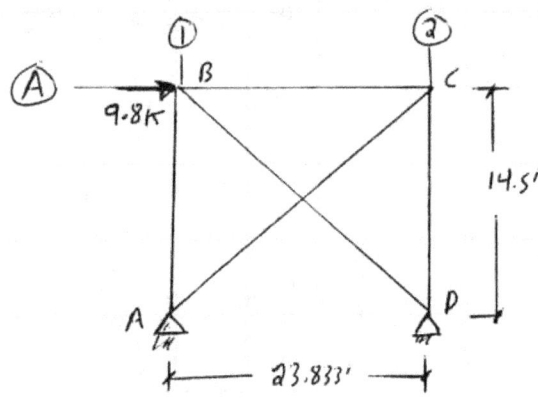

COLS. I_{xx} = 170 IN⁴
BEAMS I_{xx} = 310 IN⁴

COLS A = 9.71 IN.
BEAMS A = 11.8 IN²

BEAMS = W12×40
COLS = W10×33

BRACES A = 3.59 IN²
 HSS 4×4 × 1/4

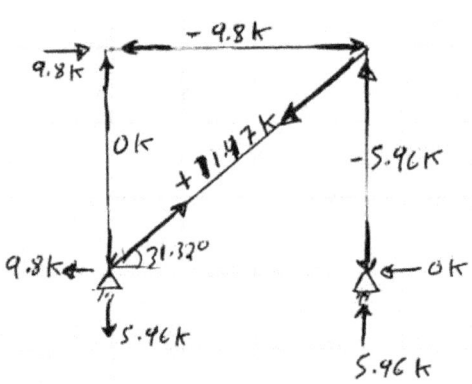

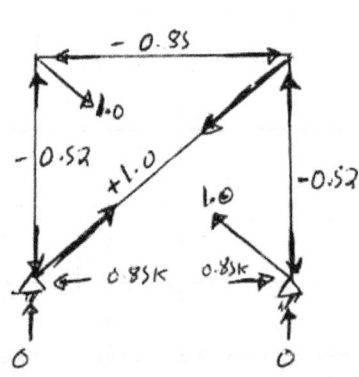

JOB NAME: *TWO-STORY OFFICE BUILDING DESIGN*	JOB NUMBER: *17-089*	
JOB LOCATION: *LINCOLN WAY WEST, MASSILLON, OHIO*	DATE: *11-10-2017*	
DESIGNED BY: *SAAD HASAN TANTAWI (M.S.CE, B.S.CE, E.I.)*	SHEET 24	OF

MEMBER	LENGTH (IN)	AE (KIPS)	P-FORCE (KIPS)	PL/AE (IN.)	U
AB	174	281590	0	0	−0.52
BC	286	342200	−9.8	−0.0082 ~~−0.000029~~	−0.85
CD	174	281590	−5.96	−0.0037 ~~−0.0000217~~	−0.52
AC	335	104110	+11.47	+0.037 ~~0.00011~~	1.0
BD	335	104110	0	0	1.0

MEMBER	$\dfrac{(PL)}{AE} \times U$ (IN.)	$\dfrac{(UL)}{AE} \times u$ (IN.)	MEMBER FORCE (KIPS) $\left(\text{P-FORCE} + \left\{ \dfrac{U \times \frac{S \cdot PL}{AE} u}{\frac{\mathcal{E} \cdot ul}{AE} u} \right\} \right)$
AB	0	0.000167	−3.23
BC	0.007 ~~0.000025~~	0.000604	(−15.1) W 12×40 Bm
CD	0.0014 ~~0.0000113~~	0.000167	−9.2
AC	0.037 ~~0.00011~~	0.00322	+17.68
BD	0	0.00322	+6.21

$$\mathcal{E} = \underset{0.0459}{D.~~\text{0.000146}} \qquad \mathcal{E} = 0.0074$$

$$\mathcal{E}\frac{PL}{AE} \times u \,/\, \mathcal{E}\frac{ul}{AE} \times u = ~~\cancel{0.40}~~ 6.21 \ KIPS \ (TENSION)$$

JOB NAME: *TWO-STORY OFFICE BUILDING DESIGN*	JOB NUMBER: *17-089*
JOB LOCATION: *LINCOLN WAY WEST, MASSILLON, OHIO*	DATE: *11-10-2017*
DESIGNED BY: *SAAD HASAN TANTAWI (M.S.CE, B.S.CE, E.I.)*	SHEET 25 OF

$V_A = -5.96 + (0) = -5.96$ KIPS $(5.96 K \downarrow)$

$V_D = 5.96$ KIPS $\uparrow$

$H_A = -9.8 K + (-0.85 \times 6.21) = 15.1$ KIPS $\leftarrow$

$H_D = 0$ KIPS $+ (0.85 \times 6.21) = 5.3$ KIPS $\rightarrow$

$\triangle$ CHECK ROOF BEAM W12×40. $P_a = 15.1$ KIPS (COMPRESSION)

$$M_u = \frac{(0.375 KLF)(23.833)^2}{8}$$

$$M_u = 23.96 \text{ KIP-FT}$$

ASSUME BEAM IS CONT. BRACED BY ROOF DIAPHRAGM.

$$\sqrt{\left(\frac{M_a}{\frac{(Z_{xx} \times F_Y)}{1.67}}\right)^2 + \left(\frac{(15.1)}{\frac{\left[\frac{\pi^2 E}{(\frac{KL}{r})^2}\right] \times A}{1.67}}\right)^2} \leq 1.0$$

$= 0.57 < 1.0$ (O.K.)

JOB NAME: *TWO-STORY OFFICE BUILDING DESIGN*	JOB NUMBER: *17-089*
JOB LOCATION: *LINCOLN WAY WEST, MASSILLON, OHIO*	DATE: *11-10-2017*
DESIGNED BY: *SAAD HASAN TANTAWI (M.S.CE, B.S.CE, E.I.)*	SHEET 4O OF

△ <u>MOMENT FRAME WITH SIDESWAY</u> : (USING SLOPE DEFL. METHOD)
(BUILDING 2)

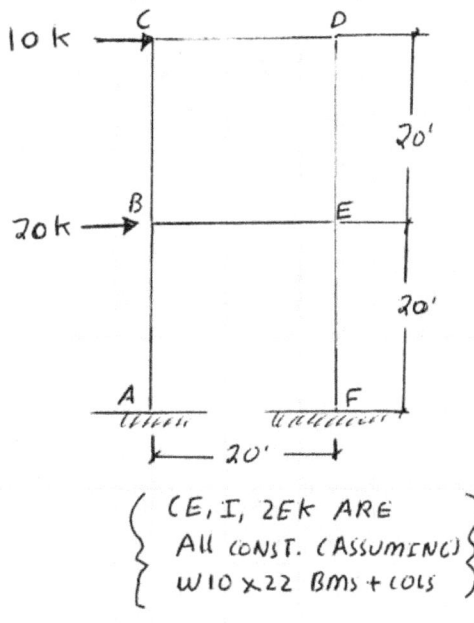

$\Psi_{AB} = \Psi_{EF} = \Psi_1$
$\Psi_{BC} = \Psi_{DE} = \Psi_2$
$\theta_A = \theta_F = 0$

$m_{AB} = \theta_B - 3\Psi_1$
$m_{BA} = 2\theta_B - 3\Psi_1$
$m_{BC} = 2\theta_B + \theta_C - 3\Psi_2$
$m_{BE} = 2\theta_B + \theta_E$
$m_{CB} = \theta_B + 2\theta_C - 3\Psi_2$
$m_{CD} = 2\theta_C + \theta_D$
$m_{DC} = \theta_C + 2\theta_D$
$m_{DE} = 2\theta_D + \theta_E - 3\Psi_2$
$m_{EB} = \theta_B + 2\theta_E$
$m_{ED} = \theta_D + 2\theta_E - 3\Psi_2$
$m_{EF} = 2\theta_E - 3\Psi_1$
$m_{FE} = \theta_E - 3\Psi_1$

10 k →
20 k →
20'
20'
20'

{ (E, I, 2EK ARE
All CONST. (ASSUMING)
W10×22 BMs + COLS }

$\mathcal{E} m_B = 0 = m_{BA} + m_{BC} + m_{BE}$
$2\theta_B - 3\Psi_1 + 2\theta_B + \theta_C - 3\Psi_2 + 2\theta_B + \theta_E = 0$
$6\theta_B + \theta_C + \theta_E - 3\Psi_1 - 3\Psi_2 = 0$ EQ. ①

$\mathcal{E} m_C = 0 = m_{CB} + m_{CD}$
$\theta_B + 2\theta_C - 3\Psi_2 + 2\theta_C + \theta_D = 0$
$\theta_B + 4\theta_C + \theta_D - 3\Psi_2 = 0$ EQ. ②

$\mathcal{E} m_D = 0 = m_{DC} + m_{DE}$
$\theta_C + 2\theta_D + 2\theta_D + \theta_E - 3\Psi_2 = 0$
$\theta_C + 4\theta_D + \theta_E - 3\Psi_2 = 0$ EQ ③

→

JOB NAME: *TWO-STORY OFFICE BUILDING DESIGN*	JOB NUMBER: *17-089*
JOB LOCATION: *LINCOLN WAY WEST, MASSILLON, OHIO*	DATE: *11-10-2017*
DESIGNED BY: *SAAD HASAN TANTAWI (M.S.CE, B.S.CE, E.I.)*	SHEET **41** OF

$\Sigma M_E = O = M_{ED} + M_{EB} + M_{EF}$

$\theta_D + 2\theta_E - 3\psi_2 + \theta_B + 2\theta_E + 2\theta_E - 3\psi_1 = 0$

$\theta_B + \theta_D + 6\theta_E - 3\psi_1 - 3\psi_2 = 0$ 　　　　　　　 EQ. ④

$\Sigma H = -10$, TOP LEVEL:

$$\frac{M_{BC} + M_{CB}}{20} + \frac{M_{DE} + M_{ED}}{20} = -10$$

$$\frac{2\theta_B + \theta_C - 3\psi_2 + \theta_B + 2\theta_C - 3\psi_2}{20} + \frac{2\theta_D + \theta_E - 3\psi_2 + \theta_D + 2\theta_E - 3\psi_2}{20} = -10$$

$3\theta_B + 3\theta_C + 3\theta_D + 3\theta_E - 12\psi_2 = -200$ 　　　　 EQ. ⑤

$\Sigma H = 30$, BOTTOM LEVEL:

$$\frac{M_{AB} + M_{BA}}{20} + \frac{M_{EF} + M_{FE}}{20} = -30$$

$3\theta_B + 3\theta_E - 12\psi_1 = -600$ 　　　　　　　 EQ. ⑥

→ SOLVING THE ⑥ EQ. SIMULTANEOUSLY:

$\theta_B = \theta_E = +52.75$ 　　　　 $\psi_1 = +76.37$

$\theta_C = \theta_D = +21.82$ 　　　　 $\psi_2 = +53.95$

FINAL MOMENTS:

$M_{AB} = M_{FE} = -176.4$ KIPS-FT	$M_{BE} = M_{EB} = +158.2$ KIP-FT
$M_{BA} = M_{EF} = -123.6$ KIP-FT	$M_{CB} = M_{DE} = -65.5$ KIP-FT
$M_{BC} = M_{ED} = -34.6$ KIP-FT	$M_{CD} = M_{DC} = +65.5$ KIP-FT

JOB NAME: *TWO-STORY OFFICE BUILDING DESIGN*	JOB NUMBER: *17-089*
JOB LOCATION: *LINCOLN WAY WEST, MASSILLON, OHIO*	DATE: *11-10-2017*
DESIGNED BY: *SAAD HASAN TANTAWI (M.S.CE, B.S.CE, E.I.)*	SHEET **47** OF

△ <u>MOMENT CONNECTION DESIGN</u> :

$V_{DL} = 7 \text{ KIPs}$
$V_L = 21 \text{ KIPs}$
$M_D = 42 \text{ KIP-FT}$
$M_L = 126 \text{ KIP-FT}$

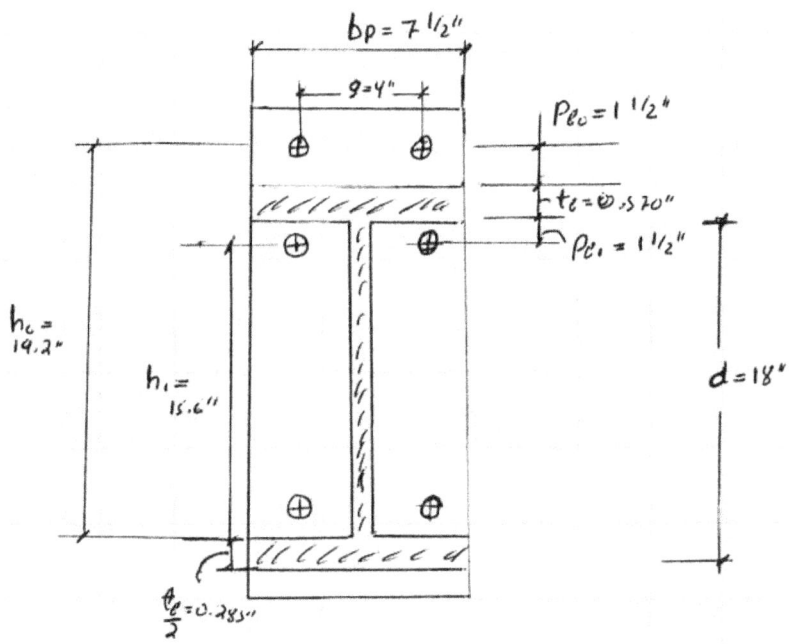

((ALL BOLTS ARE 1" DIA. ASTM A325-N SNUG-TIGHTENED BOLTS))

→ BEAM <u>W18×50</u> : $d = 18.0$ IN. $t_w = 0.355$ IN
  $b_\ell = 7.50$ IN. $S_x = 88.9$ IN³.
  $t_\ell = 0.570$ IN

→ COLUM <u>W14×99</u> : $d = 14.2$ IN $b_\ell = 14.6$ IN.
  $t_\ell = 0.780$ IN. $t_w = 0.485$ IN.
  $K_{des} = 1.38$ IN.

→

JOB NAME: *TWO-STORY OFFICE BUILDING DESIGN*	JOB NUMBER: *17-089*
JOB LOCATION: *LINCOLN WAY WEST, MASSILLON, OHIO*	DATE: *11-10-2017*
DESIGNED BY: *SAAD HASAN TANTAWI (M.S.CE, B.S.CE, E.I.)*	SHEET 48 OF

$R_a = 7 + 21 = 28$ KIPS

$M_a = 42 + 126 = 168$ KIP-FT.

➡ EXTENDED END PL. PROPERTIES:

$$h_o = d + P_{eo} - \frac{t_e}{2}$$

$$= 18 + 1.5 - \frac{0.570}{2} = 19.2 \text{ IN.}$$

$$h_1 = d - P_{ei} - t_e - \frac{t_e}{2}$$

$$= 18 - 1.5 - 0.57 - \frac{0.57}{2} = 15.6 \text{ IN.}$$

➡ REQ'D BOLT DIA. ASSUMING NO PRYING ACTION.

$$d_b \text{ REQ'D} = \sqrt{\frac{2M_a \Omega}{\pi F_{nt}(h_o + h_i)}} = \sqrt{\frac{(2)(168)(12)(2)}{\pi(90)(19.2 + 15.6)}}$$

$$= 0.405 \text{ IN.}$$

USE 1-IN. DIA. ASTM A325-N SNUG TIGHT BOLTS

➡ REQ'D END PLATE THICKNESS:

$$S = \frac{\sqrt{b_p g}}{2} = \frac{\sqrt{7 \frac{1}{2} \text{ IN.} (4 \text{ IN})}}{2} = 2.74 \text{ IN.} > P_{ei} = 1.5 \text{ IN.} \underline{\text{O.K}}$$

$$Y_p = \frac{b_p}{2}\left[h_i\left(\frac{1}{P_{ei}} + \frac{1}{s}\right) + h_o\left(\frac{1}{P_{eo}}\right) - \frac{1}{2}\right] + \frac{2}{g}\left[h_i(P_{ei} + s)\right]$$

SOLVING FOR Y_p. $Y_p = 140$ IN.

$$P_t = F_{nt}\left(\frac{\pi d_b^2}{4}\right) = \frac{90(\pi)}{4} = 70.7 \text{ KIPS}$$

➡

JOB NAME: *TWO-STORY OFFICE BUILDING DESIGN*	JOB NUMBER: *17-089*
JOB LOCATION: *LINCOLN WAY WEST, MASSILLON, OHIO*	DATE: *11-10-2017*
DESIGNED BY: *SAAD HASAN TANTAWI (M.S.CE, B.S.CE, E.I.)*	SHEET **49** OF

$\Rightarrow \quad M_{np} = 2P_t(h_0 + h_1)$
$= 2(70.7)(19.2 + 15.6)$
$= 4920 \text{ KIP-IN.}$

THE NO PRYING BOLT AVAILABLE FLEXURAL STRENGTH IS:

$\Omega = 2.0 \rightarrow \dfrac{M_{np}}{\Omega} = \dfrac{4920 \text{ K-IN}}{2} = 2460 \text{ KIP-IN.}$

$t_{p\ REQ'D} = \sqrt{\dfrac{1.11\left(\dfrac{M_{np}}{\Omega}\right)}{\left(\dfrac{F_{yp}}{\Omega}\right)y_p}} = \sqrt{\dfrac{1.11(2460)}{\left(\dfrac{36}{1.67}\right)(140)}} = 0.951 \text{ IN.}$

USE A 1" THICK GR. 36 END PLATE.

$M_{all.} = \dfrac{F_{yp}\,t_p^2\,y_p}{1.11\,\Omega b}$

$= \dfrac{36(1)^2(140)}{1.11(1.67)} = 2720 \text{ K-IN.}$ o.k.

$\Rightarrow$ BEAM FLANGE FORCE: THE REQ'D FORCE APPLIED TO THE END PLATE THROUGH THE BEAM FLANGE IS:

$F_{fa} = \dfrac{M_a}{d - t_b}$

$= \dfrac{168 \text{ KIP-FT}(12 \text{ IN./FT})}{(18.0 - 0.570) \text{ IN}}$

$= 116 \text{ KIPS}$

$\Rightarrow$ SHEAR YIELDING OF THE EXTENDED END-PLATE:

39

JOB NAME: *TWO-STORY OFFICE BUILDING DESIGN*	JOB NUMBER: *17-089*
JOB LOCATION: *LINCOLN WAY WEST, MASSILLON, OHIO*	DATE: *11-10-2017*
DESIGNED BY: *SAAD HASAN TANTAWI (M.S.CE, B.S.CE, E.I.)*	SHEET **50** OF

$$\frac{R_n}{\Omega} = \frac{0.6\,F_{yp}\,b_p\,t_p}{\Omega} > \frac{F_{fa}}{2}$$

$$= \frac{0.6\,(36)\,(7.5)\,(1.0)}{1.67} > \frac{116\;KIPS}{2}$$

$$= 97.0\;KIPS > 58.0\;KIPS \quad (O.K)$$

→ SHEAR RUPTURE OF THE EXTENDED END-PLATE:

$$A_n = \left[7.5 - 2\left(1\frac{1}{16}\;IN. + \frac{1}{16}\;IN.\right)\right](1.0\;IN.) = 5.25\;IN^2$$

$$\frac{R_n}{\Omega} = \frac{0.6\,F_{up}\,A_n}{\Omega} > \frac{F_{fa}}{2} = \frac{0.6\,(58\;KSI)\,(5.25\;IN^2)}{2} > \frac{116\;KIPS}{2}$$

$$\frac{R_n}{\Omega} = 91.4\;KIPS > 58.0\;KIPS. \qquad O.K.$$

→ BOLT SHEAR AND BEARING:

TRY THE MIN. OF FOUR BOLTS AT THE TENSION FLANGE AND TWO BOLTS AT THE COMPRESSION FLANGE.

BOLT SHEAR STRENGTH : $\dfrac{R_n}{\Omega} = \dfrac{n\,r_n}{\Omega} = 2\;BOLTS\,(21.2\;KIPS/BOLT)$
$$= 42.4\;KIPS > 28.0\;KIPS \quad O.K$$

BOLT BEARING ON THE END-PLATE $(Le \geq Le_{full})$:

$$\frac{R_n}{\Omega} = \frac{n\,r_n}{\Omega} = (69.6\;KIPS/IN./BOLT)(1.0\;IN.)$$
$$= \frac{69.6\;KIPS}{BOLT} > \frac{21.2\;KIPS}{BOLT} \qquad O.K.$$

BOLT BEARING ON COLUMN FLANGE :

$$\frac{R_n}{\Omega} = \frac{n\,r_n}{\Omega} = (78.0)(0.780) = \frac{60.8\;KIPS}{BOLT} > \frac{21.2\;KIPS}{BOLT}$$

$$\underline{BOLT\;SHER\;CONTROLS = 42.4\;KIPS}$$

JOB NAME: *TWO-STORY OFFICE BUILDING DESIGN*	JOB NUMBER: *17-089*
JOB LOCATION: *LINCOLN WAY WEST, MASSILLON, OHIO*	DATE: *11-10-2017*
DESIGNED BY: *SAAD HASAN TANTAWI (M.S.CE, B.S.CE, E.I.)*	SHEET *51* OF

⚠ MIN. WELD SIZE REQ'D TO MATCH THE SHEAR RUPTURE STRENGTH OF THE WELD TO THE TENSION YIELD STRENGTH OF THE BEAM WEB IS:

$$D_{MIN} = \frac{F_y t_w (1.0 \ IN.)}{1.2 \ (2)(1.5)(0.928)(1.0 \ IN)}$$

$$= \frac{(50)(0.355)(1.0)}{1.67 \ (2)(1.5)(0.928)(1.0)} = 3.82$$

USE $\frac{1}{4}$ - IN. FILLET WELDS ON BOTH SIDES.

NOTE: 1.5 FACTOR IS TO ACCOUNT FOR THE INCREASED STRENGTH OF A TRANSVERELY LOADED FILLET WELD.

→ WELD SIZE REQ'D FOR THE END REACTION

THE END REACTION R_a, IS RESISTED BY THE LESSER OF THE OF THE BEAM WEB-TO-END-PLATE WELD:

1) BETWEEN THE MID-DEPTH OF THE BEAM AND THE INSIDE FACE OF THE COMPRESSION FLANGE. ← CONTROLS.

2) BETWEEN THE INNE ROW OF TENSION BOLTS PLUS TWO BOLT DIA. AND THE INSIDE FACE OF THE BEAM COMPRESSION FLANGE

$$\ell = \frac{d}{2} - t_\ell = \frac{18.0 \ IN.}{2} - 0.570 \ IN.$$

$$\ell = 8.43 \ IN.$$

$$D_{MIN} = \frac{R_a}{2(0.928)(8.43)} = \frac{1.79}{16^{th} IN} \ 0$$

USE $\frac{3}{16}$ - IN. FILLET WELD ON BOTH SIDES OF THE BEAM WEB BELOW THE TENSION - BOLT REGION.

JOB NAME: *TWO-STORY OFFICE BUILDING DESIGN*	JOB NUMBER: *17-089*
JOB LOCATION: *LINCOLN WAY WEST, MASSILLON, OHIO*	DATE: *11-10-2017*
DESIGNED BY: *SAAD HASAN TANTAWI (M.S.CE, B.S.CE, E.I.)*	SHEET **52** OF

➤ CONNECTING ELEMENTS RUPTURE STRENGTH @ WELDS:

$$t_{min.} = \frac{6.19D}{F_u} = \frac{6.19 \,(1.79 \text{ SIXTEENTHS})}{65 \text{ KSI}}$$
$$= 0.170 \text{ IN.} < 0.355 \text{ IN. BEAM WEB O.K}$$

$$t_{min.} = \frac{3.09D}{F_u} = \frac{3.09 \,(1.79 \text{ SIXTEENTHS})}{58 \text{ KSI}}$$
$$= 0.09454 \text{ IN.} < 1.00 \text{ IN. END-PLATE O.K.}$$

➤ REQ'D FILLET WELD SIZE FOR THE BEAM FLANGE TO END-PLATE CONN.:

$$l = 2(b_f) - t_w$$
$$= 2\,(7.50 \text{ IN}) - 0.355 \text{ IN.}$$
$$= 14.6 \text{ IN.}$$

$$F_{fa} = 116 \text{ KIPS} \rightarrow D_{min} = \frac{F_{fa}}{1.5\,(0.928)\,l} = \frac{116 \text{ KIPS}}{1.5\,(0.928)(14.6 \text{ IN.})}$$
$$D_{min.} = 5.71 \rightarrow \underline{6 \text{ SIXTEENTHS.}}$$

USE $\frac{3}{8}$ IN. FILLET WELDS @ THE BEAM TENSION FLANGE

USE $\frac{1}{4}$ IN. FILLET WELDS @ COMPRESSION FLANGE
(MIN. SIZE PER AISC TABLE J2.4)

➤ CONNECTING ELEMENTS RUPTURE STRENGTH @ WELDS:

SHEAR RUPTURE STRENGTH OF BASE METAL:

$$t_{min} = \frac{3.09D}{F_u} = \frac{3.09\,(5.71 \text{ SIXTEENTHS})}{58 \text{ KSI}}$$
$$= 0.304 \text{ IN.} < 1.00 \text{ IN. END-PLATE} \underline{O.K.}$$

2-7 References

1. *Minimum Design Loads for Buildings and Other Structures,* American Society of Civil Engineers (ASCE/SEI 7-10). Reston, VA, 2010.

2. *Manual of Steel Construction,* 14th ed. American Institute of Steel Construction, Inc., Chicago, IL, 2010.

3. *Standard Specifications, Load Tables, and Weight Tables for Steel Joists and Joist Girders.* Steel Joist Institute, Forest, VA, 2005.

4. *Diaphragm Design Manual,* 3rd ed. Steel Deck Institute, Fox River Grove, IL, 2004.

5. West, M., Fisher, J. and Griffis, L.A. *Serviceability Design Considerations for Steel Buildings, Design Guide 3,* 2nd ed. AISC, Chicago, IL, 2003.

6. *Design Examples,* 14th ed. American Institute of Steel Construction, Inc., Chicago, IL, 2011.

7. *Construction Industry Digest* (OSHA 2202). U.S. Department of Labor, Washington, D.C.

8. *Code of Federal Regulations,* Title 29, Chapter XVII, Part 1926. Office of the Federal Register, National Archives and Records Administration, Washington, D.C.

3 EXISTING OPEN WEB ROOF JOISTS REINFORCEMENT

3-1 Evaluation of Joists and Their Typical Reinforcement

Unforeseen changes often occur in the design loads on a building, either during construction or afterwards in the existing structure. One of the structural components that is highly impacted by any change in loading is the open-web steel joist. The most common design change in the field is due to concentrated loads such as HVAC units. Therefore, it is imperative that the building owner or general contractor notify the joist fabricator or design engineer immediately of any changes to the loads before proceeding with any installations or reinforcements

3-2 Loading Pattern

- The top chord members (composed of two angles welded together back-to-back) are subjected to compressive forces.

- The bottom chord members (composed of two angles welded together back-toback) are subjected to tensile forces.

- The first or end diagonal is primarily subjected to tensile forces.

- The second diagonal and the other diagonals sloped in the same direction from the bottom chord (generally towards the centre of the joist) are primarily subjected to compressive forces.

- The third diagonal and the other diagonals sloped in the same direction from the bottom chord (generally towards the end of the joist) are primarily subjected to tensile forces.

- The vertical members are normally subjected to compressive forces. (The first vertical, usually located between the first and second diagonals, may be inclined from the bottom chord towards the end of the joist.)

3-3 Round Bars Reinforcement

The most commonly used type of reinforcement for top chords and bottom chords is the addition of round rods having a diameter according to the capacity of each existing member subjected to the added applied forces (Figure 3-1).

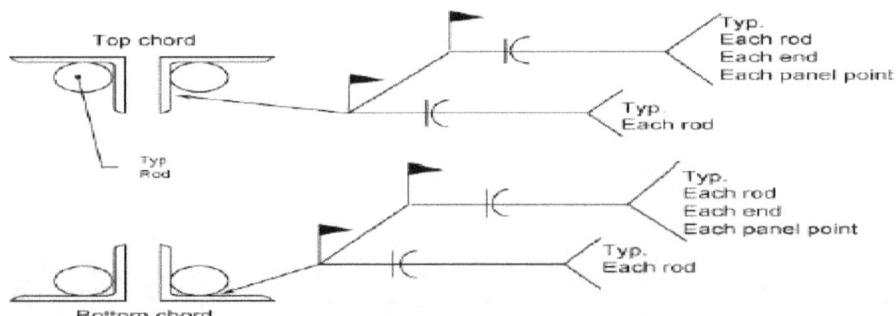

Figure 3-1: Joist Chord Reinforcement with Round Bars

3-4   Angle Reinforcement

For diagonal or vertical members subjected to overstressing of the compressive or tensile forces, the type of repair procedure may vary. To reinforce these members, the best option is to add two angles, one on either side of the affected member (Figure 3-2). The weld, which is essential in distributing the load, is also applied to the top and bottom chord members. However, if the top and bottom chords also need to be reinforced, a solution must be chosen so as not to interfere with the chord members (Figures 3-2 and 3-3).

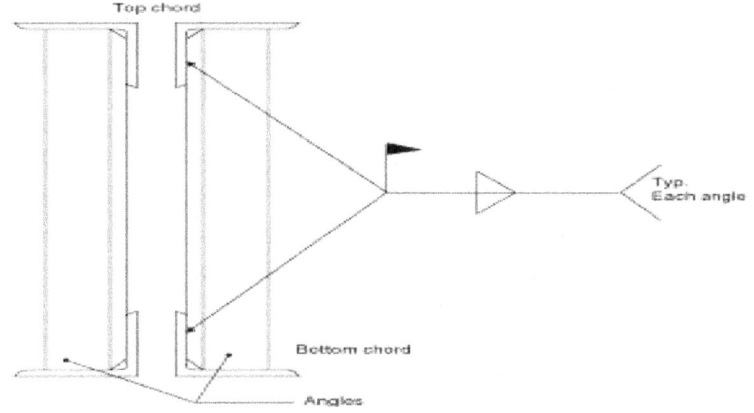

Figure 3-2: Angle Reinforcement

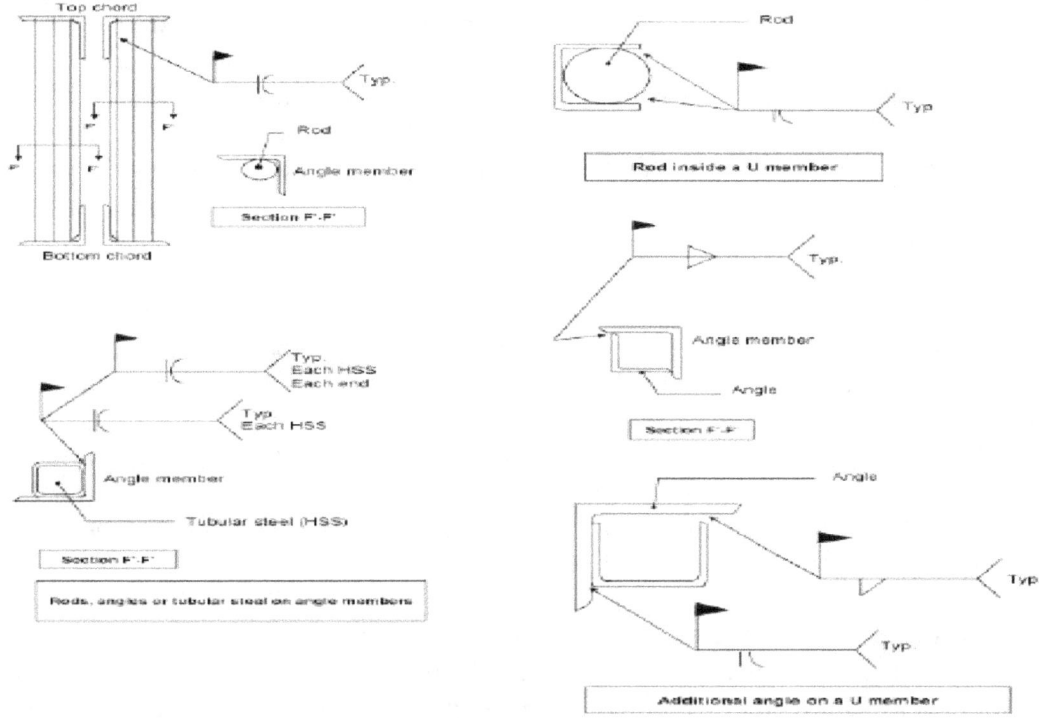

Figure 3-3: Angle Reinforcement

3-5   References

1.   *Construction Industry Digest* (OSHA 2202). U.S. Department of Labor, Washington, D.C.

4   DETAILS

4-1   Why the need of accurate structural details is important

The Structural Engineer plays a vital part to the successful completion of a project through construction and their involvement during construction with the Owner, Architect, other design consultants, and contractor is critical.

A Structural Engineer provides a key link for the Contractor during construction, but the SE's role and connection with construction of the project starts well before the first shovel hits dirt. Coordination of a well-planned and detailed set of construction documents is a critical first step. Unforeseen site conditions are costly enough, however, without this first step, a project can get lost in request for information (RFI's) or confusing drawings. A well-coordinated set of documents includes coordination among the many design disciplines. This involves checks between different consultant drawings and collaboration of design team members during meetings to resolve challenges during design and before construction.
On many projects, a general contractor may be involved early in a pre-construction role or on a Design-Build project. This provides a great opportunity to coordinate and collaborate during design with the entity that is going to build the structure. There are many different ideas on the best methods to construct and working together during design should greatly enhance the ability of the project to be built.

There may be no more critical time period for a project schedule than during construction and within the construction period one could argue that completion of the building structure is quite possibly the most critical. While other phases of construction may allow for multiple trades to be working on site, the building structure physically supports all other elements and must be completed first before others begin. Keeping the structure schedule on track is critically important and the SE needs to keep up with the pace of the Contractor. Beyond responding to RFI's and submittals as quickly as possible (which should go without saying) and at most no more than the time allotted by the project specifications, projects that allow for regular site visits with the Contractor to answer questions and provide clarification go a long way to keeping job sites busy and productive.

This last item is a key and critical point to providing construction administration for the Structural Engineer. Site visits and Code required Structural Observations are necessary for the success of virtually all projects. Below is the description of Structural Observation from the City of Los Angeles DBS Structural Observation Form:
"STRUCTURAL OBSERVATION means the visual observation of the structural system, for general conformance to the approved plans and specifications, at significant construction stages and at completion of the structural system. Structural observation does not include or waive the responsibility for the inspections required by Section 108, 1701 or other sections of the code."

Site visits and observations go hand in hand with other disciplines and help to provide checks and balances during construction as noted above. Inspections by third party inspectors provide specific review of the contractors work and provide the SE with information on the quantity and quality of constructed material. The SE works in conjunction with the inspection team and the Contractors to verify the structure is being constructed in conformance with the approved construction documents.

4-2   Abbreviations

Each detail in this book has been carefully considered and tested in an actual construction operation, please note T.S. referenced through the detailing part is in reference to HSS beams and columns.

Whenever you see a section cut with the page marked "NEXT" that means the actual detail will be in the next series, for example light gauge metal bracing is been put with the light gauge metal detailing part of the series. If you see the word "PREV." in a section cut, that means the actual detail is found in this part of the series in a previous detail.

Details have been carefully considered to include roof cuts with masonry walls, light gauge metal walls with brick veneer, wood stud wall, and metal siding. Some section cuts details has been omitted for simplicity.

The following is a list of different symbols and meaning associated with it as it is found throughout the book: (See glossary for a complete list of abbreviations)

STL.	-	STEEL
L.G.M	-	LIGHT GAUGE METALS
T.S.	-	HSS SECTIONS
FFE	-	FINISH FLOOR ELEVATION
J.B.E	-	JOIST BEARING ELEVATION
T.O.S	-	TOP OF STEEL
PL.	-	PLATE
F.V.	-	FIELD VERIFY
LLH	-	LONG LEG HORIZONTAL
LLV	-	LONG LEG VERTICAL
U.N.O	-	UNLESS NOTED OTHERWISE
TYP.	-	TYPICAL

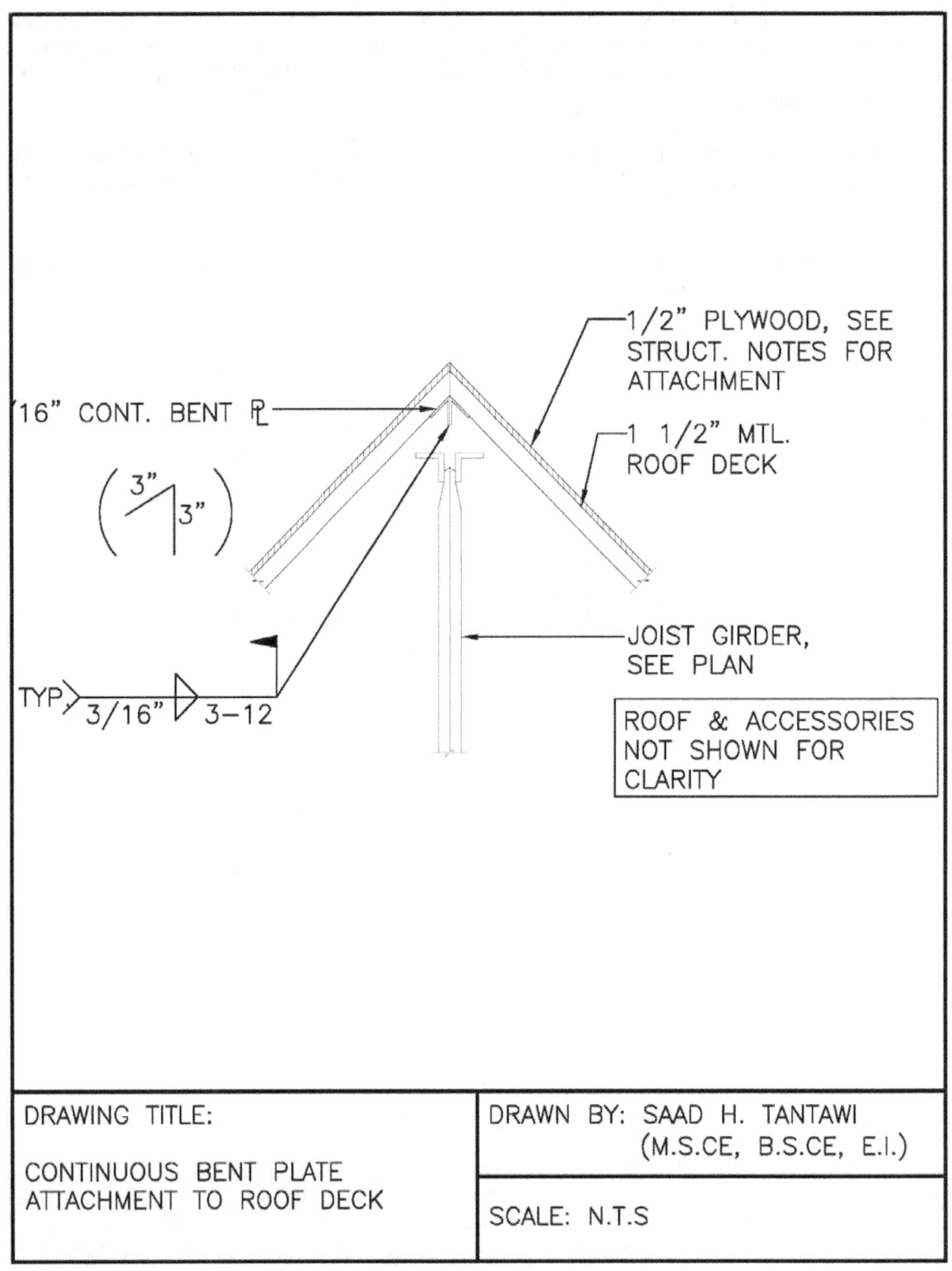

16" CONT. BENT ℄

1/2" PLYWOOD, SEE STRUCT. NOTES FOR ATTACHMENT

1 1/2" MTL. ROOF DECK

JOIST GIRDER, SEE PLAN

ROOF & ACCESSORIES NOT SHOWN FOR CLARITY

TYP 3/16" 3-12

DRAWING TITLE:	DRAWN BY: SAAD H. TANTAWI
	(M.S.CE, B.S.CE, E.I.)
CONTINUOUS BENT PLATE ATTACHMENT TO ROOF DECK	
	SCALE: N.T.S

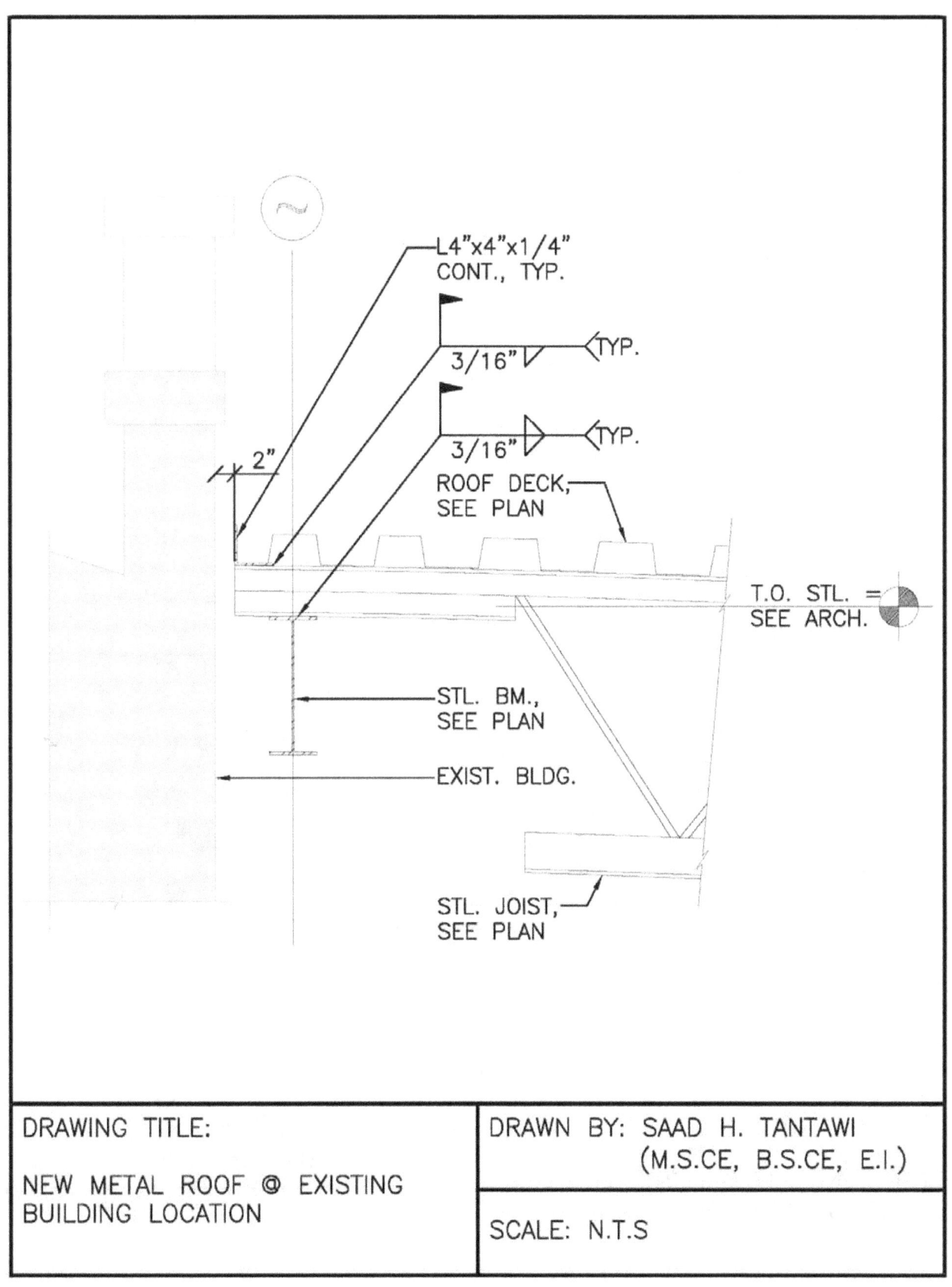

L4"x4"x1/4"
CONT., TYP.

3/16" TYP.

3/16" TYP.

2"

ROOF DECK,
SEE PLAN

T.O. STL. =
SEE ARCH.

STL. BM.,
SEE PLAN

EXIST. BLDG.

STL. JOIST,
SEE PLAN

DRAWING TITLE:	DRAWN BY: SAAD H. TANTAWI (M.S.CE, B.S.CE, E.I.)
NEW METAL ROOF @ EXISTING BUILDING LOCATION	SCALE: N.T.S

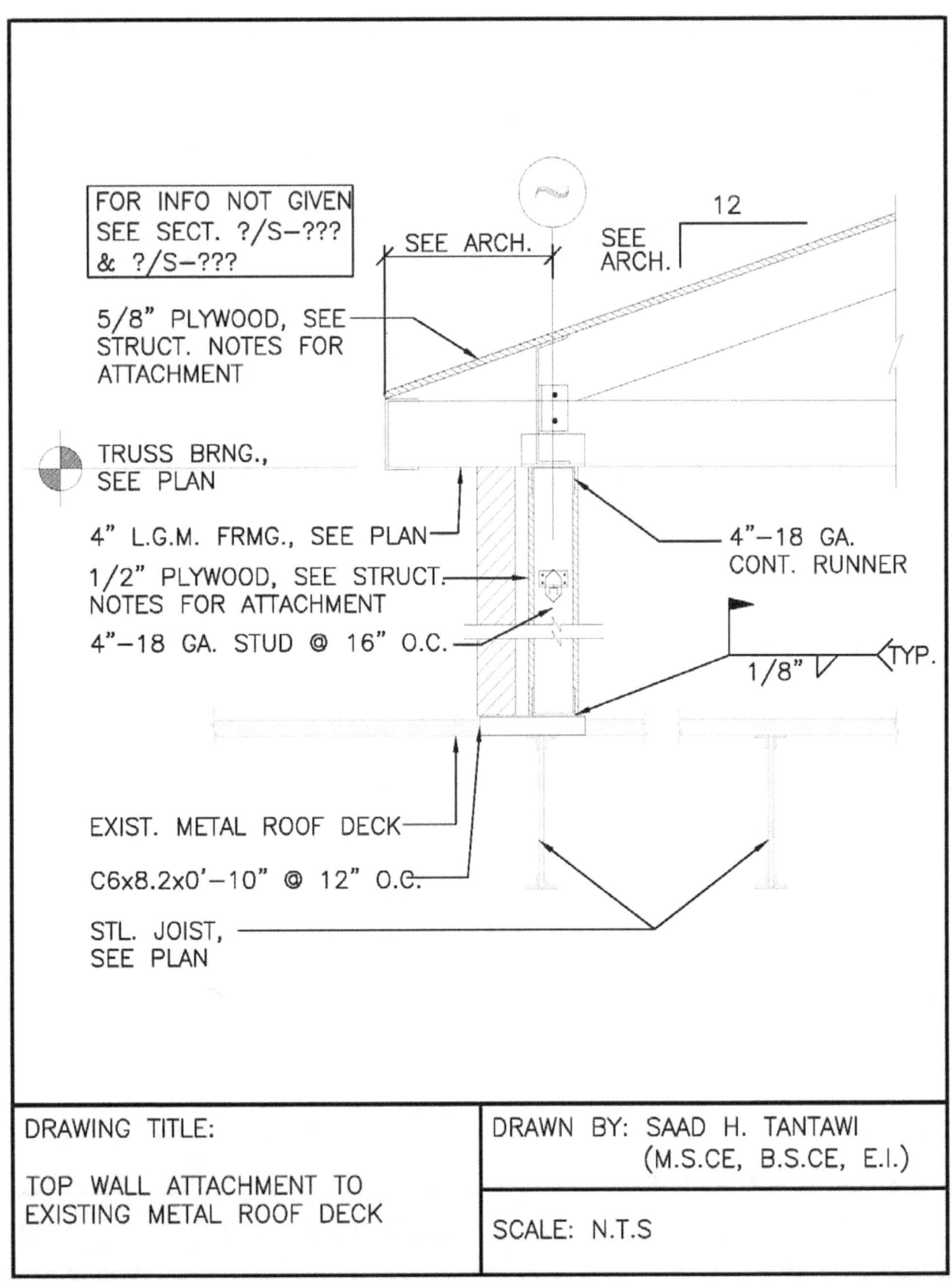

FOR INFO NOT GIVEN SEE SECT. ?/S-??? & ?/S-???

SEE ARCH.

SEE ARCH.

12

5/8" PLYWOOD, SEE STRUCT. NOTES FOR ATTACHMENT

TRUSS BRNG., SEE PLAN

4" L.G.M. FRMG., SEE PLAN

1/2" PLYWOOD, SEE STRUCT. NOTES FOR ATTACHMENT

4"-18 GA. STUD @ 16" O.C.

4"-18 GA. CONT. RUNNER

1/8" TYP.

EXIST. METAL ROOF DECK

C6x8.2x0'-10" @ 12" O.C.

STL. JOIST, SEE PLAN

DRAWING TITLE:	DRAWN BY: SAAD H. TANTAWI (M.S.CE, B.S.CE, E.I.)
TOP WALL ATTACHMENT TO EXISTING METAL ROOF DECK	SCALE: N.T.S

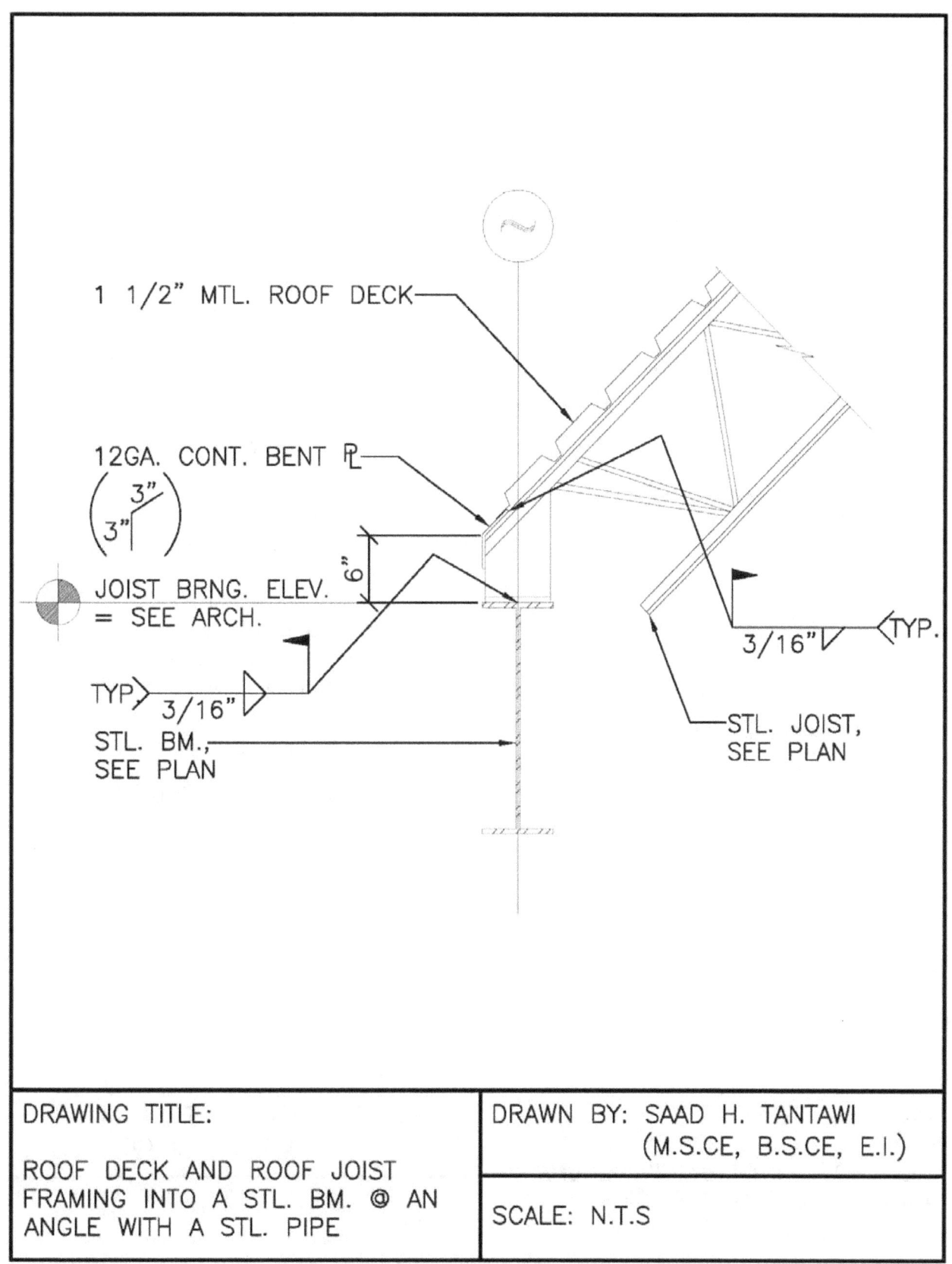

1 1/2" MTL. ROOF DECK

12GA. CONT. BENT PL
(3" / 3")

JOIST BRNG. ELEV.
= SEE ARCH.

6"

TYP> 3/16"

STL. BM.,
SEE PLAN

3/16" TYP.

STL. JOIST,
SEE PLAN

DRAWING TITLE:	DRAWN BY: SAAD H. TANTAWI (M.S.CE, B.S.CE, E.I.)
ROOF DECK AND ROOF JOIST FRAMING INTO A STL. BM. @ AN ANGLE WITH A STL. PIPE	SCALE: N.T.S

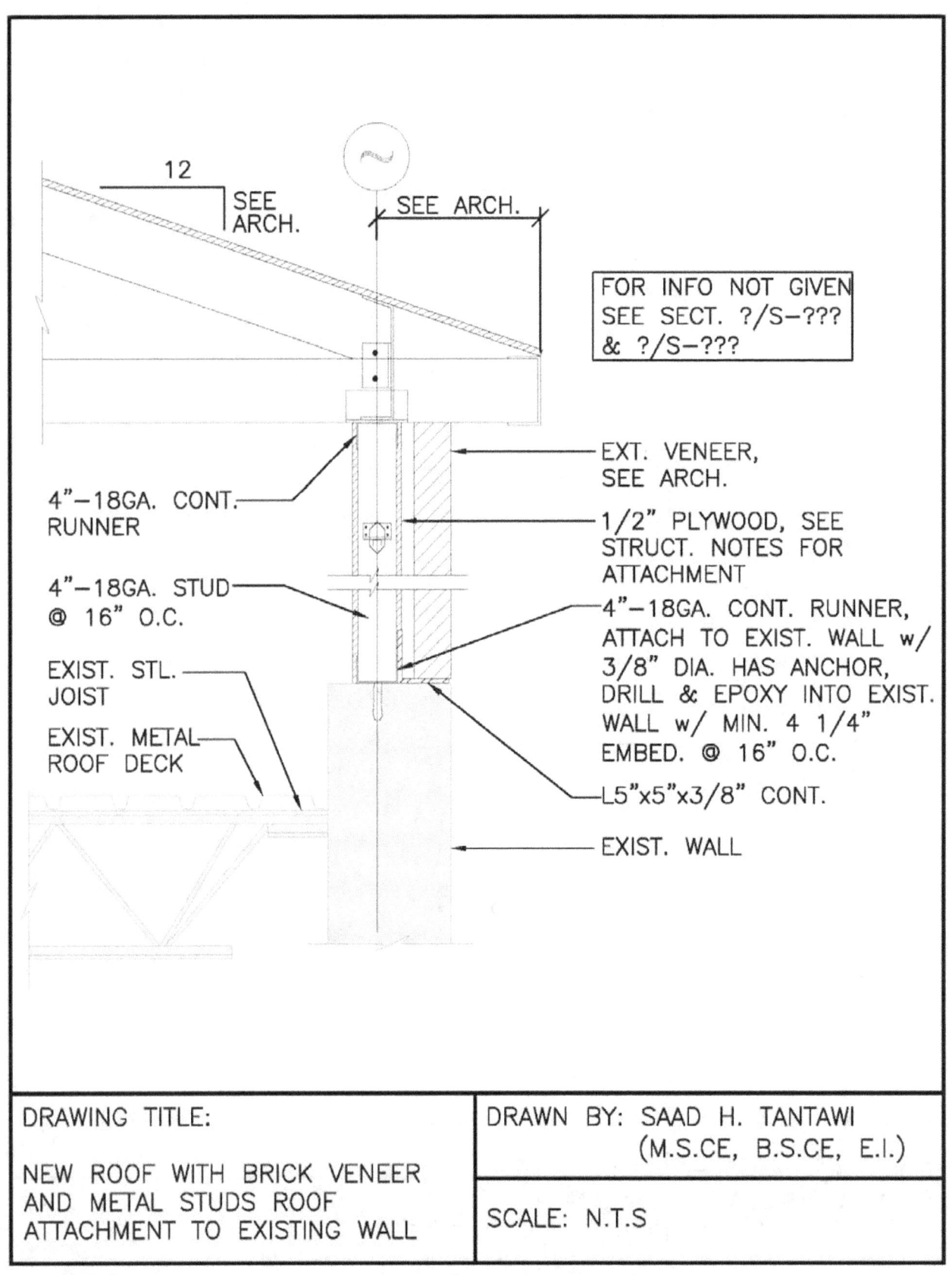

12

SEE ARCH.

SEE ARCH.

FOR INFO NOT GIVEN
SEE SECT. ?/S-???
& ?/S-???

EXT. VENEER,
SEE ARCH.

4"-18GA. CONT.
RUNNER

1/2" PLYWOOD, SEE
STRUCT. NOTES FOR
ATTACHMENT

4"-18GA. STUD
@ 16" O.C.

4"-18GA. CONT. RUNNER,
ATTACH TO EXIST. WALL w/
3/8" DIA. HAS ANCHOR,
DRILL & EPOXY INTO EXIST.
WALL w/ MIN. 4 1/4"
EMBED. @ 16" O.C.

EXIST. STL.
JOIST

EXIST. METAL
ROOF DECK

L5"x5"x3/8" CONT.

EXIST. WALL

DRAWING TITLE:	DRAWN BY: SAAD H. TANTAWI
	(M.S.CE, B.S.CE, E.I.)
NEW ROOF WITH BRICK VENEER AND METAL STUDS ROOF ATTACHMENT TO EXISTING WALL	SCALE: N.T.S

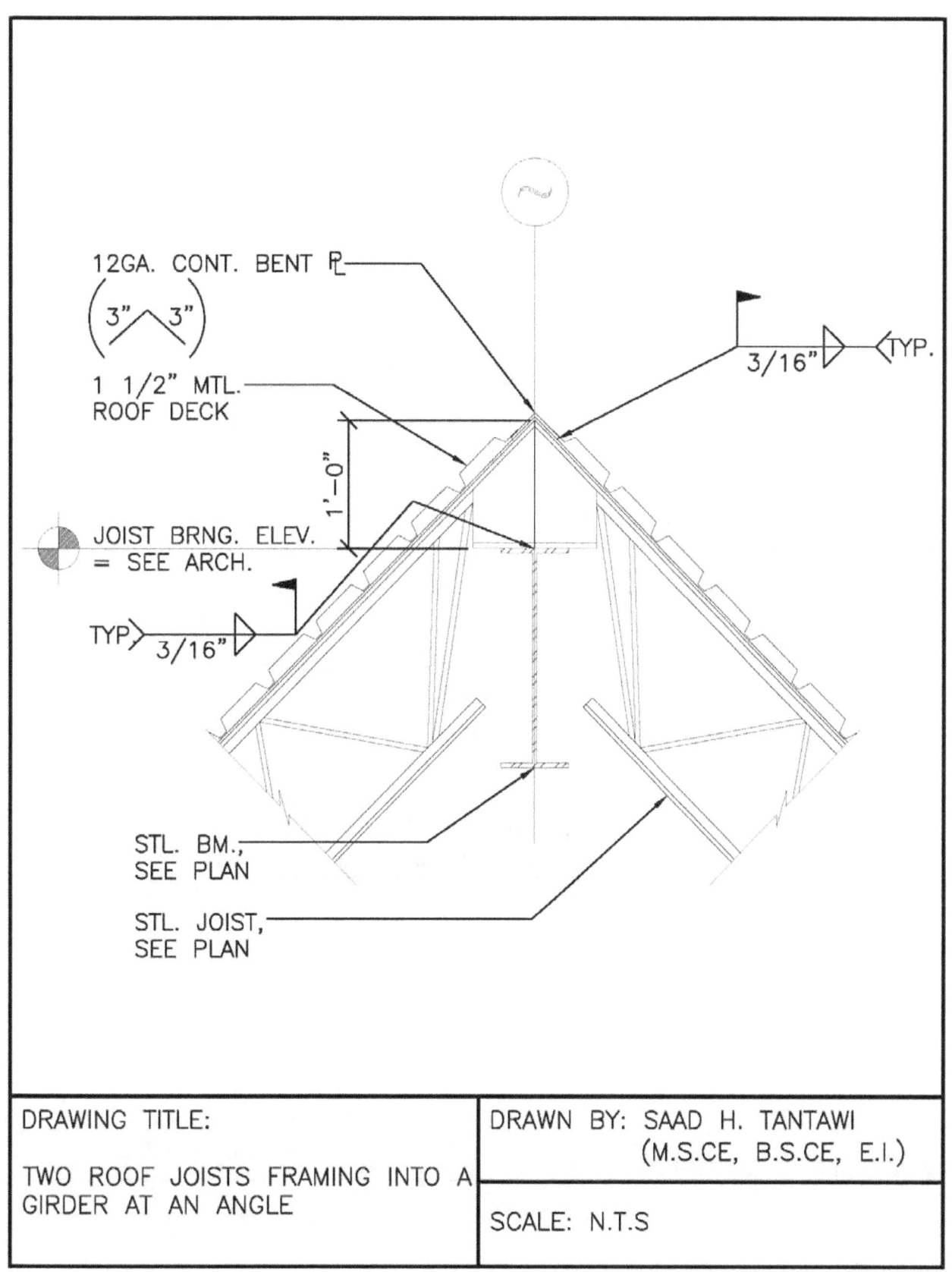

12GA. CONT. BENT ℙ

(3" ⋀ 3")

1 1/2" MTL.
ROOF DECK

1'-0"

3/16" ▷ TYP.

JOIST BRNG. ELEV.
= SEE ARCH.

TYP ▷ 3/16"

STL. BM.,
SEE PLAN

STL. JOIST,
SEE PLAN

DRAWING TITLE:	DRAWN BY: SAAD H. TANTAWI
	(M.S.CE, B.S.CE, E.I.)
TWO ROOF JOISTS FRAMING INTO A GIRDER AT AN ANGLE	
	SCALE: N.T.S

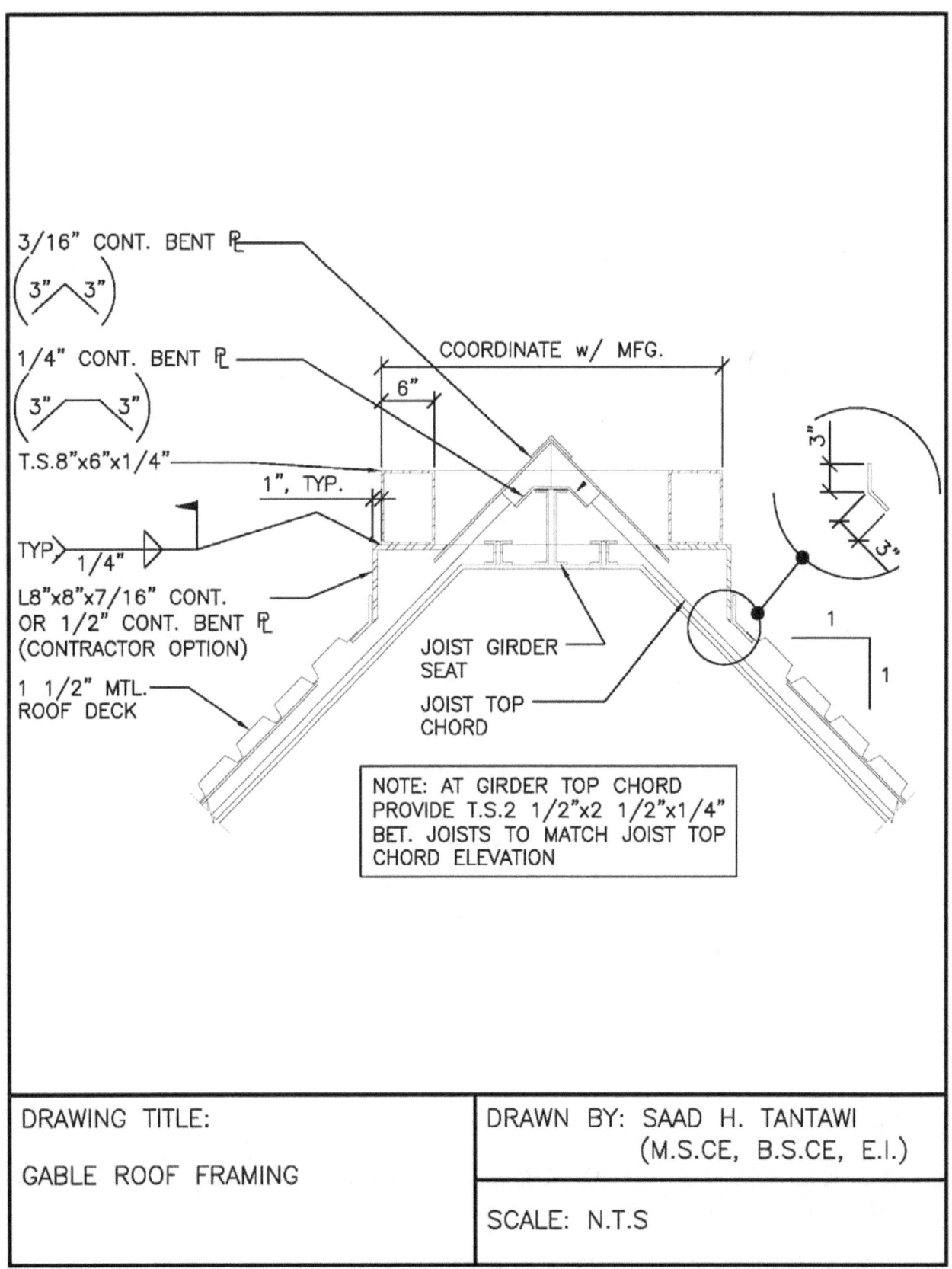

3/16" CONT. BENT ℓ
(3" ∧ 3")

1/4" CONT. BENT ℓ
(3" ⌐⌐ 3")

T.S.8"x6"x1/4"

1", TYP.

TYP 1/4"

L8"x8"x7/16" CONT.
OR 1/2" CONT. BENT ℓ
(CONTRACTOR OPTION)

1 1/2" MTL.
ROOF DECK

COORDINATE w/ MFG.

6"

3"

3"

1

1

JOIST GIRDER
SEAT

JOIST TOP
CHORD

NOTE: AT GIRDER TOP CHORD
PROVIDE T.S.2 1/2"x2 1/2"x1/4"
BET. JOISTS TO MATCH JOIST TOP
CHORD ELEVATION

DRAWING TITLE:	DRAWN BY: SAAD H. TANTAWI
	(M.S.CE, B.S.CE, E.I.)
GABLE ROOF FRAMING	SCALE: N.T.S

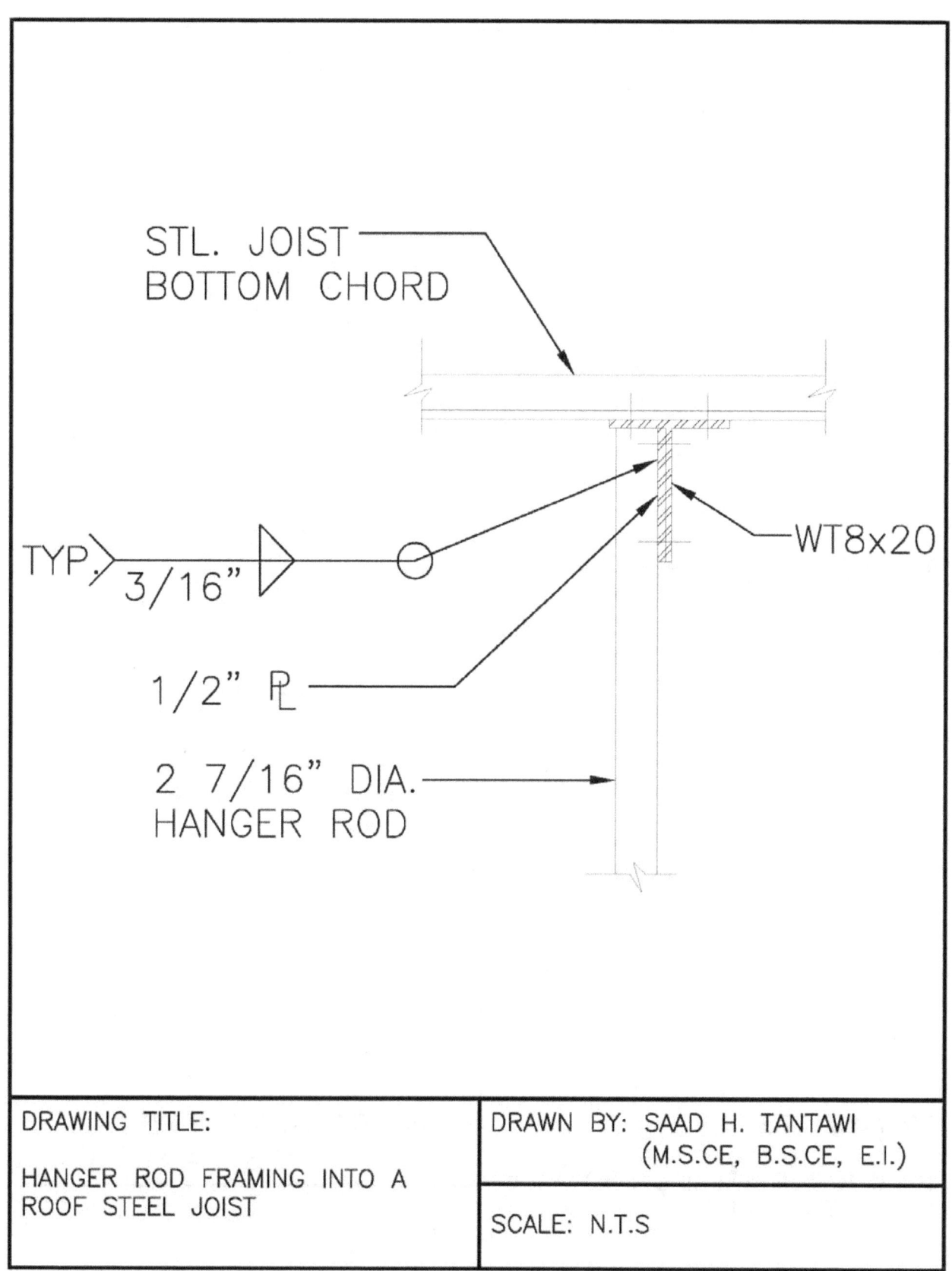

STL. JOIST
BOTTOM CHORD

TYP. 3/16"

1/2" PL

2 7/16" DIA.
HANGER ROD

WT8x20

DRAWING TITLE:	DRAWN BY: SAAD H. TANTAWI
HANGER ROD FRAMING INTO A ROOF STEEL JOIST	(M.S.CE, B.S.CE, E.I.)
	SCALE: N.T.S

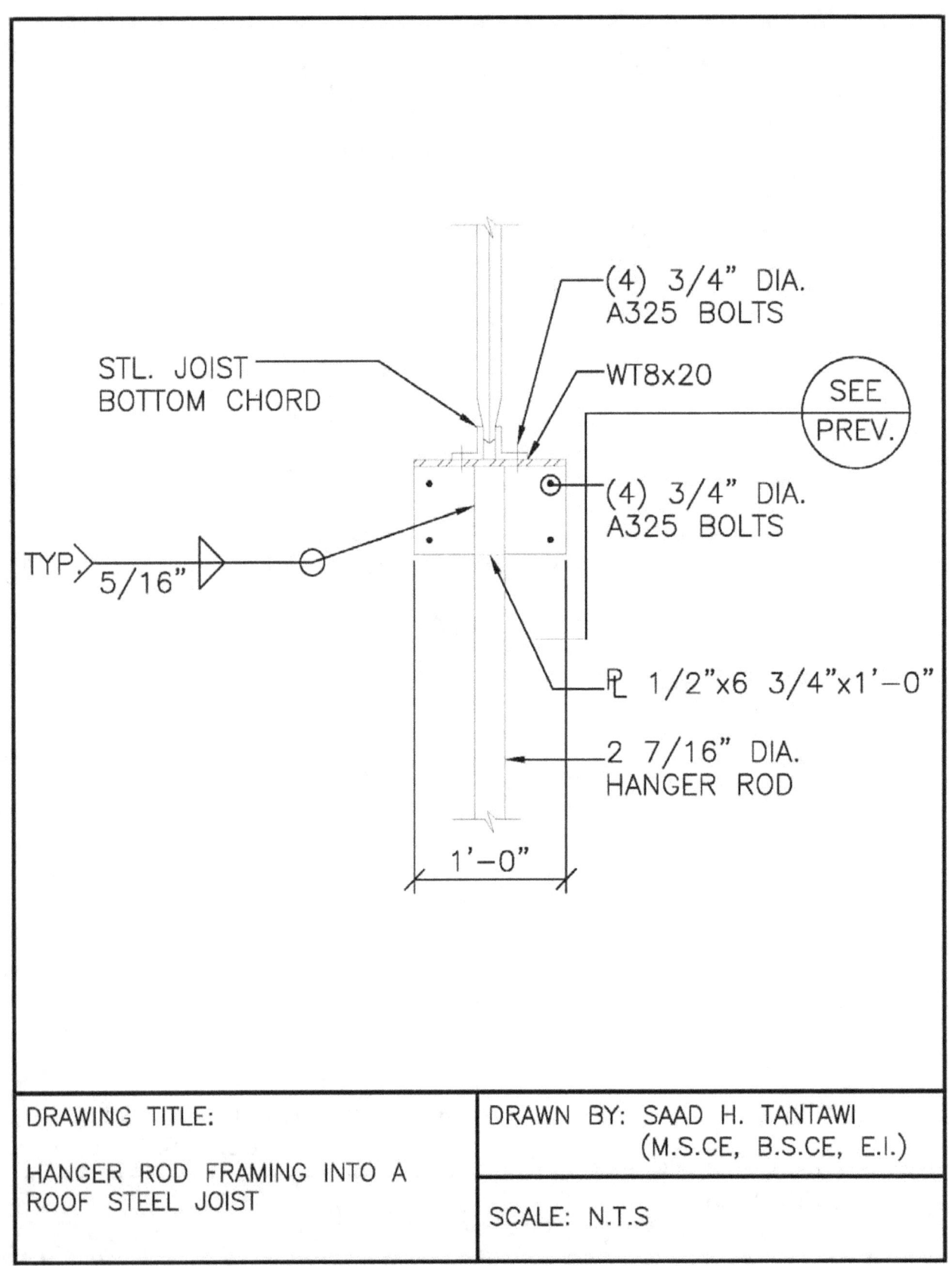

STL. JOIST
BOTTOM CHORD

(4) 3/4" DIA.
A325 BOLTS

WT8x20

SEE
PREV.

(4) 3/4" DIA.
A325 BOLTS

TYP. 5/16"

PL 1/2"x6 3/4"x1'-0"

2 7/16" DIA.
HANGER ROD

1'-0"

DRAWING TITLE:	DRAWN BY: SAAD H. TANTAWI (M.S.CE, B.S.CE, E.I.)
HANGER ROD FRAMING INTO A ROOF STEEL JOIST	SCALE: N.T.S

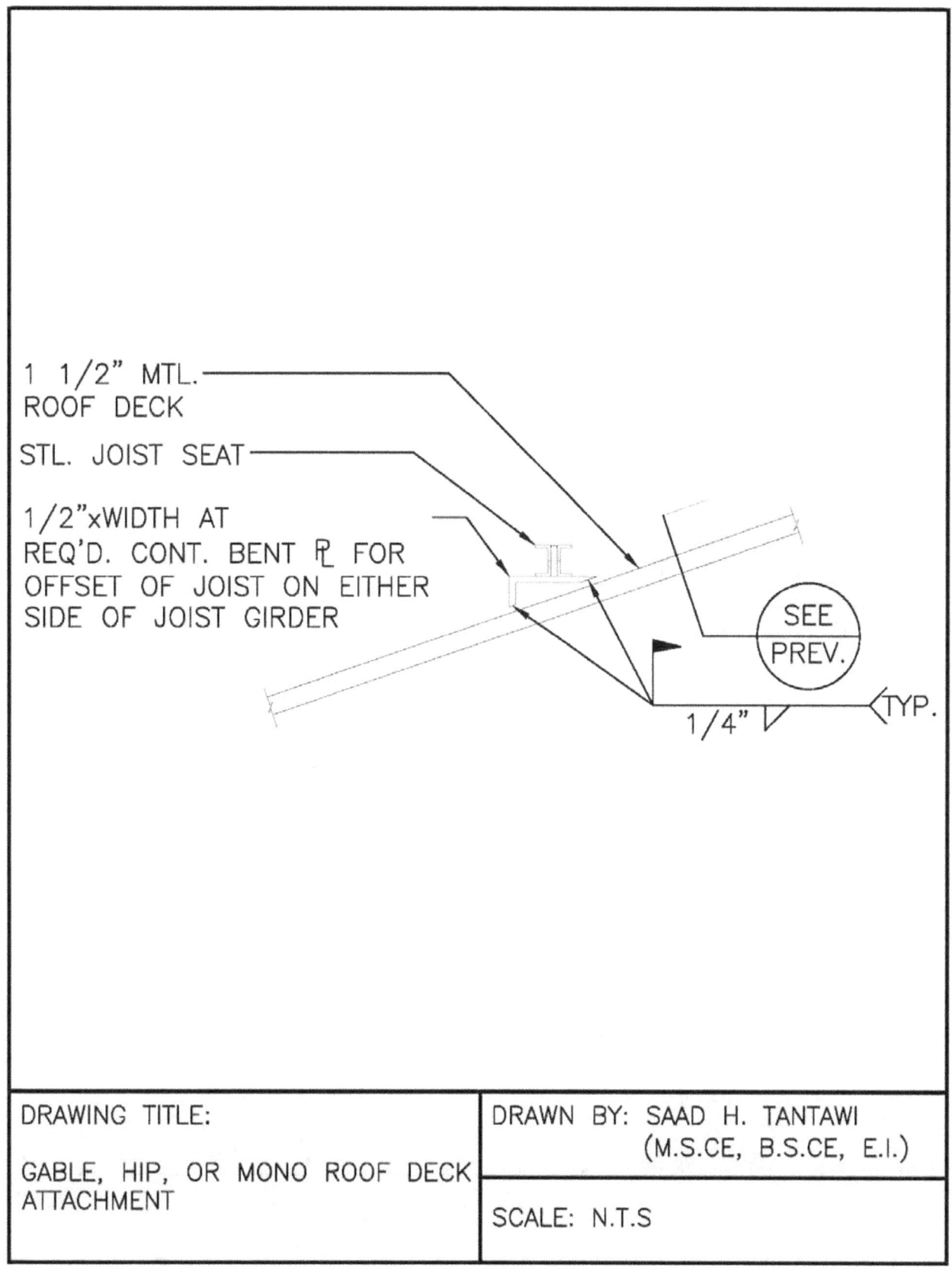

1 1/2" MTL. ROOF DECK

STL. JOIST SEAT

1/2"xWIDTH AT REQ'D. CONT. BENT ℞ FOR OFFSET OF JOIST ON EITHER SIDE OF JOIST GIRDER

SEE PREV.

1/4"

TYP.

DRAWING TITLE:	DRAWN BY: SAAD H. TANTAWI
GABLE, HIP, OR MONO ROOF DECK ATTACHMENT	(M.S.CE, B.S.CE, E.I.)
	SCALE: N.T.S

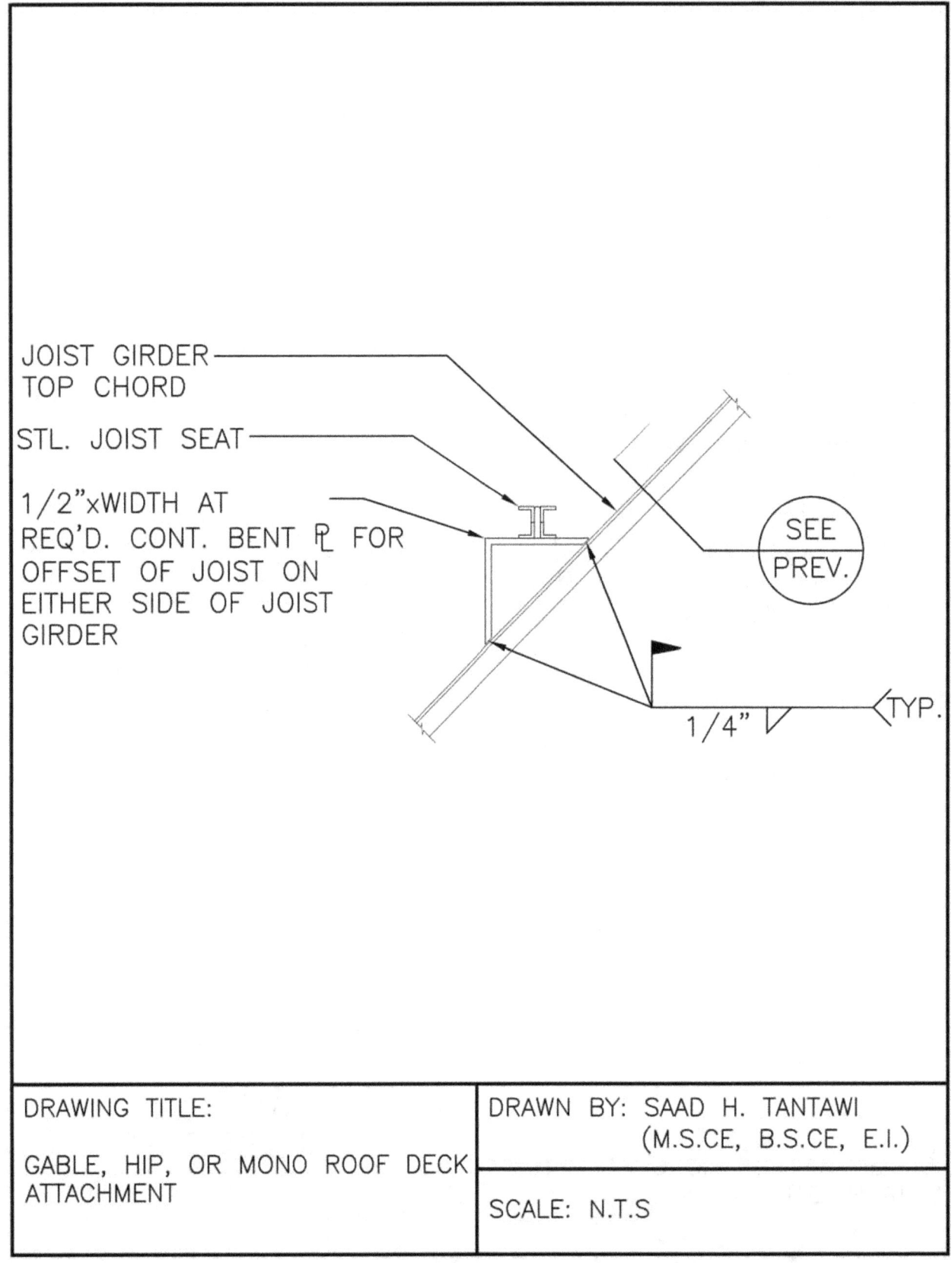

JOIST GIRDER
TOP CHORD

STL. JOIST SEAT

1/2"xWIDTH AT
REQ'D. CONT. BENT ℄ FOR
OFFSET OF JOIST ON
EITHER SIDE OF JOIST
GIRDER

SEE
PREV.

1/4"   TYP.

DRAWING TITLE:	DRAWN BY: SAAD H. TANTAWI
GABLE, HIP, OR MONO ROOF DECK ATTACHMENT	(M.S.CE, B.S.CE, E.I.)
	SCALE: N.T.S

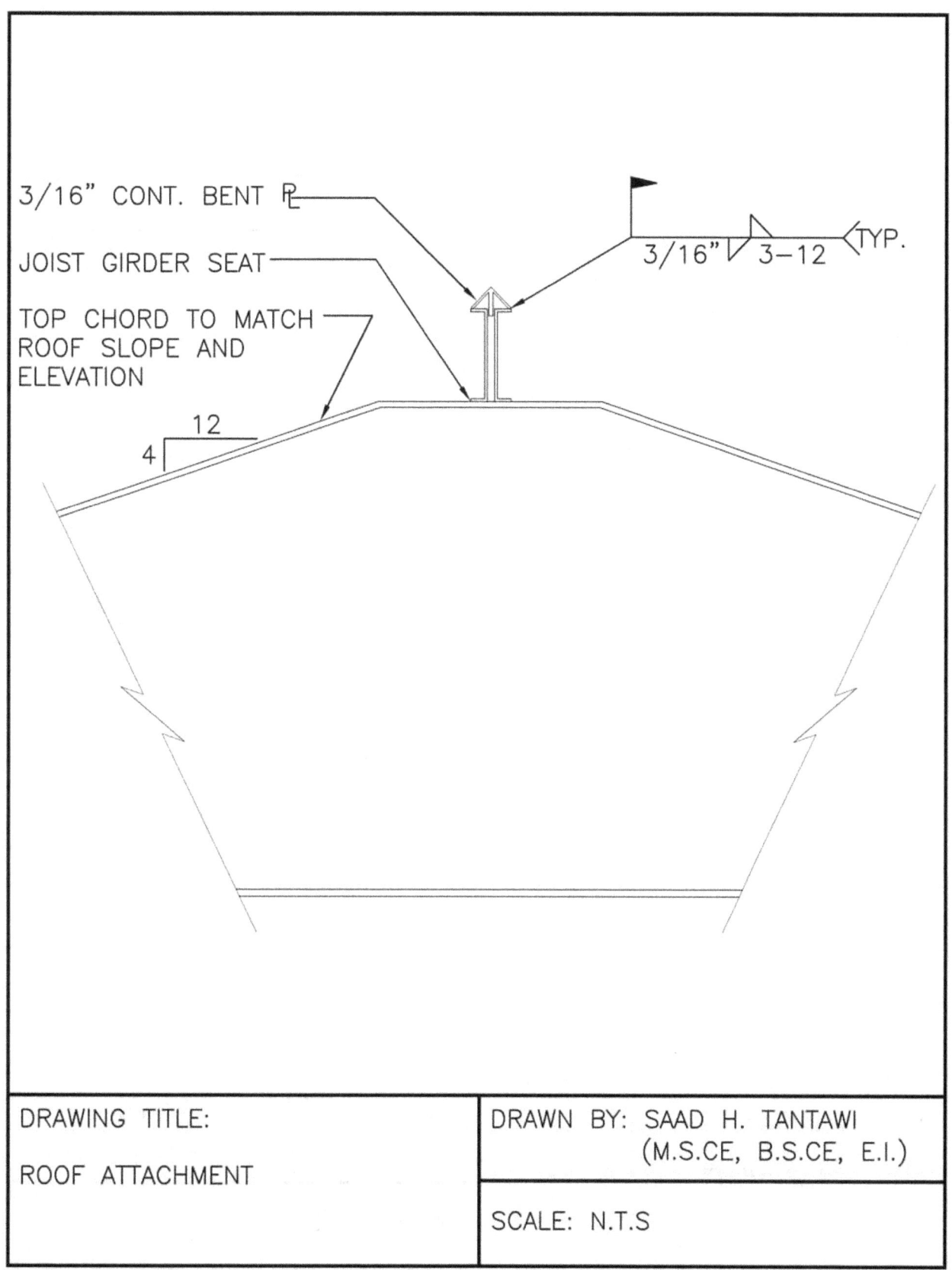

3/16" CONT. BENT ℄

JOIST GIRDER SEAT

TOP CHORD TO MATCH
ROOF SLOPE AND
ELEVATION

3/16" 3-12 TYP.

12
4

DRAWING TITLE:	DRAWN BY: SAAD H. TANTAWI
ROOF ATTACHMENT	(M.S.CE, B.S.CE, E.I.)
	SCALE: N.T.S

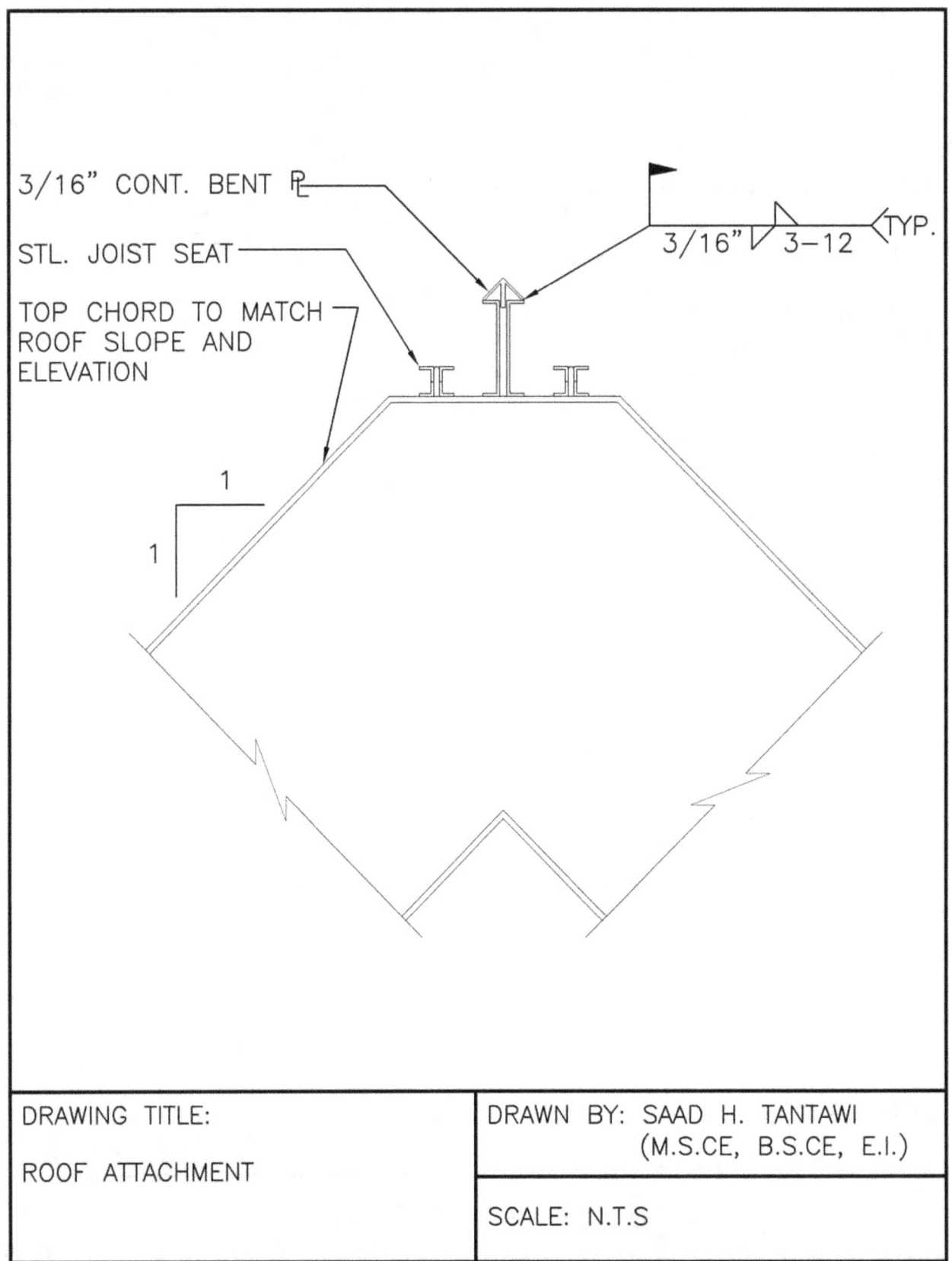

3/16" CONT. BENT ℄

STL. JOIST SEAT

TOP CHORD TO MATCH
ROOF SLOPE AND
ELEVATION

3/16" 3-12 TYP.

1
1

DRAWING TITLE:	DRAWN BY: SAAD H. TANTAWI
	(M.S.CE, B.S.CE, E.I.)
ROOF ATTACHMENT	
	SCALE: N.T.S

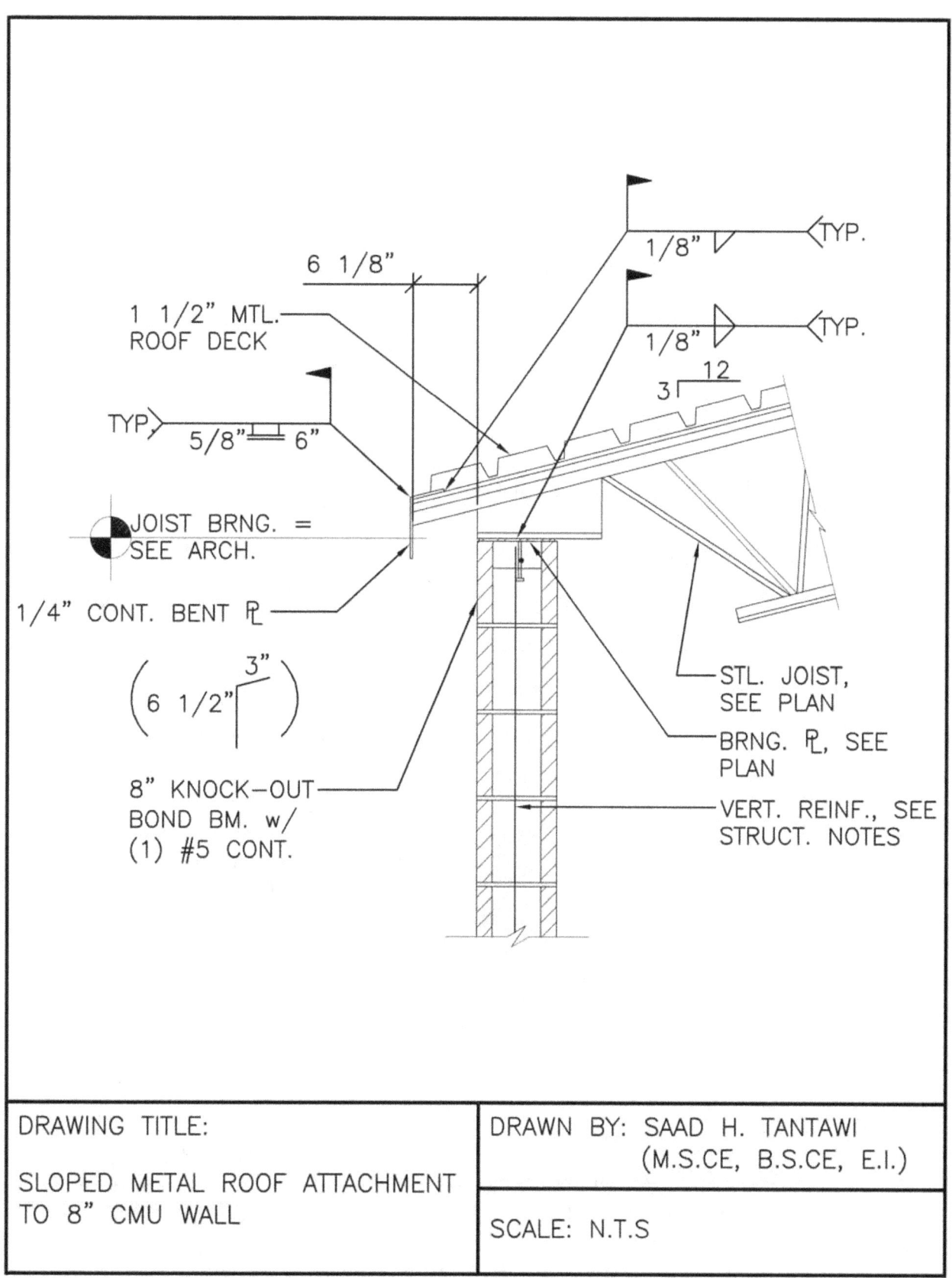

6 1/8"

1 1/2" MTL.
ROOF DECK

1/8" TYP.

1/8" TYP.

3 ⌐12

TYP.
5/8" 6"

JOIST BRNG. =
SEE ARCH.

1/4" CONT. BENT ℞

$\left(6 \ 1/2" \overbrace{}^{3"} \right)$

8" KNOCK-OUT
BOND BM. w/
(1) #5 CONT.

STL. JOIST,
SEE PLAN

BRNG. ℞, SEE
PLAN

VERT. REINF., SEE
STRUCT. NOTES

DRAWING TITLE:	DRAWN BY: SAAD H. TANTAWI
SLOPED METAL ROOF ATTACHMENT TO 8" CMU WALL	(M.S.CE, B.S.CE, E.I.)
	SCALE: N.T.S

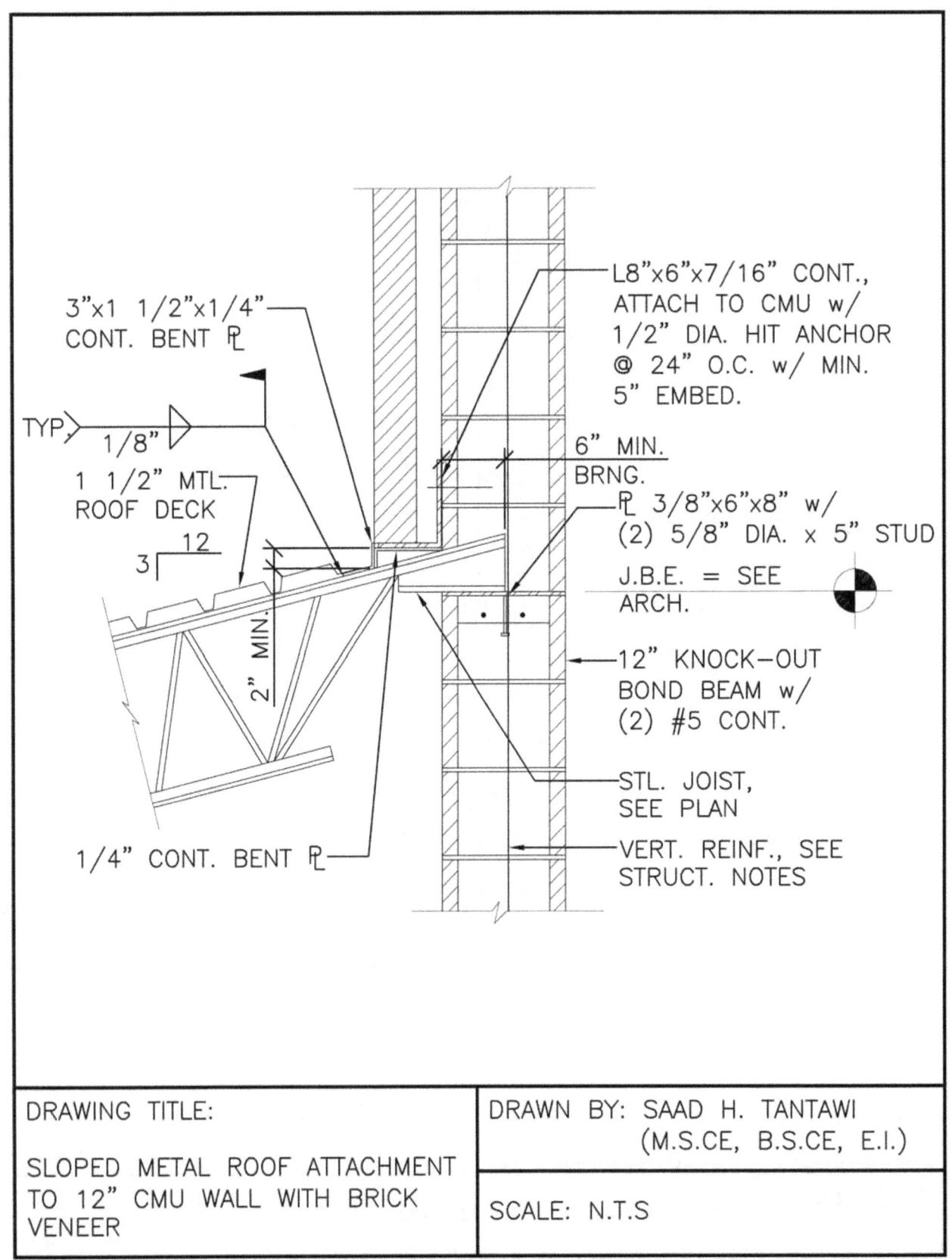

3"x1 1/2"x1/4" CONT. BENT ℙL

L8"x6"x7/16" CONT., ATTACH TO CMU w/ 1/2" DIA. HIT ANCHOR @ 24" O.C. w/ MIN. 5" EMBED.

TYP 1/8"

6" MIN. BRNG.

1 1/2" MTL. ROOF DECK

ℙL 3/8"x6"x8" w/ (2) 5/8" DIA. x 5" STUD

3 ⌐ 12

J.B.E. = SEE ARCH.

2" MIN.

12" KNOCK—OUT BOND BEAM w/ (2) #5 CONT.

STL. JOIST, SEE PLAN

1/4" CONT. BENT ℙL

VERT. REINF., SEE STRUCT. NOTES

DRAWING TITLE:	DRAWN BY: SAAD H. TANTAWI (M.S.CE, B.S.CE, E.I.)
SLOPED METAL ROOF ATTACHMENT TO 12" CMU WALL WITH BRICK VENEER	SCALE: N.T.S

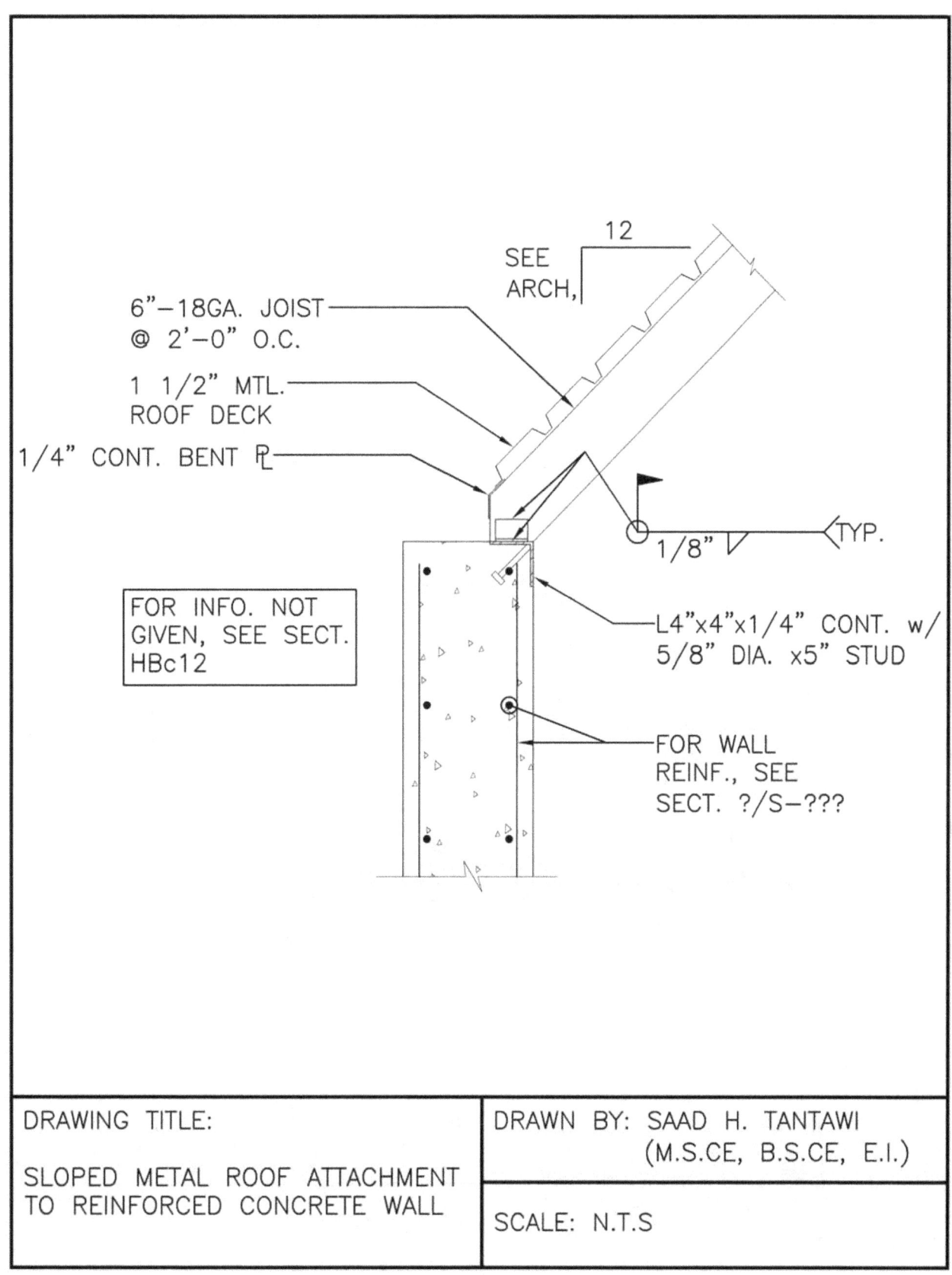

6"-18GA. JOIST
@ 2'-0" O.C.

1 1/2" MTL.
ROOF DECK

1/4" CONT. BENT PL

SEE
ARCH,

12

1/8"   TYP.

L4"x4"x1/4" CONT. w/
5/8" DIA. x5" STUD

FOR INFO. NOT
GIVEN, SEE SECT.
HBc12

FOR WALL
REINF., SEE
SECT. ?/S-???

DRAWING TITLE:	DRAWN BY: SAAD H. TANTAWI
	(M.S.CE, B.S.CE, E.I.)
SLOPED METAL ROOF ATTACHMENT TO REINFORCED CONCRETE WALL	
	SCALE: N.T.S

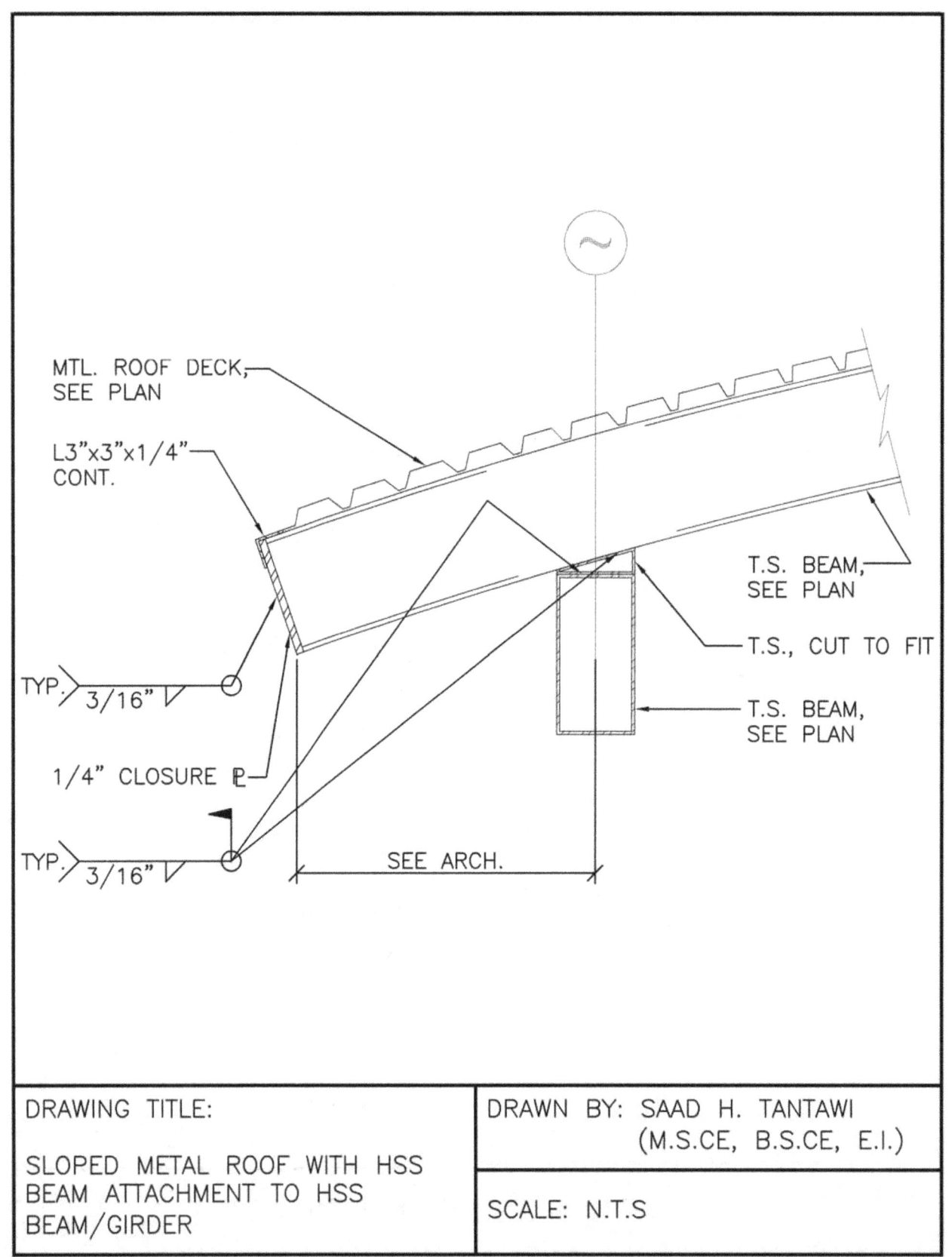

MTL. ROOF DECK,
SEE PLAN

L3"x3"x1/4"
CONT.

TYP. 3/16"

1/4" CLOSURE ⅊

TYP. 3/16"

SEE ARCH.

T.S. BEAM,
SEE PLAN

T.S., CUT TO FIT

T.S. BEAM,
SEE PLAN

DRAWING TITLE:	DRAWN BY: SAAD H. TANTAWI (M.S.CE, B.S.CE, E.I.)
SLOPED METAL ROOF WITH HSS BEAM ATTACHMENT TO HSS BEAM/GIRDER	SCALE: N.T.S

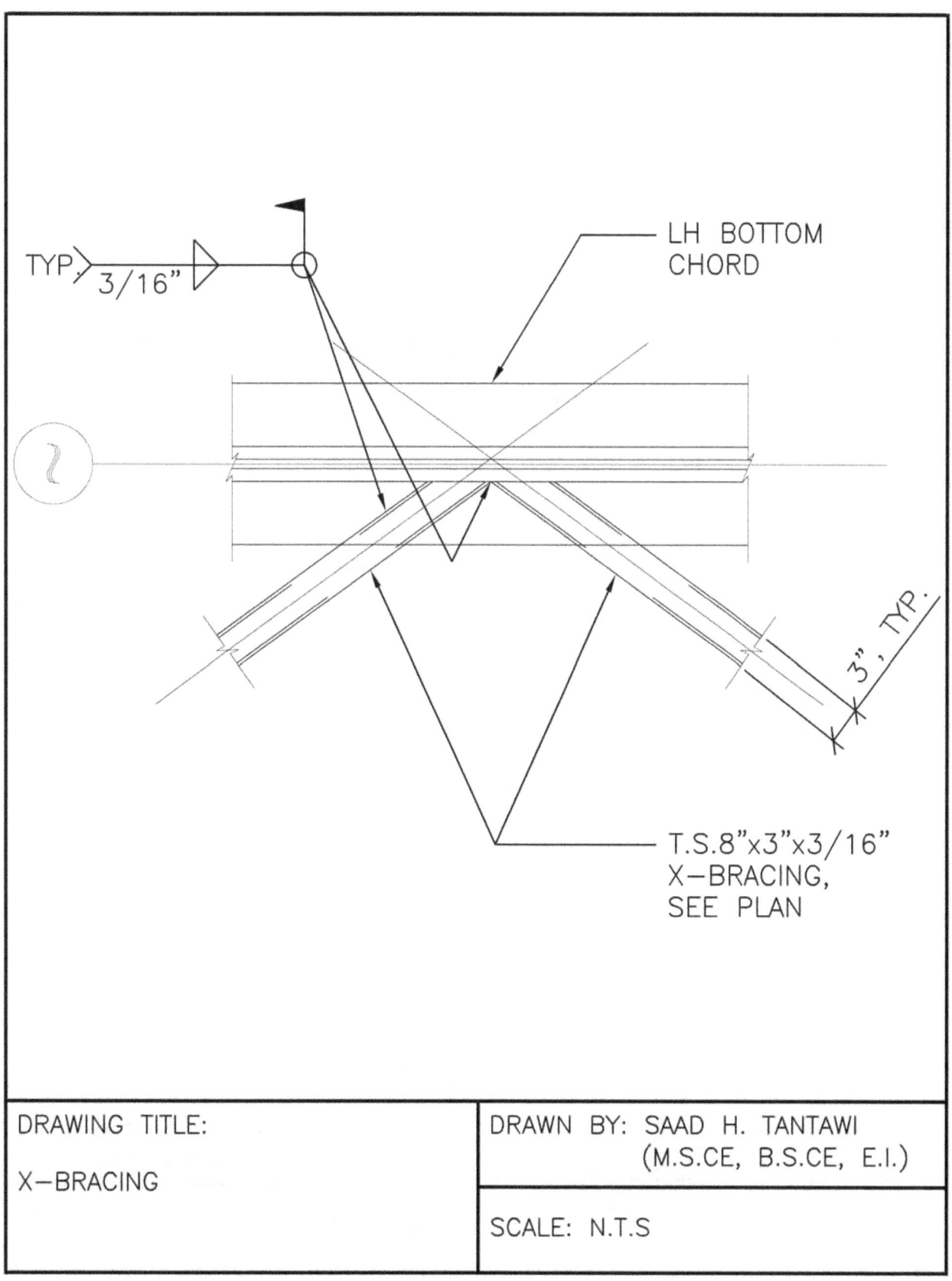

TYP. 3/16"

LH BOTTOM
CHORD

3", TYP.

T.S.8"x3"x3/16"
X-BRACING,
SEE PLAN

DRAWING TITLE: X-BRACING	DRAWN BY: SAAD H. TANTAWI (M.S.CE, B.S.CE, E.I.)
	SCALE: N.T.S

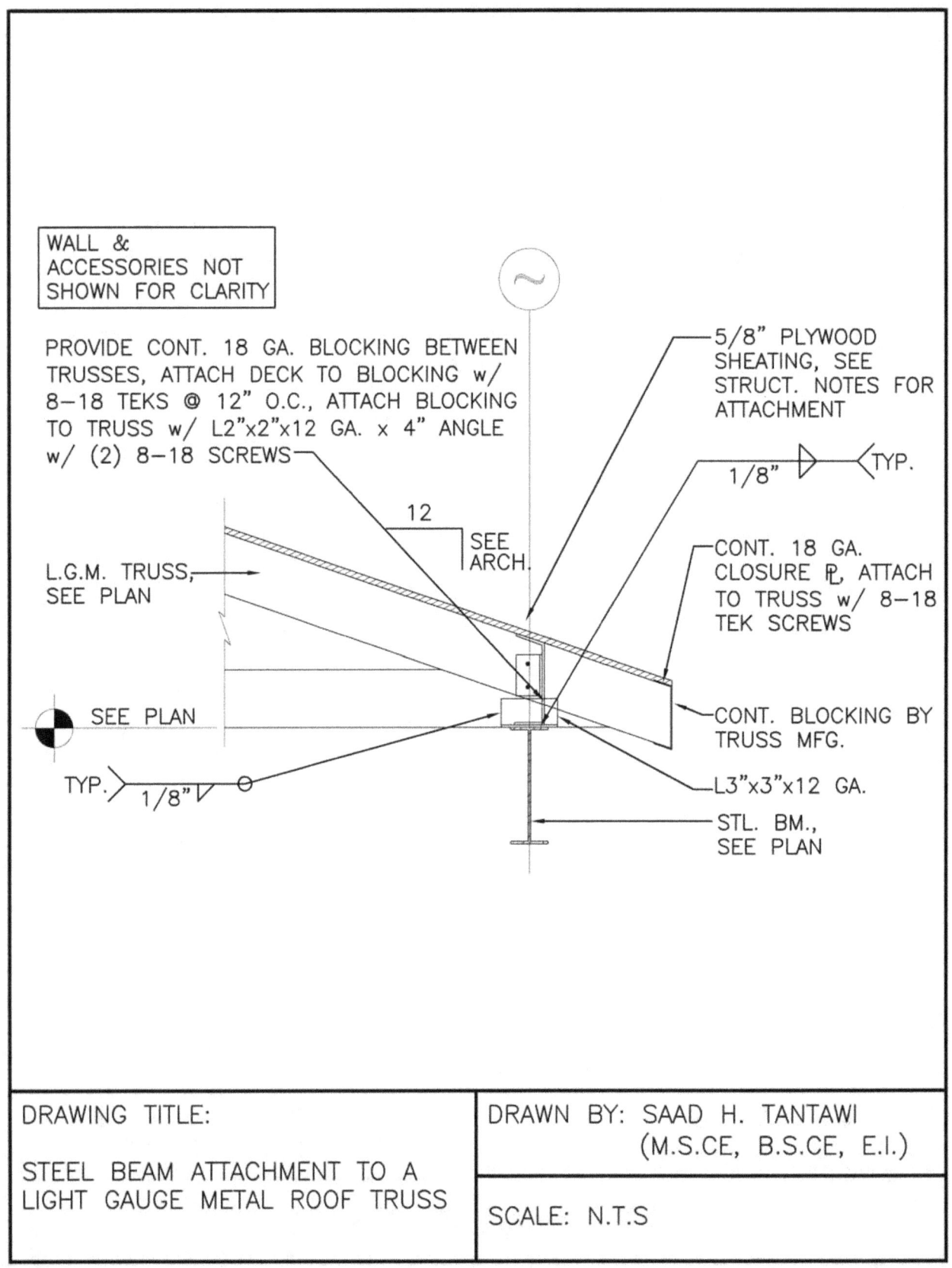

WALL &
ACCESSORIES NOT
SHOWN FOR CLARITY

PROVIDE CONT. 18 GA. BLOCKING BETWEEN
TRUSSES, ATTACH DECK TO BLOCKING w/
8-18 TEKS @ 12" O.C., ATTACH BLOCKING
TO TRUSS w/ L2"x2"x12 GA. x 4" ANGLE
w/ (2) 8-18 SCREWS

5/8" PLYWOOD
SHEATING, SEE
STRUCT. NOTES FOR
ATTACHMENT

1/8"   TYP.

12
SEE
ARCH.

CONT. 18 GA.
CLOSURE ℙ, ATTACH
TO TRUSS w/ 8-18
TEK SCREWS

L.G.M. TRUSS,
SEE PLAN

SEE PLAN

CONT. BLOCKING BY
TRUSS MFG.

TYP.   1/8"

L3"x3"x12 GA.

STL. BM.,
SEE PLAN

DRAWING TITLE:

STEEL BEAM ATTACHMENT TO A
LIGHT GAUGE METAL ROOF TRUSS

DRAWN BY: SAAD H. TANTAWI
(M.S.CE, B.S.CE, E.I.)

SCALE: N.T.S

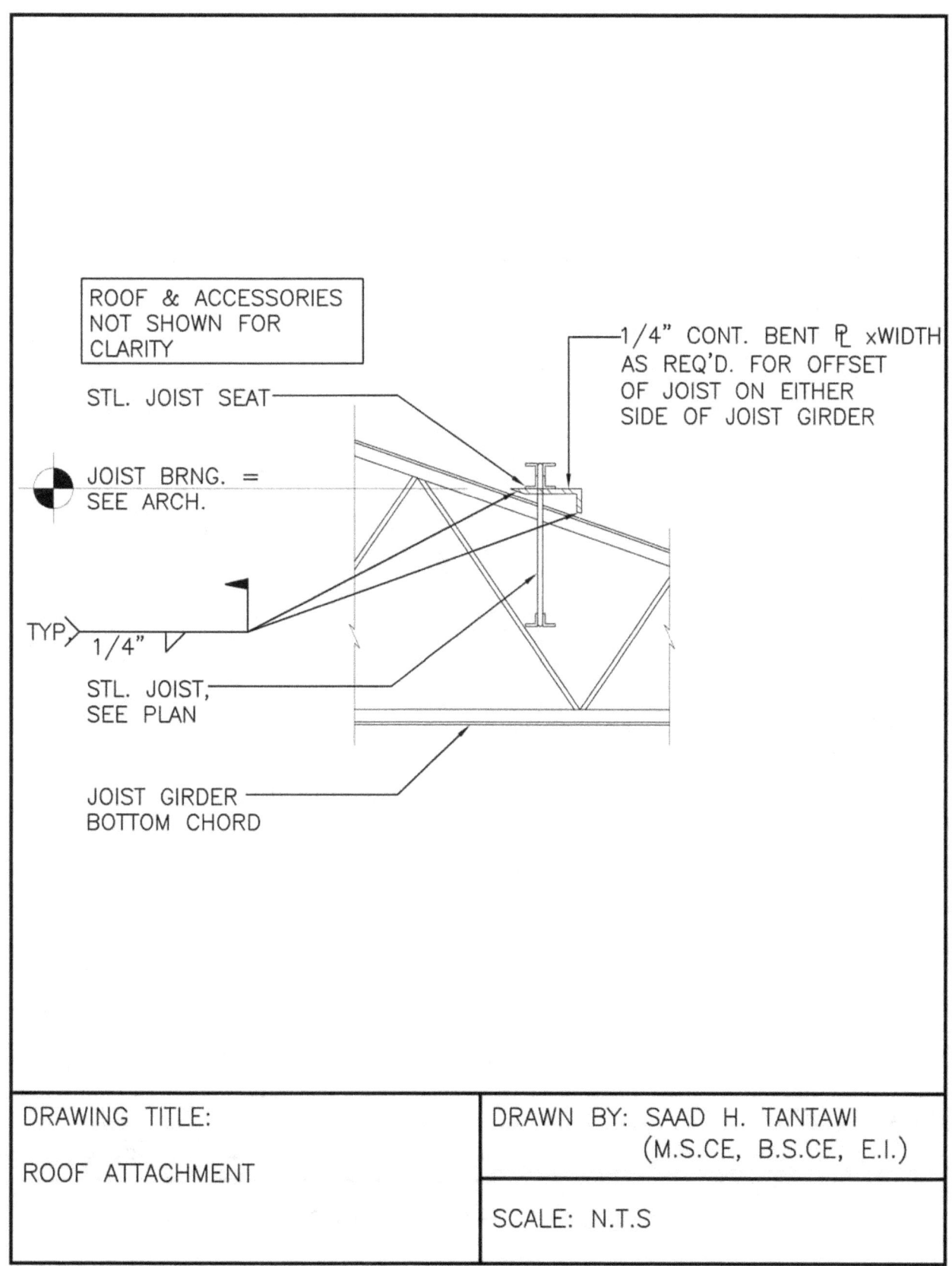

ROOF & ACCESSORIES
NOT SHOWN FOR
CLARITY

STL. JOIST SEAT

JOIST BRNG. =
SEE ARCH.

1/4" CONT. BENT ℞ xWIDTH
AS REQ'D. FOR OFFSET
OF JOIST ON EITHER
SIDE OF JOIST GIRDER

TYP
1/4"

STL. JOIST,
SEE PLAN

JOIST GIRDER
BOTTOM CHORD

DRAWING TITLE: ROOF ATTACHMENT	DRAWN BY: SAAD H. TANTAWI (M.S.CE, B.S.CE, E.I.)
	SCALE: N.T.S

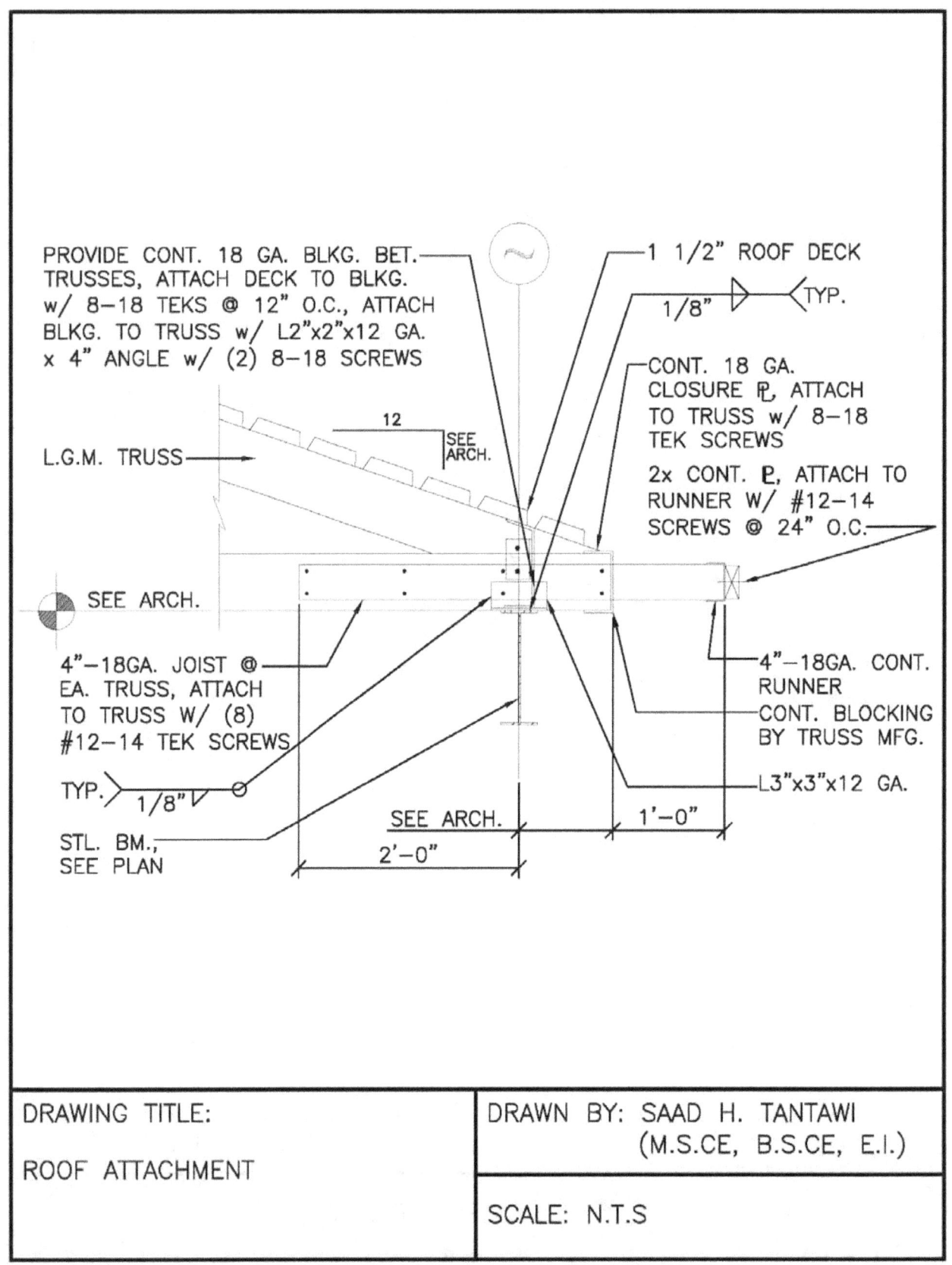

PROVIDE CONT. 18 GA. BLKG. BET.
TRUSSES, ATTACH DECK TO BLKG.
w/ 8-18 TEKS @ 12" O.C., ATTACH
BLKG. TO TRUSS w/ L2"x2"x12 GA.
x 4" ANGLE w/ (2) 8-18 SCREWS

1 1/2" ROOF DECK

1/8" TYP.

CONT. 18 GA.
CLOSURE Ⴐ, ATTACH
TO TRUSS w/ 8-18
TEK SCREWS

2x CONT. Ⴐ, ATTACH TO
RUNNER W/ #12-14
SCREWS @ 24" O.C.

12 SEE ARCH.

L.G.M. TRUSS

SEE ARCH.

4"-18GA. JOIST @
EA. TRUSS, ATTACH
TO TRUSS w/ (8)
#12-14 TEK SCREWS

TYP. 1/8"

STL. BM.,
SEE PLAN

4"-18GA. CONT.
RUNNER

CONT. BLOCKING
BY TRUSS MFG.

L3"x3"x12 GA.

SEE ARCH. 1'-0"

2'-0"

DRAWING TITLE:

ROOF ATTACHMENT

DRAWN BY: SAAD H. TANTAWI
(M.S.CE, B.S.CE, E.I.)

SCALE: N.T.S

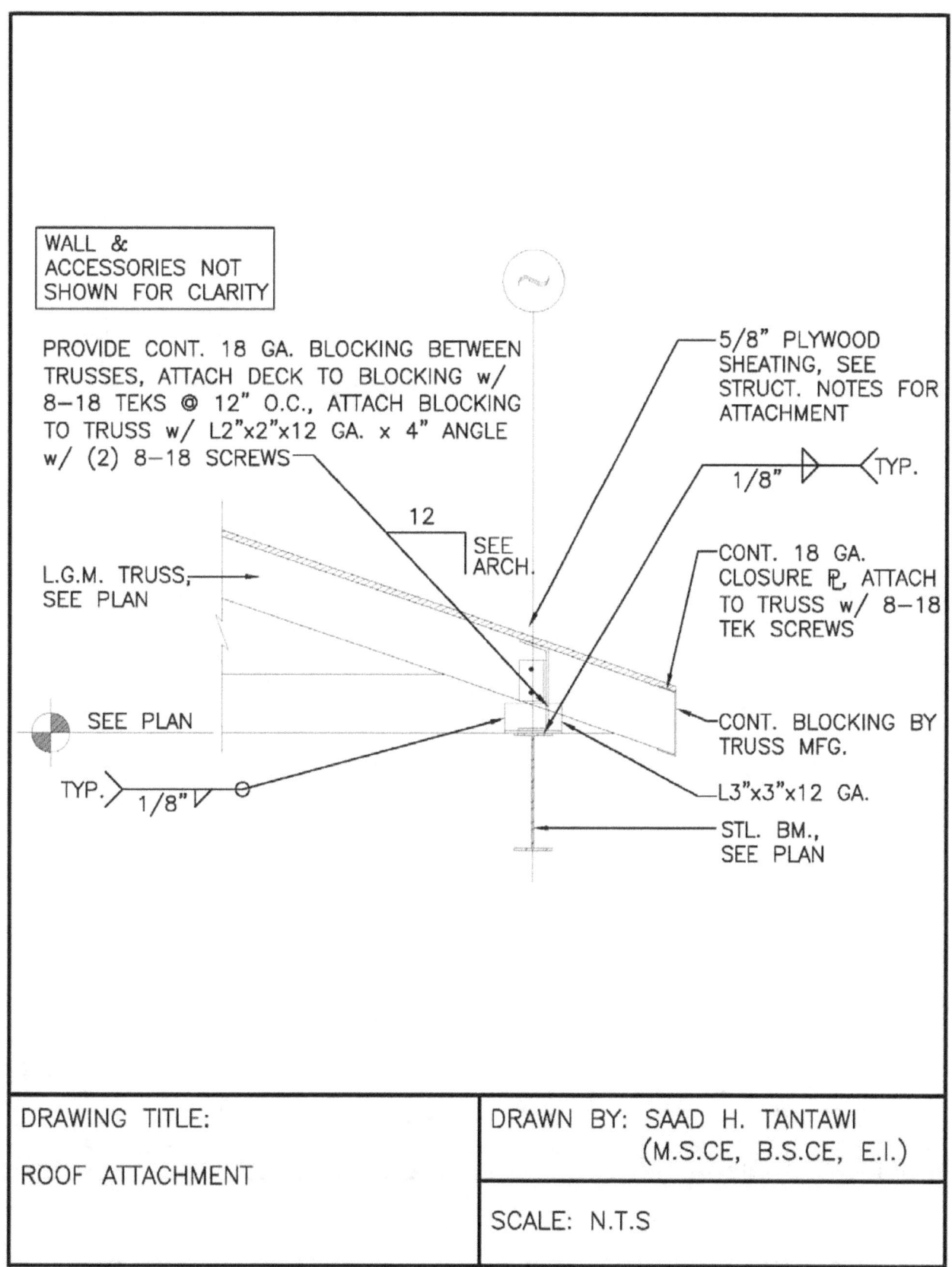

WALL &
ACCESSORIES NOT
SHOWN FOR CLARITY

PROVIDE CONT. 18 GA. BLOCKING BETWEEN
TRUSSES, ATTACH DECK TO BLOCKING w/
8-18 TEKS @ 12" O.C., ATTACH BLOCKING
TO TRUSS w/ L2"x2"x12 GA. x 4" ANGLE
w/ (2) 8-18 SCREWS

12

SEE
ARCH.

5/8" PLYWOOD
SHEATING, SEE
STRUCT. NOTES FOR
ATTACHMENT

1/8"   TYP.

L.G.M. TRUSS,
SEE PLAN

CONT. 18 GA.
CLOSURE P, ATTACH
TO TRUSS w/ 8-18
TEK SCREWS

SEE PLAN

CONT. BLOCKING BY
TRUSS MFG.

TYP.   1/8"

L3"x3"x12 GA.

STL. BM.,
SEE PLAN

DRAWING TITLE:	DRAWN BY: SAAD H. TANTAWI
	(M.S.CE, B.S.CE, E.I.)
ROOF ATTACHMENT	
	SCALE: N.T.S

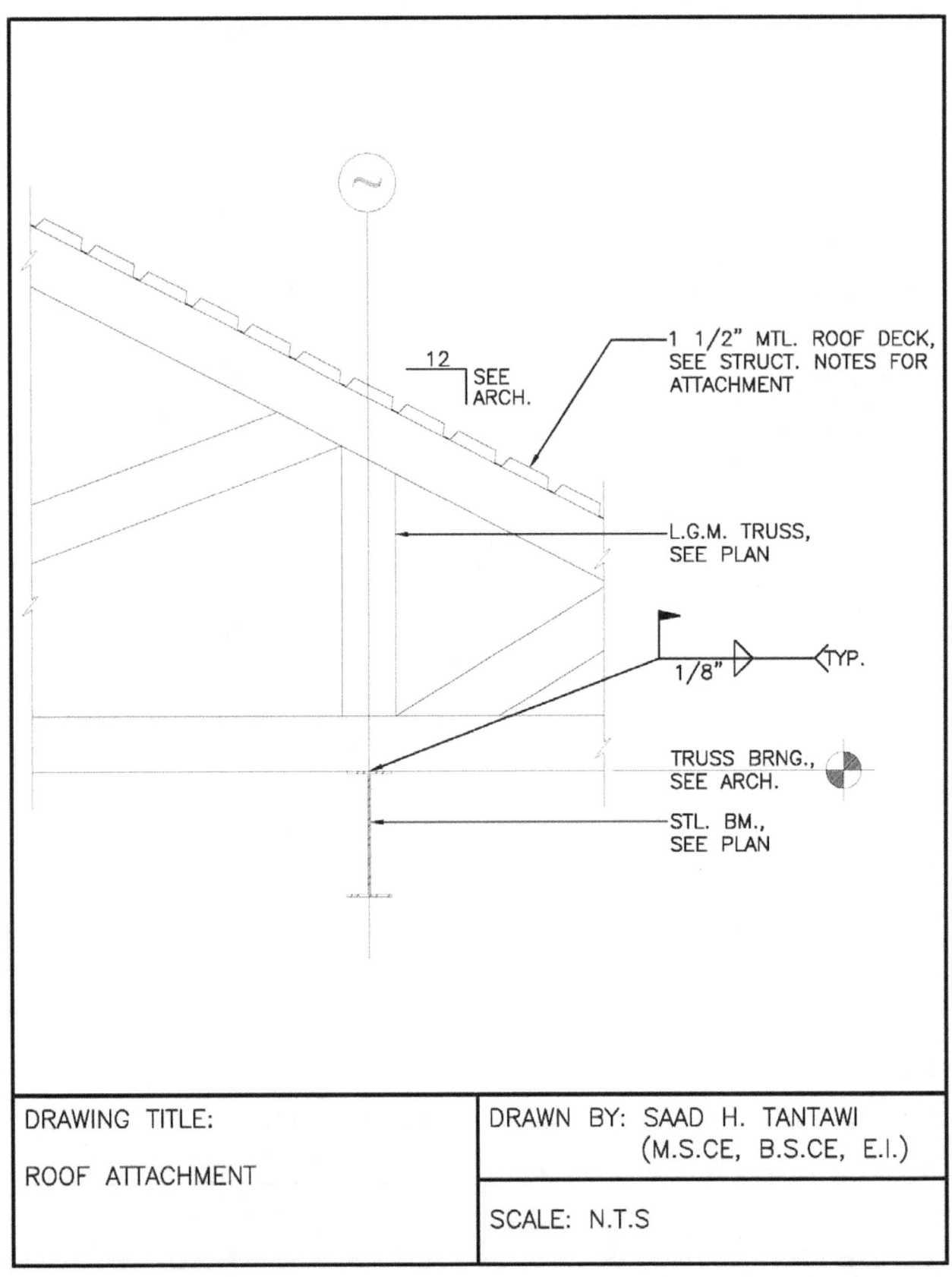

DRAWING TITLE:	DRAWN BY: SAAD H. TANTAWI
	(M.S.CE, B.S.CE, E.I.)
ROOF ATTACHMENT	SCALE: N.T.S

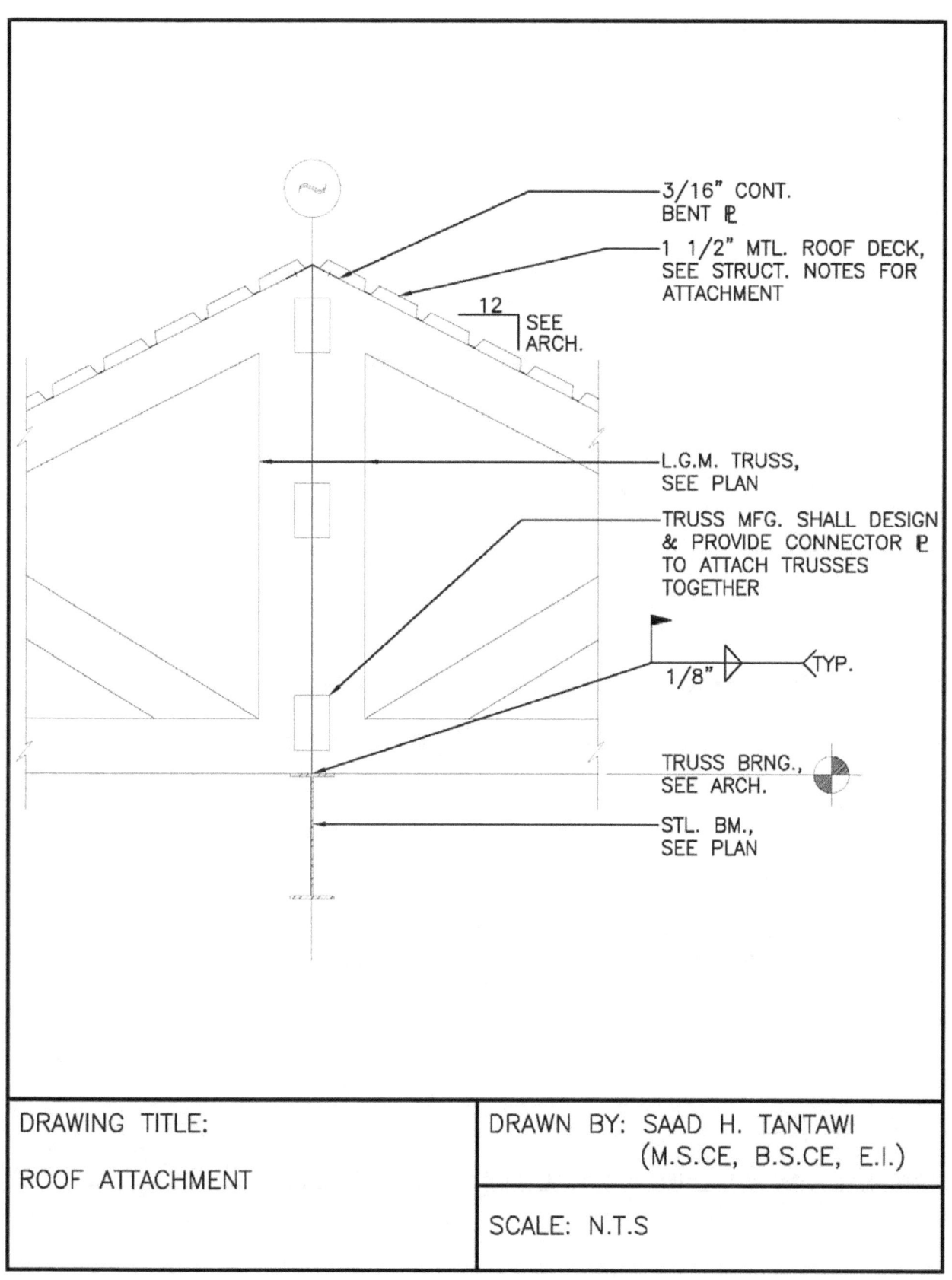

3/16" CONT.
BENT ℄

1 1/2" MTL. ROOF DECK,
SEE STRUCT. NOTES FOR
ATTACHMENT

12
|‾ SEE
| ARCH.

L.G.M. TRUSS,
SEE PLAN

TRUSS MFG. SHALL DESIGN
& PROVIDE CONNECTOR ℄
TO ATTACH TRUSSES
TOGETHER

1/8"   ⊳   ⟨TYP.

TRUSS BRNG.,
SEE ARCH.

STL. BM.,
SEE PLAN

DRAWING TITLE:	DRAWN BY: SAAD H. TANTAWI
	(M.S.CE, B.S.CE, E.I.)
ROOF ATTACHMENT	SCALE: N.T.S

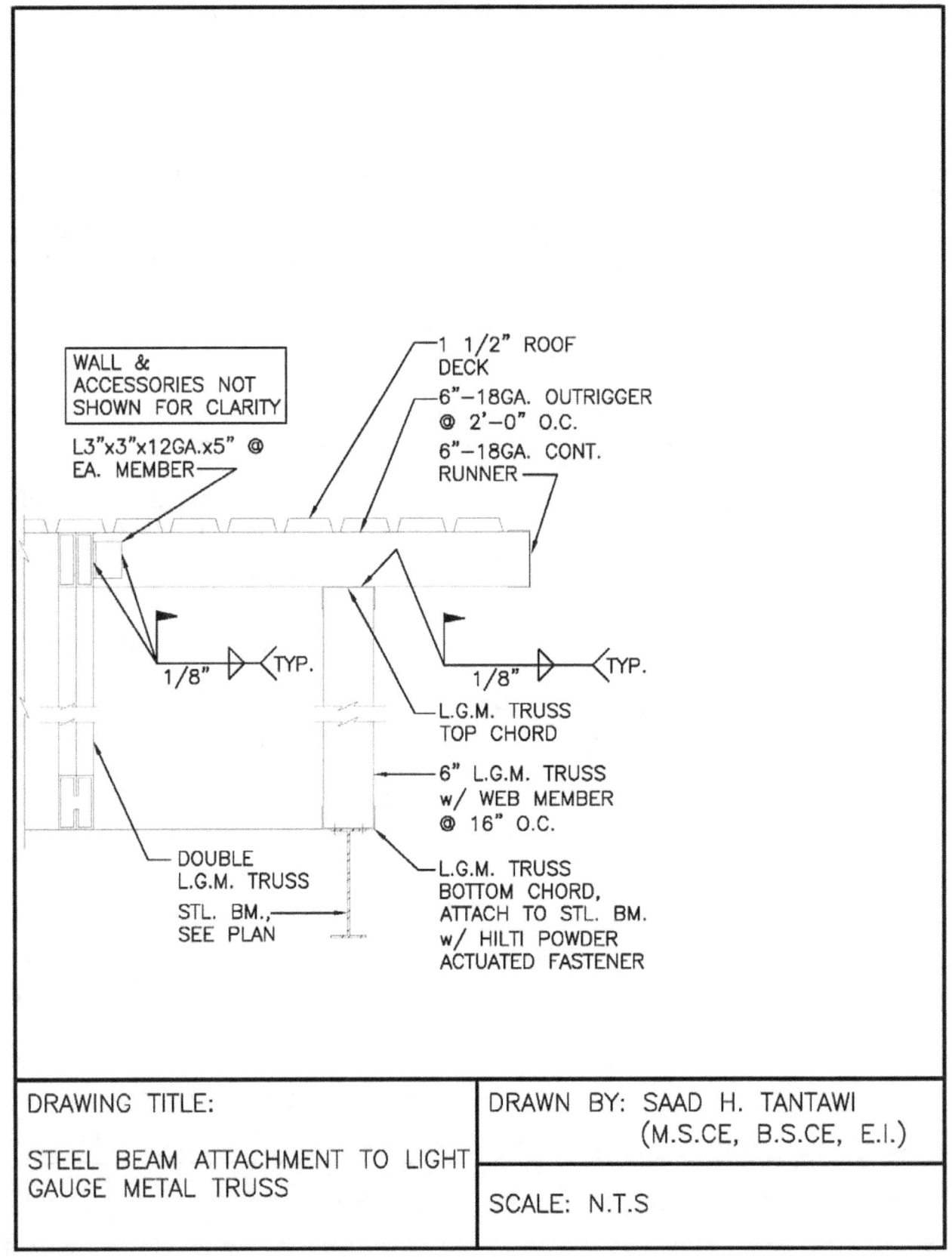

WALL &
ACCESSORIES NOT
SHOWN FOR CLARITY

L3"x3"x12GA.x5" @
EA. MEMBER

1 1/2" ROOF
DECK

6"-18GA. OUTRIGGER
@ 2'-0" O.C.

6"-18GA. CONT.
RUNNER

1/8"   TYP.

1/8"   TYP.

L.G.M. TRUSS
TOP CHORD

6" L.G.M. TRUSS
w/ WEB MEMBER
@ 16" O.C.

DOUBLE
L.G.M. TRUSS

STL. BM.,
SEE PLAN

L.G.M. TRUSS
BOTTOM CHORD,
ATTACH TO STL. BM.
w/ HILTI POWDER
ACTUATED FASTENER

DRAWING TITLE:	DRAWN BY: SAAD H. TANTAWI
STEEL BEAM ATTACHMENT TO LIGHT GAUGE METAL TRUSS	(M.S.CE, B.S.CE, E.I.)
	SCALE: N.T.S

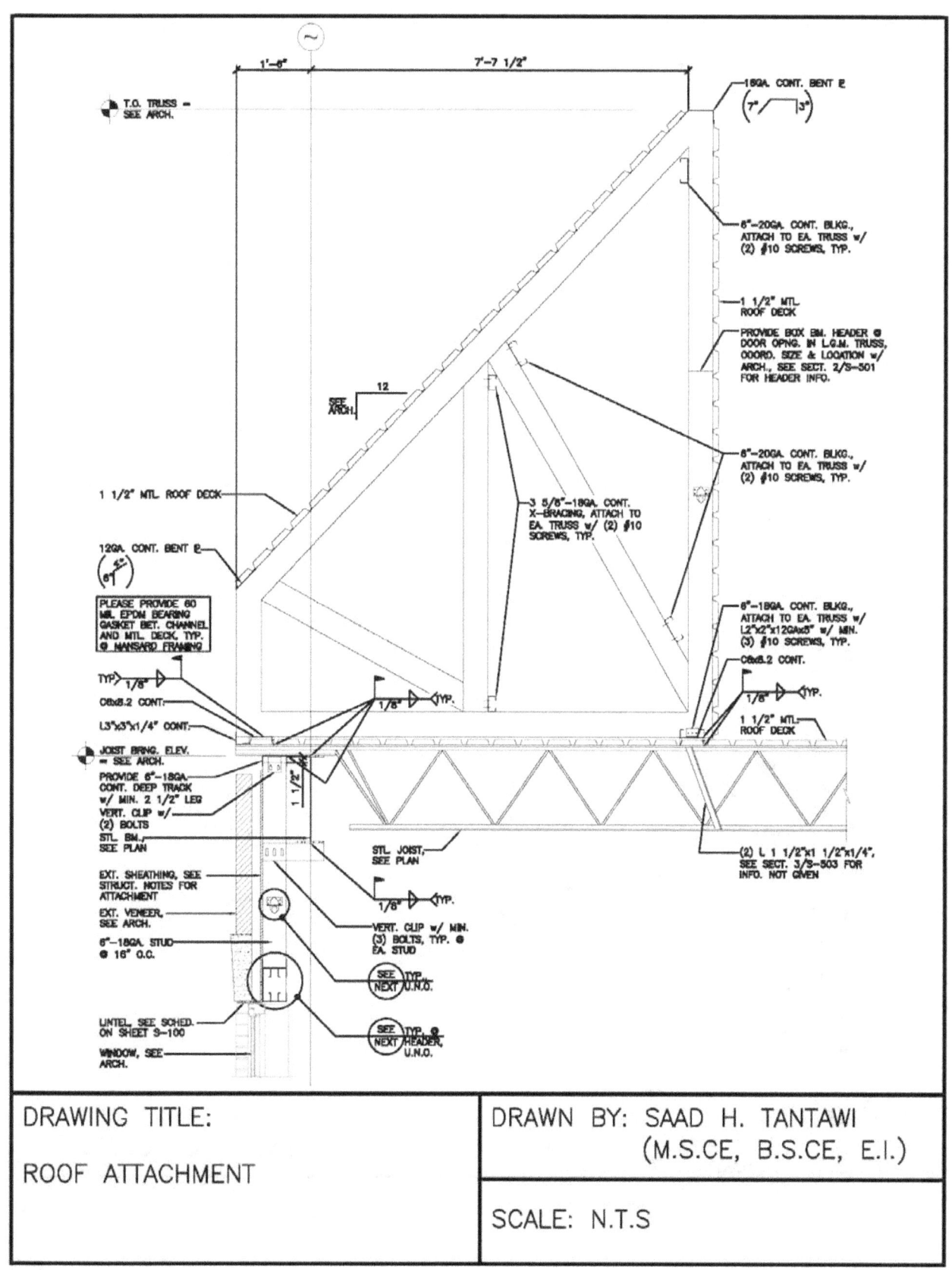

T.O. TRUSS — SEE ARCH.

1'-6"

7'-7 1/2"

18GA. CONT. BENT ℙ (7" / 3")

6"-20GA. CONT. BLKG., ATTACH TO EA. TRUSS w/ (2) #10 SCREWS, TYP.

1 1/2" MTL. ROOF DECK

PROVIDE BOX BM. HEADER @ DOOR OPNG. IN L.G.M. TRUSS, COORD. SIZE & LOCATION w/ ARCH., SEE SECT. 2/S-301 FOR HEADER INFO.

6"-20GA. CONT. BLKG., ATTACH TO EA. TRUSS w/ (2) #10 SCREWS, TYP.

12

SEE ARCH.

1 1/2" MTL. ROOF DECK

3 5/8"-18GA. CONT. X-BRACING, ATTACH TO EA. TRUSS w/ (2) #10 SCREWS, TYP.

6"-18GA. CONT. BLKG., ATTACH TO EA. TRUSS w/ L2"x2"x12GAx5" w/ MIN. (3) #10 SCREWS, TYP.

12GA. CONT. BENT ℙ (4" / 8")

C6x8.2 CONT.

1 1/2" MTL. ROOF DECK

PLEASE PROVIDE 60 MIL EPDM BEARING GASKET BET. CHANNEL AND MTL. DECK, TYP. @ MANSARD FRAMING

1/8" TYP.

C6x8.2 CONT.

L3"x3"x1/4" CONT.

JOIST BRNG. ELEV. = SEE ARCH.

PROVIDE 6"-18GA. CONT. DEEP TRACK w/ MIN. 2 1/2" LEG

VERT. CLIP w/ (2) BOLTS

STL. BM., SEE PLAN

EXT. SHEATHING, SEE STRUCT. NOTES FOR ATTACHMENT

EXT. VENEER, SEE ARCH.

6"-18GA. STUD @ 16" O.C.

LINTEL, SEE SCHED. ON SHEET S-100

WINDOW, SEE ARCH.

1 1/2"

1/8" TYP.

1/8" TYP.

STL. JOIST, SEE PLAN

1/8" TYP.

VERT. CLIP w/ MIN. (3) BOLTS, TYP. @ EA. STUD

SEE NEXT TYP. U.N.O.

SEE NEXT HEADER TYP. @ U.N.O.

(2) L 1 1/2"x1 1/2"x1/4", SEE SECT. 3/S-503 FOR INFO. NOT GIVEN

DRAWING TITLE:	DRAWN BY: SAAD H. TANTAWI
ROOF ATTACHMENT	(M.S.CE, B.S.CE, E.I.)
	SCALE: N.T.S

Saad Hasan Tantawi (M.S.CE, B.S.CE, E.I., A.M.ASCE)

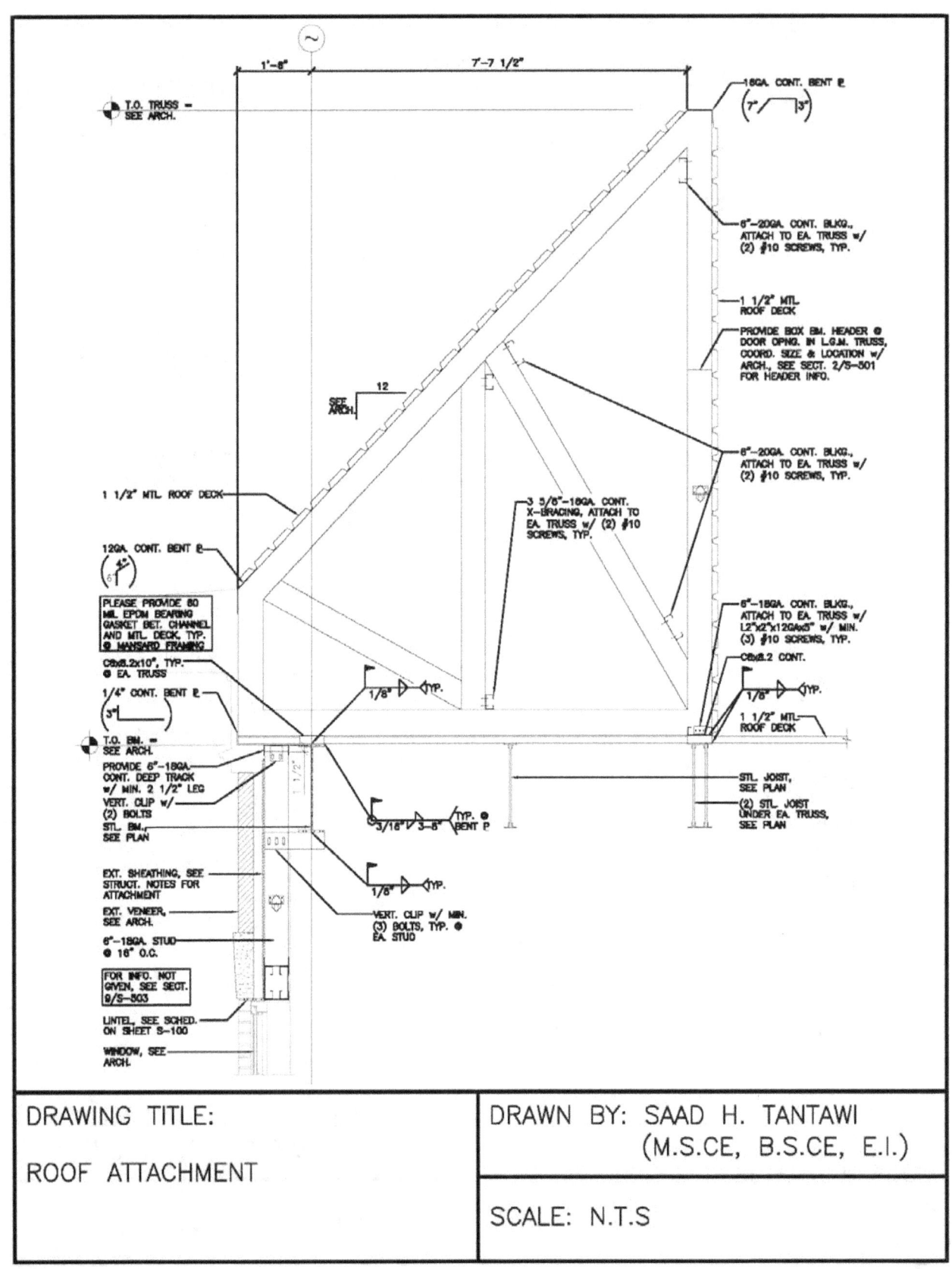

DRAWING TITLE:	DRAWN BY: SAAD H. TANTAWI
ROOF ATTACHMENT	(M.S.CE, B.S.CE, E.I.)
	SCALE: N.T.S

74

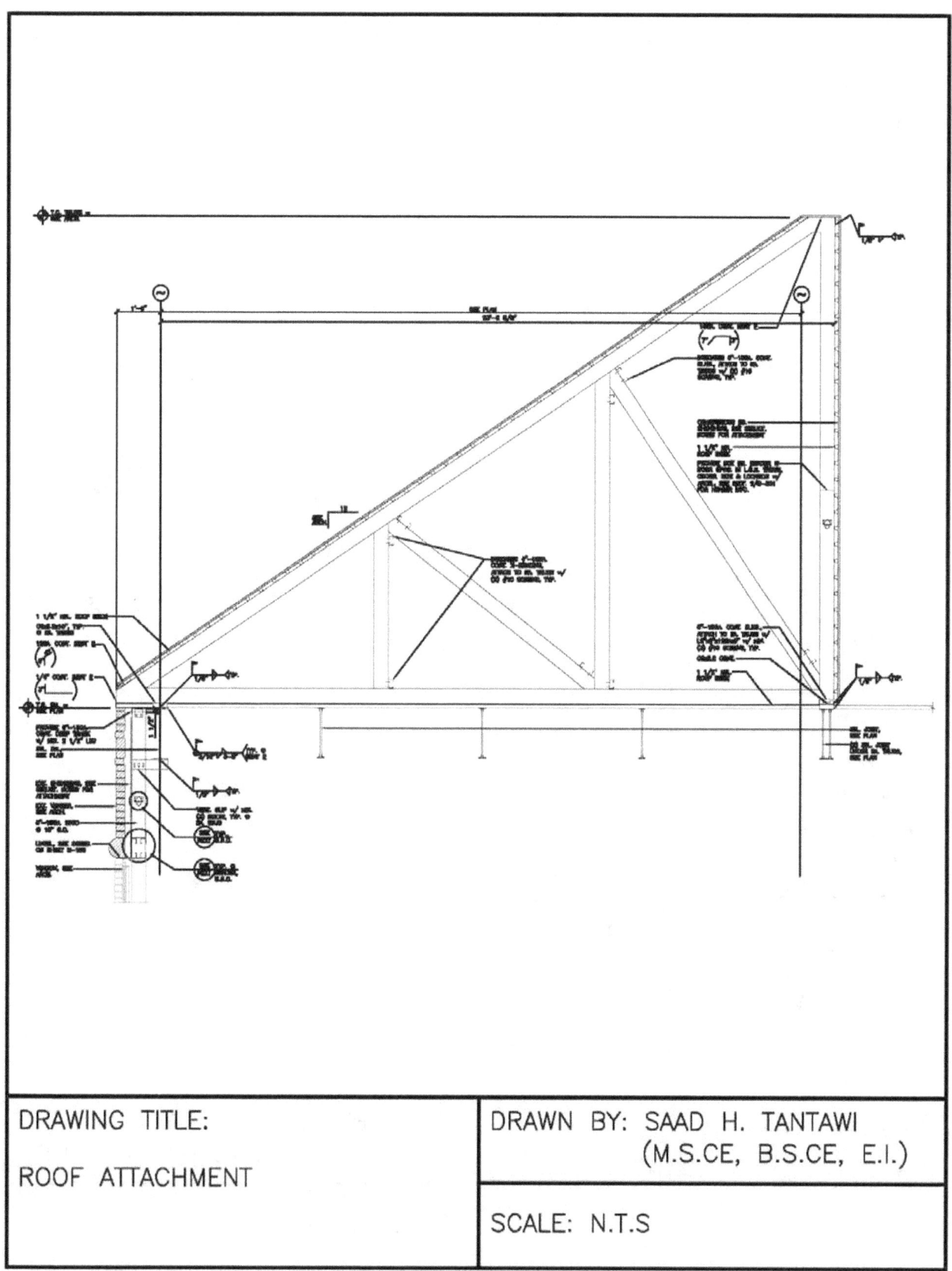

DRAWING TITLE:	DRAWN BY: SAAD H. TANTAWI
ROOF ATTACHMENT	(M.S.CE, B.S.CE, E.I.)
	SCALE: N.T.S

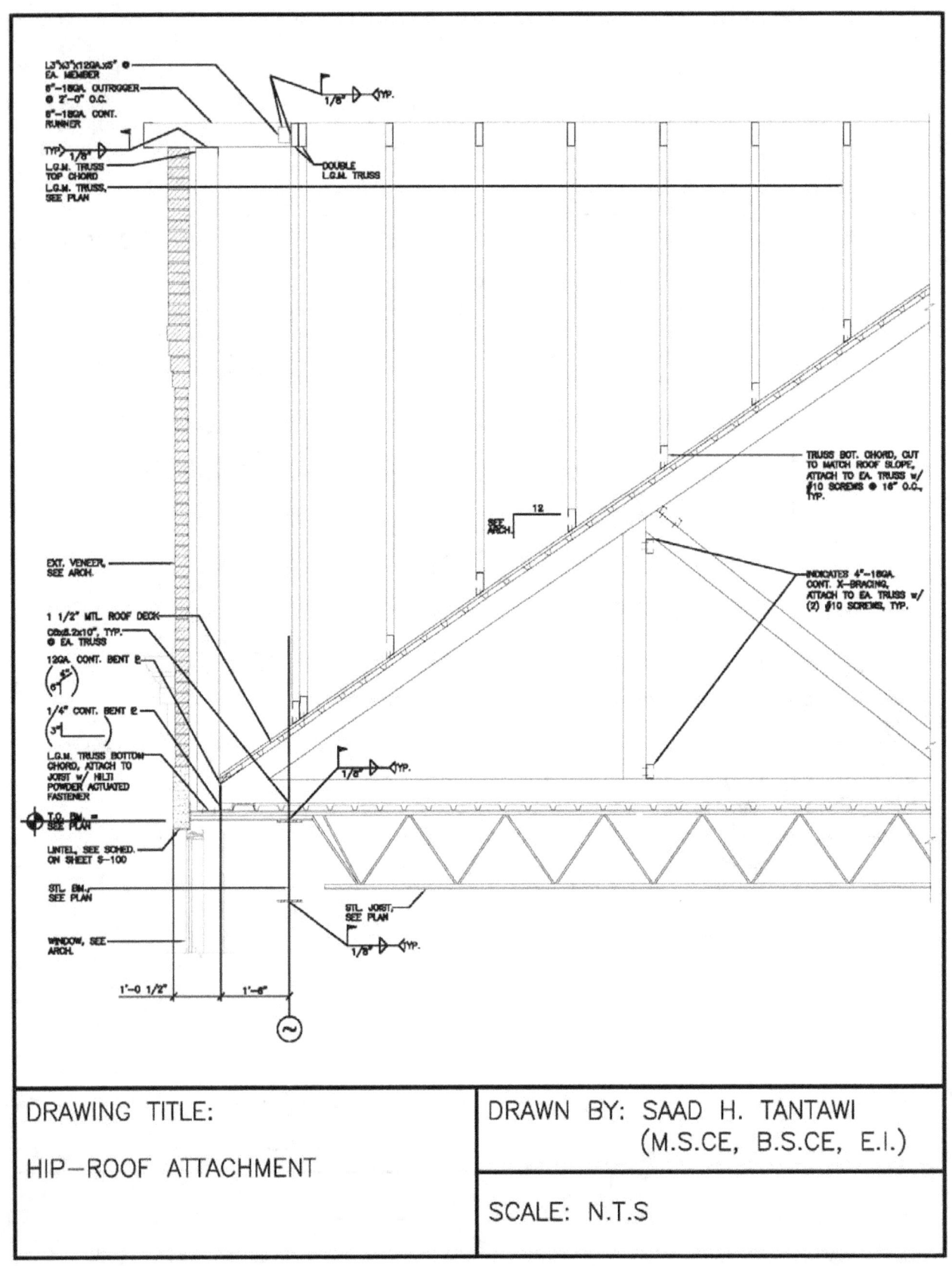

DRAWING TITLE: HIP—ROOF ATTACHMENT	DRAWN BY: SAAD H. TANTAWI (M.S.CE, B.S.CE, E.I.)
	SCALE: N.T.S

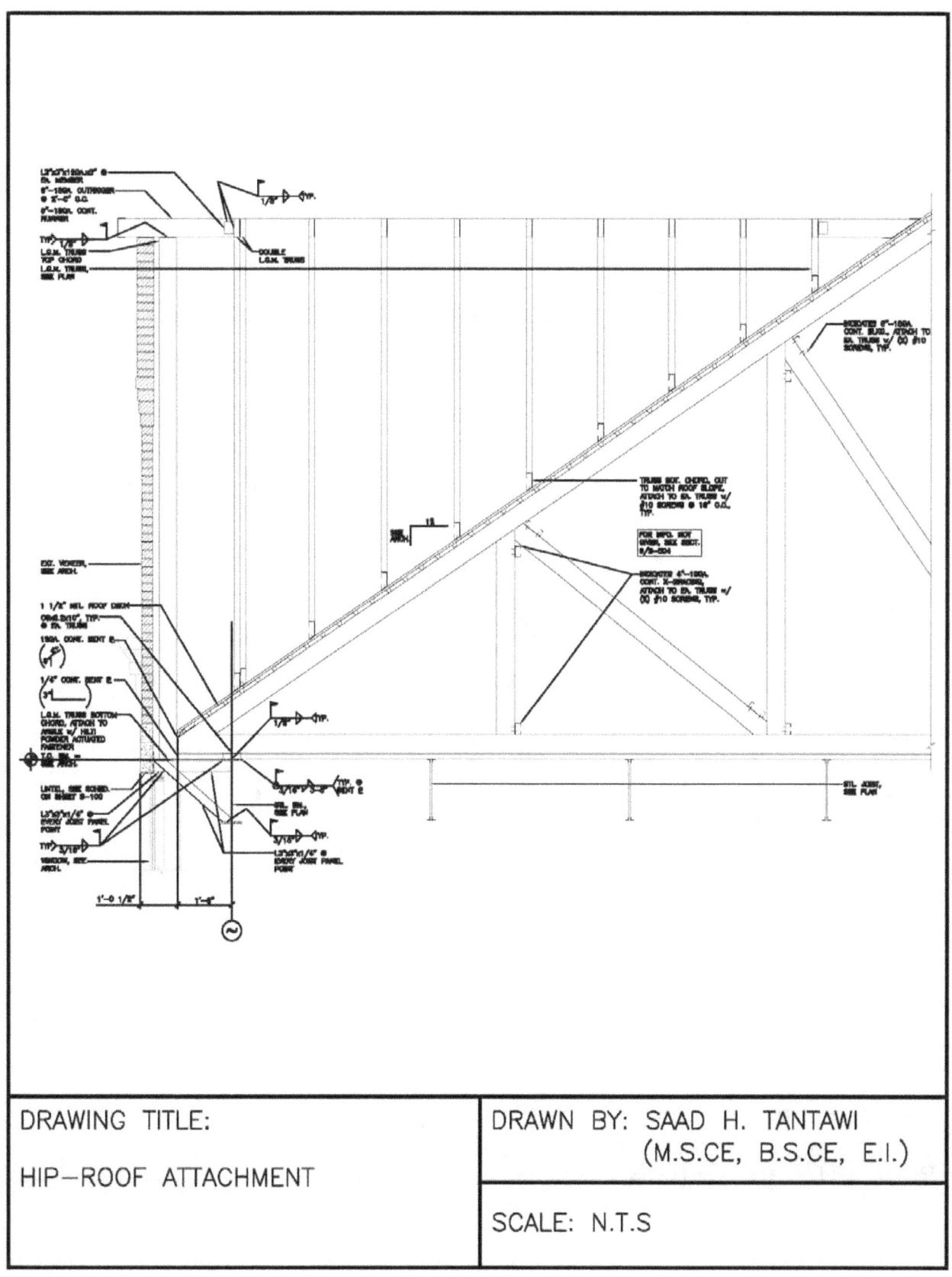

DRAWING TITLE:	DRAWN BY: SAAD H. TANTAWI
HIP—ROOF ATTACHMENT	(M.S.CE, B.S.CE, E.I.)
	SCALE: N.T.S

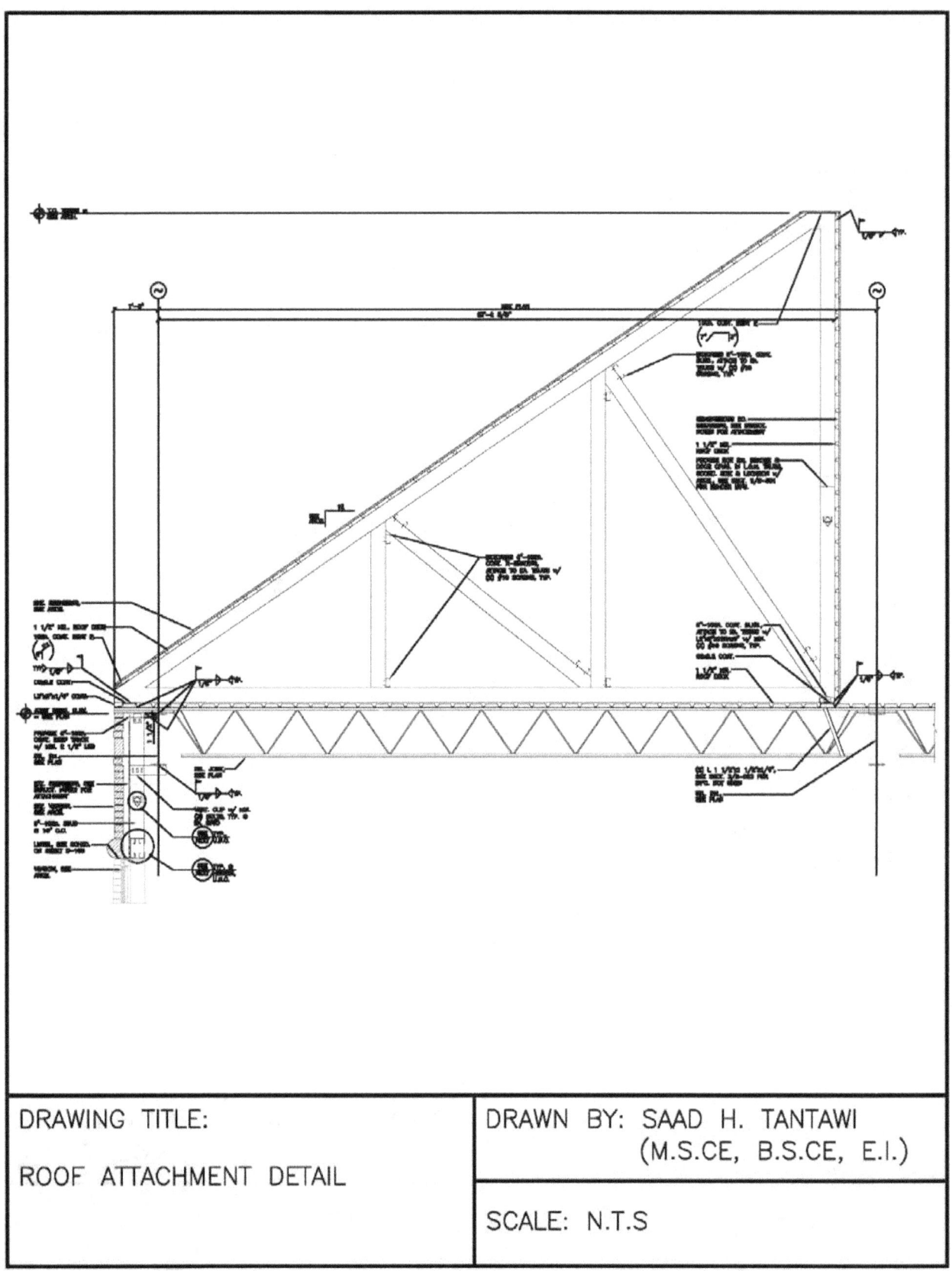

DRAWING TITLE:	DRAWN BY: SAAD H. TANTAWI
ROOF ATTACHMENT DETAIL	(M.S.CE, B.S.CE, E.I.)
	SCALE: N.T.S

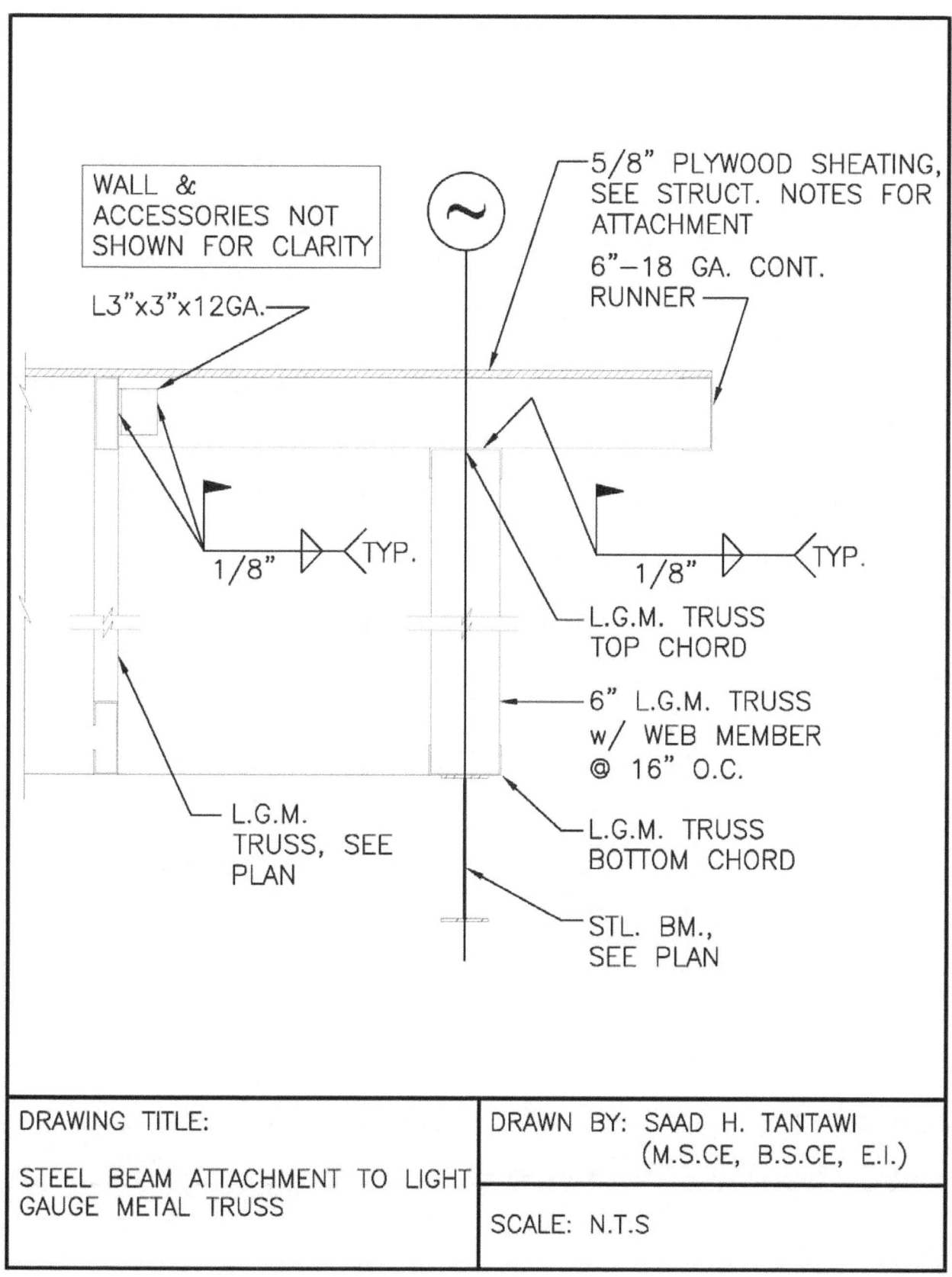

WALL &
ACCESSORIES NOT
SHOWN FOR CLARITY

L3"x3"x12GA.

5/8" PLYWOOD SHEATING,
SEE STRUCT. NOTES FOR
ATTACHMENT

6"-18 GA. CONT.
RUNNER

1/8"   TYP.

1/8"   TYP.

L.G.M. TRUSS
TOP CHORD

6" L.G.M. TRUSS
w/ WEB MEMBER
@ 16" O.C.

L.G.M.
TRUSS, SEE
PLAN

L.G.M. TRUSS
BOTTOM CHORD

STL. BM.,
SEE PLAN

DRAWING TITLE:	DRAWN BY: SAAD H. TANTAWI
	(M.S.CE, B.S.CE, E.I.)
STEEL BEAM ATTACHMENT TO LIGHT GAUGE METAL TRUSS	
	SCALE: N.T.S

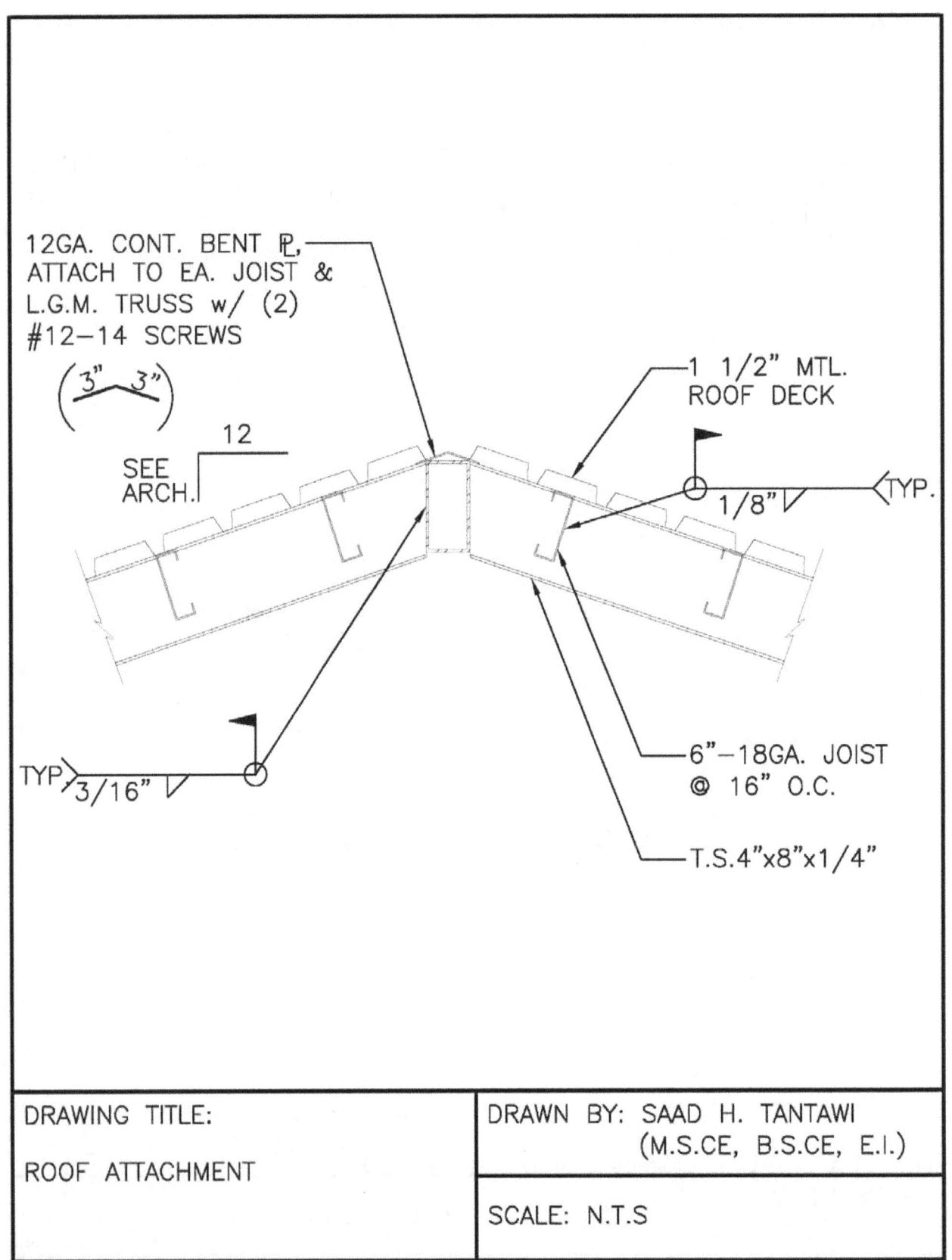

12GA. CONT. BENT ℙ,
ATTACH TO EA. JOIST &
L.G.M. TRUSS w/ (2)
#12−14 SCREWS

1 1/2" MTL.
ROOF DECK

12

SEE
ARCH.

TYP.

1/8"

TYP. 3/16"

6"−18GA. JOIST
@ 16" O.C.

T.S.4"x8"x1/4"

DRAWING TITLE:	DRAWN BY: SAAD H. TANTAWI
	(M.S.CE, B.S.CE, E.I.)
ROOF ATTACHMENT	SCALE: N.T.S

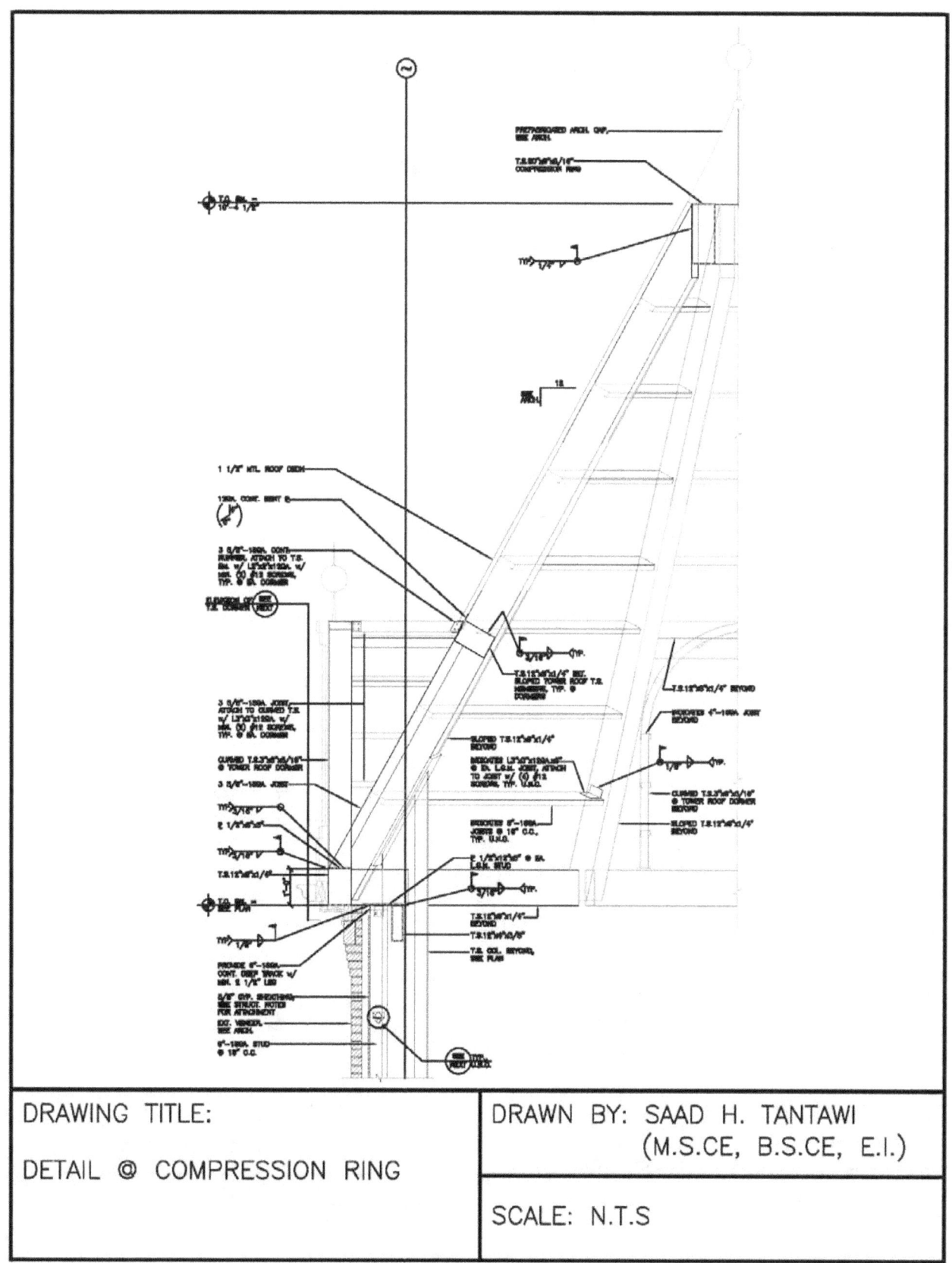

DRAWING TITLE:	DRAWN BY: SAAD H. TANTAWI
DETAIL @ COMPRESSION RING	(M.S.CE, B.S.CE, E.I.)
	SCALE: N.T.S

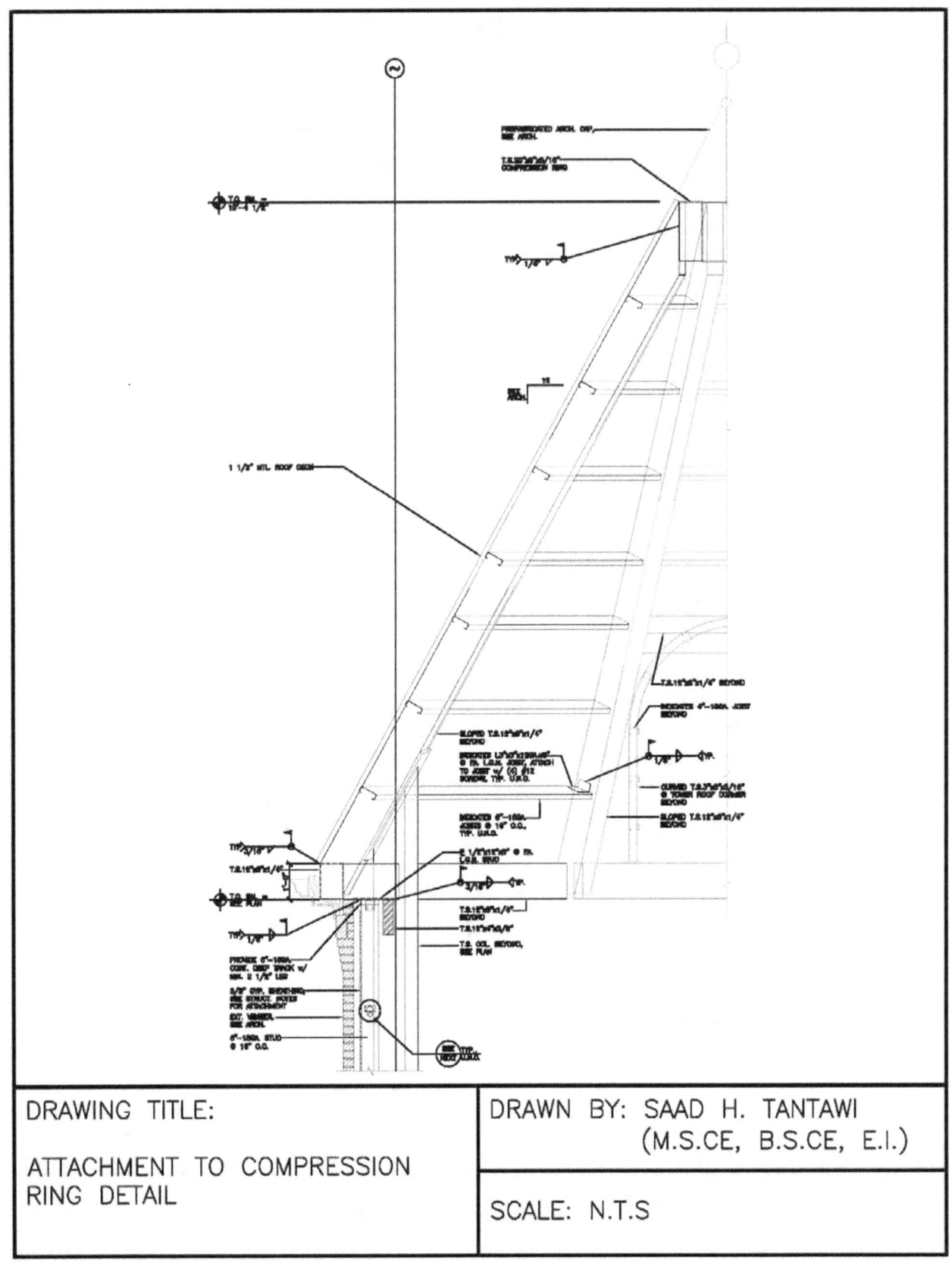

DRAWING TITLE:	DRAWN BY: SAAD H. TANTAWI
ATTACHMENT TO COMPRESSION RING DETAIL	(M.S.CE, B.S.CE, E.I.)
	SCALE: N.T.S

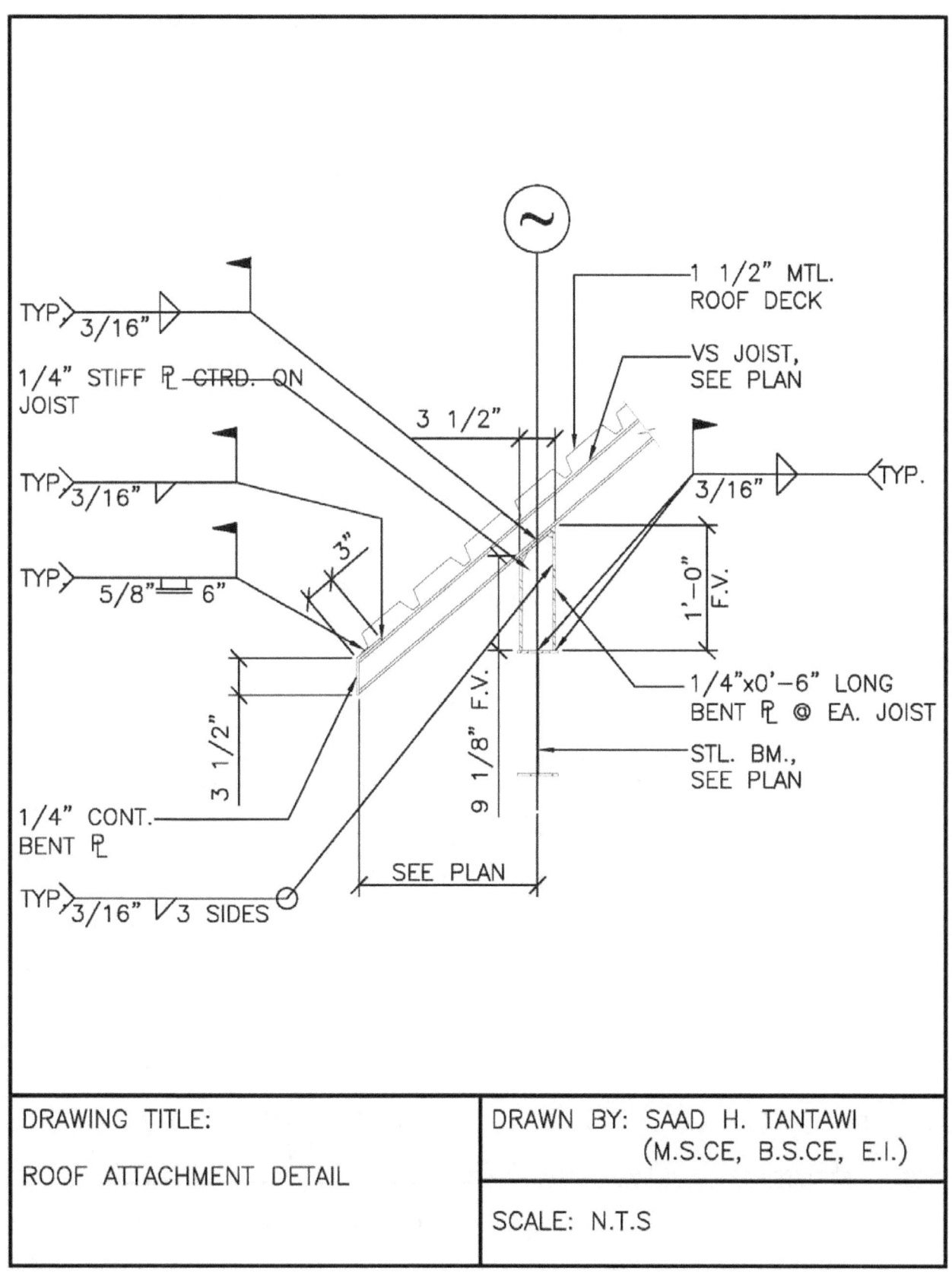

DRAWING TITLE:	DRAWN BY: SAAD H. TANTAWI
	(M.S.CE, B.S.CE, E.I.)
ROOF ATTACHMENT DETAIL	
	SCALE: N.T.S

Saad Hasan Tantawi (M.S.CE, B.S.CE, E.I., A.M.ASCE)

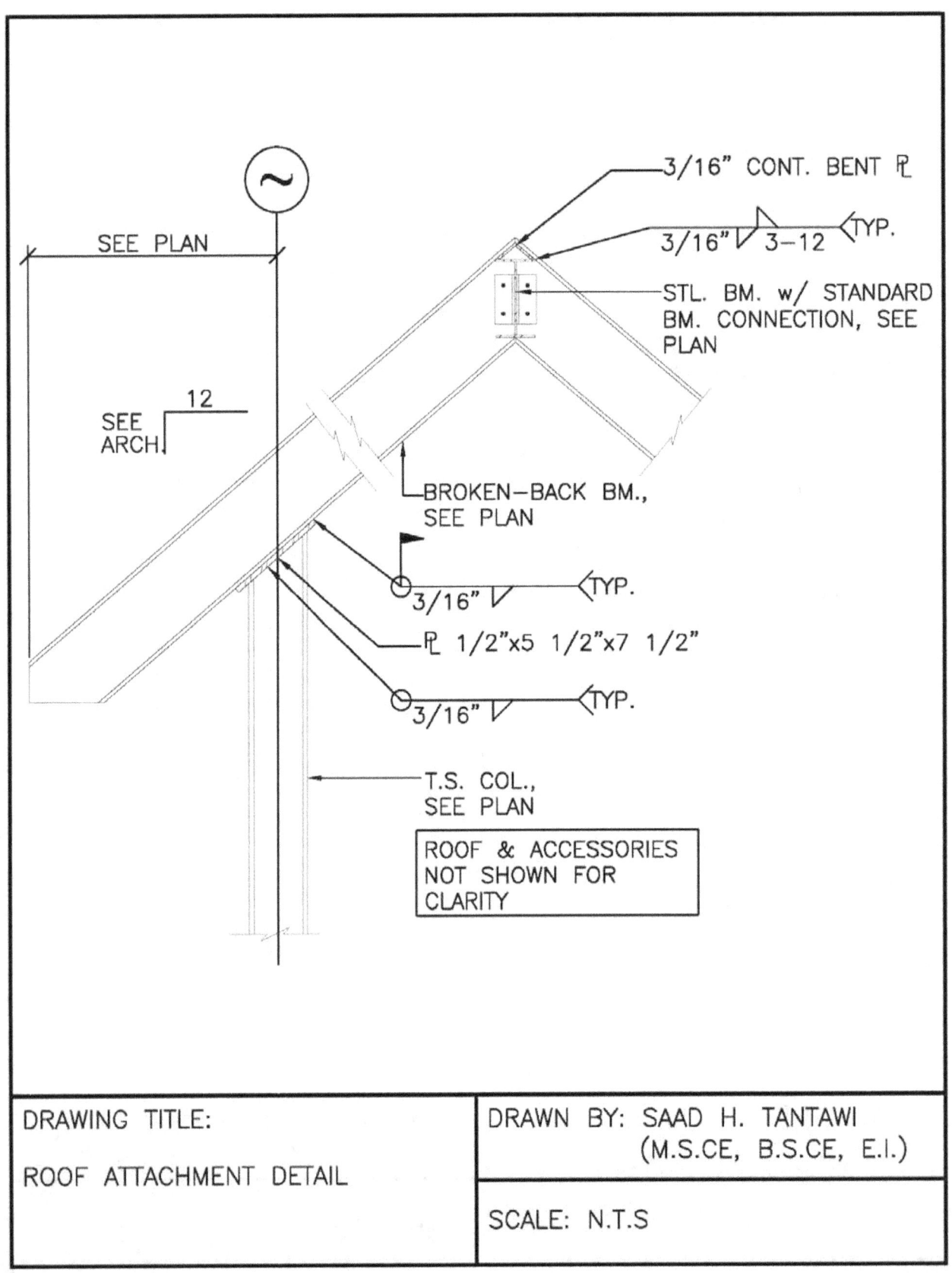

DRAWING TITLE: ROOF ATTACHMENT DETAIL	DRAWN BY: SAAD H. TANTAWI (M.S.CE, B.S.CE, E.I.)
	SCALE: N.T.S

84

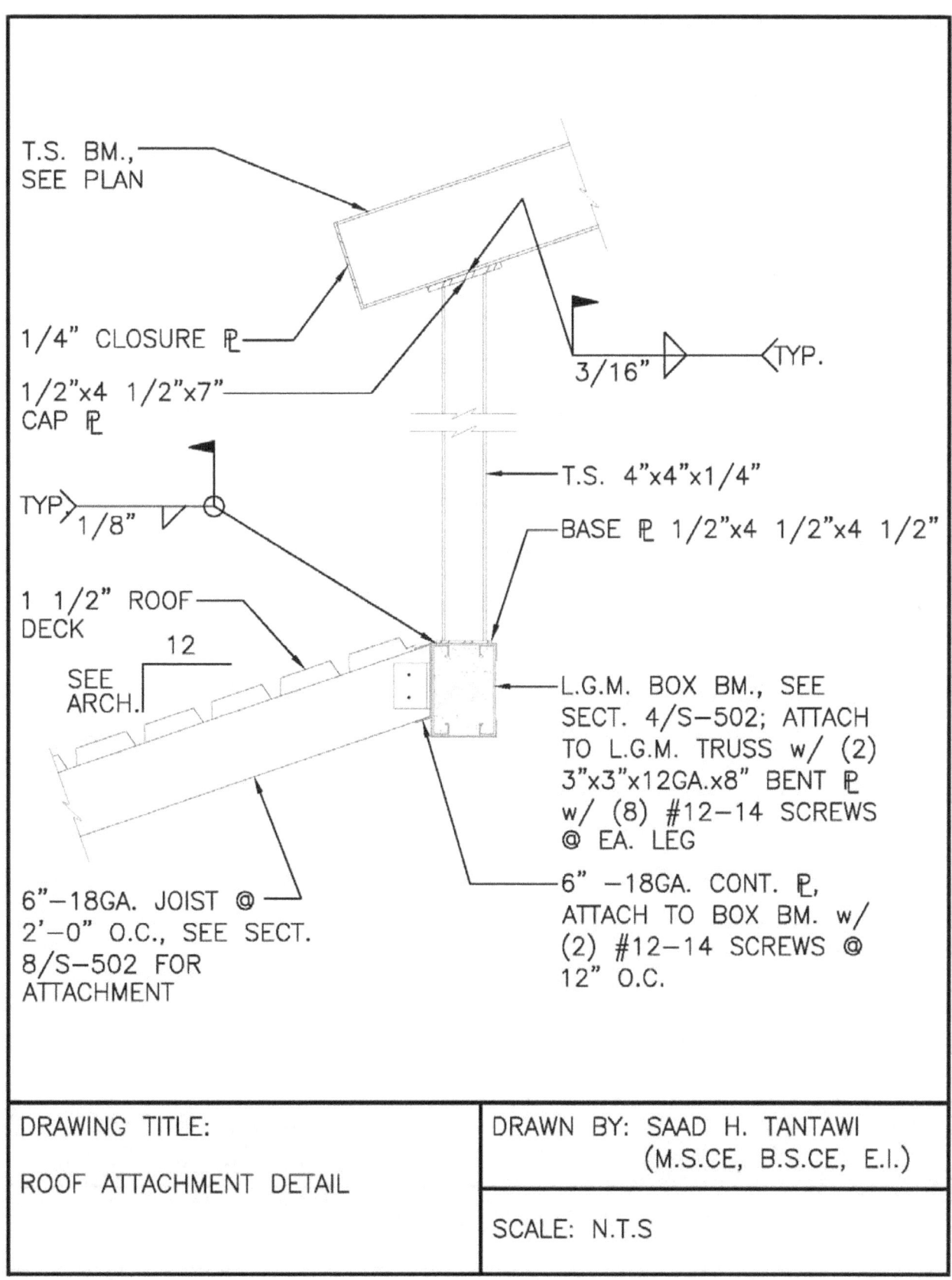

T.S. BM.,
SEE PLAN

1/4" CLOSURE ℙ

1/2"x4 1/2"x7"
CAP ℙ

TYP. 1/8"

1 1/2" ROOF
DECK
12
SEE
ARCH.

6"-18GA. JOIST @
2'-0" O.C., SEE SECT.
8/S-502 FOR
ATTACHMENT

3/16" TYP.

T.S. 4"x4"x1/4"

BASE ℙ 1/2"x4 1/2"x4 1/2"

L.G.M. BOX BM., SEE
SECT. 4/S-502; ATTACH
TO L.G.M. TRUSS w/ (2)
3"x3"x12GA.x8" BENT ℙ
w/ (8) #12-14 SCREWS
@ EA. LEG

6" -18GA. CONT. ℙ,
ATTACH TO BOX BM. w/
(2) #12-14 SCREWS @
12" O.C.

DRAWING TITLE:	DRAWN BY: SAAD H. TANTAWI
ROOF ATTACHMENT DETAIL	(M.S.CE, B.S.CE, E.I.)
	SCALE: N.T.S

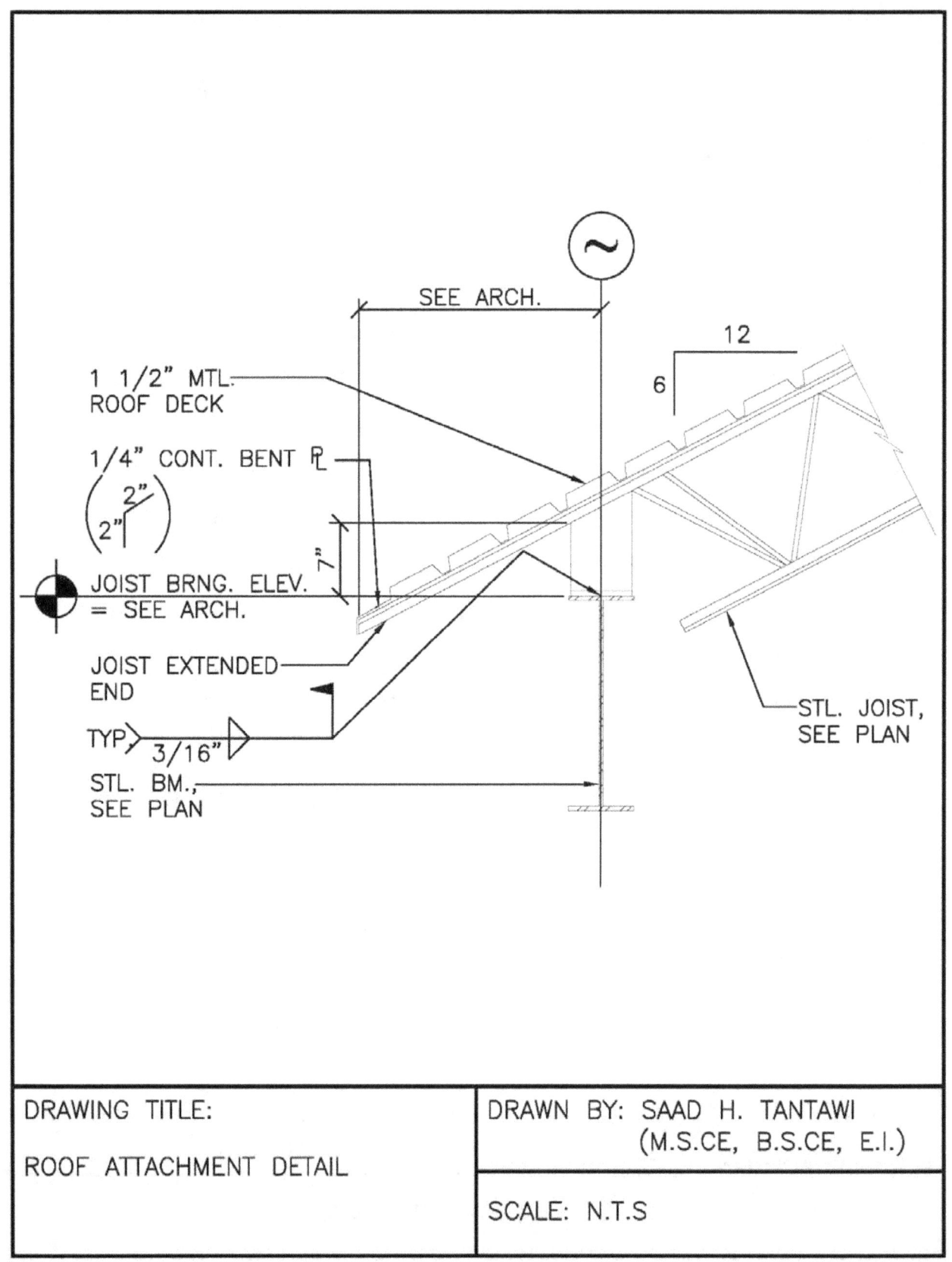

DRAWING TITLE: ROOF ATTACHMENT DETAIL	DRAWN BY: SAAD H. TANTAWI (M.S.CE, B.S.CE, E.I.)
	SCALE: N.T.S

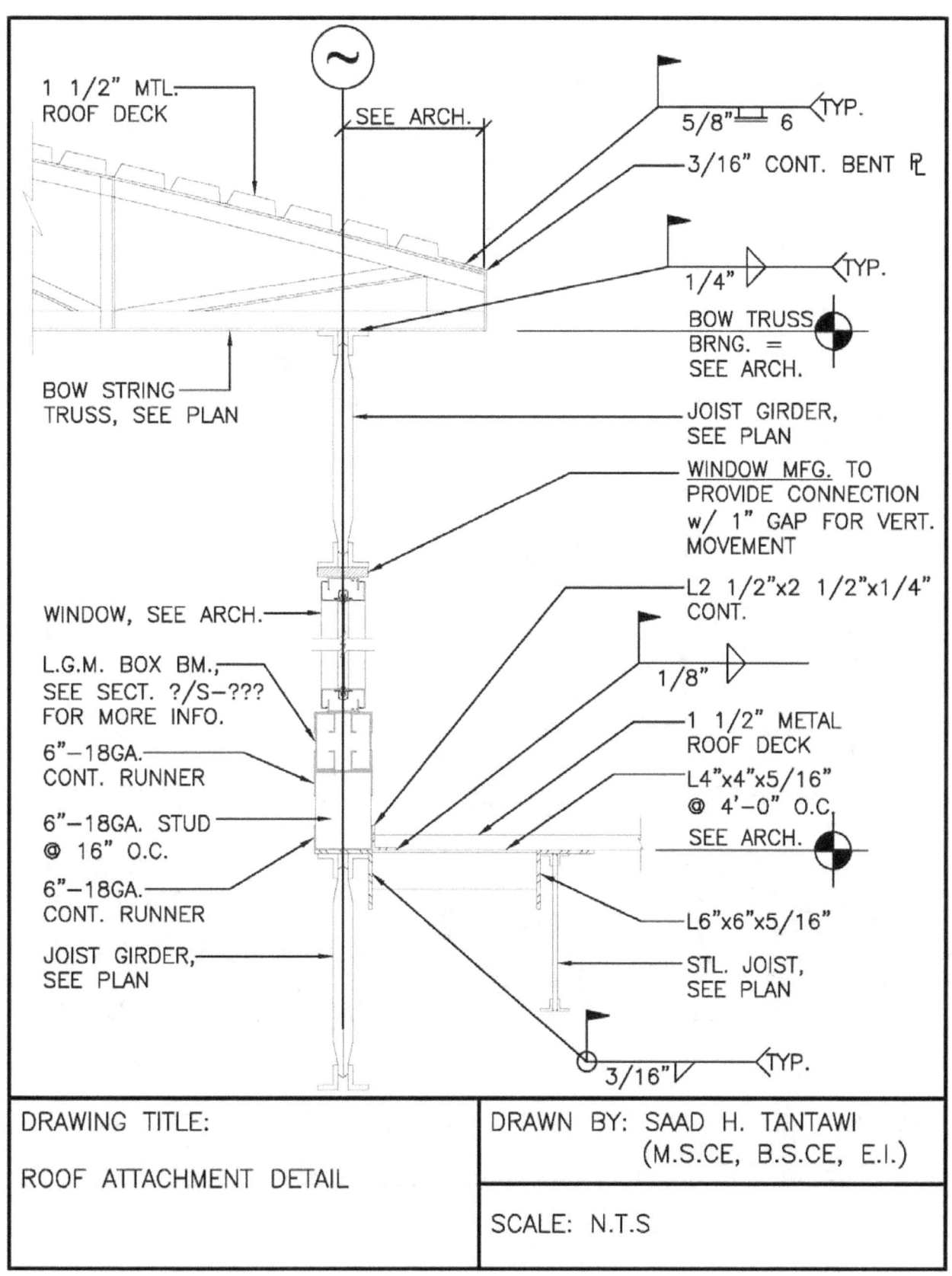

1 1/2" MTL.
ROOF DECK

SEE ARCH.

5/8" ___ 6   TYP.

3/16" CONT. BENT PL

1/4"   TYP.

BOW TRUSS
BRNG. =
SEE ARCH.

BOW STRING
TRUSS, SEE PLAN

JOIST GIRDER,
SEE PLAN

WINDOW MFG. TO
PROVIDE CONNECTION
w/ 1" GAP FOR VERT.
MOVEMENT

L2 1/2"x2 1/2"x1/4"
CONT.

WINDOW, SEE ARCH.

1/8"

L.G.M. BOX BM.,
SEE SECT. ?/S-???
FOR MORE INFO.

1 1/2" METAL
ROOF DECK

6"-18GA.
CONT. RUNNER

L4"x4"x5/16"
@ 4'-0" O.C.
SEE ARCH.

6"-18GA. STUD
@ 16" O.C.

6"-18GA.
CONT. RUNNER

L6"x6"x5/16"

JOIST GIRDER,
SEE PLAN

STL. JOIST,
SEE PLAN

3/16"   TYP.

DRAWING TITLE:	DRAWN BY: SAAD H. TANTAWI (M.S.CE, B.S.CE, E.I.)
ROOF ATTACHMENT DETAIL	
	SCALE: N.T.S

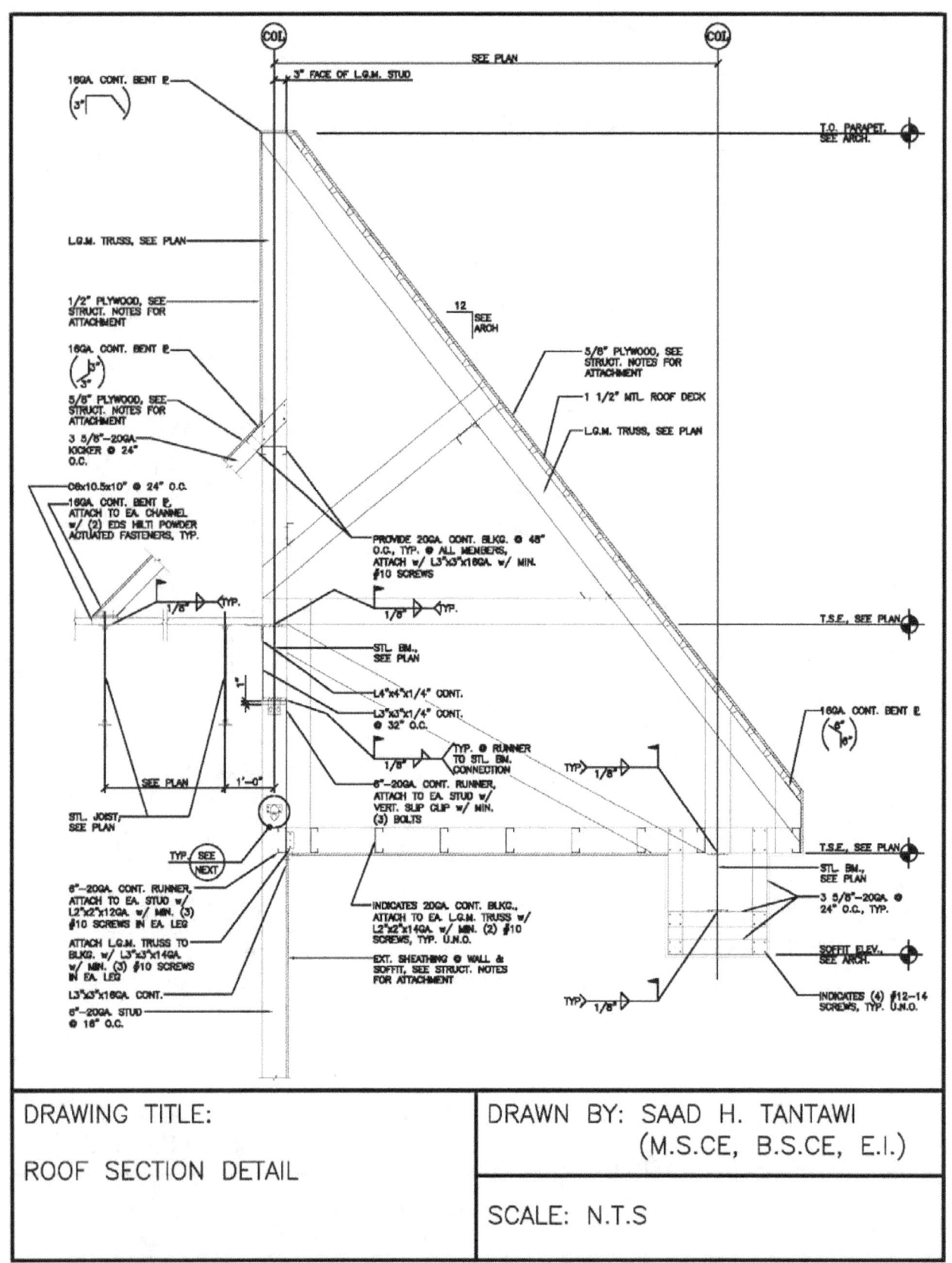

DRAWING TITLE:	DRAWN BY: SAAD H. TANTAWI
ROOF SECTION DETAIL	(M.S.CE, B.S.CE, E.I.)
	SCALE: N.T.S

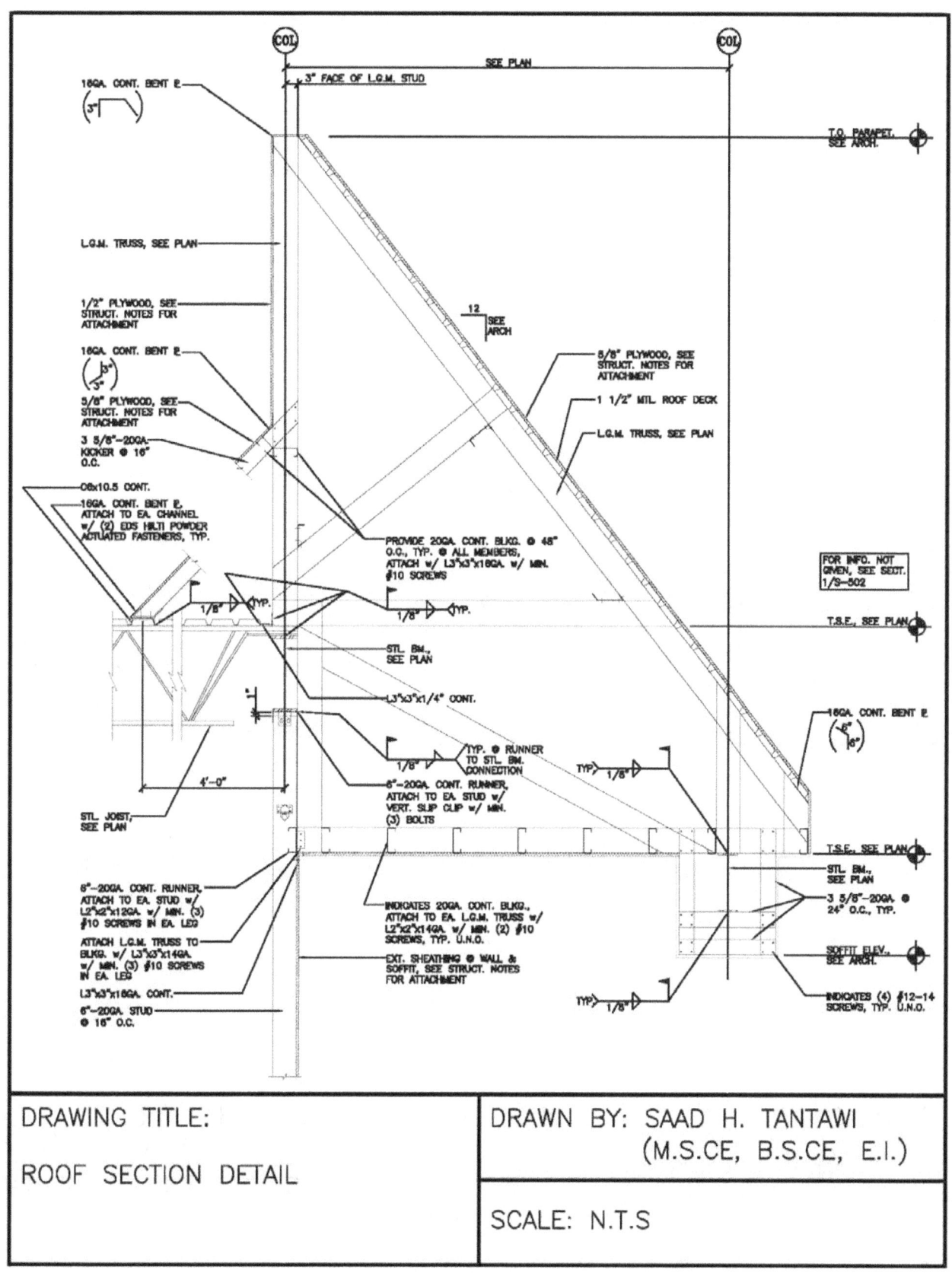

DRAWING TITLE:	DRAWN BY: SAAD H. TANTAWI
ROOF SECTION DETAIL	(M.S.CE, B.S.CE, E.I.)
	SCALE: N.T.S

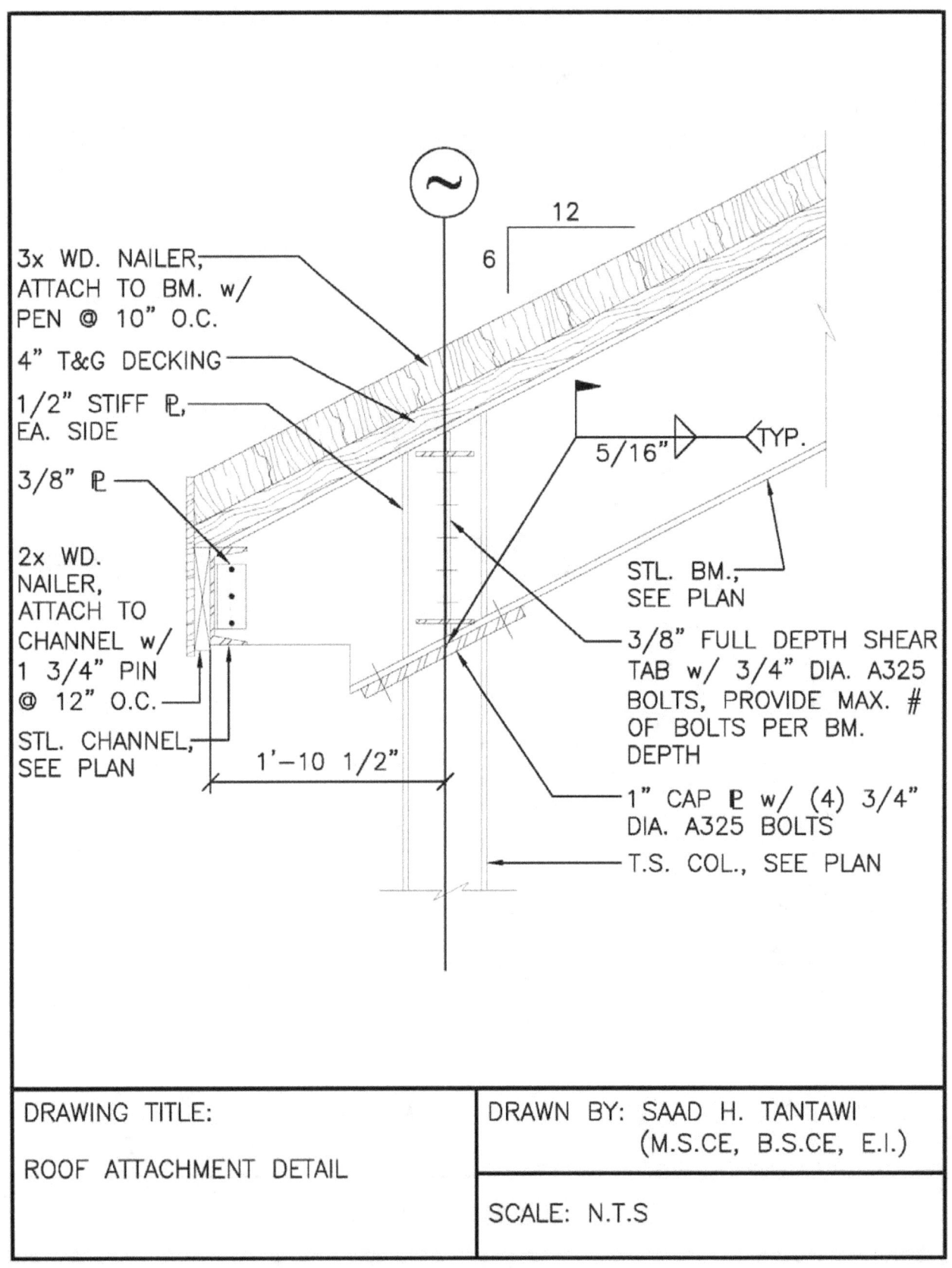

DRAWING TITLE:	DRAWN BY: SAAD H. TANTAWI
ROOF ATTACHMENT DETAIL	(M.S.CE, B.S.CE, E.I.)
	SCALE: N.T.S

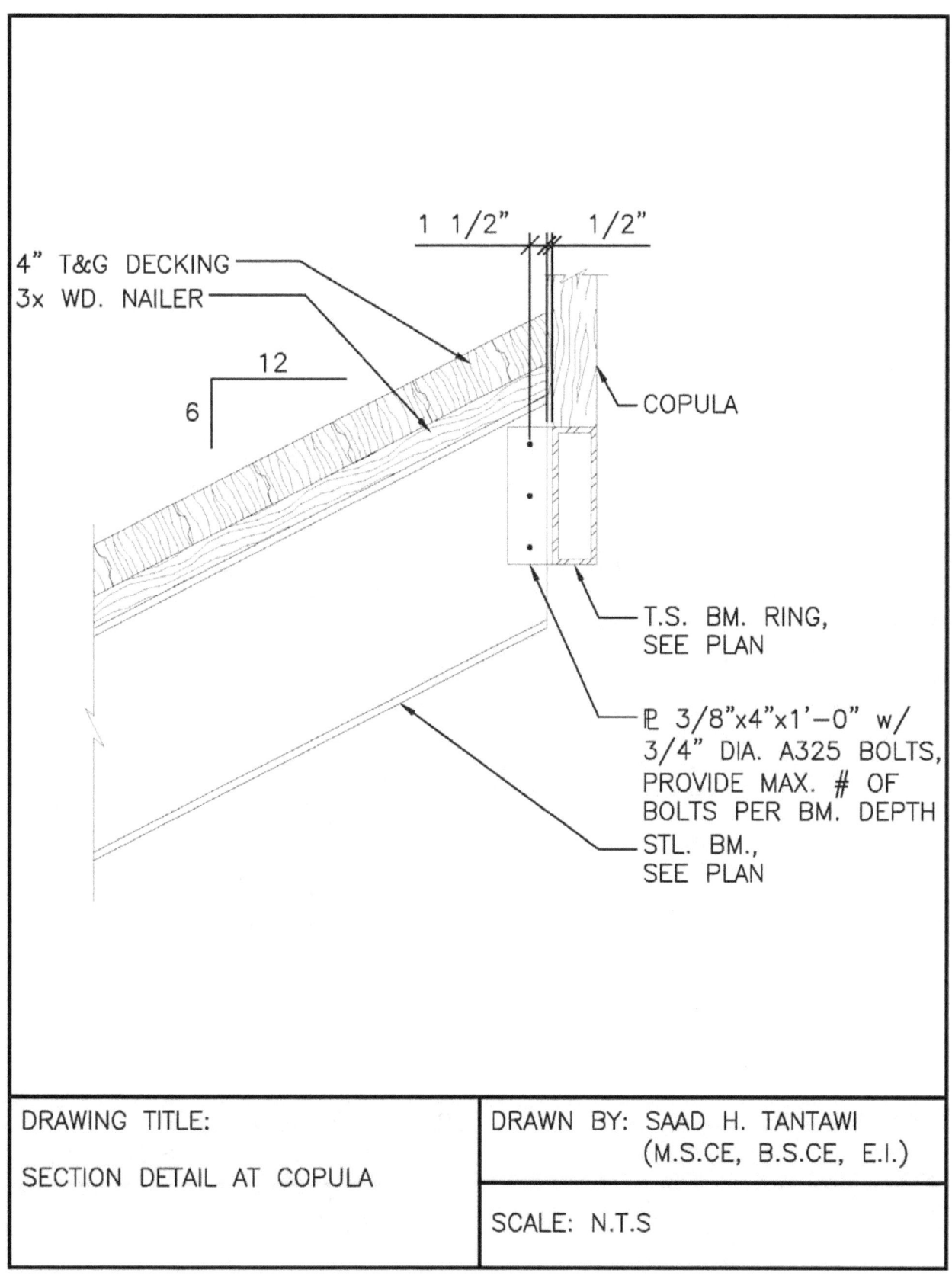

4" T&G DECKING

3x WD. NAILER

12

6

1 1/2"   1/2"

COPULA

T.S. BM. RING,
SEE PLAN

℄ 3/8"x4"x1'-0" w/
3/4" DIA. A325 BOLTS,
PROVIDE MAX. # OF
BOLTS PER BM. DEPTH

STL. BM.,
SEE PLAN

DRAWING TITLE:	DRAWN BY: SAAD H. TANTAWI
	(M.S.CE, B.S.CE, E.I.)
SECTION DETAIL AT COPULA	
	SCALE: N.T.S

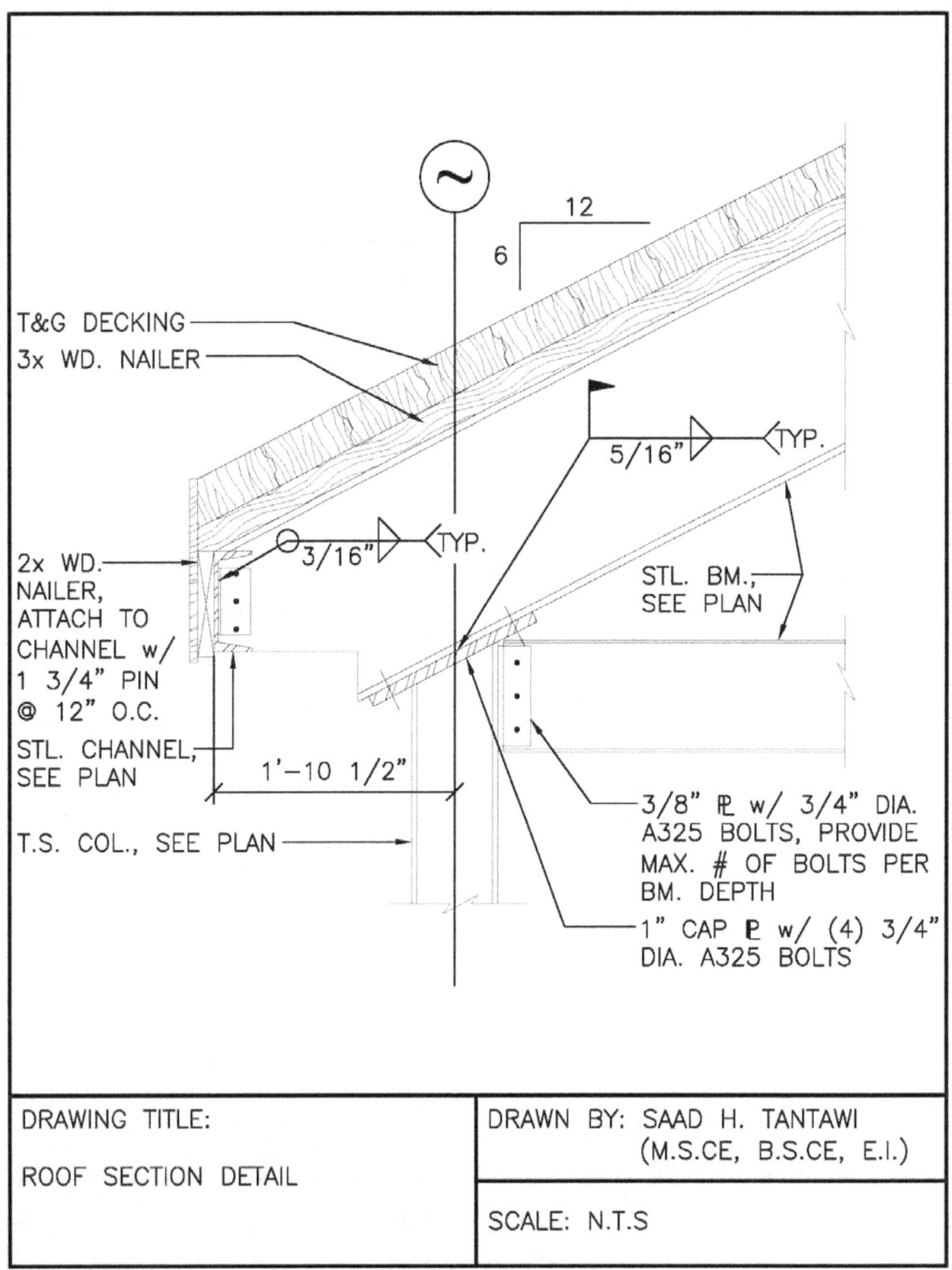

T&G DECKING

3x WD. NAILER

12

6

5/16" TYP.

2x WD. NAILER, ATTACH TO CHANNEL w/ 1 3/4" PIN @ 12" O.C.

3/16" TYP.

STL. BM., SEE PLAN

STL. CHANNEL, SEE PLAN

1'-10 1/2"

T.S. COL., SEE PLAN

3/8" ℙ w/ 3/4" DIA. A325 BOLTS, PROVIDE MAX. # OF BOLTS PER BM. DEPTH

1" CAP ℙ w/ (4) 3/4" DIA. A325 BOLTS

DRAWING TITLE: ROOF SECTION DETAIL	DRAWN BY: SAAD H. TANTAWI (M.S.CE, B.S.CE, E.I.)
	SCALE: N.T.S

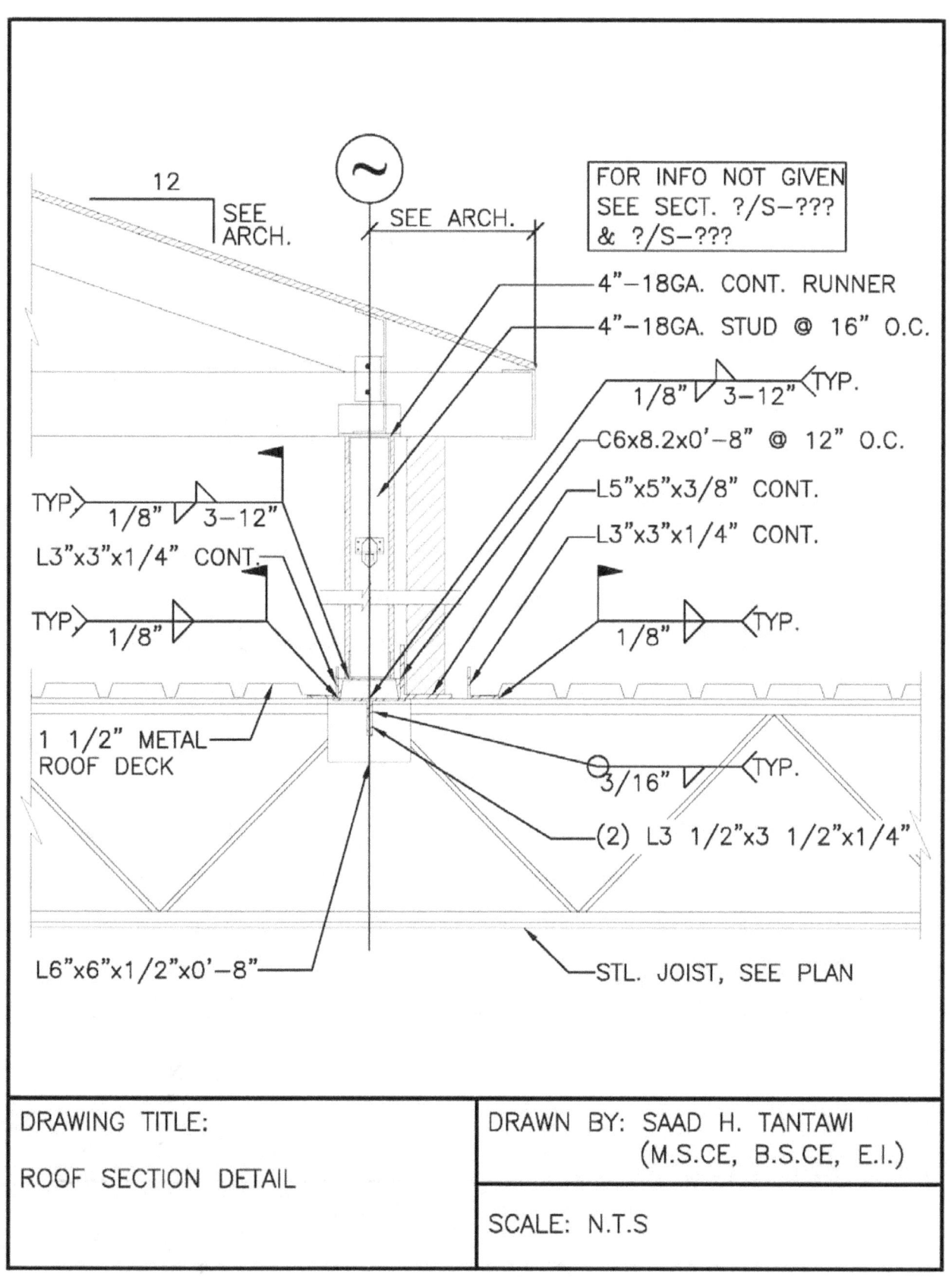

12
SEE ARCH.
SEE ARCH.

~

FOR INFO NOT GIVEN
SEE SECT. ?/S-???
& ?/S-???

4"-18GA. CONT. RUNNER

4"-18GA. STUD @ 16" O.C.

1/8" 3-12" TYP.

C6x8.2x0'-8" @ 12" O.C.

L5"x5"x3/8" CONT.

L3"x3"x1/4" CONT.

TYP. 1/8" 3-12"

L3"x3"x1/4" CONT.

TYP. 1/8"

1/8" TYP.

1 1/2" METAL ROOF DECK

3/16" TYP.

(2) L3 1/2"x3 1/2"x1/4"

L6"x6"x1/2"x0'-8"

STL. JOIST, SEE PLAN

DRAWING TITLE:

ROOF SECTION DETAIL

DRAWN BY: SAAD H. TANTAWI
(M.S.CE, B.S.CE, E.I.)

SCALE: N.T.S

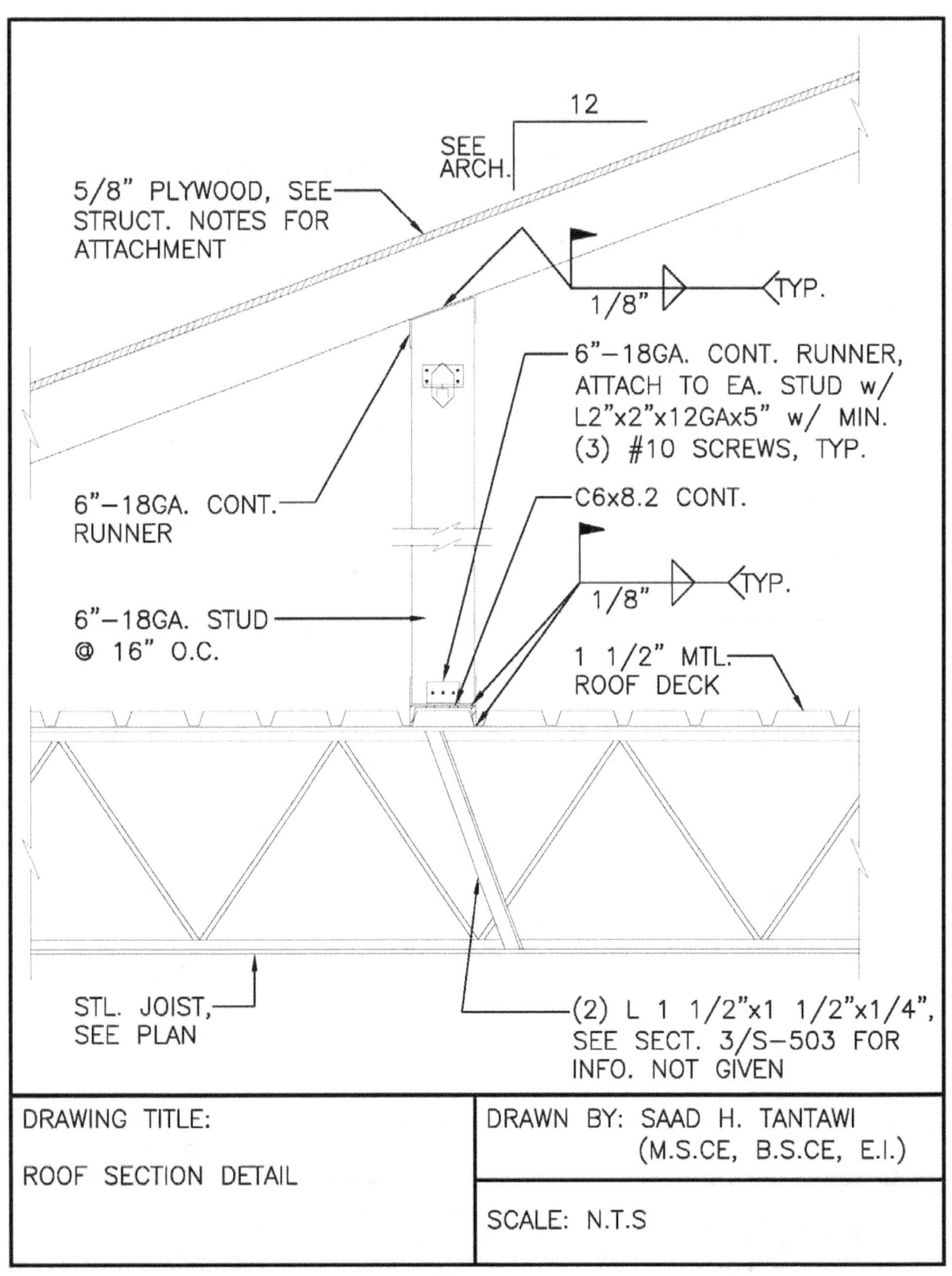

5/8" PLYWOOD, SEE
STRUCT. NOTES FOR
ATTACHMENT

SEE
ARCH.

12

1/8" TYP.

6"-18GA. CONT.
RUNNER

6"-18GA. STUD
@ 16" O.C.

6"-18GA. CONT. RUNNER,
ATTACH TO EA. STUD w/
L2"x2"x12GAx5" w/ MIN.
(3) #10 SCREWS, TYP.

C6x8.2 CONT.

1/8" TYP.

1 1/2" MTL.
ROOF DECK

STL. JOIST,
SEE PLAN

(2) L 1 1/2"x1 1/2"x1/4",
SEE SECT. 3/S-503 FOR
INFO. NOT GIVEN

DRAWING TITLE:	DRAWN BY: SAAD H. TANTAWI
ROOF SECTION DETAIL	(M.S.CE, B.S.CE, E.I.)
	SCALE: N.T.S

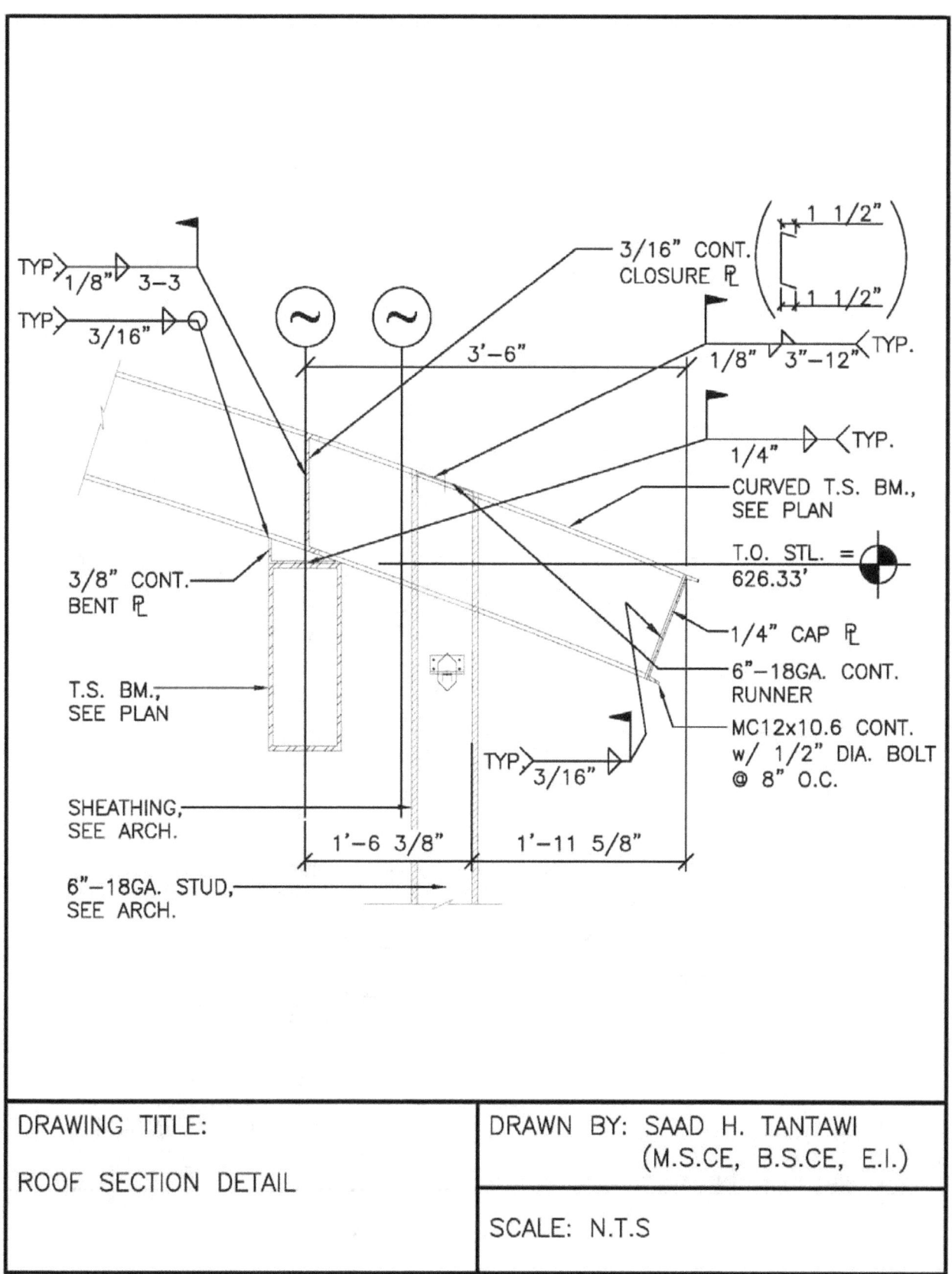

DRAWING TITLE:

ROOF SECTION DETAIL

DRAWN BY: SAAD H. TANTAWI
(M.S.CE, B.S.CE, E.I.)

SCALE: N.T.S

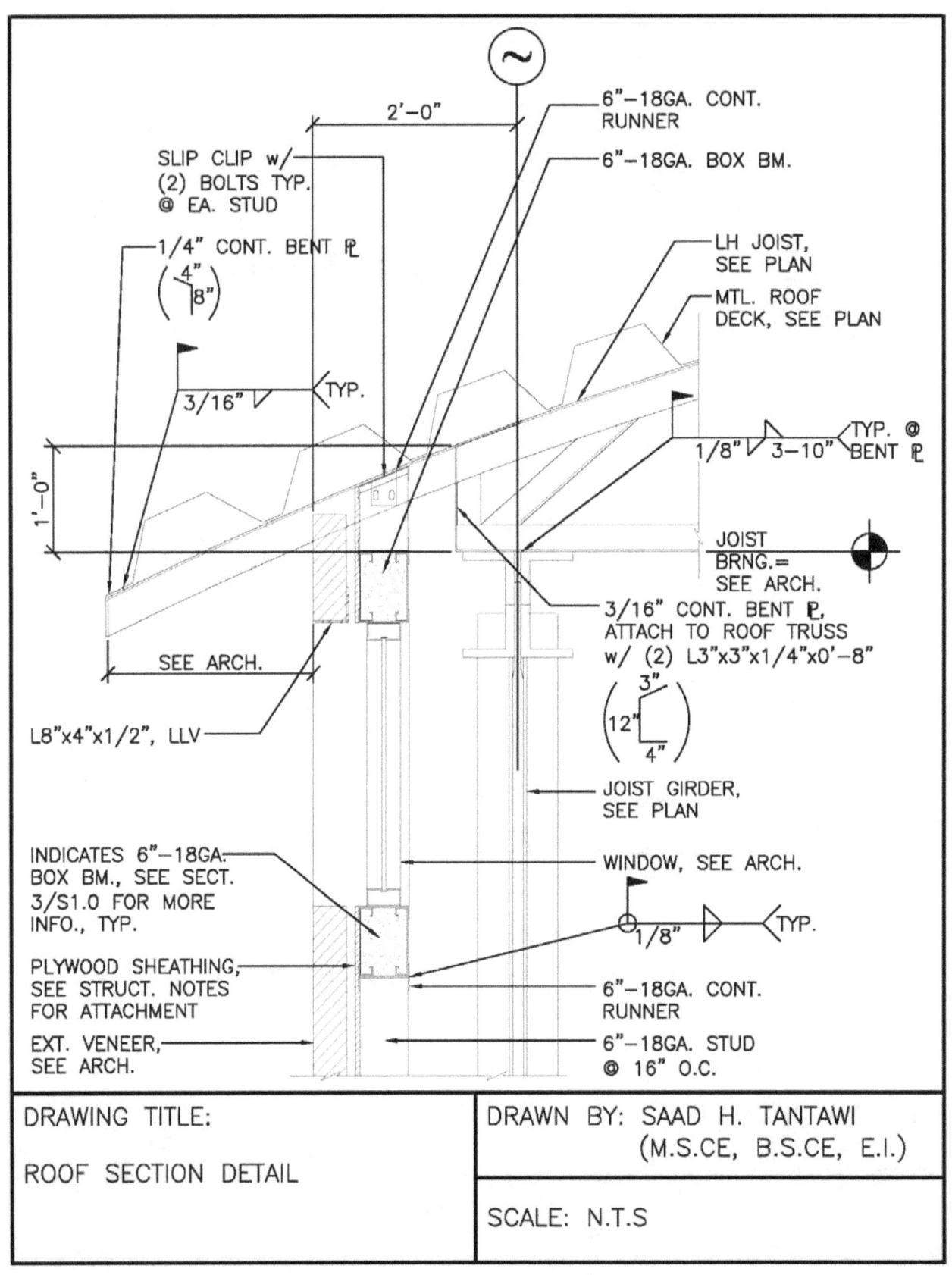

DRAWING TITLE:	DRAWN BY: SAAD H. TANTAWI
ROOF SECTION DETAIL	(M.S.CE, B.S.CE, E.I.)
	SCALE: N.T.S

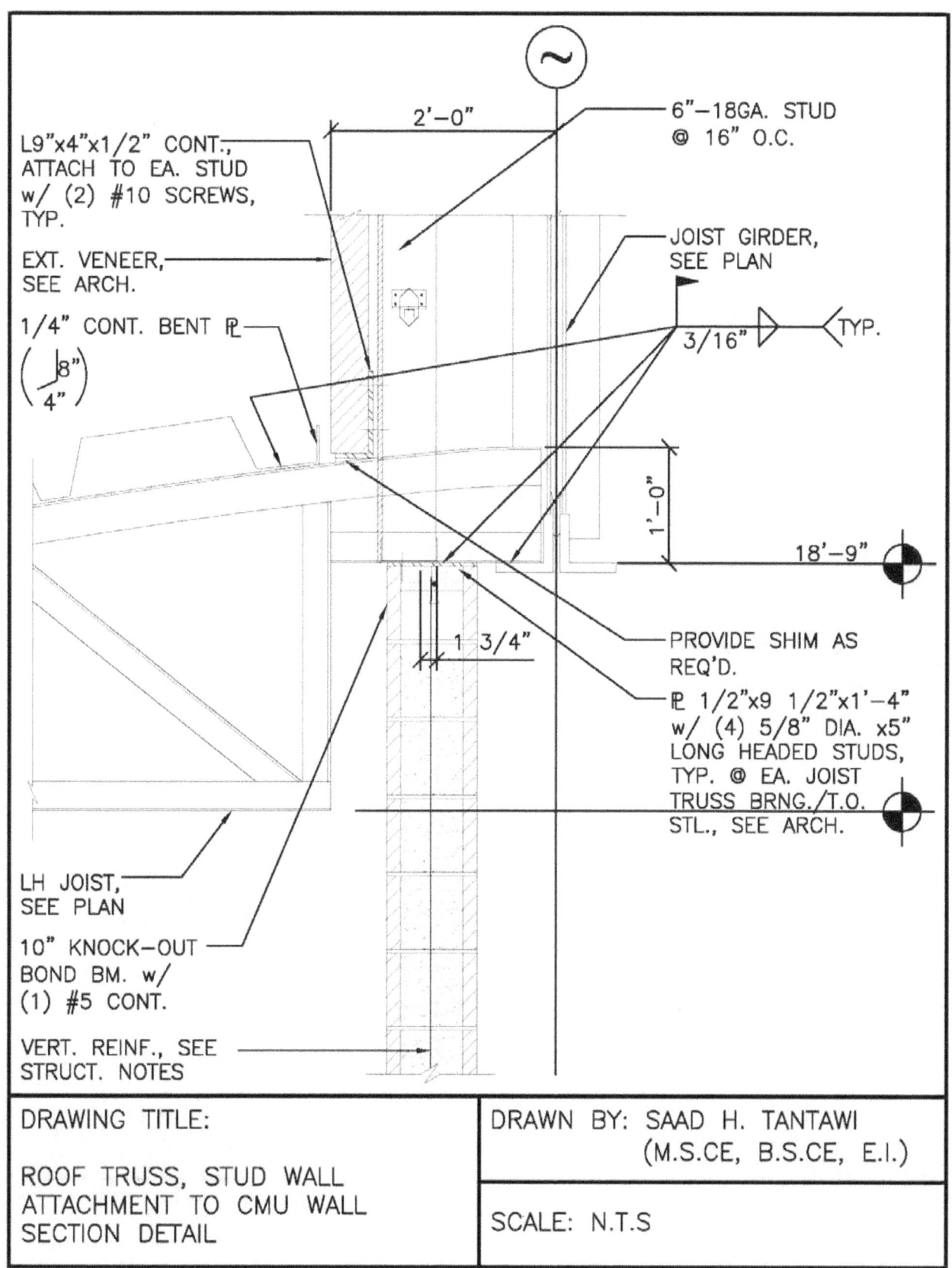

L9"x4"x1/2" CONT., ATTACH TO EA. STUD w/ (2) #10 SCREWS, TYP.

EXT. VENEER, SEE ARCH.

1/4" CONT. BENT ℙ

$\left(\dfrac{8"}{4"}\right)$

2'-0"

6"-18GA. STUD @ 16" O.C.

JOIST GIRDER, SEE PLAN

3/16" TYP.

1'-0"

1 3/4"

18'-9"

PROVIDE SHIM AS REQ'D.

ℙ 1/2"x9 1/2"x1'-4" w/ (4) 5/8" DIA. x5" LONG HEADED STUDS, TYP. @ EA. JOIST TRUSS BRNG./T.O. STL., SEE ARCH.

LH JOIST, SEE PLAN

10" KNOCK-OUT BOND BM. w/ (1) #5 CONT.

VERT. REINF., SEE STRUCT. NOTES

DRAWING TITLE:	DRAWN BY: SAAD H. TANTAWI (M.S.CE, B.S.CE, E.I.)
ROOF TRUSS, STUD WALL ATTACHMENT TO CMU WALL SECTION DETAIL	SCALE: N.T.S

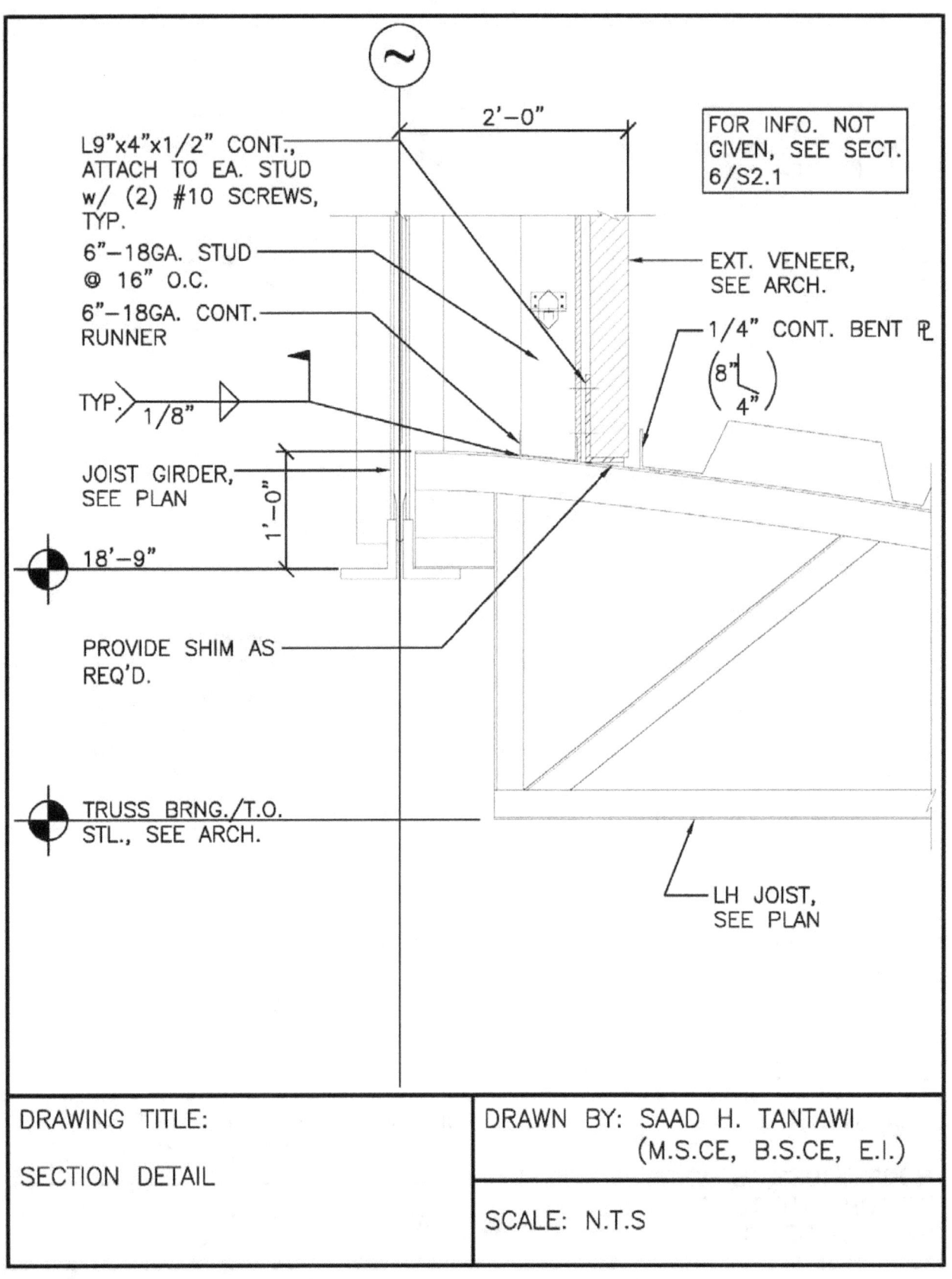

DRAWING TITLE: SECTION DETAIL	DRAWN BY: SAAD H. TANTAWI (M.S.CE, B.S.CE, E.I.)
	SCALE: N.T.S

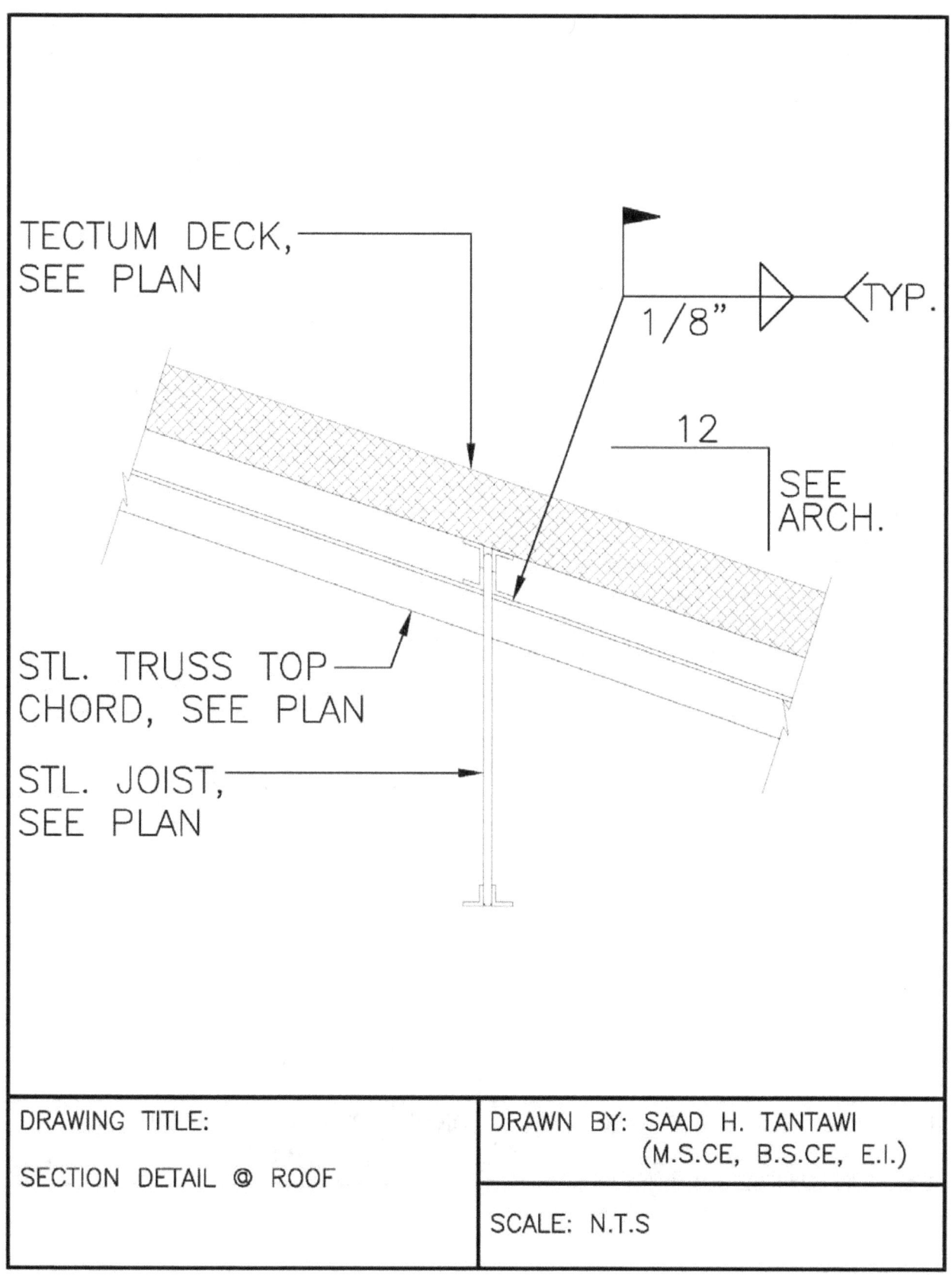

TECTUM DECK,
SEE PLAN

1/8" TYP.

12
SEE
ARCH.

STL. TRUSS TOP
CHORD, SEE PLAN

STL. JOIST,
SEE PLAN

DRAWING TITLE:	DRAWN BY: SAAD H. TANTAWI
	(M.S.CE, B.S.CE, E.I.)
SECTION DETAIL @ ROOF	
	SCALE: N.T.S

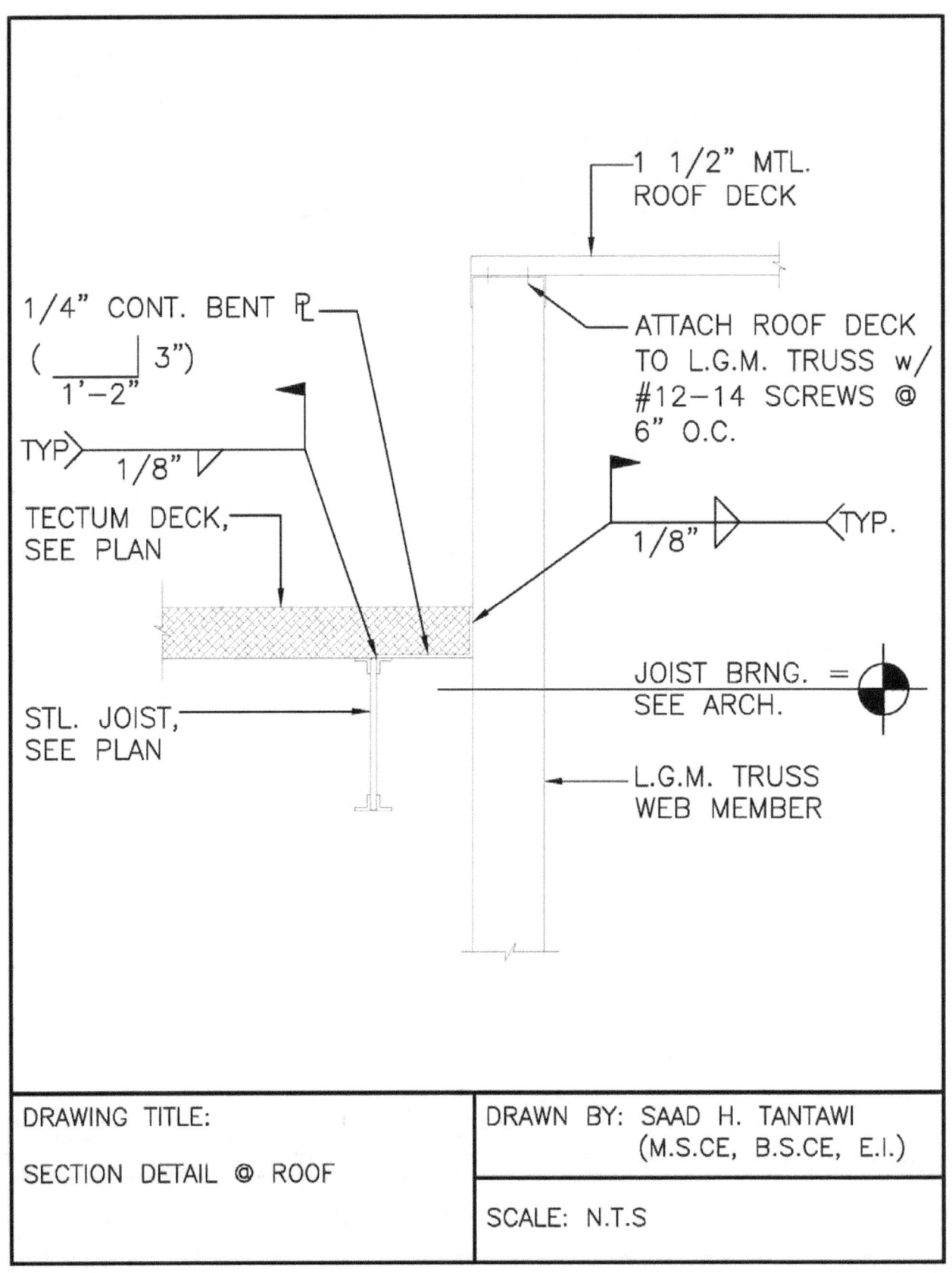

1 1/2" MTL. ROOF DECK

1/4" CONT. BENT PL (⌐ 3" / 1'-2")

TYP 1/8"

TECTUM DECK, SEE PLAN

STL. JOIST, SEE PLAN

ATTACH ROOF DECK TO L.G.M. TRUSS w/ #12-14 SCREWS @ 6" O.C.

1/8" TYP.

JOIST BRNG. = SEE ARCH.

L.G.M. TRUSS WEB MEMBER

DRAWING TITLE: SECTION DETAIL @ ROOF	DRAWN BY: SAAD H. TANTAWI (M.S.CE, B.S.CE, E.I.)
	SCALE: N.T.S

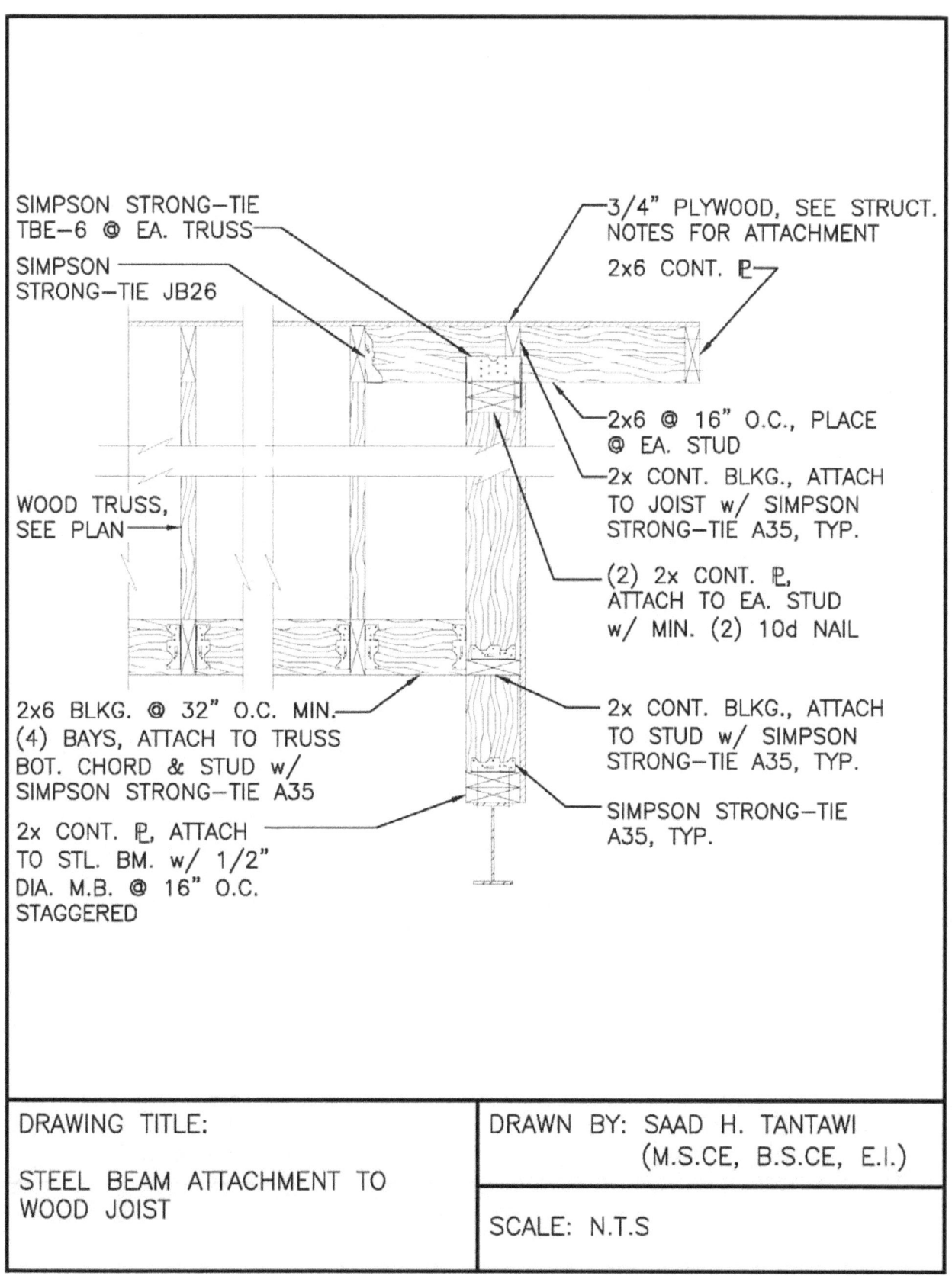

SIMPSON STRONG—TIE
TBE—6 @ EA. TRUSS

SIMPSON
STRONG—TIE JB26

WOOD TRUSS,
SEE PLAN

2x6 BLKG. @ 32" O.C. MIN.
(4) BAYS, ATTACH TO TRUSS
BOT. CHORD & STUD w/
SIMPSON STRONG—TIE A35

2x CONT. ℄, ATTACH
TO STL. BM. w/ 1/2"
DIA. M.B. @ 16" O.C.
STAGGERED

3/4" PLYWOOD, SEE STRUCT.
NOTES FOR ATTACHMENT

2x6 CONT. ℄

2x6 @ 16" O.C., PLACE
@ EA. STUD

2x CONT. BLKG., ATTACH
TO JOIST w/ SIMPSON
STRONG—TIE A35, TYP.

(2) 2x CONT. ℄,
ATTACH TO EA. STUD
w/ MIN. (2) 10d NAIL

2x CONT. BLKG., ATTACH
TO STUD w/ SIMPSON
STRONG—TIE A35, TYP.

SIMPSON STRONG—TIE
A35, TYP.

DRAWING TITLE: STEEL BEAM ATTACHMENT TO WOOD JOIST	DRAWN BY: SAAD H. TANTAWI (M.S.CE, B.S.CE, E.I.)
	SCALE: N.T.S

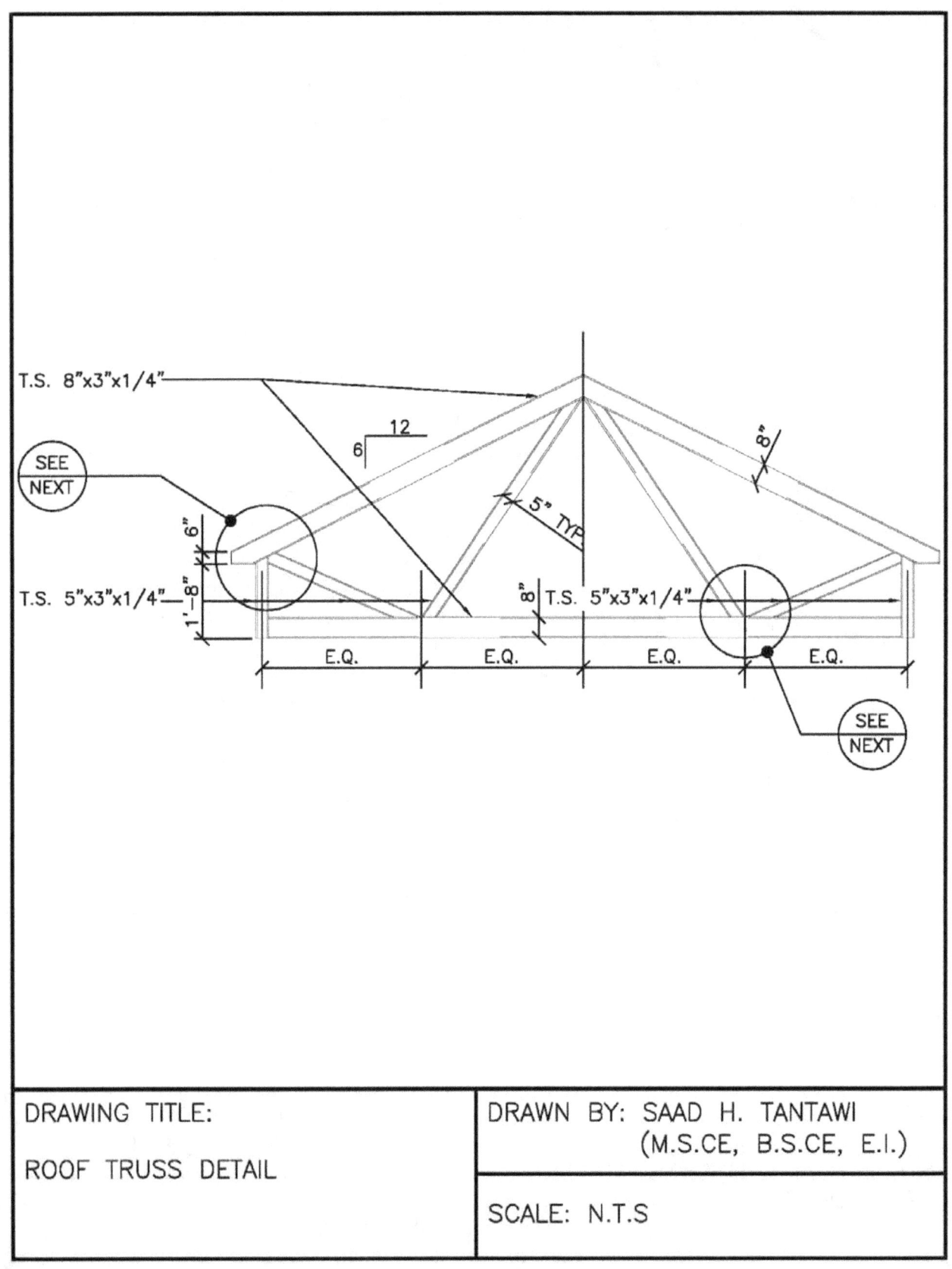

DRAWING TITLE:	DRAWN BY: SAAD H. TANTAWI
ROOF TRUSS DETAIL	(M.S.CE, B.S.CE, E.I.)
	SCALE: N.T.S

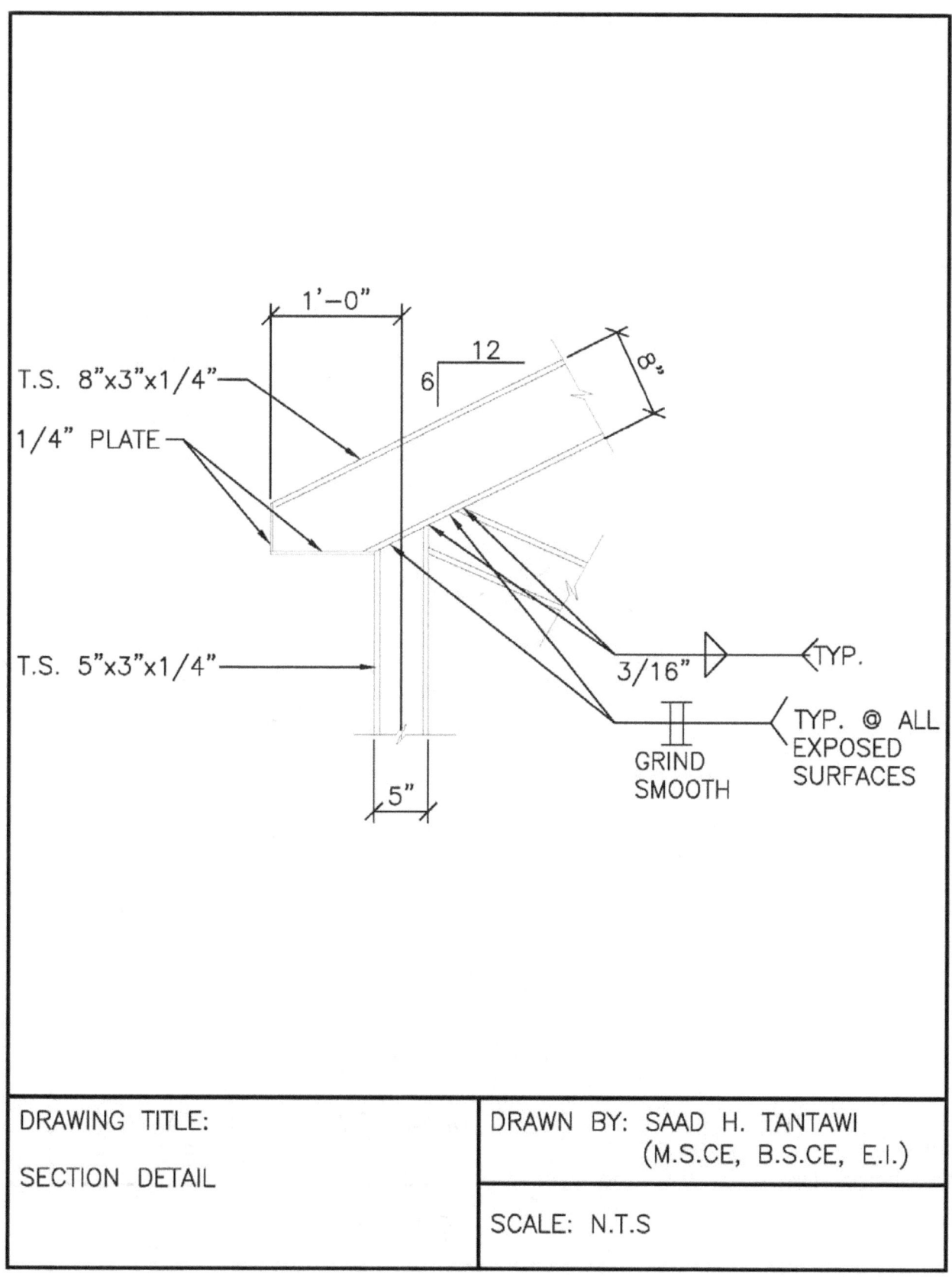

T.S. 8"x3"x1/4"

1/4" PLATE

T.S. 5"x3"x1/4"

1'-0"

12

6

8"

3/16"   TYP.

GRIND
SMOOTH

TYP. @ ALL
EXPOSED
SURFACES

5"

DRAWING TITLE:	DRAWN BY: SAAD H. TANTAWI
	(M.S.CE, B.S.CE, E.I.)
SECTION DETAIL	
	SCALE: N.T.S

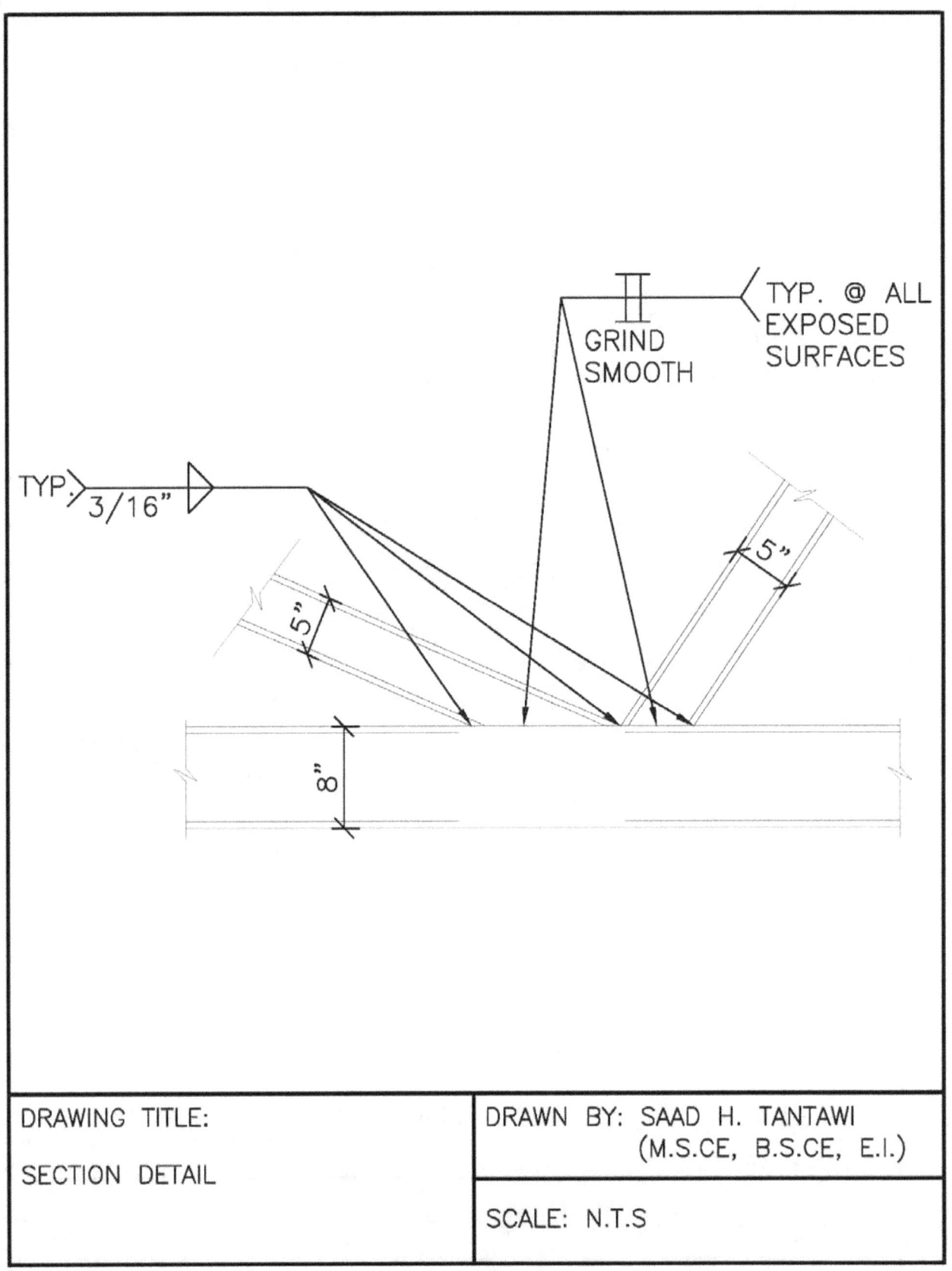

TYP. @ ALL
EXPOSED
SURFACES

GRIND
SMOOTH

TYP.
3/16"

5"

5"

8"

DRAWING TITLE:	DRAWN BY: SAAD H. TANTAWI
SECTION DETAIL	(M.S.CE, B.S.CE, E.I.)
	SCALE: N.T.S

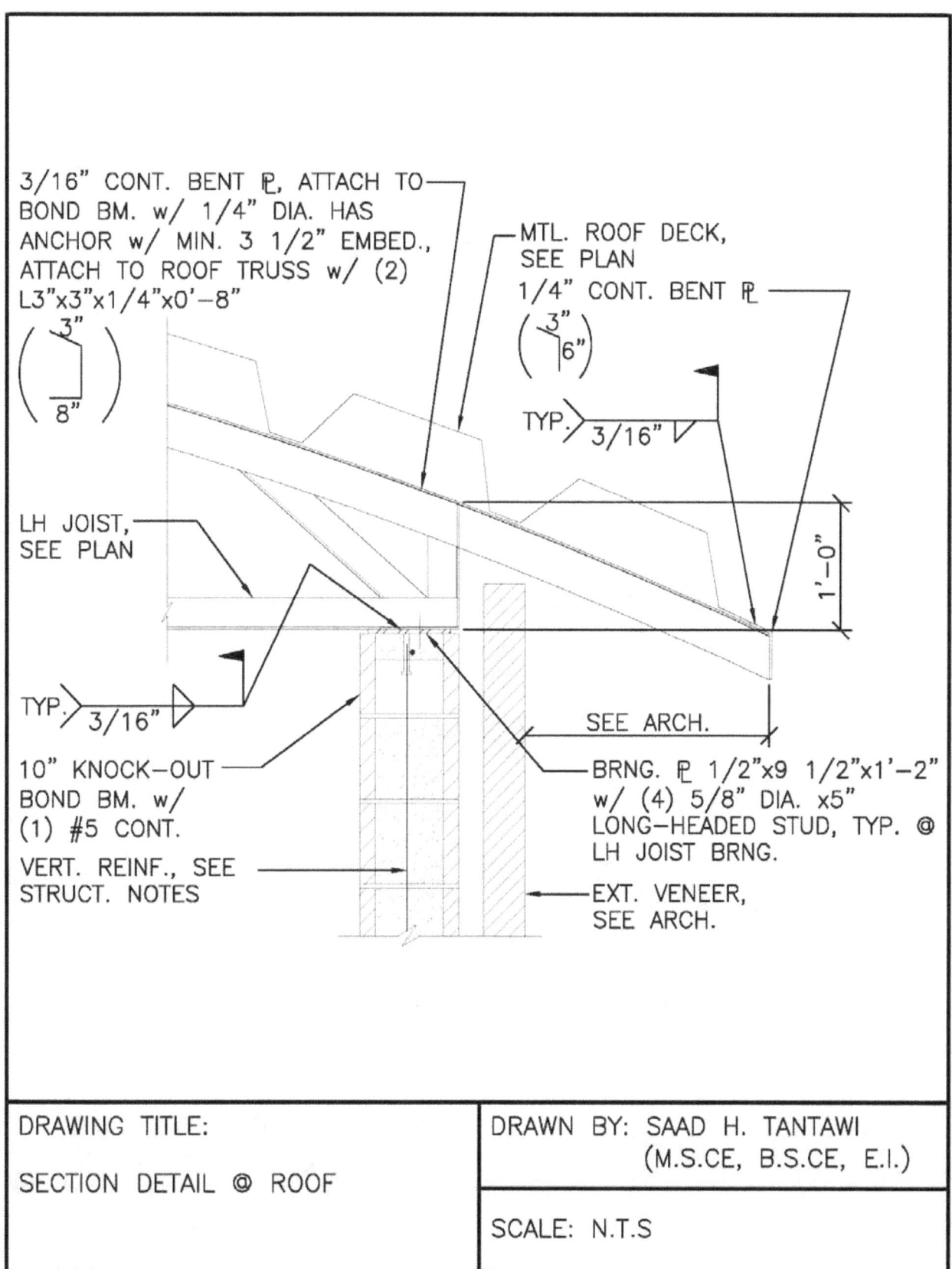

3/16" CONT. BENT ℔, ATTACH TO
BOND BM. w/ 1/4" DIA. HAS
ANCHOR w/ MIN. 3 1/2" EMBED.,
ATTACH TO ROOF TRUSS w/ (2)
L3"x3"x1/4"x0'-8"

MTL. ROOF DECK,
SEE PLAN
1/4" CONT. BENT ℔

TYP. 3/16"

LH JOIST,
SEE PLAN

1'-0"

TYP. 3/16"

SEE ARCH.

10" KNOCK-OUT
BOND BM. w/
(1) #5 CONT.

VERT. REINF., SEE
STRUCT. NOTES

BRNG. ℔ 1/2"x9 1/2"x1'-2"
w/ (4) 5/8" DIA. x5"
LONG-HEADED STUD, TYP. @
LH JOIST BRNG.

EXT. VENEER,
SEE ARCH.

DRAWING TITLE:	DRAWN BY: SAAD H. TANTAWI
	(M.S.CE, B.S.CE, E.I.)
SECTION DETAIL @ ROOF	
	SCALE: N.T.S

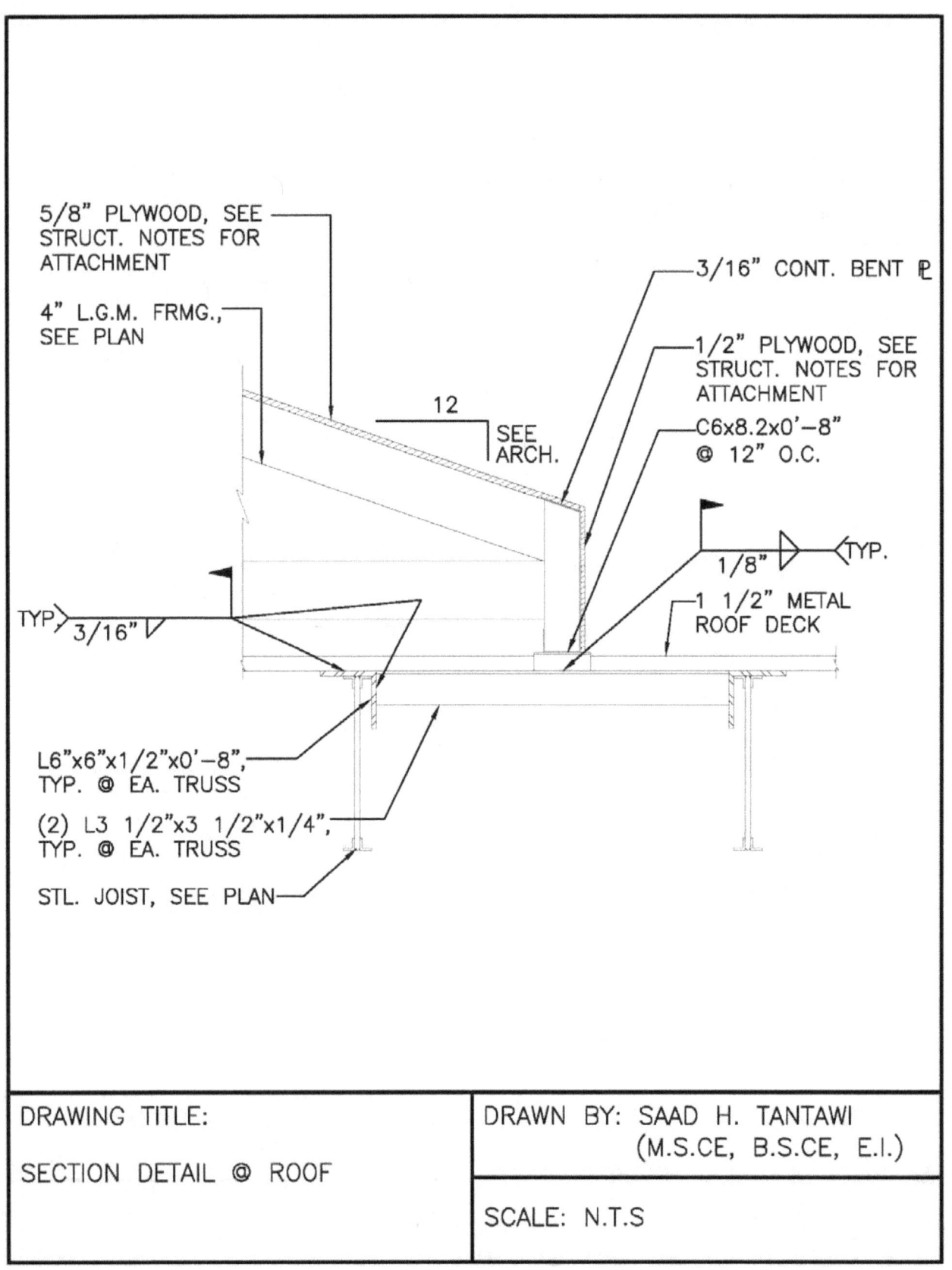

5/8" PLYWOOD, SEE STRUCT. NOTES FOR ATTACHMENT

4" L.G.M. FRMG., SEE PLAN

3/16" CONT. BENT ℙ

1/2" PLYWOOD, SEE STRUCT. NOTES FOR ATTACHMENT

C6x8.2x0'-8" @ 12" O.C.

12

SEE ARCH.

1/8"

TYP.

1 1/2" METAL ROOF DECK

TYP. 3/16"

L6"x6"x1/2"x0'-8", TYP. @ EA. TRUSS

(2) L3 1/2"x3 1/2"x1/4", TYP. @ EA. TRUSS

STL. JOIST, SEE PLAN

DRAWING TITLE:	DRAWN BY: SAAD H. TANTAWI
SECTION DETAIL @ ROOF	(M.S.CE, B.S.CE, E.I.)
	SCALE: N.T.S

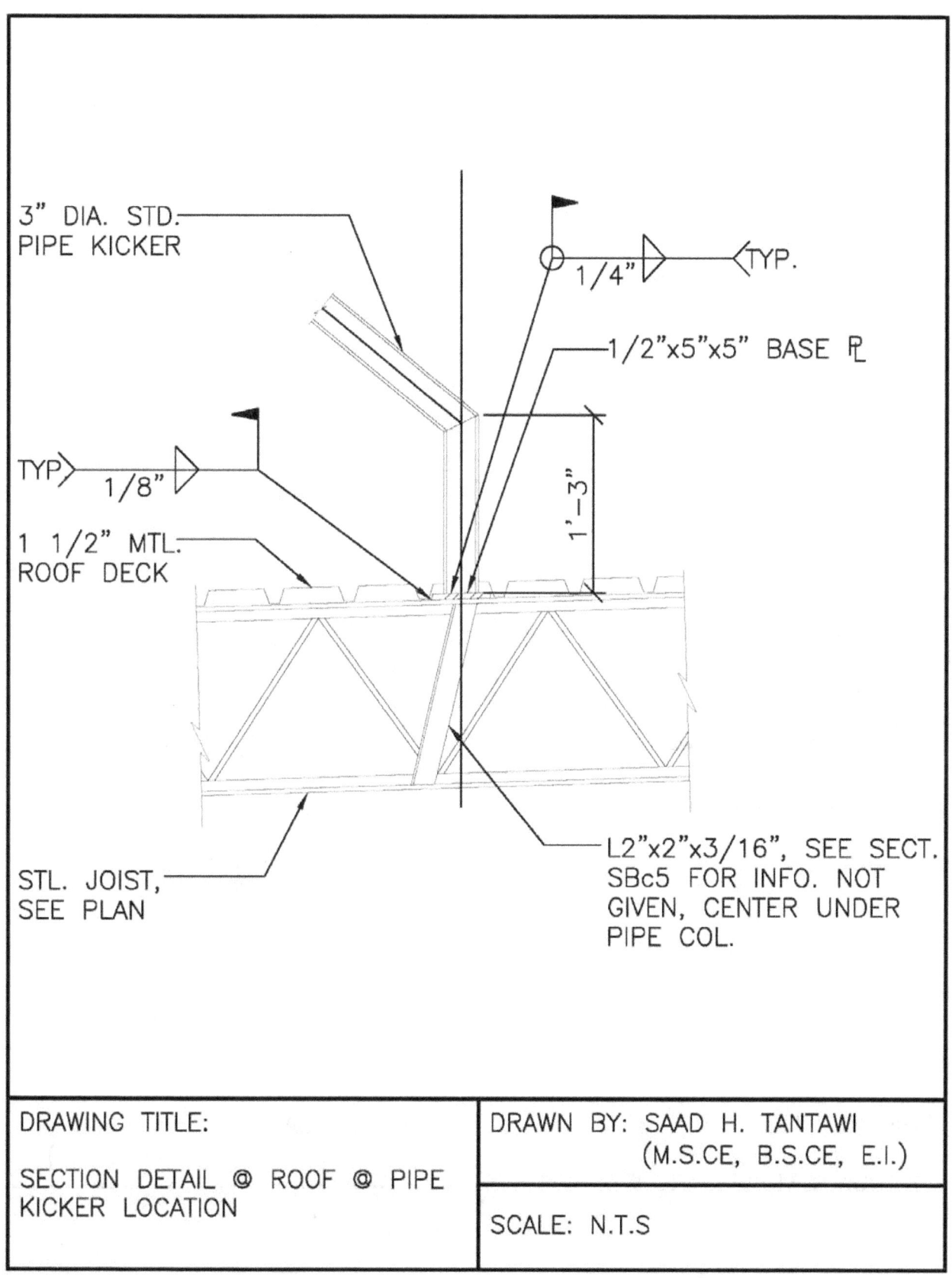

3" DIA. STD. PIPE KICKER

1/4" TYP.

1/2"x5"x5" BASE PL

TYP 1/8"

1'-3"

1 1/2" MTL. ROOF DECK

STL. JOIST, SEE PLAN

L2"x2"x3/16", SEE SECT. SBc5 FOR INFO. NOT GIVEN, CENTER UNDER PIPE COL.

DRAWING TITLE:	DRAWN BY: SAAD H. TANTAWI (M.S.CE, B.S.CE, E.I.)
SECTION DETAIL @ ROOF @ PIPE KICKER LOCATION	SCALE: N.T.S

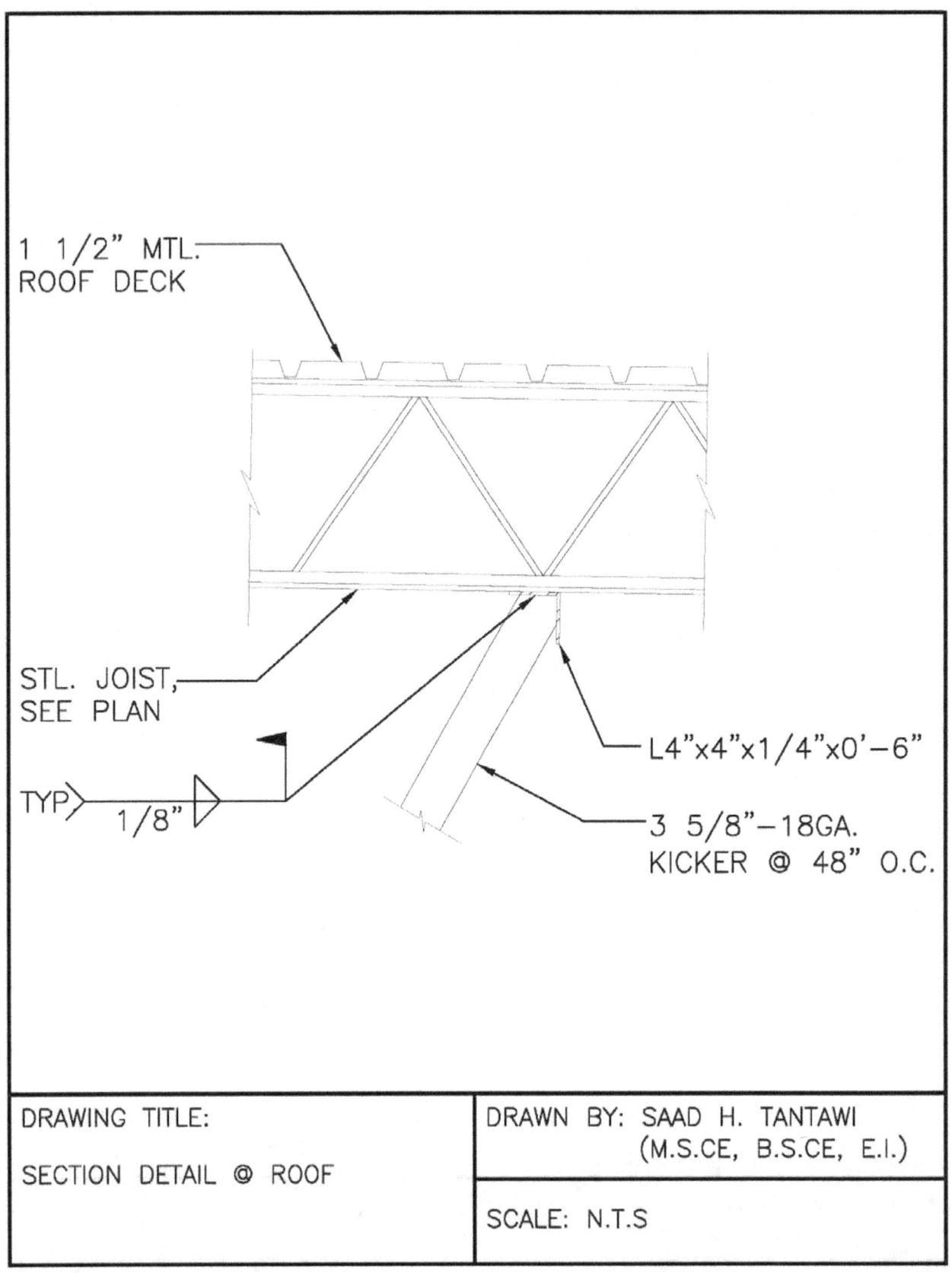

1 1/2" MTL. ROOF DECK

STL. JOIST, SEE PLAN

TYP

1/8"

L4"x4"x1/4"x0'-6"

3 5/8"-18GA. KICKER @ 48" O.C.

DRAWING TITLE:	DRAWN BY: SAAD H. TANTAWI
SECTION DETAIL @ ROOF	(M.S.CE, B.S.CE, E.I.)
	SCALE: N.T.S

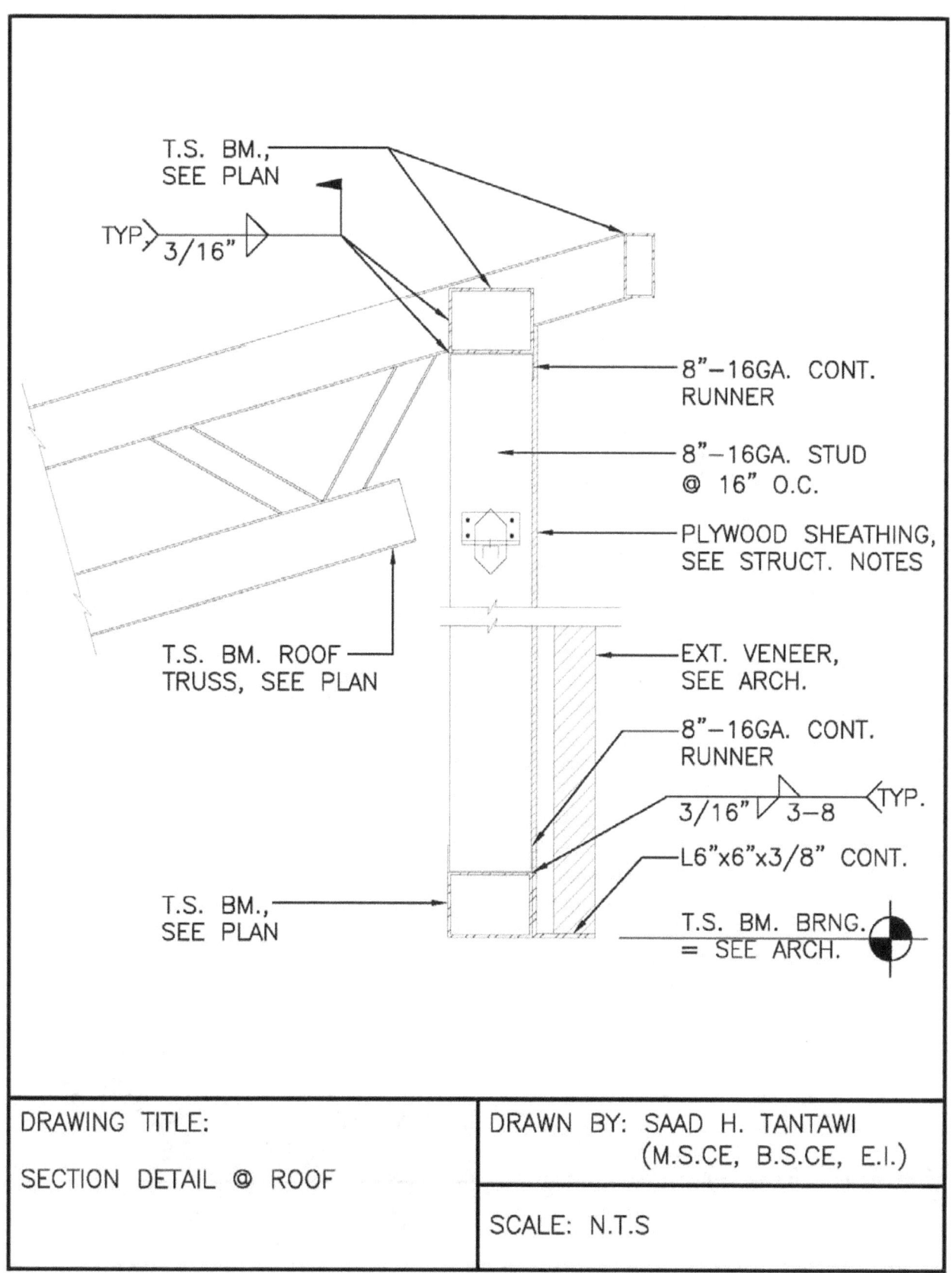

T.S. BM.,
SEE PLAN

TYP 3/16"

8"-16GA. CONT.
RUNNER

8"-16GA. STUD
@ 16" O.C.

PLYWOOD SHEATHING,
SEE STRUCT. NOTES

T.S. BM. ROOF
TRUSS, SEE PLAN

EXT. VENEER,
SEE ARCH.

8"-16GA. CONT.
RUNNER

3/16" 3-8 TYP.

L6"x6"x3/8" CONT.

T.S. BM.,
SEE PLAN

T.S. BM. BRNG.
= SEE ARCH.

DRAWING TITLE:	DRAWN BY: SAAD H. TANTAWI
	(M.S.CE, B.S.CE, E.I.)
SECTION DETAIL @ ROOF	
	SCALE: N.T.S

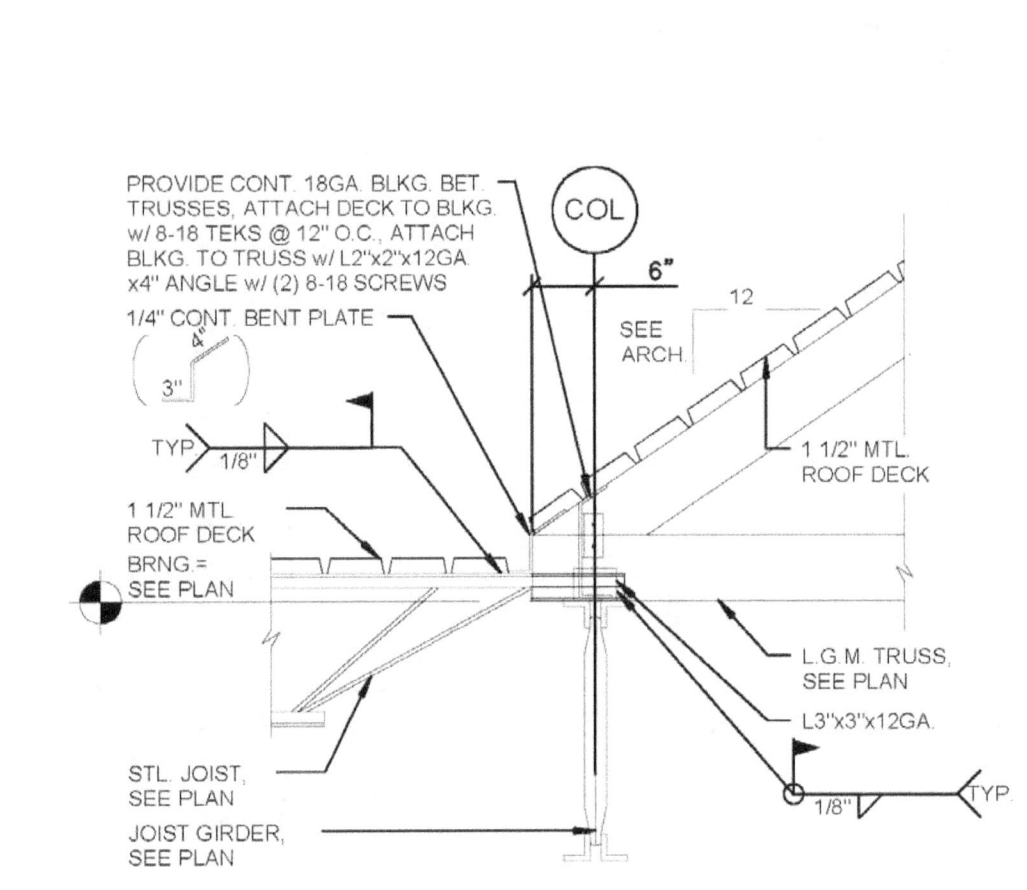

PROVIDE CONT. 18GA. BLKG. BET.
TRUSSES, ATTACH DECK TO BLKG.
w/ 8-18 TEKS @ 12" O.C., ATTACH
BLKG. TO TRUSS w/ L2"x2"x12GA.
x4" ANGLE w/ (2) 8-18 SCREWS

1/4" CONT. BENT PLATE

4"

3"

TYP.   1/8"

COL

6"

12

SEE
ARCH.

1 1/2" MTL.
ROOF DECK

1 1/2" MTL
ROOF DECK
BRNG.=
SEE PLAN

L.G.M. TRUSS,
SEE PLAN

L3"x3"x12GA.

STL. JOIST,
SEE PLAN

1/8"   TYP.

JOIST GIRDER,
SEE PLAN

TYP. @ L.G.M. TRUSS/STL. JOIST ON STL. GIRDER CONNECTION, U.N.O.

DRAWING TITLE:	DRAWN BY: SAAD H. TANTAWI
	(M.S.CE, B.S.CE, E.I.)
SECTION DETAIL @ ROOF	SCALE: N.T.S

110

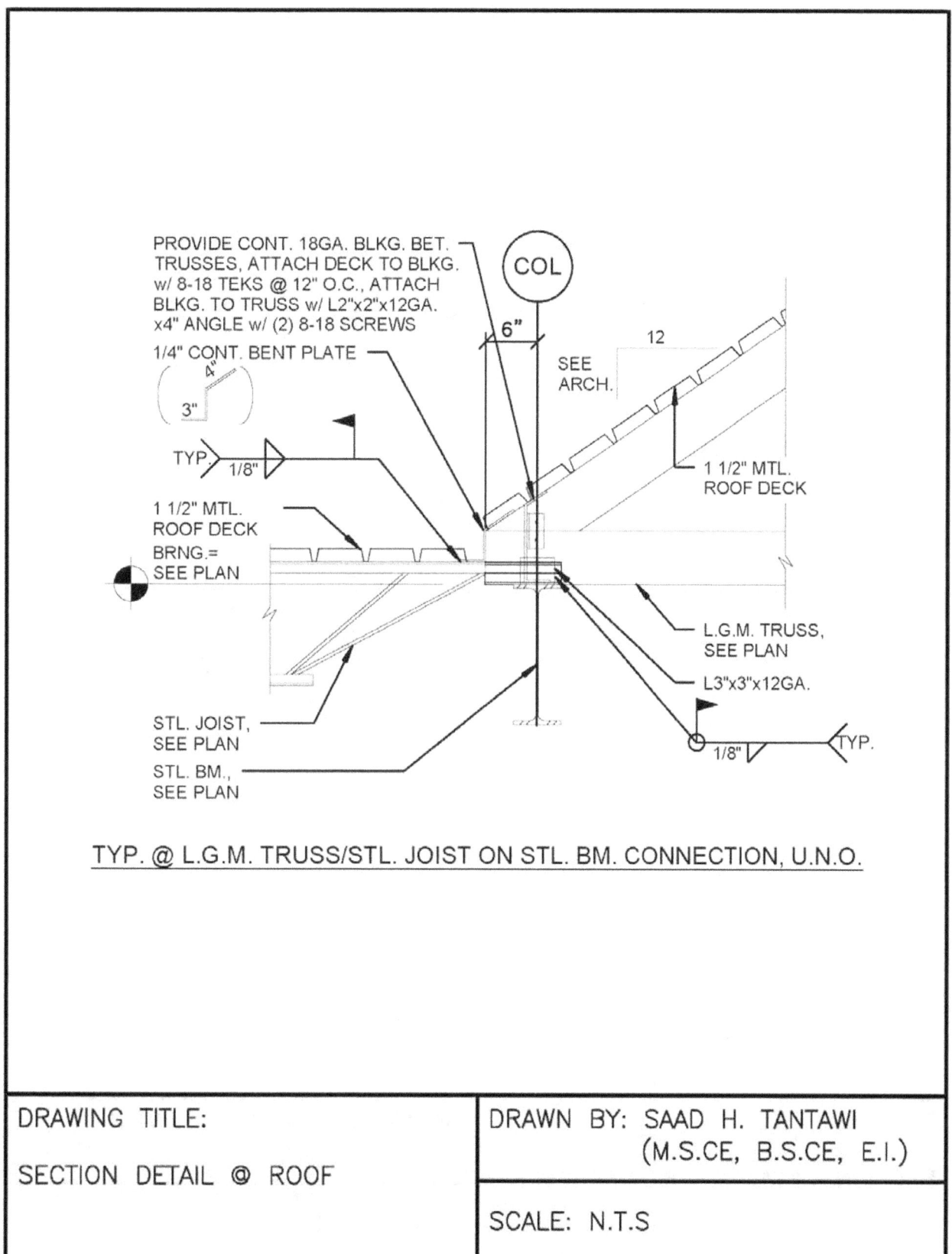

PROVIDE CONT. 18GA. BLKG. BET. TRUSSES, ATTACH DECK TO BLKG. w/ 8-18 TEKS @ 12" O.C., ATTACH BLKG. TO TRUSS w/ L2"x2"x12GA. x4" ANGLE w/ (2) 8-18 SCREWS

1/4" CONT. BENT PLATE

4"

3"

TYP. 1/8"

COL

6"

12

SEE ARCH.

1 1/2" MTL. ROOF DECK

1 1/2" MTL. ROOF DECK BRNG.= SEE PLAN

L.G.M. TRUSS, SEE PLAN

L3"x3"x12GA.

STL. JOIST, SEE PLAN

STL. BM., SEE PLAN

1/8" TYP.

TYP. @ L.G.M. TRUSS/STL. JOIST ON STL. BM. CONNECTION, U.N.O.

DRAWING TITLE:	DRAWN BY: SAAD H. TANTAWI
SECTION DETAIL @ ROOF	(M.S.CE, B.S.CE, E.I.)
	SCALE: N.T.S

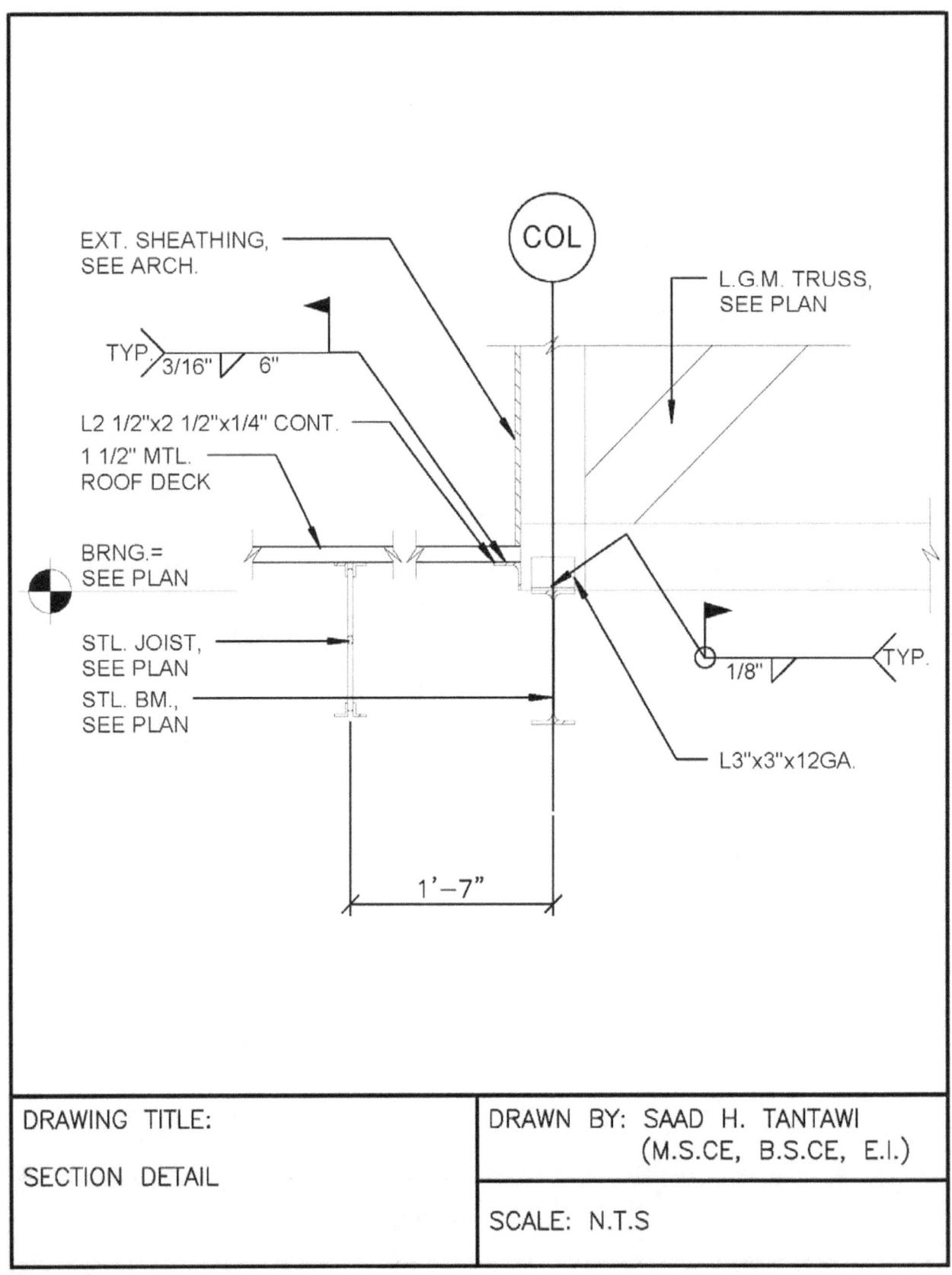

EXT. SHEATHING, SEE ARCH.

COL

L.G.M. TRUSS, SEE PLAN

TYP. 3/16" 6"

L2 1/2"x2 1/2"x1/4" CONT.

1 1/2" MTL. ROOF DECK

BRNG.= SEE PLAN

STL. JOIST, SEE PLAN

STL. BM., SEE PLAN

1/8" TYP.

L3"x3"x12GA.

1'-7"

DRAWING TITLE:	DRAWN BY: SAAD H. TANTAWI
	(M.S.CE, B.S.CE, E.I.)
SECTION DETAIL	
	SCALE: N.T.S

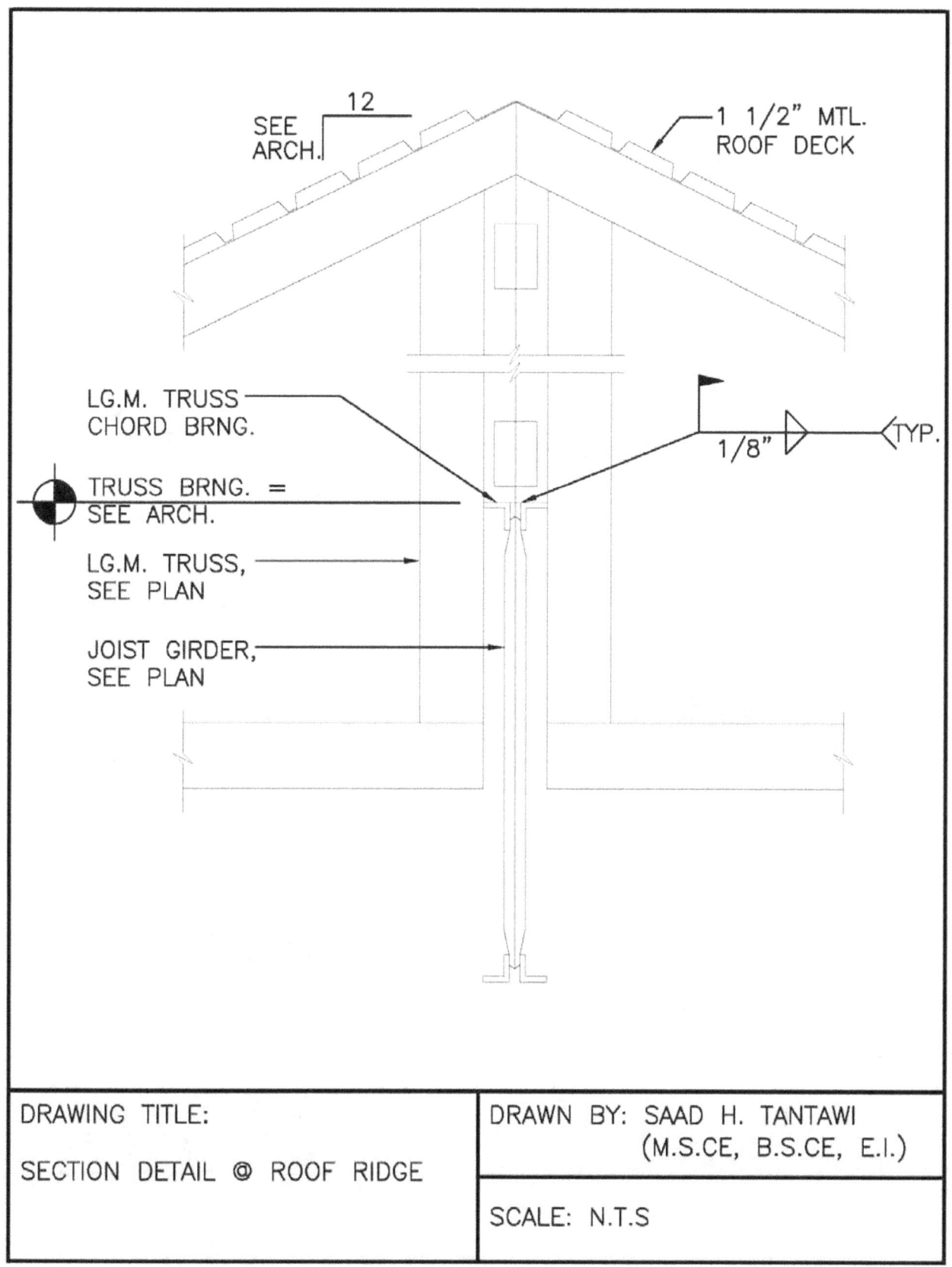

SEE
ARCH. 12

1 1/2" MTL.
ROOF DECK

LG.M. TRUSS
CHORD BRNG.

TRUSS BRNG. =
SEE ARCH.

LG.M. TRUSS,
SEE PLAN

JOIST GIRDER,
SEE PLAN

1/8"   TYP.

DRAWING TITLE:	DRAWN BY: SAAD H. TANTAWI
	(M.S.CE, B.S.CE, E.I.)
SECTION DETAIL @ ROOF RIDGE	
	SCALE: N.T.S

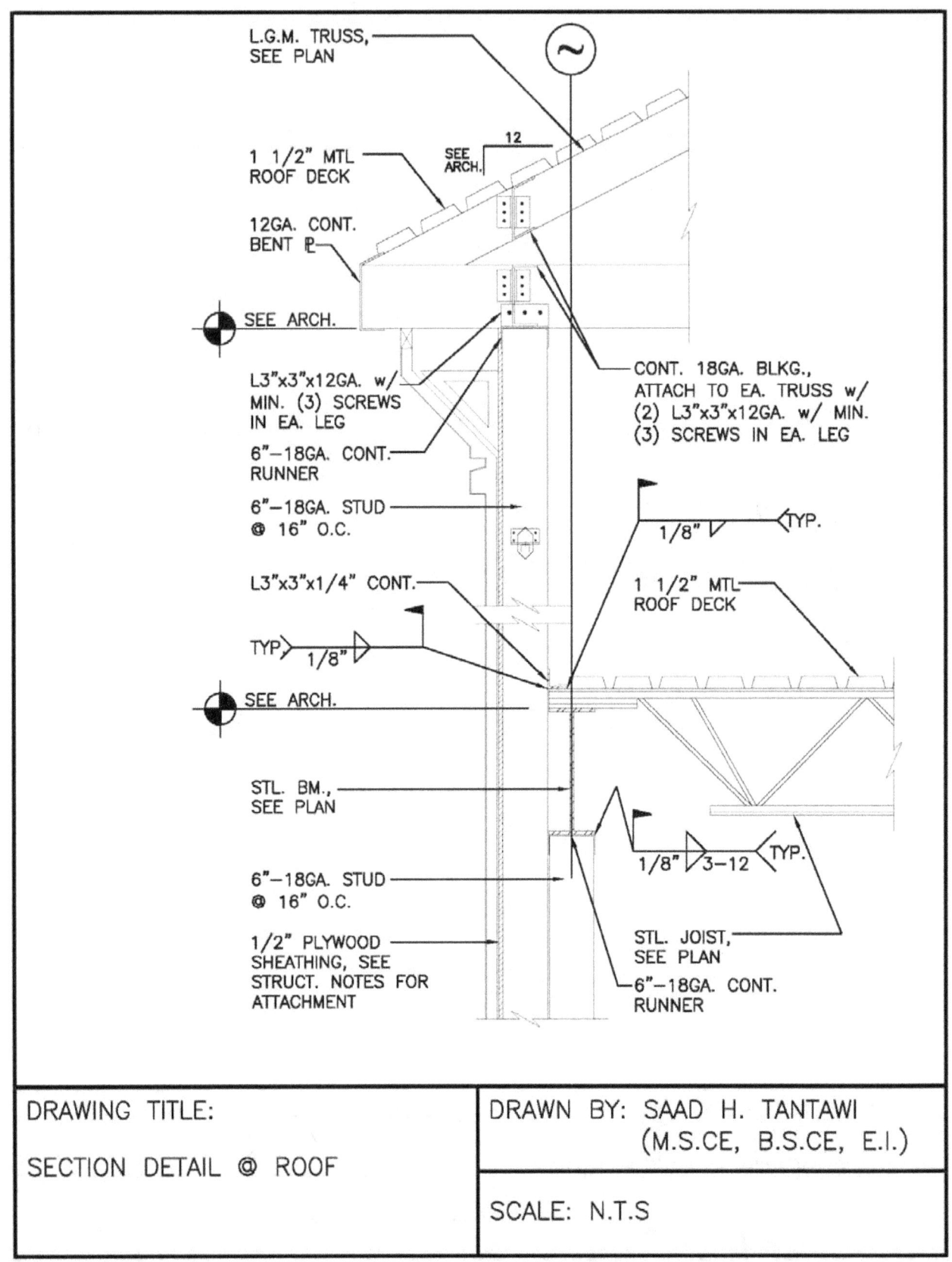

DRAWING TITLE:

SECTION DETAIL @ ROOF

DRAWN BY: SAAD H. TANTAWI
(M.S.CE, B.S.CE, E.I.)

SCALE: N.T.S

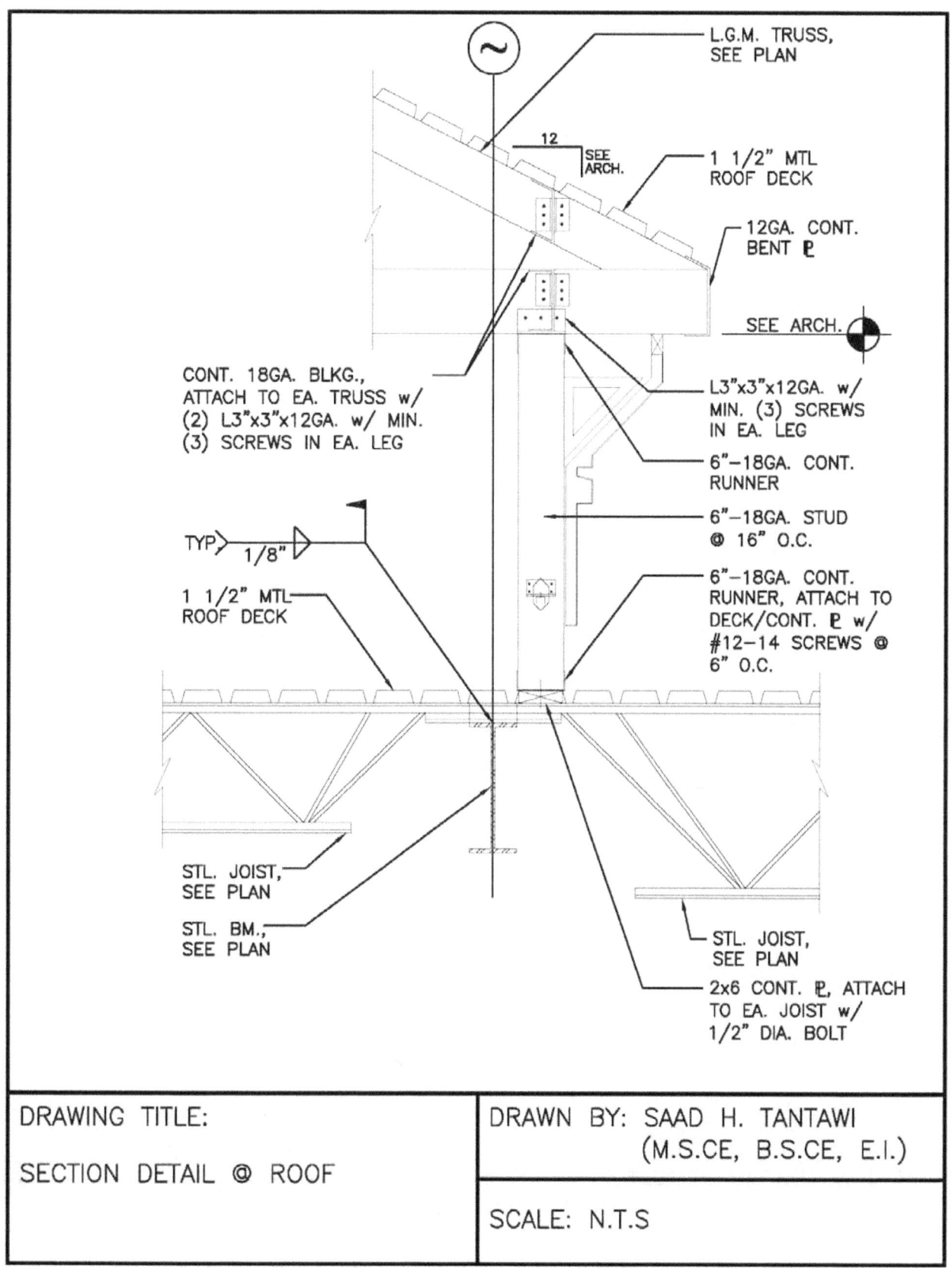

L.G.M. TRUSS,
SEE PLAN

1 1/2" MTL
ROOF DECK

12 GA. CONT.
BENT ℗

SEE ARCH.

L3"x3"x12GA. w/
MIN. (3) SCREWS
IN EA. LEG

6"-18GA. CONT.
RUNNER

6"-18GA. STUD
@ 16" O.C.

6"-18GA. CONT.
RUNNER, ATTACH TO
DECK/CONT. ℗ w/
#12-14 SCREWS @
6" O.C.

CONT. 18GA. BLKG.,
ATTACH TO EA. TRUSS w/
(2) L3"x3"x12GA. w/ MIN.
(3) SCREWS IN EA. LEG

12
SEE
ARCH.

TYP
1/8"

1 1/2" MTL
ROOF DECK

STL. JOIST,
SEE PLAN

STL. BM.,
SEE PLAN

STL. JOIST,
SEE PLAN

2x6 CONT. ℗, ATTACH
TO EA. JOIST w/
1/2" DIA. BOLT

DRAWING TITLE:	DRAWN BY: SAAD H. TANTAWI
	(M.S.CE, B.S.CE, E.I.)
SECTION DETAIL @ ROOF	
	SCALE: N.T.S

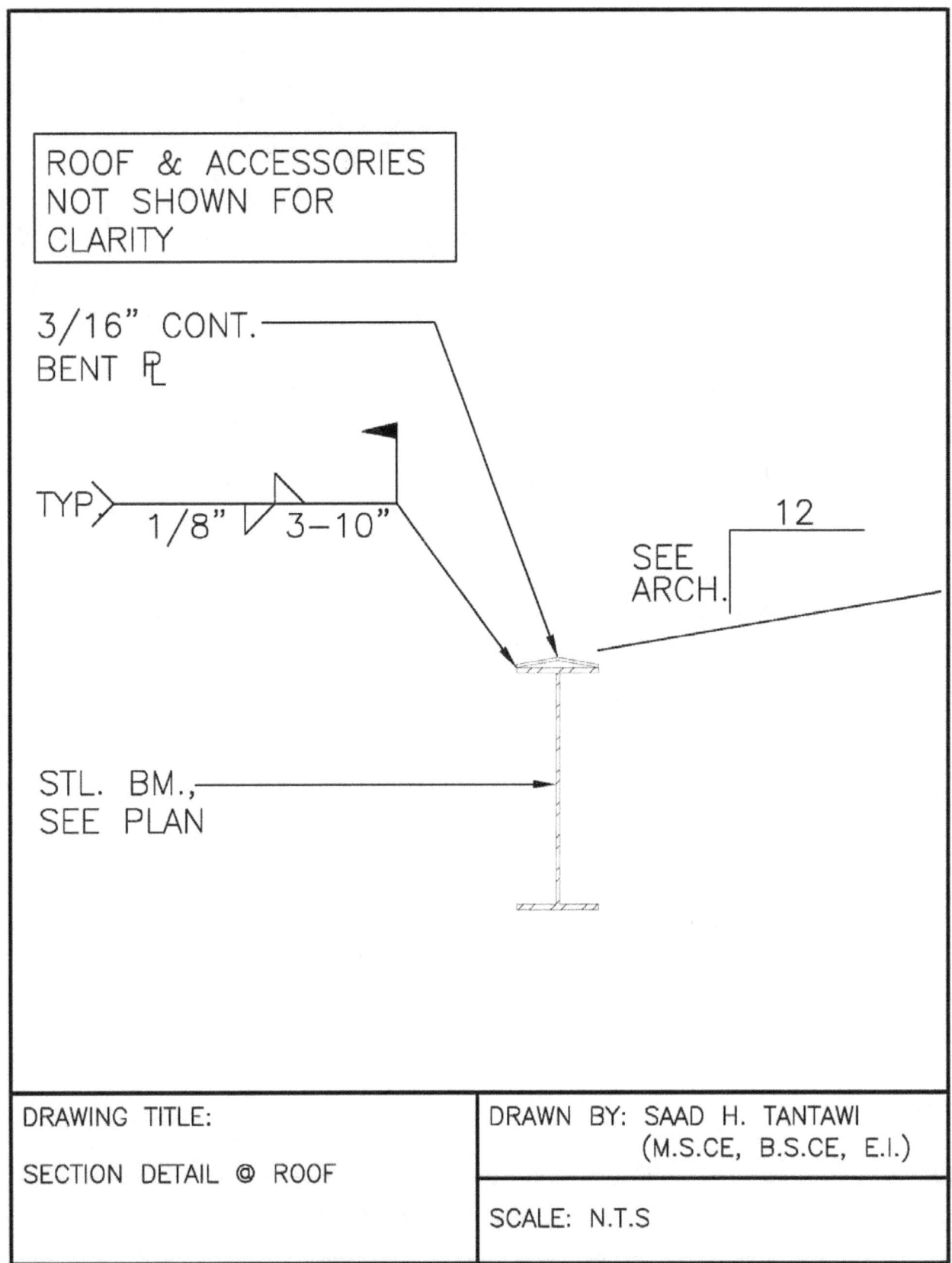

ROOF & ACCESSORIES
NOT SHOWN FOR
CLARITY

3/16" CONT.
BENT ℞

TYP

1/8"   3-10"

12

SEE
ARCH.

STL. BM.,
SEE PLAN

DRAWING TITLE:	DRAWN BY: SAAD H. TANTAWI
SECTION DETAIL @ ROOF	(M.S.CE, B.S.CE, E.I.)
	SCALE: N.T.S

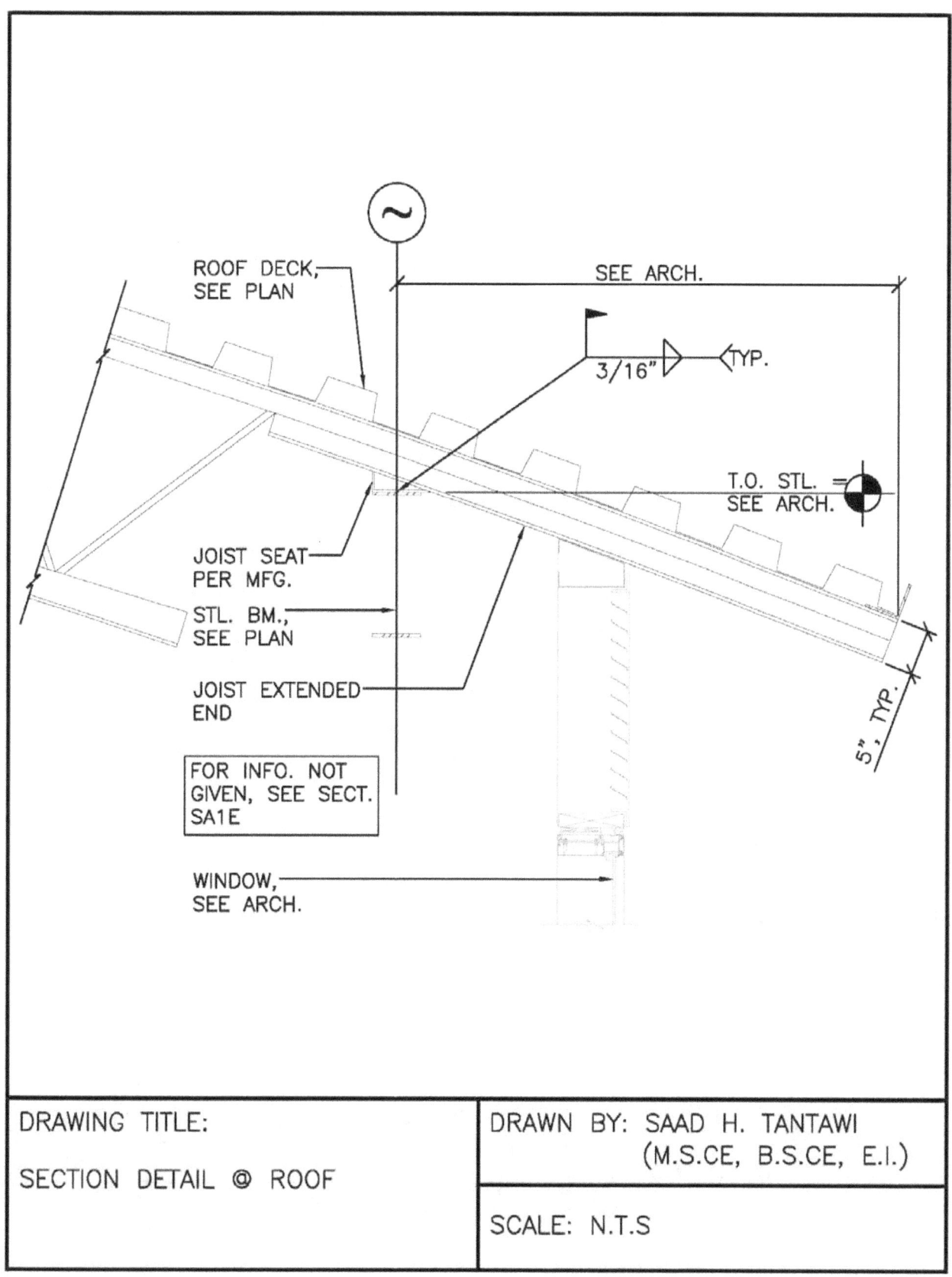

ROOF DECK,
SEE PLAN

SEE ARCH.

3/16" TYP.

T.O. STL. =
SEE ARCH.

JOIST SEAT
PER MFG.

STL. BM.,
SEE PLAN

JOIST EXTENDED
END

5", TYP.

FOR INFO. NOT
GIVEN, SEE SECT.
SA1E

WINDOW,
SEE ARCH.

DRAWING TITLE:	DRAWN BY: SAAD H. TANTAWI
	(M.S.CE, B.S.CE, E.I.)
SECTION DETAIL @ ROOF	
	SCALE: N.T.S

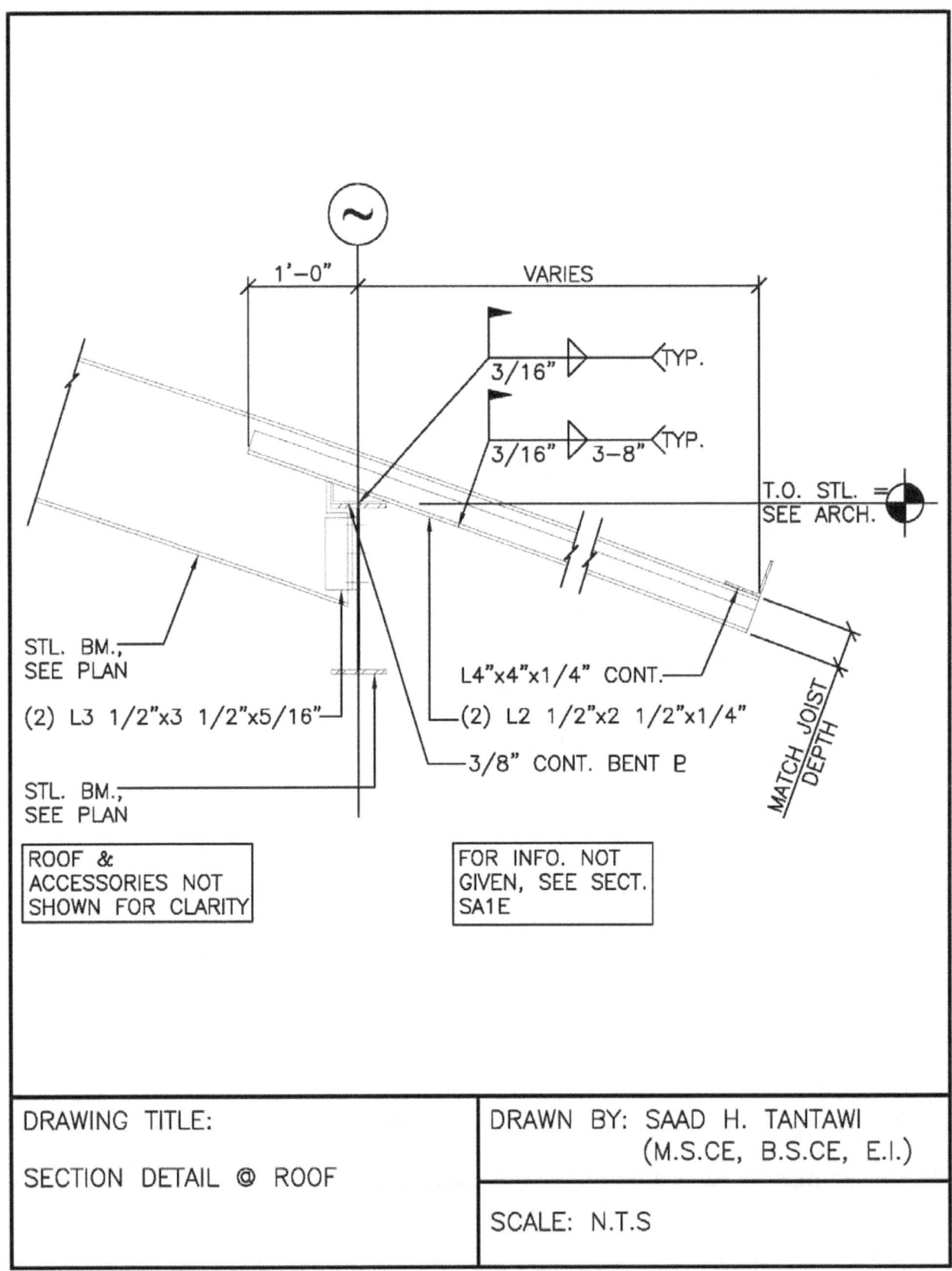

1'-0"

VARIES

3/16"

TYP.

3/16"   3-8"

TYP.

T.O. STL. =
SEE ARCH.

STL. BM.,
SEE PLAN

(2) L3 1/2"x3 1/2"x5/16"

STL. BM.,
SEE PLAN

L4"x4"x1/4" CONT.

(2) L2 1/2"x2 1/2"x1/4"

3/8" CONT. BENT P

MATCH JOIST
DEPTH

| ROOF &
ACCESSORIES NOT
SHOWN FOR CLARITY | FOR INFO. NOT
GIVEN, SEE SECT.
SA1E |

DRAWING TITLE: SECTION DETAIL @ ROOF	DRAWN BY: SAAD H. TANTAWI (M.S.CE, B.S.CE, E.I.)
	SCALE: N.T.S

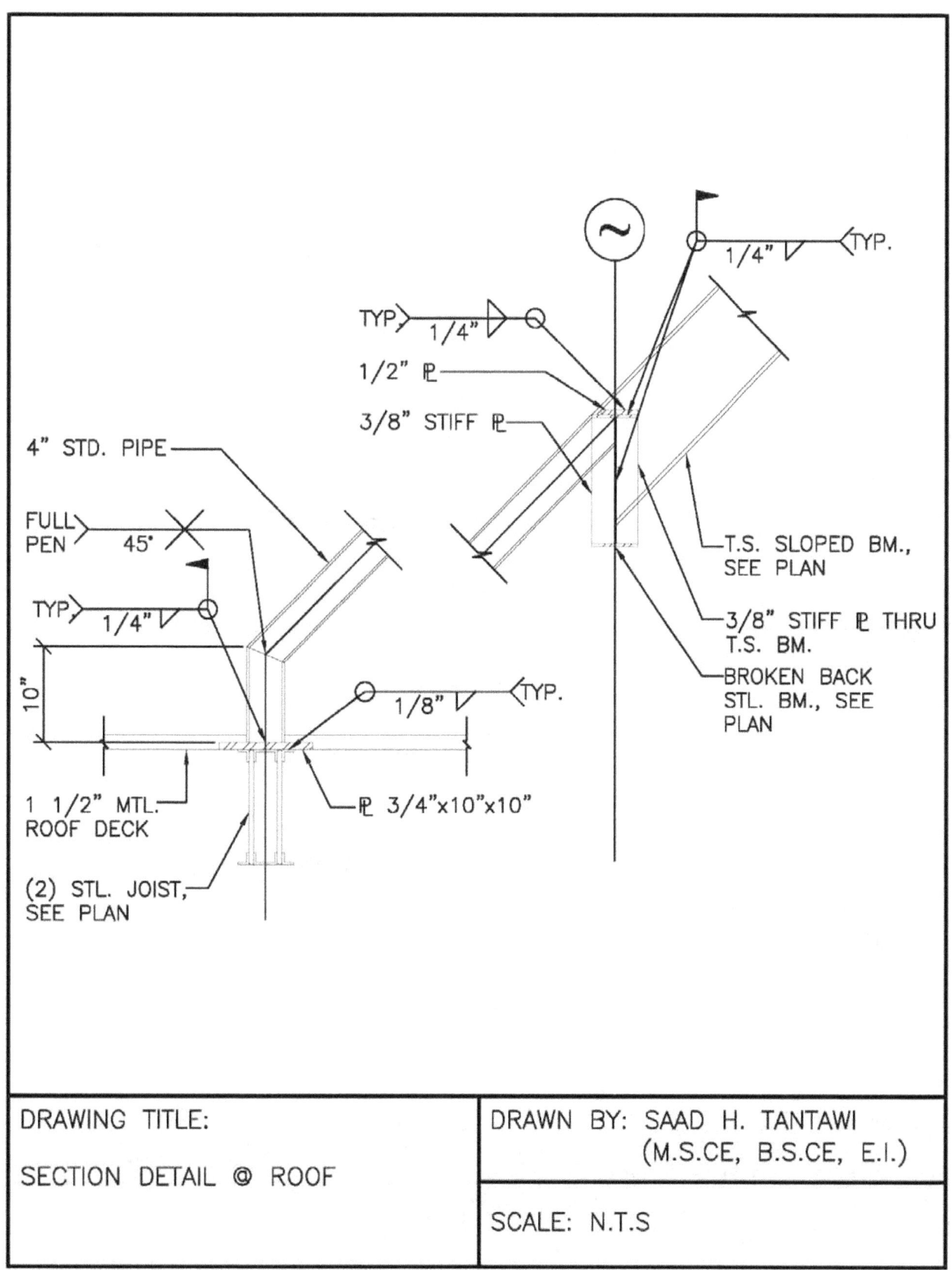

DRAWING TITLE:	DRAWN BY: SAAD H. TANTAWI
	(M.S.CE, B.S.CE, E.I.)
SECTION DETAIL @ ROOF	
	SCALE: N.T.S

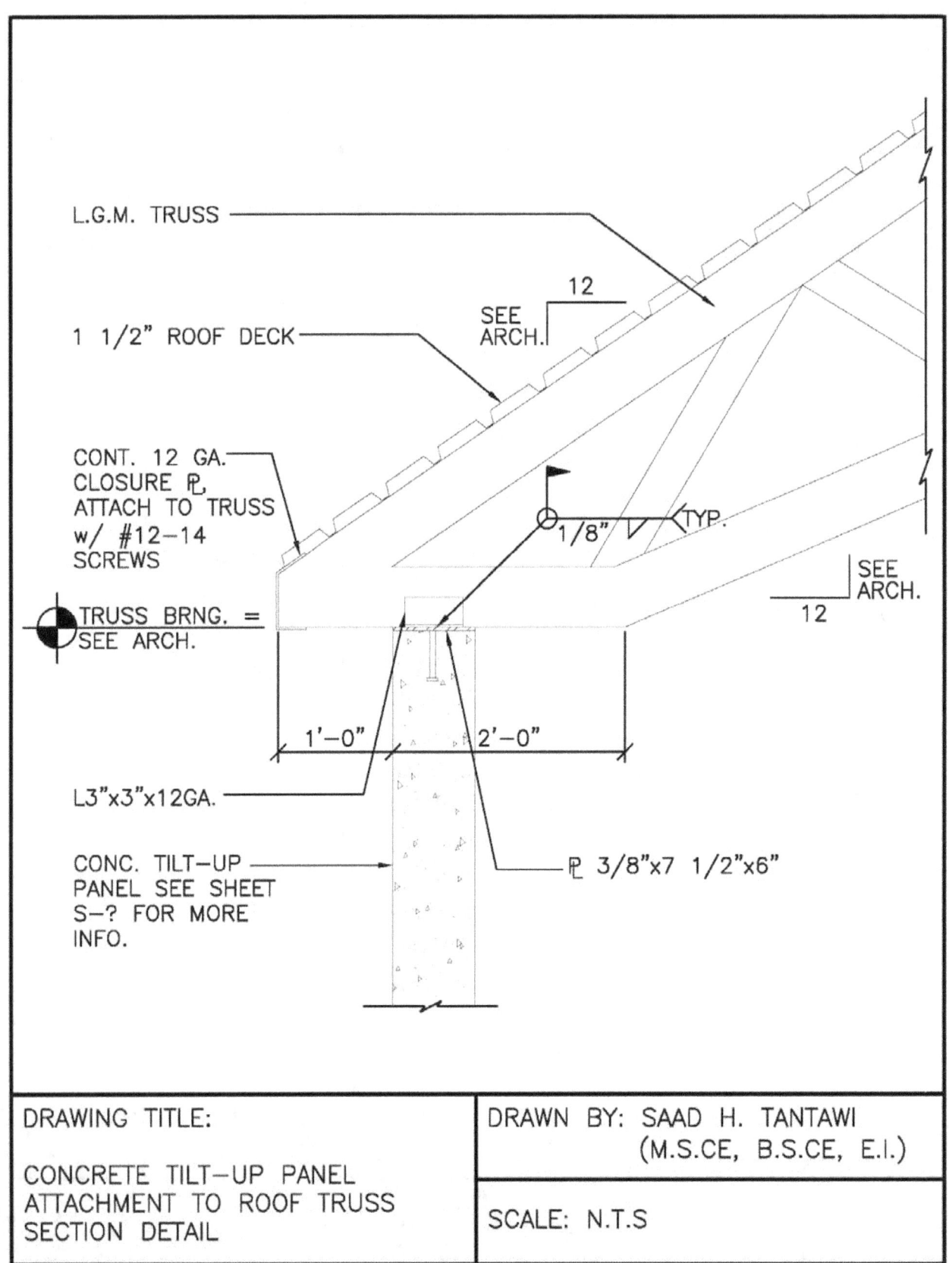

L.G.M. TRUSS

1 1/2" ROOF DECK

SEE ARCH. 12

CONT. 12 GA. CLOSURE ℙ, ATTACH TO TRUSS w/ #12-14 SCREWS

1/8" TYP.

SEE ARCH. 12

TRUSS BRNG. = SEE ARCH.

1'-0"   2'-0"

L3"x3"x12GA.

CONC. TILT-UP PANEL SEE SHEET S-? FOR MORE INFO.

ℙ 3/8"x7 1/2"x6"

DRAWING TITLE:	DRAWN BY: SAAD H. TANTAWI
	(M.S.CE, B.S.CE, E.I.)
CONCRETE TILT-UP PANEL ATTACHMENT TO ROOF TRUSS SECTION DETAIL	SCALE: N.T.S

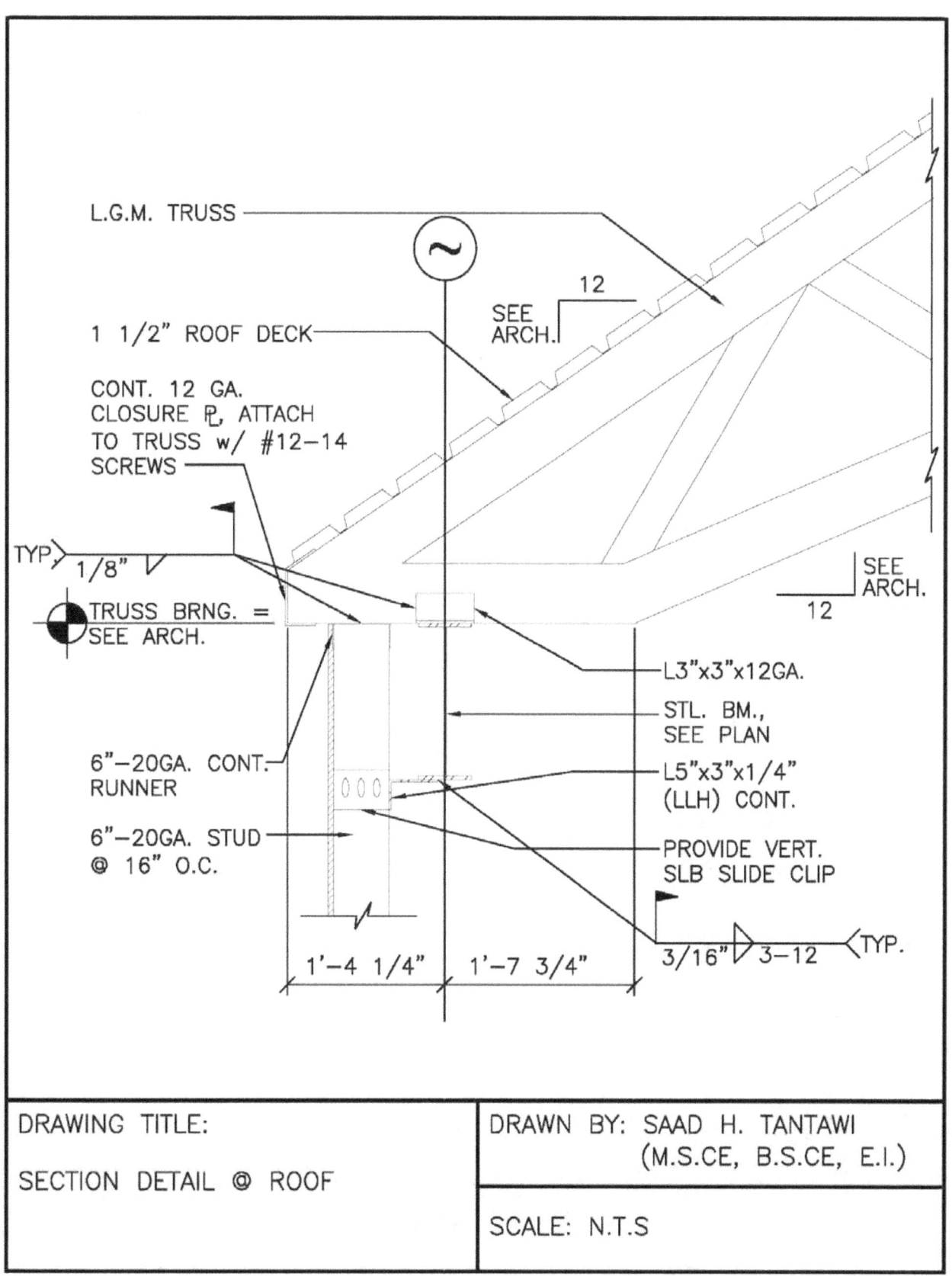

L.G.M. TRUSS

1 1/2" ROOF DECK

CONT. 12 GA.
CLOSURE ℙ, ATTACH
TO TRUSS w/ #12−14
SCREWS

SEE
ARCH.

12

TYP.
1/8"

TRUSS BRNG. =
SEE ARCH.

SEE
ARCH.
12

6"−20GA. CONT.
RUNNER

6"−20GA. STUD
@ 16" O.C.

L3"x3"x12GA.

STL. BM.,
SEE PLAN

L5"x3"x1/4"
(LLH) CONT.

PROVIDE VERT.
SLB SLIDE CLIP

1'−4 1/4"   1'−7 3/4"   3/16" 3−12   TYP.

DRAWING TITLE:	DRAWN BY: SAAD H. TANTAWI
	(M.S.CE, B.S.CE, E.I.)
SECTION DETAIL @ ROOF	
	SCALE: N.T.S

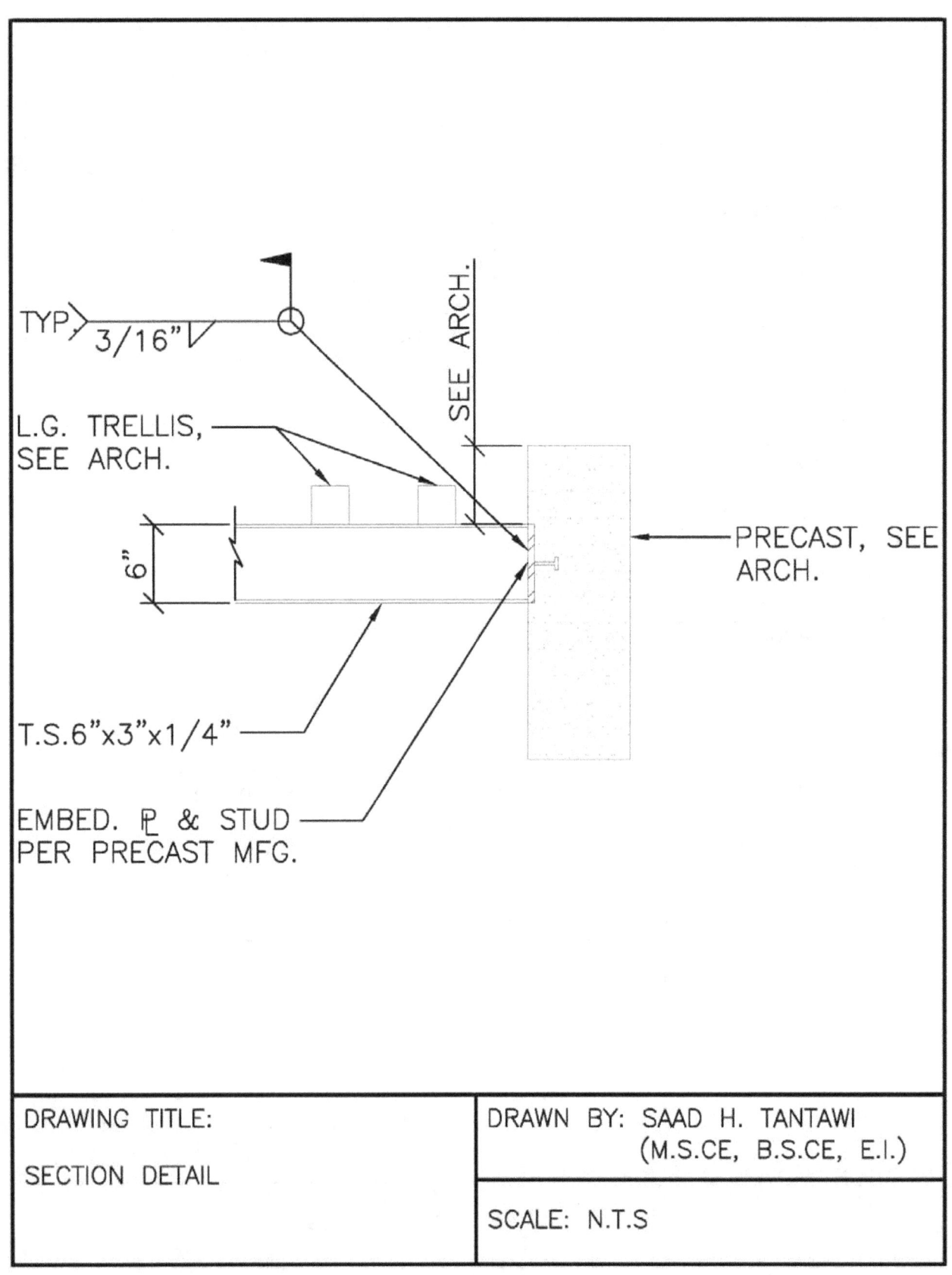

TYP〉
3/16"

L.G. TRELLIS,
SEE ARCH.

SEE ARCH.

6"

PRECAST, SEE
ARCH.

T.S.6"x3"x1/4"

EMBED. ℗ & STUD
PER PRECAST MFG.

DRAWING TITLE:	DRAWN BY: SAAD H. TANTAWI
	(M.S.CE, B.S.CE, E.I.)
SECTION DETAIL	
	SCALE: N.T.S

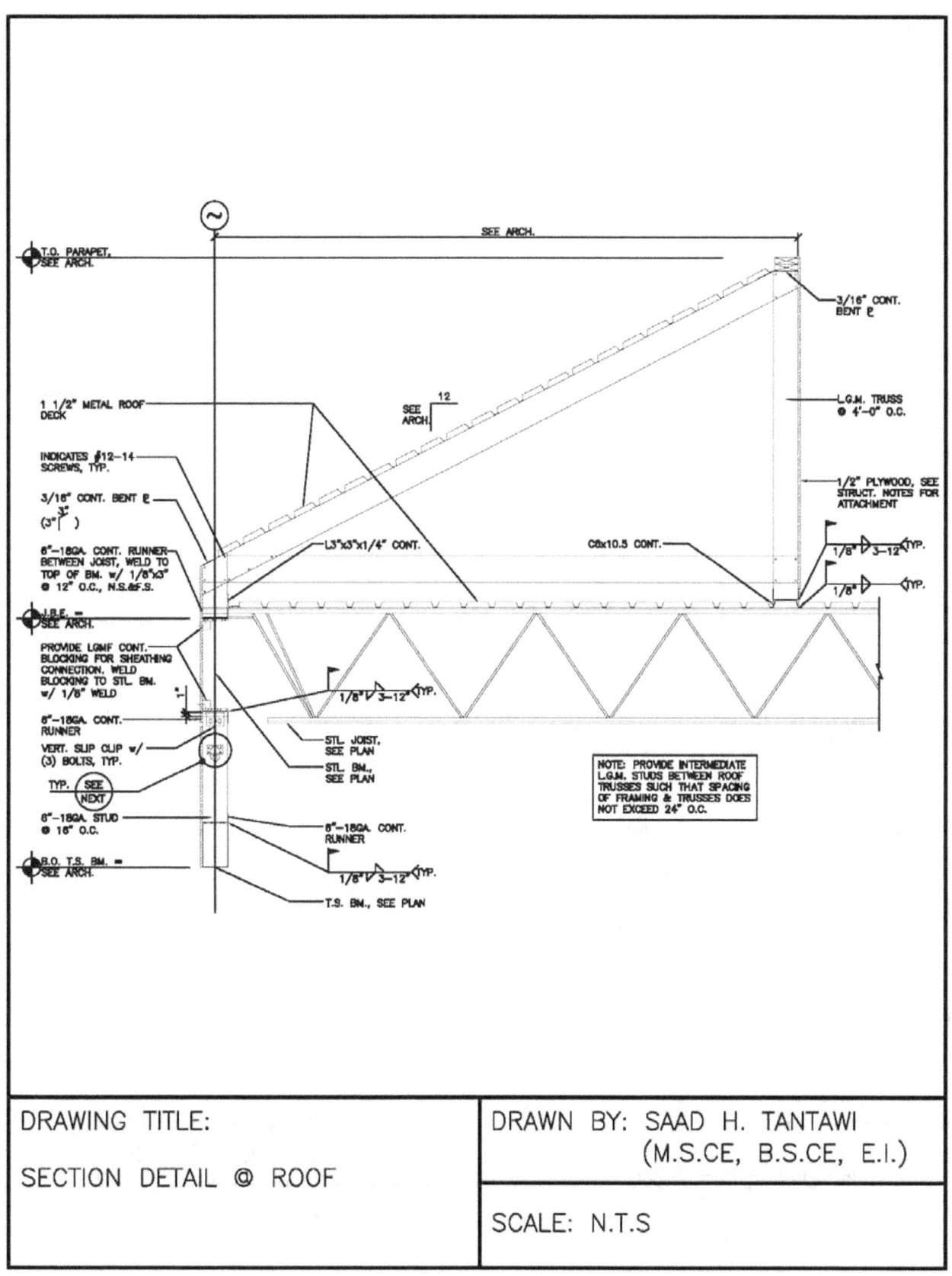

DRAWING TITLE:	DRAWN BY: SAAD H. TANTAWI
SECTION DETAIL @ ROOF	(M.S.CE, B.S.CE, E.I.)
	SCALE: N.T.S

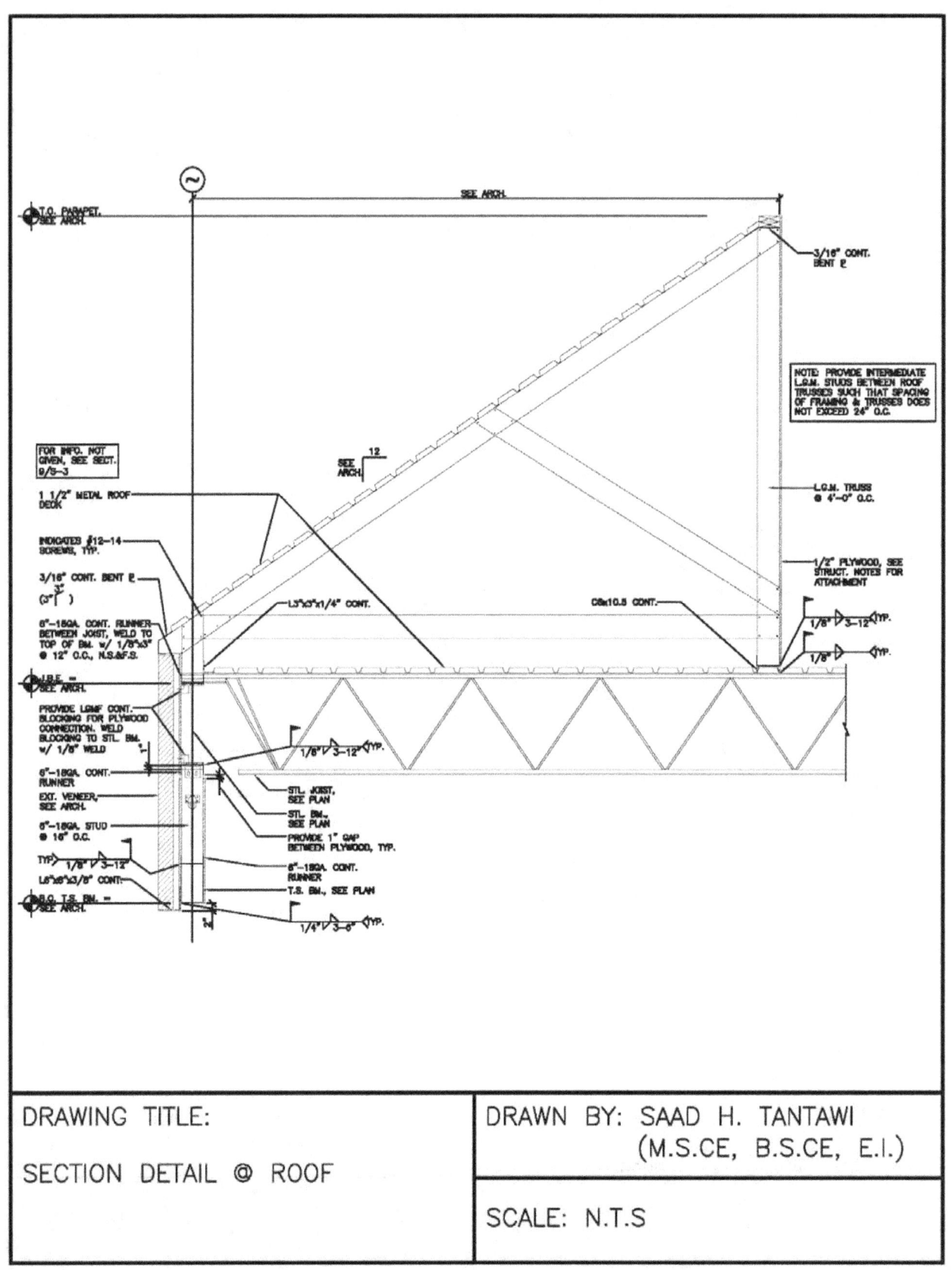

DRAWING TITLE:	DRAWN BY: SAAD H. TANTAWI
	(M.S.CE, B.S.CE, E.I.)
SECTION DETAIL @ ROOF	SCALE: N.T.S

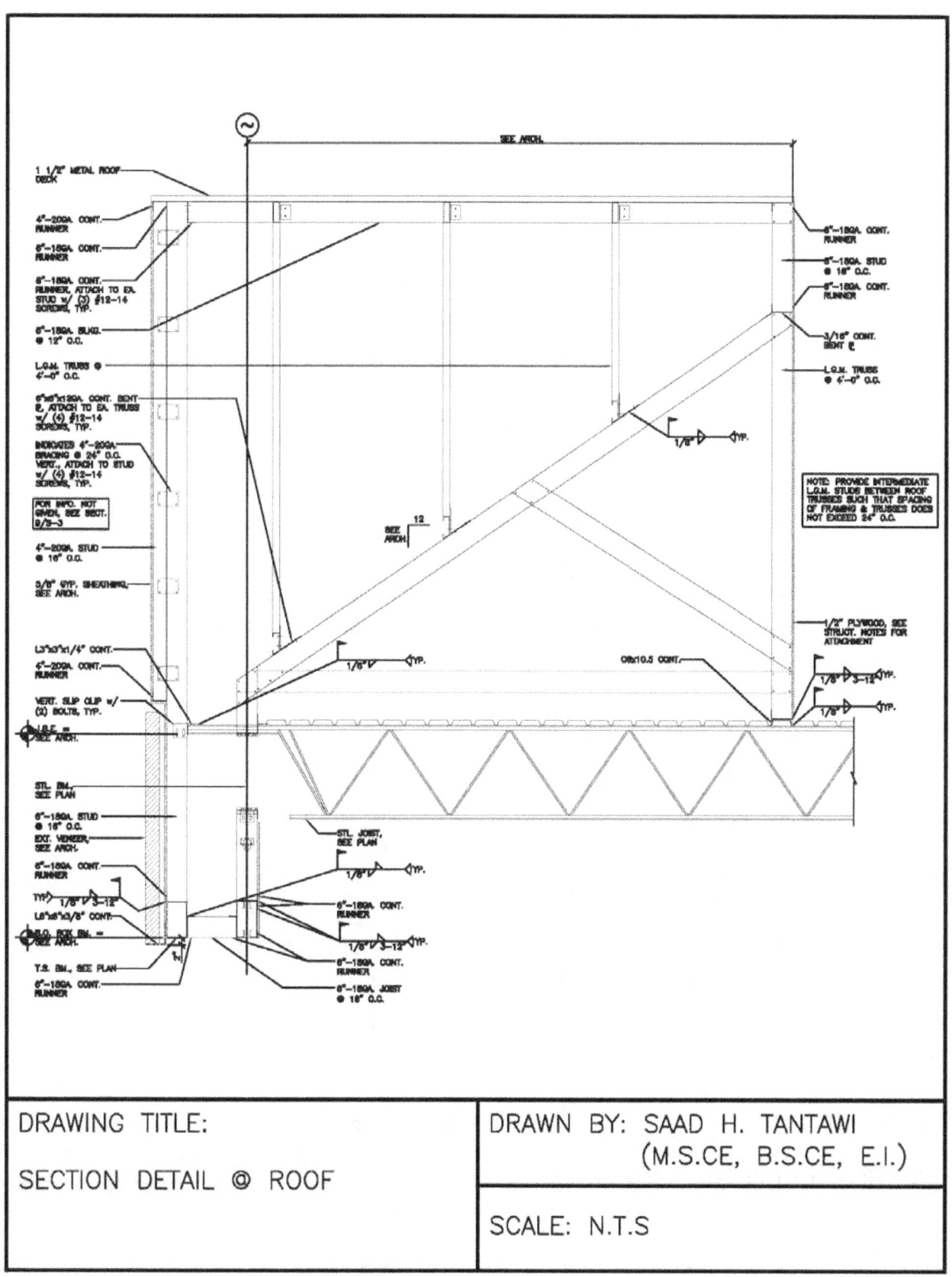

DRAWING TITLE:	DRAWN BY: SAAD H. TANTAWI
	(M.S.CE, B.S.CE, E.I.)
SECTION DETAIL @ ROOF	
	SCALE: N.T.S

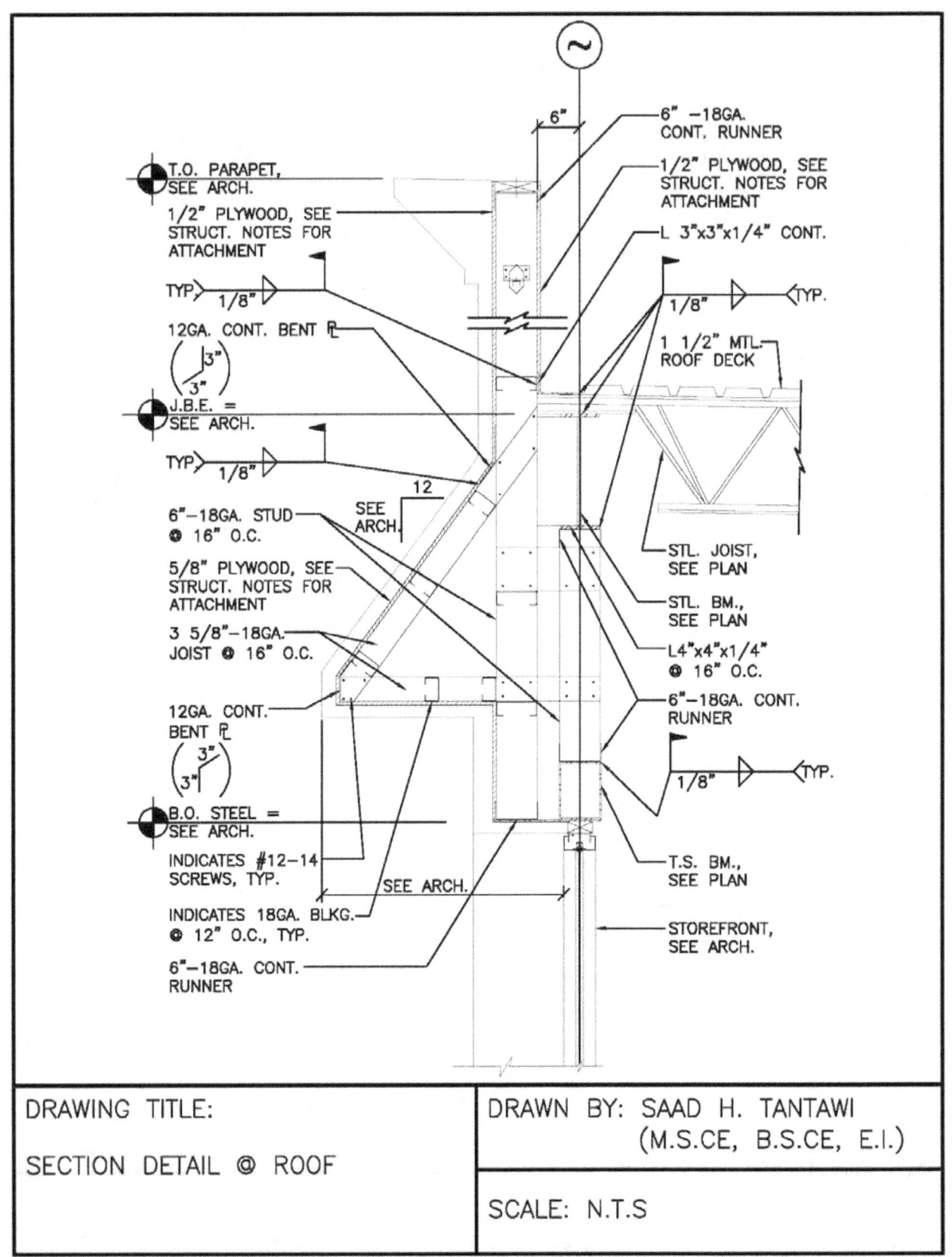

DRAWING TITLE:	DRAWN BY: SAAD H. TANTAWI
SECTION DETAIL @ ROOF	(M.S.CE, B.S.CE, E.I.)
	SCALE: N.T.S

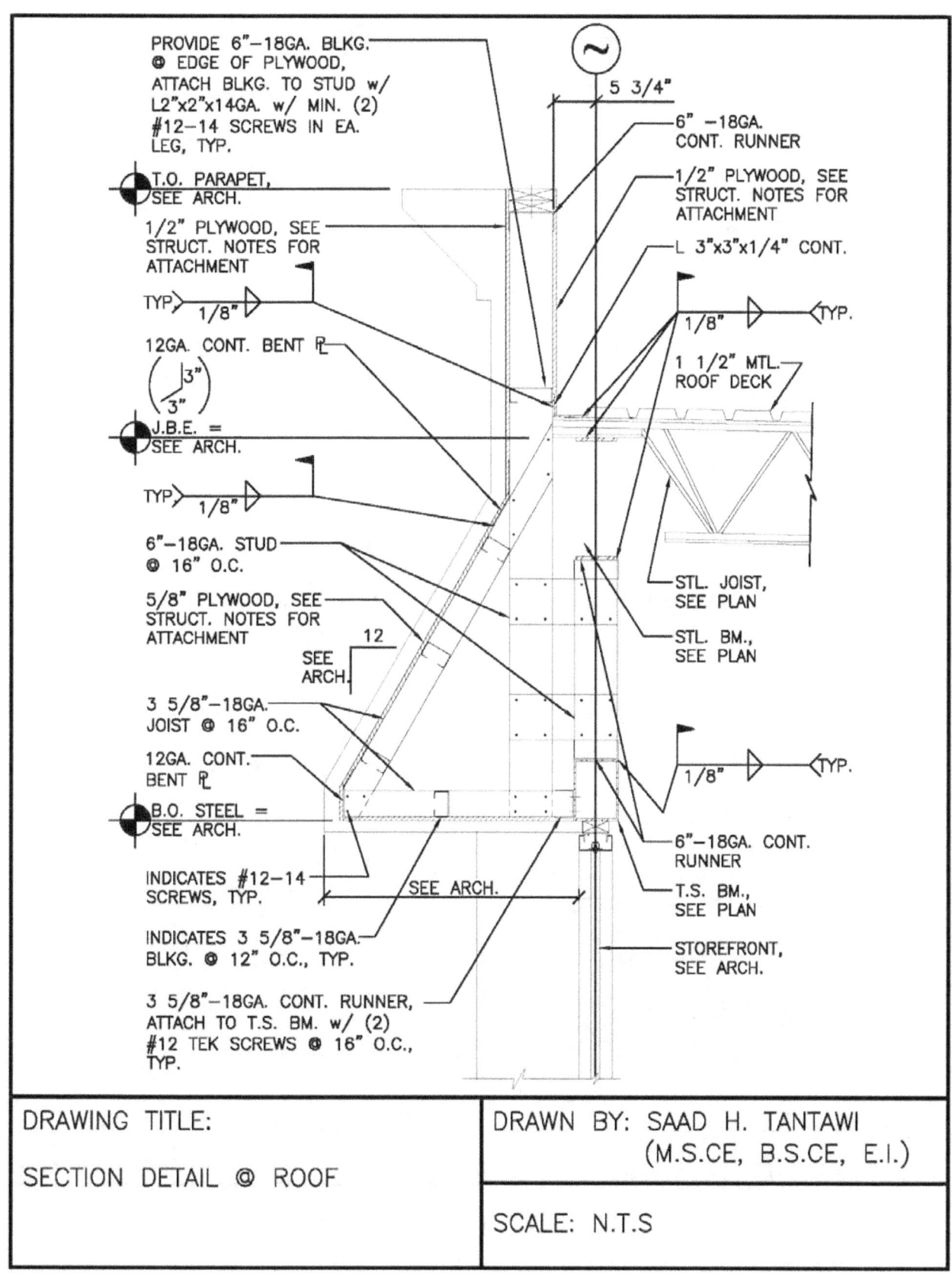

PROVIDE 6"–18GA. BLKG.
@ EDGE OF PLYWOOD,
ATTACH BLKG. TO STUD w/
L2"x2"x14GA. w/ MIN. (2)
#12–14 SCREWS IN EA.
LEG, TYP.

5 3/4"

6" –18GA.
CONT. RUNNER

T.O. PARAPET,
SEE ARCH.

1/2" PLYWOOD, SEE
STRUCT. NOTES FOR
ATTACHMENT

1/2" PLYWOOD, SEE
STRUCT. NOTES FOR
ATTACHMENT

L 3"x3"x1/4" CONT.

TYP 1/8" TYP.

12GA. CONT. BENT PL

1 1/2" MTL.
ROOF DECK

J.B.E. =
SEE ARCH.

TYP 1/8"

6"–18GA. STUD
@ 16" O.C.

STL. JOIST,
SEE PLAN

5/8" PLYWOOD, SEE
STRUCT. NOTES FOR
ATTACHMENT

STL. BM.,
SEE PLAN

12

SEE
ARCH.

3 5/8"–18GA.
JOIST @ 16" O.C.

12GA. CONT.
BENT PL

1/8" TYP.

B.O. STEEL =
SEE ARCH.

INDICATES #12–14
SCREWS, TYP.

SEE ARCH.

6"–18GA. CONT.
RUNNER

T.S. BM.,
SEE PLAN

INDICATES 3 5/8"–18GA.
BLKG. @ 12" O.C., TYP.

STOREFRONT,
SEE ARCH.

3 5/8"–18GA. CONT. RUNNER,
ATTACH TO T.S. BM. w/ (2)
#12 TEK SCREWS @ 16" O.C.,
TYP.

DRAWING TITLE:

SECTION DETAIL @ ROOF

DRAWN BY: SAAD H. TANTAWI
(M.S.CE, B.S.CE, E.I.)

SCALE: N.T.S

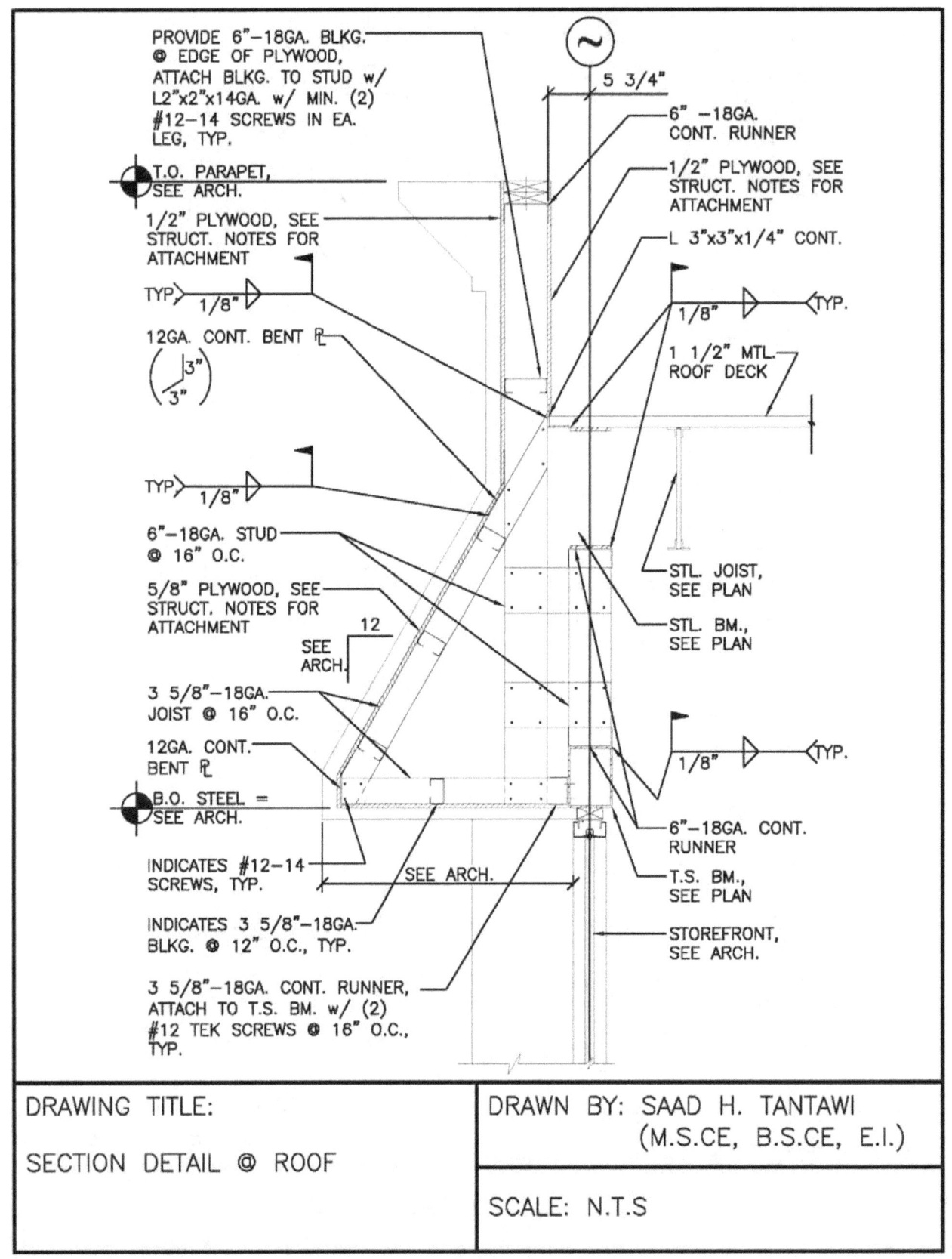

PROVIDE 6"–18GA. BLKG.
@ EDGE OF PLYWOOD,
ATTACH BLKG. TO STUD w/
L2"x2"x14GA. w/ MIN. (2)
#12–14 SCREWS IN EA.
LEG, TYP.

T.O. PARAPET,
SEE ARCH.

1/2" PLYWOOD, SEE
STRUCT. NOTES FOR
ATTACHMENT

TYP 1/8"

12GA. CONT. BENT ℞

6"–18GA. STUD
@ 16" O.C.

5/8" PLYWOOD, SEE
STRUCT. NOTES FOR
ATTACHMENT

SEE
ARCH.

3 5/8"–18GA.
JOIST @ 16" O.C.

12GA. CONT.
BENT ℞

B.O. STEEL =
SEE ARCH.

INDICATES #12–14
SCREWS, TYP.

INDICATES 3 5/8"–18GA.
BLKG. @ 12" O.C., TYP.

3 5/8"–18GA. CONT. RUNNER,
ATTACH TO T.S. BM. w/ (2)
#12 TEK SCREWS @ 16" O.C.,
TYP.

5 3/4"

6" –18GA.
CONT. RUNNER

1/2" PLYWOOD, SEE
STRUCT. NOTES FOR
ATTACHMENT

L 3"x3"x1/4" CONT.

1/8" TYP.

1 1/2" MTL.
ROOF DECK

STL. JOIST,
SEE PLAN

STL. BM.,
SEE PLAN

1/8" TYP.

6"–18GA. CONT.
RUNNER

T.S. BM.,
SEE PLAN

STOREFRONT,
SEE ARCH.

SEE ARCH.

12

DRAWING TITLE:	DRAWN BY: SAAD H. TANTAWI
	(M.S.CE, B.S.CE, E.I.)
SECTION DETAIL @ ROOF	
	SCALE: N.T.S

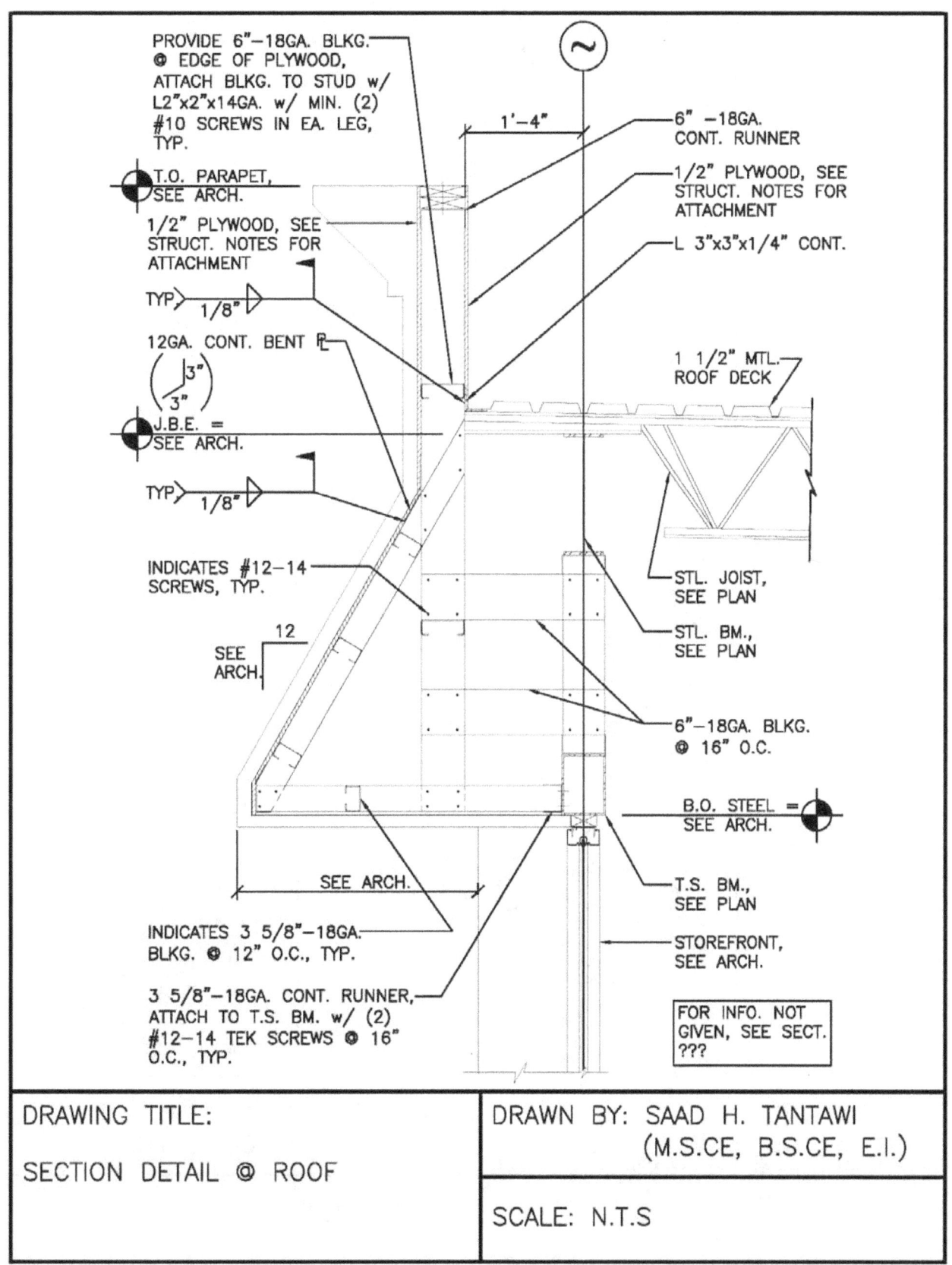

PROVIDE 6"−18GA. BLKG.
@ EDGE OF PLYWOOD,
ATTACH BLKG. TO STUD w/
L2"x2"x14GA. w/ MIN. (2)
#10 SCREWS IN EA. LEG,
TYP.

T.O. PARAPET,
SEE ARCH.

1/2" PLYWOOD, SEE
STRUCT. NOTES FOR
ATTACHMENT

TYP 1/8"

12GA. CONT. BENT PL.
3"
3"

J.B.E. =
SEE ARCH.

TYP 1/8"

INDICATES #12−14
SCREWS, TYP.

SEE
ARCH.
12

INDICATES 3 5/8"−18GA.
BLKG. @ 12" O.C., TYP.

3 5/8"−18GA. CONT. RUNNER,
ATTACH TO T.S. BM. w/ (2)
#12−14 TEK SCREWS @ 16"
O.C., TYP.

1'−4"

6" −18GA.
CONT. RUNNER

1/2" PLYWOOD, SEE
STRUCT. NOTES FOR
ATTACHMENT

L 3"x3"x1/4" CONT.

1 1/2" MTL.
ROOF DECK

STL. JOIST,
SEE PLAN

STL. BM.,
SEE PLAN

6"−18GA. BLKG.
@ 16" O.C.

B.O. STEEL =
SEE ARCH.

T.S. BM.,
SEE PLAN

STOREFRONT,
SEE ARCH.

SEE ARCH.

FOR INFO. NOT
GIVEN, SEE SECT.
???

DRAWING TITLE:	DRAWN BY: SAAD H. TANTAWI
	(M.S.CE, B.S.CE, E.I.)
SECTION DETAIL @ ROOF	
	SCALE: N.T.S

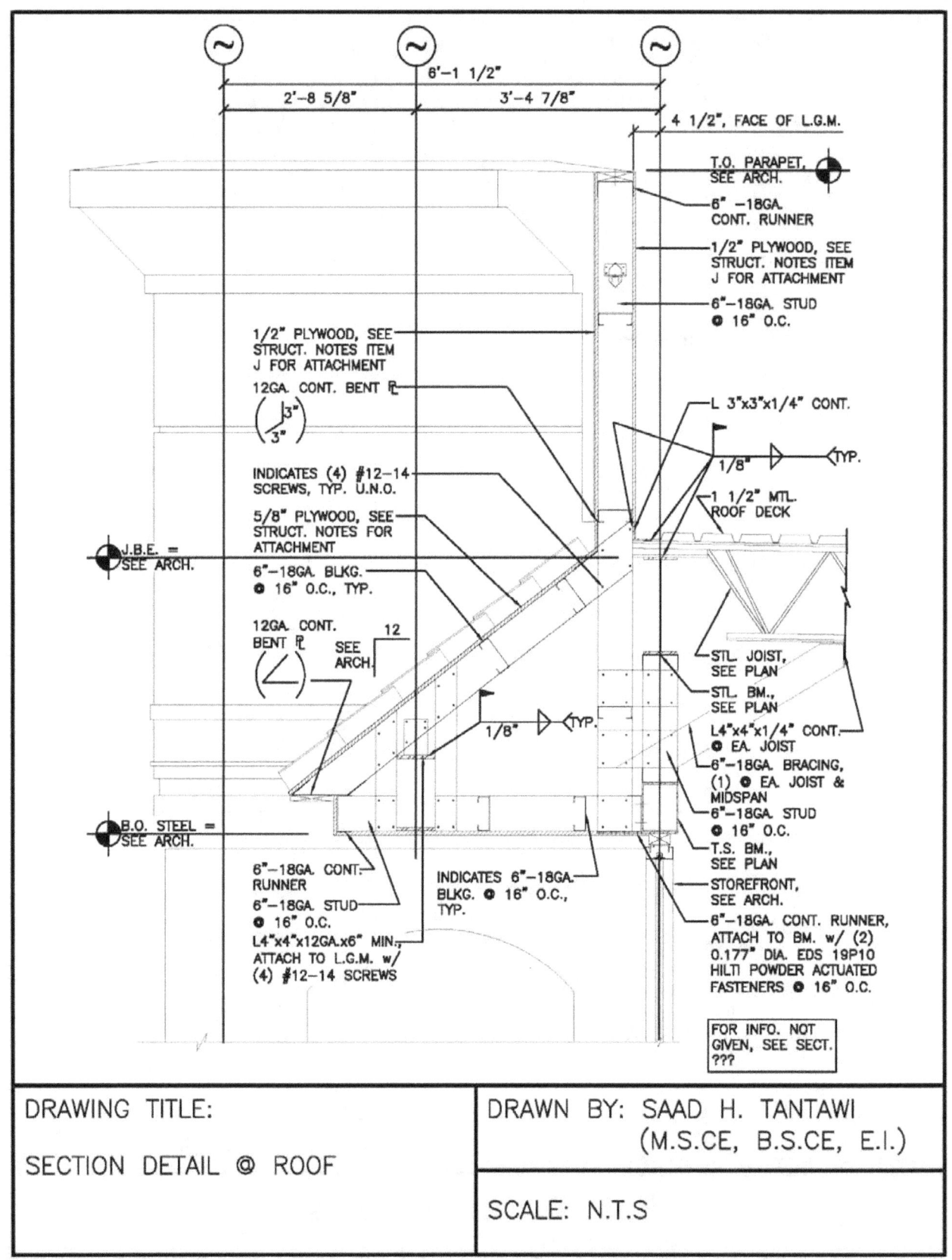

4 1/2", FACE OF L.G.M.

T.O. PARAPET,
SEE ARCH.

6" -18GA.
CONT. RUNNER

1/2" PLYWOOD, SEE
STRUCT. NOTES ITEM
J FOR ATTACHMENT

6"-18GA. STUD
@ 16" O.C.

1/2" PLYWOOD, SEE
STRUCT. NOTES ITEM
J FOR ATTACHMENT

12GA. CONT. BENT PL

INDICATES (4) #12-14
SCREWS, TYP. U.N.O.

5/8" PLYWOOD, SEE
STRUCT. NOTES FOR
ATTACHMENT

J.B.E. =
SEE ARCH.

6"-18GA. BLKG.
@ 16" O.C., TYP.

12GA. CONT.
BENT PL

SEE
ARCH.

12

L 3"x3"x1/4" CONT.

1/8"   TYP.

1 1/2" MTL.
ROOF DECK

STL. JOIST,
SEE PLAN

STL. BM.,
SEE PLAN

L4"x4"x1/4" CONT.
@ EA. JOIST

6"-18GA. BRACING,
(1) @ EA. JOIST &
MIDSPAN

6"-18GA. STUD
@ 16" O.C.

T.S. BM.,
SEE PLAN

STOREFRONT,
SEE ARCH.

1/8"   TYP.

B.O. STEEL =
SEE ARCH.

6"-18GA. CONT.
RUNNER

6"-18GA. STUD
@ 16" O.C.

L4"x4"x12GA.x6" MIN.,
ATTACH TO L.G.M. w/
(4) #12-14 SCREWS

INDICATES 6"-18GA.
BLKG. @ 16" O.C.,
TYP.

6"-18GA. CONT. RUNNER,
ATTACH TO BM. w/ (2)
0.177" DIA. EDS 19P10
HILTI POWDER ACTUATED
FASTENERS @ 16" O.C.

FOR INFO. NOT
GIVEN, SEE SECT.
???

6'-1 1/2"

2'-8 5/8"   3'-4 7/8"

DRAWING TITLE:	DRAWN BY: SAAD H. TANTAWI
	(M.S.CE, B.S.CE, E.I.)
SECTION DETAIL @ ROOF	
	SCALE: N.T.S

130

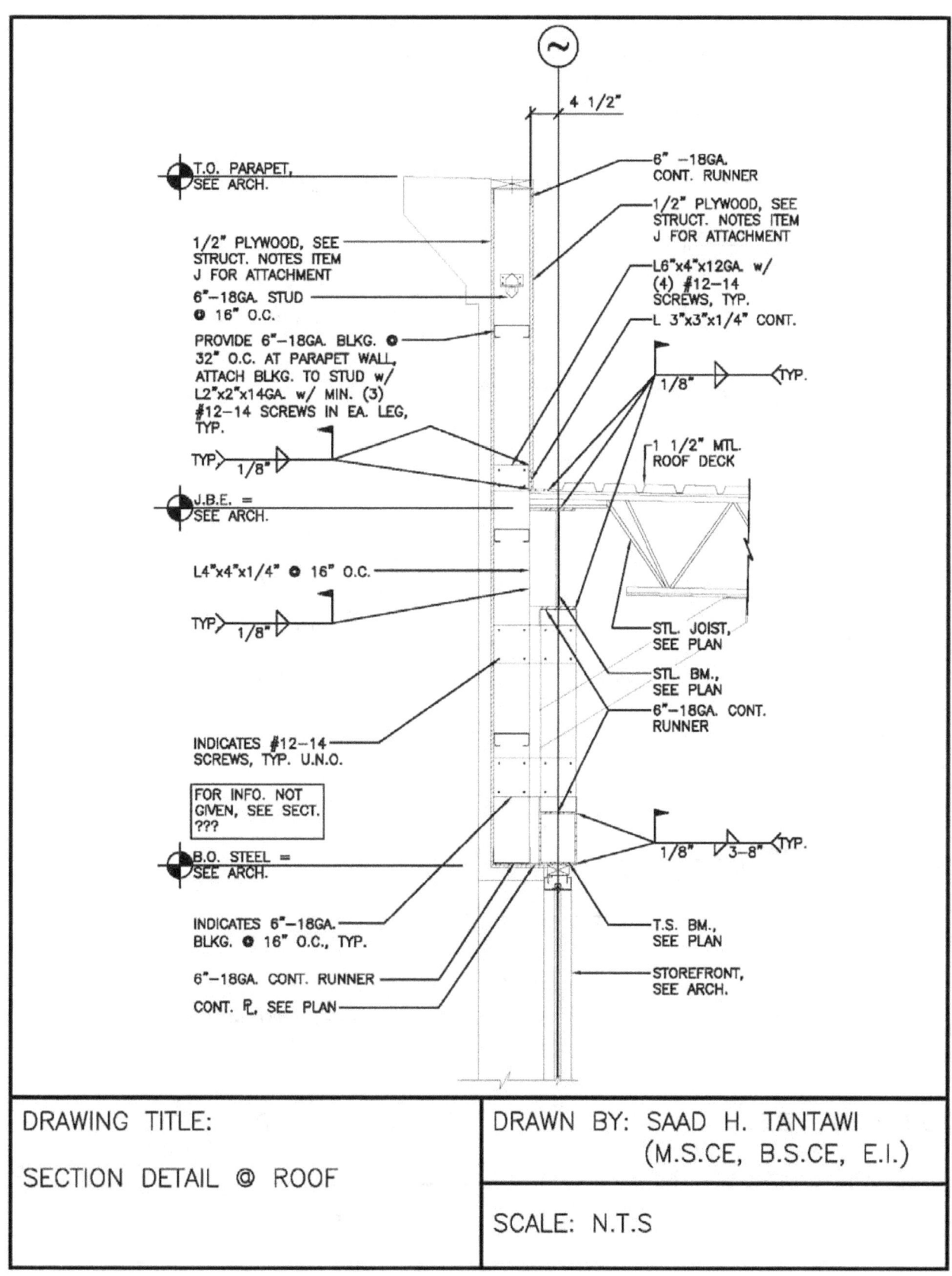

T.O. PARAPET,
SEE ARCH.

1/2" PLYWOOD, SEE
STRUCT. NOTES ITEM
J FOR ATTACHMENT

6"-18GA. STUD
@ 16" O.C.

PROVIDE 6"-18GA. BLKG. @
32" O.C. AT PARAPET WALL,
ATTACH BLKG. TO STUD w/
L2"x2"x14GA. w/ MIN. (3)
#12-14 SCREWS IN EA. LEG,
TYP.

TYP 1/8"

J.B.E. =
SEE ARCH.

L4"x4"x1/4" @ 16" O.C.

TYP 1/8"

INDICATES #12-14
SCREWS, TYP. U.N.O.

FOR INFO. NOT
GIVEN, SEE SECT.
???

B.O. STEEL =
SEE ARCH.

INDICATES 6"-18GA.
BLKG. @ 16" O.C., TYP.

6"-18GA. CONT. RUNNER

CONT. ₽, SEE PLAN

4 1/2"

6" -18GA.
CONT. RUNNER

1/2" PLYWOOD, SEE
STRUCT. NOTES ITEM
J FOR ATTACHMENT

L6"x4"x12GA. w/
(4) #12-14
SCREWS, TYP.

L 3"x3"x1/4" CONT.

1/8"   TYP.

1 1/2" MTL.
ROOF DECK

STL. JOIST,
SEE PLAN

STL. BM.,
SEE PLAN

6"-18GA. CONT.
RUNNER

1/8"   3-8"   TYP.

T.S. BM.,
SEE PLAN

STOREFRONT,
SEE ARCH.

DRAWING TITLE:	DRAWN BY: SAAD H. TANTAWI
	(M.S.CE, B.S.CE, E.I.)
SECTION DETAIL @ ROOF	
	SCALE: N.T.S

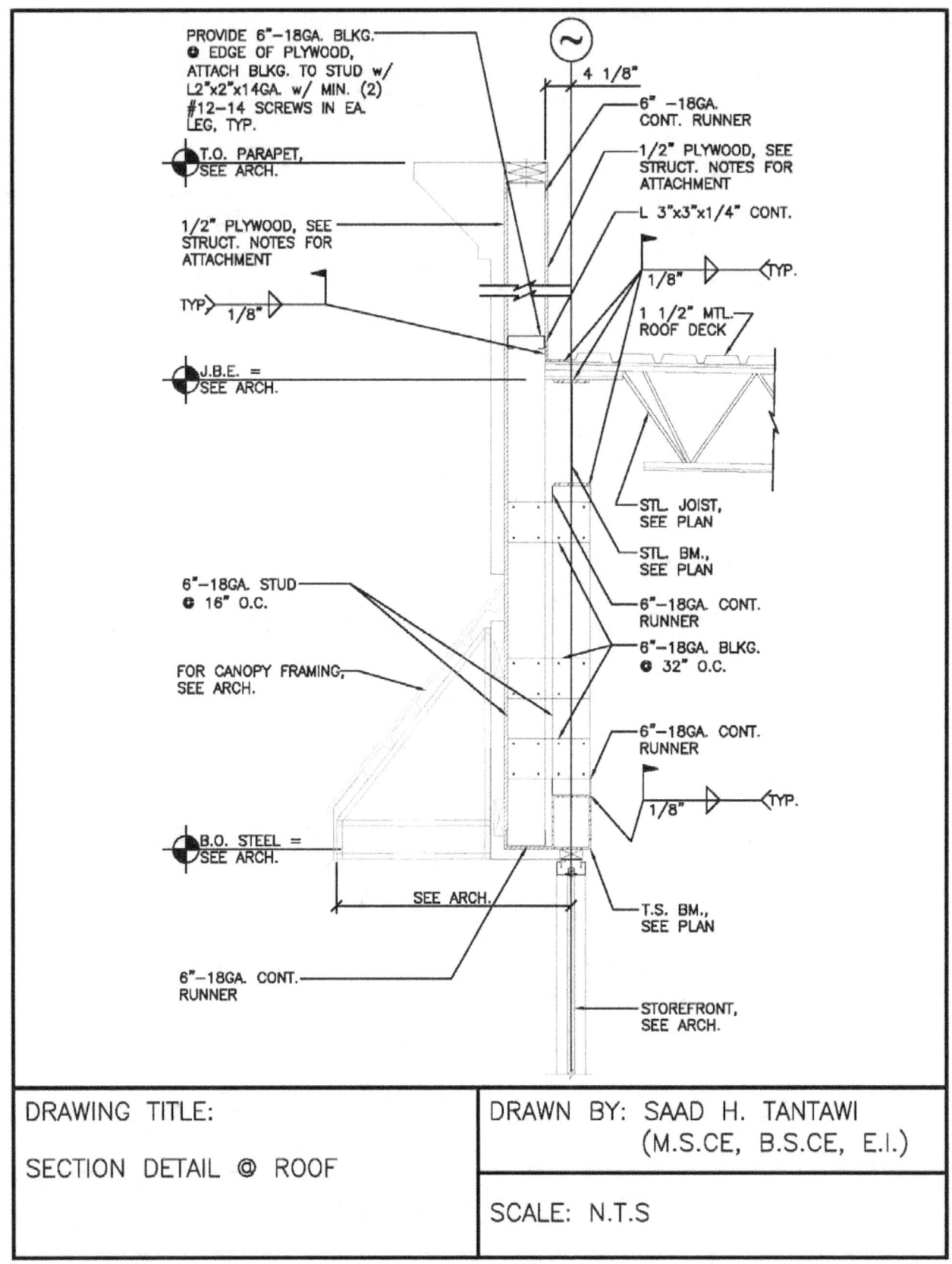

PROVIDE 6"–18GA. BLKG. @ EDGE OF PLYWOOD, ATTACH BLKG. TO STUD w/ L2"x2"x14GA. w/ MIN. (2) #12–14 SCREWS IN EA. LEG, TYP.

T.O. PARAPET, SEE ARCH.

1/2" PLYWOOD, SEE STRUCT. NOTES FOR ATTACHMENT

TYP. 1/8"

J.B.E. = SEE ARCH.

6"–18GA. STUD @ 16" O.C.

FOR CANOPY FRAMING, SEE ARCH.

B.O. STEEL = SEE ARCH.

SEE ARCH.

6"–18GA. CONT. RUNNER

4 1/8"

6" –18GA. CONT. RUNNER

1/2" PLYWOOD, SEE STRUCT. NOTES FOR ATTACHMENT

L 3"x3"x1/4" CONT.

1/8" TYP.

1 1/2" MTL. ROOF DECK

STL. JOIST, SEE PLAN

STL. BM., SEE PLAN

6"–18GA. CONT. RUNNER

6"–18GA. BLKG. @ 32" O.C.

6"–18GA. CONT. RUNNER

1/8" TYP.

T.S. BM., SEE PLAN

STOREFRONT, SEE ARCH.

DRAWING TITLE:	DRAWN BY: SAAD H. TANTAWI
SECTION DETAIL @ ROOF	(M.S.CE, B.S.CE, E.I.)
	SCALE: N.T.S

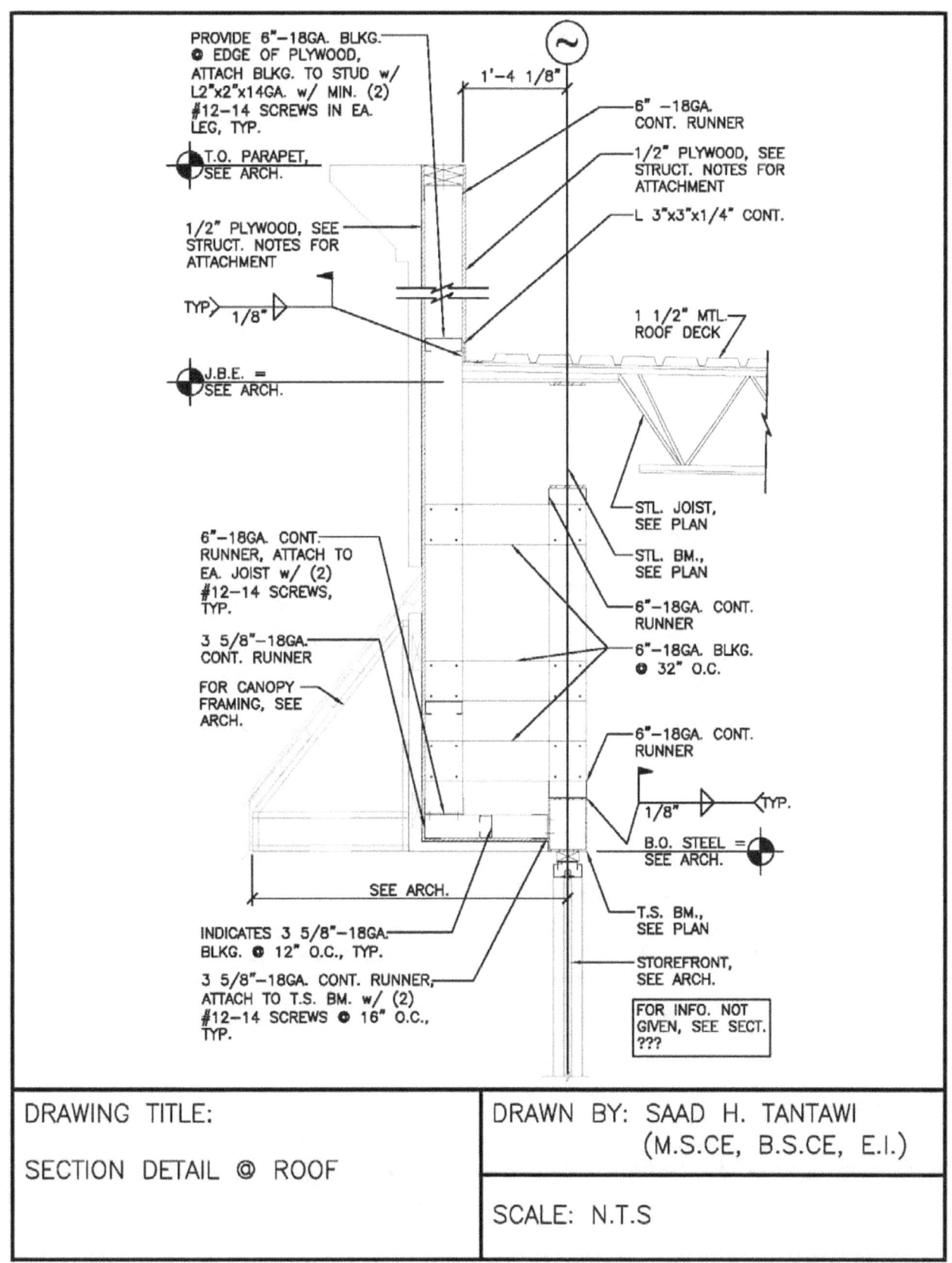

PROVIDE 6"–18GA. BLKG.
@ EDGE OF PLYWOOD,
ATTACH BLKG. TO STUD w/
L2"x2"x14GA. w/ MIN. (2)
#12–14 SCREWS IN EA.
LEG, TYP.

T.O. PARAPET,
SEE ARCH.

1/2" PLYWOOD, SEE
STRUCT. NOTES FOR
ATTACHMENT

TYP 1/8"

J.B.E. =
SEE ARCH.

6"–18GA. CONT.
RUNNER, ATTACH TO
EA. JOIST w/ (2)
#12–14 SCREWS,
TYP.

3 5/8"–18GA.
CONT. RUNNER

FOR CANOPY
FRAMING, SEE
ARCH.

SEE ARCH.

INDICATES 3 5/8"–18GA.
BLKG. @ 12" O.C., TYP.

3 5/8"–18GA. CONT. RUNNER,
ATTACH TO T.S. BM. w/ (2)
#12–14 SCREWS @ 16" O.C.,
TYP.

1'–4 1/8"

6" –18GA.
CONT. RUNNER

1/2" PLYWOOD, SEE
STRUCT. NOTES FOR
ATTACHMENT

L 3"x3"x1/4" CONT.

1 1/2" MTL.
ROOF DECK

STL. JOIST,
SEE PLAN

STL. BM.,
SEE PLAN

6"–18GA. CONT.
RUNNER

6"–18GA. BLKG.
@ 32" O.C.

6"–18GA. CONT.
RUNNER

1/8" TYP.

B.O. STEEL =
SEE ARCH.

T.S. BM.,
SEE PLAN

STOREFRONT,
SEE ARCH.

FOR INFO. NOT
GIVEN, SEE SECT.
???

DRAWING TITLE:	DRAWN BY: SAAD H. TANTAWI
	(M.S.CE, B.S.CE, E.I.)
SECTION DETAIL @ ROOF	
	SCALE: N.T.S

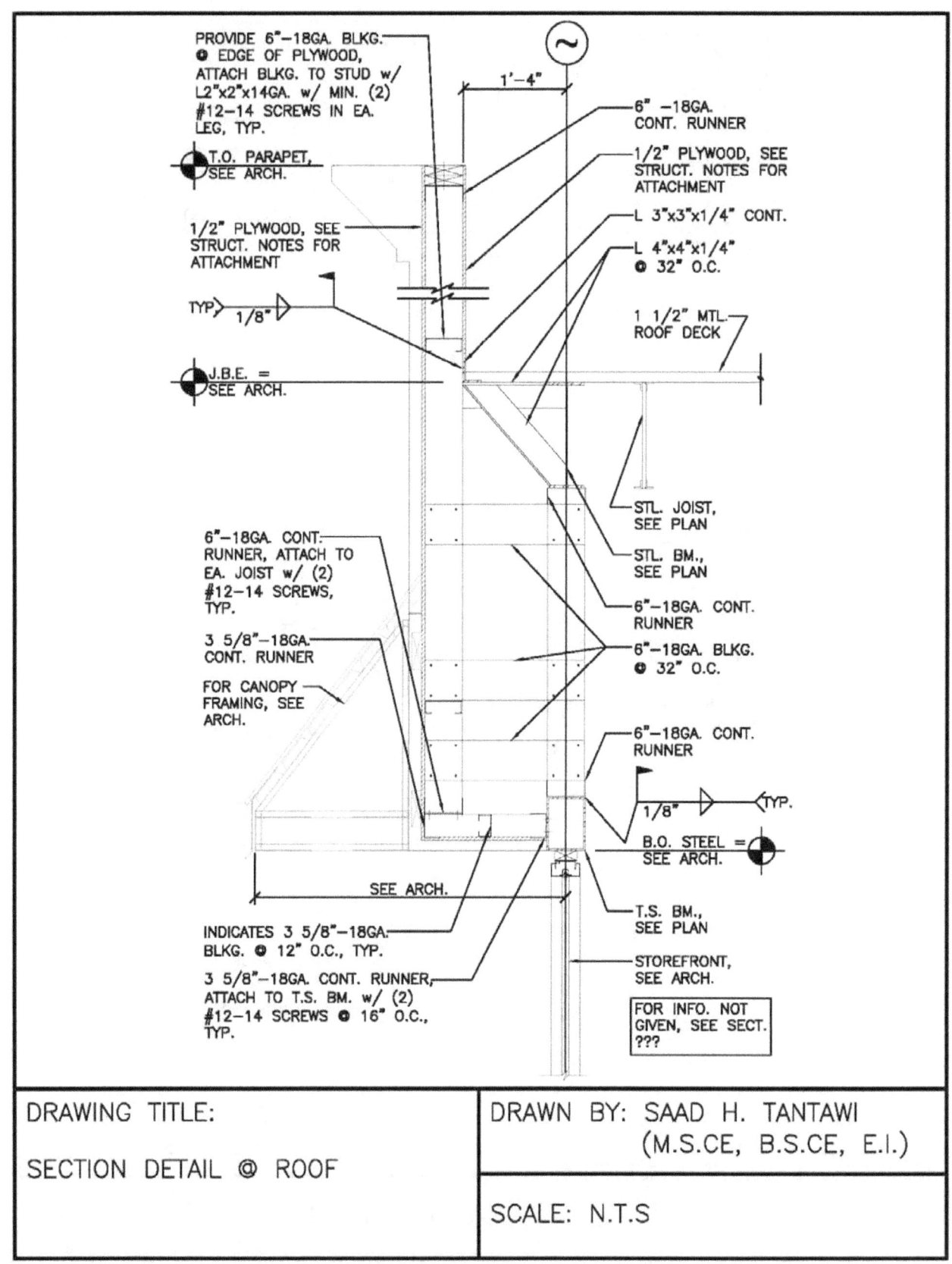

PROVIDE 6"-18GA. BLKG.
@ EDGE OF PLYWOOD,
ATTACH BLKG. TO STUD w/
L2"x2"x14GA. w/ MIN. (2)
#12-14 SCREWS IN EA.
LEG, TYP.

1'-4"

6" -18GA.
CONT. RUNNER

1/2" PLYWOOD, SEE
STRUCT. NOTES FOR
ATTACHMENT

T.O. PARAPET,
SEE ARCH.

L 3"x3"x1/4" CONT.

1/2" PLYWOOD, SEE
STRUCT. NOTES FOR
ATTACHMENT

L 4"x4"x1/4"
@ 32" O.C.

1 1/2" MTL.
ROOF DECK

TYP. 1/8"

J.B.E. =
SEE ARCH.

STL. JOIST,
SEE PLAN

STL. BM.,
SEE PLAN

6"-18GA. CONT.
RUNNER, ATTACH TO
EA. JOIST w/ (2)
#12-14 SCREWS,
TYP.

6"-18GA. CONT.
RUNNER

6"-18GA. BLKG.
@ 32" O.C.

3 5/8"-18GA.
CONT. RUNNER

FOR CANOPY
FRAMING, SEE
ARCH.

6"-18GA. CONT.
RUNNER

1/8" TYP.

B.O. STEEL =
SEE ARCH.

T.S. BM.,
SEE PLAN

SEE ARCH.

INDICATES 3 5/8"-18GA.
BLKG. @ 12" O.C., TYP.

3 5/8"-18GA. CONT. RUNNER,
ATTACH TO T.S. BM. w/ (2)
#12-14 SCREWS @ 16" O.C.,
TYP.

STOREFRONT,
SEE ARCH.

FOR INFO. NOT
GIVEN, SEE SECT.
???

DRAWING TITLE:	DRAWN BY: SAAD H. TANTAWI
	(M.S.CE, B.S.CE, E.I.)
SECTION DETAIL @ ROOF	
	SCALE: N.T.S

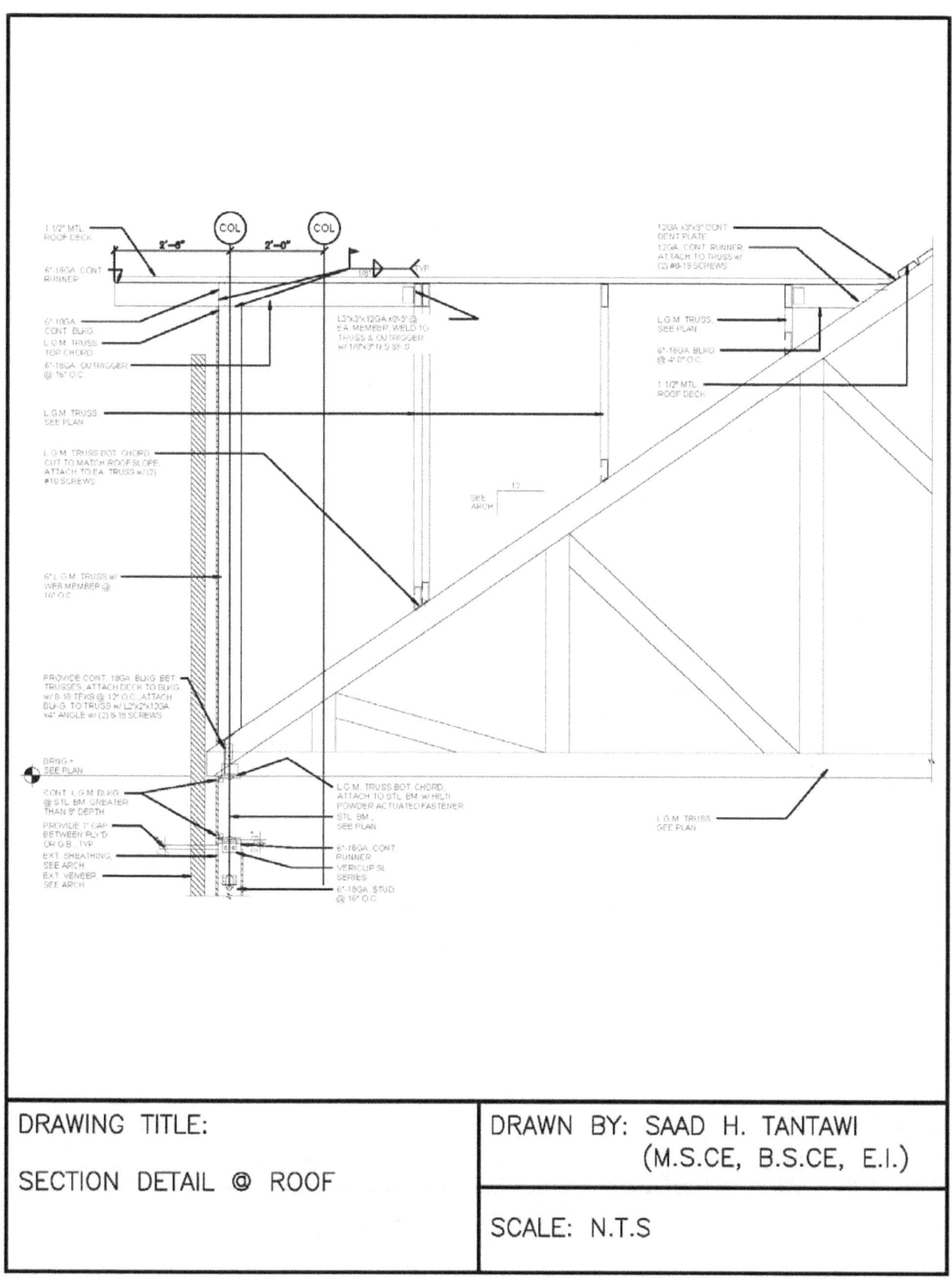

DRAWING TITLE:	DRAWN BY: SAAD H. TANTAWI
SECTION DETAIL @ ROOF	(M.S.CE, B.S.CE, E.I.)
	SCALE: N.T.S

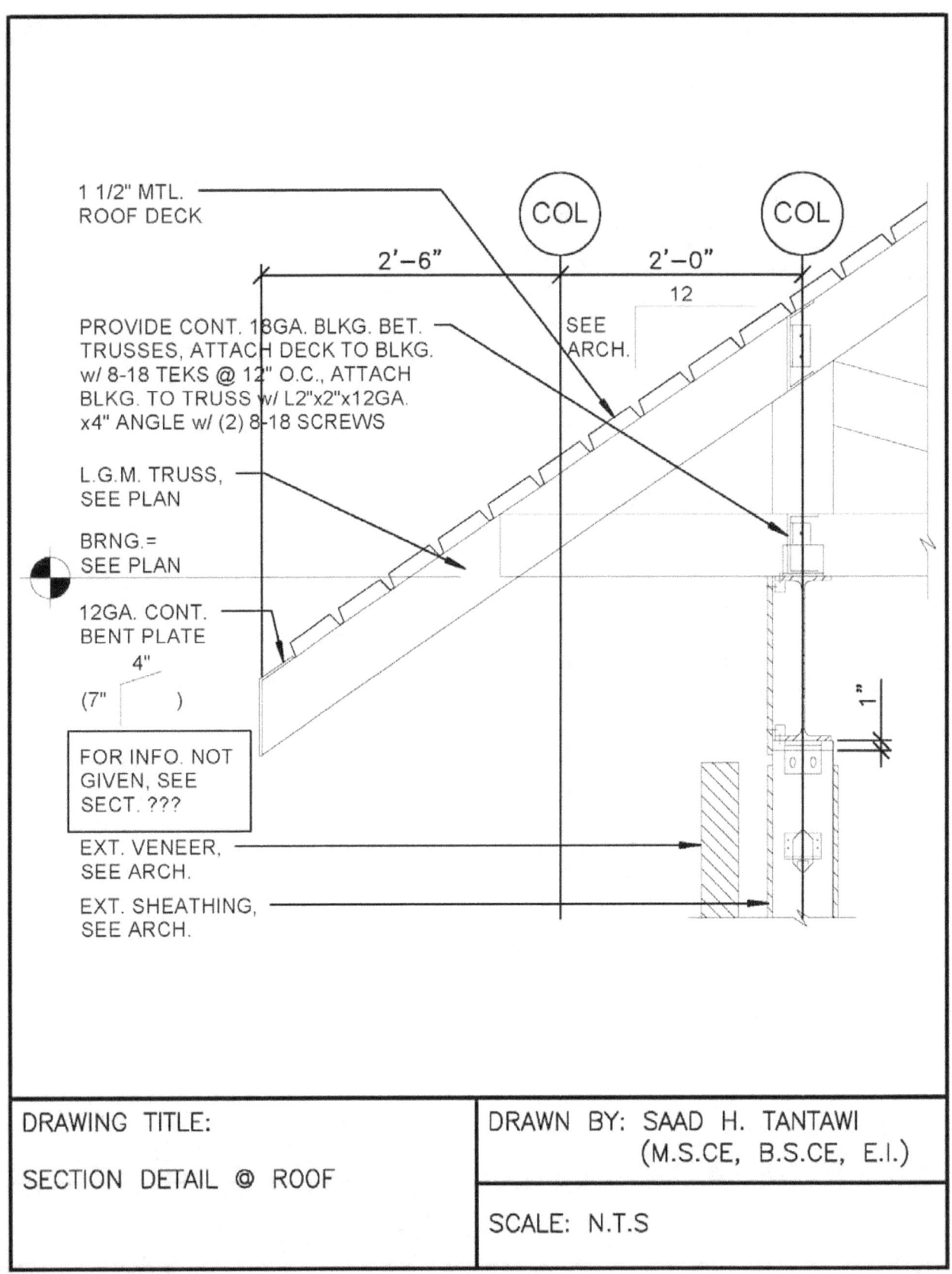

1 1/2" MTL.
ROOF DECK

COL

COL

2'-6"

2'-0"

12

PROVIDE CONT. 18GA. BLKG. BET.
TRUSSES, ATTACH DECK TO BLKG.
w/ 8-18 TEKS @ 12" O.C., ATTACH
BLKG. TO TRUSS w/ L2"x2"x12GA.
x4" ANGLE w/ (2) 8-18 SCREWS

SEE
ARCH.

L.G.M. TRUSS,
SEE PLAN

BRNG.=
SEE PLAN

12GA. CONT.
BENT PLATE
4"
(7"   )

1"

FOR INFO. NOT
GIVEN, SEE
SECT. ???

EXT. VENEER,
SEE ARCH.

EXT. SHEATHING,
SEE ARCH.

DRAWING TITLE:	DRAWN BY: SAAD H. TANTAWI
	(M.S.CE, B.S.CE, E.I.)
SECTION DETAIL @ ROOF	
	SCALE: N.T.S

136

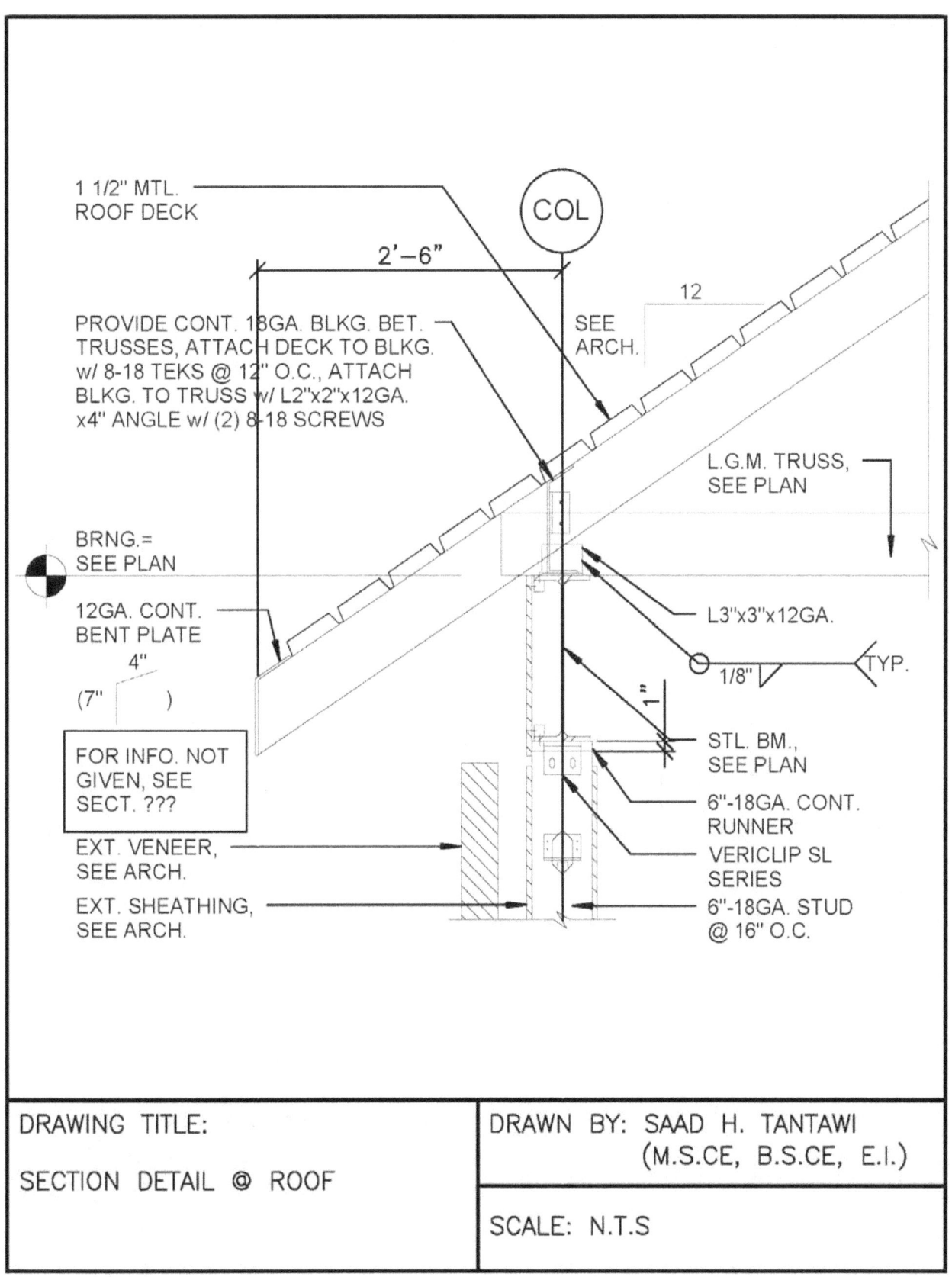

DRAWING TITLE:	DRAWN BY: SAAD H. TANTAWI
SECTION DETAIL @ ROOF	(M.S.CE, B.S.CE, E.I.)
	SCALE: N.T.S

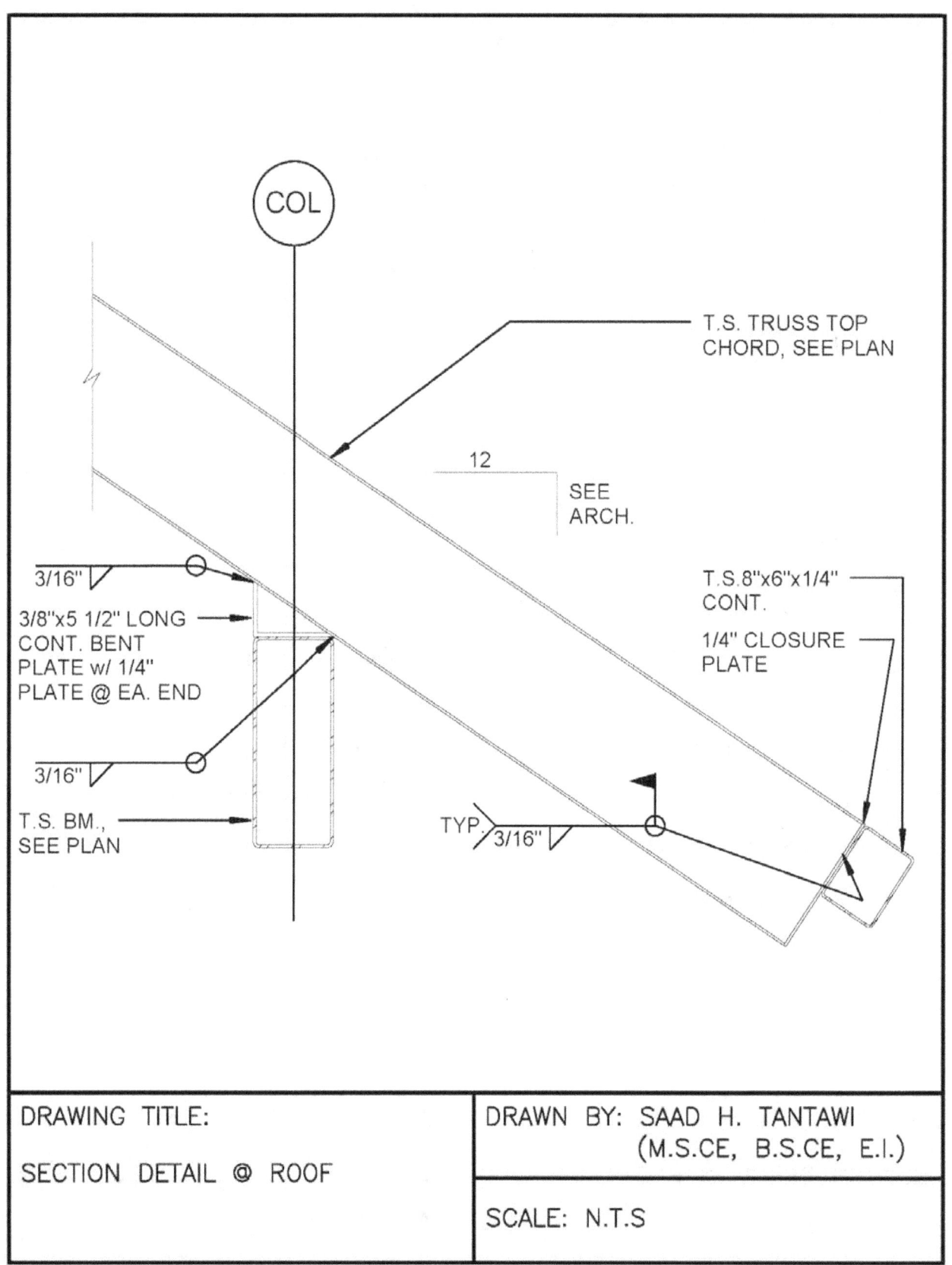

DRAWING TITLE:	DRAWN BY: SAAD H. TANTAWI
SECTION DETAIL @ ROOF	(M.S.CE, B.S.CE, E.I.)
	SCALE: N.T.S

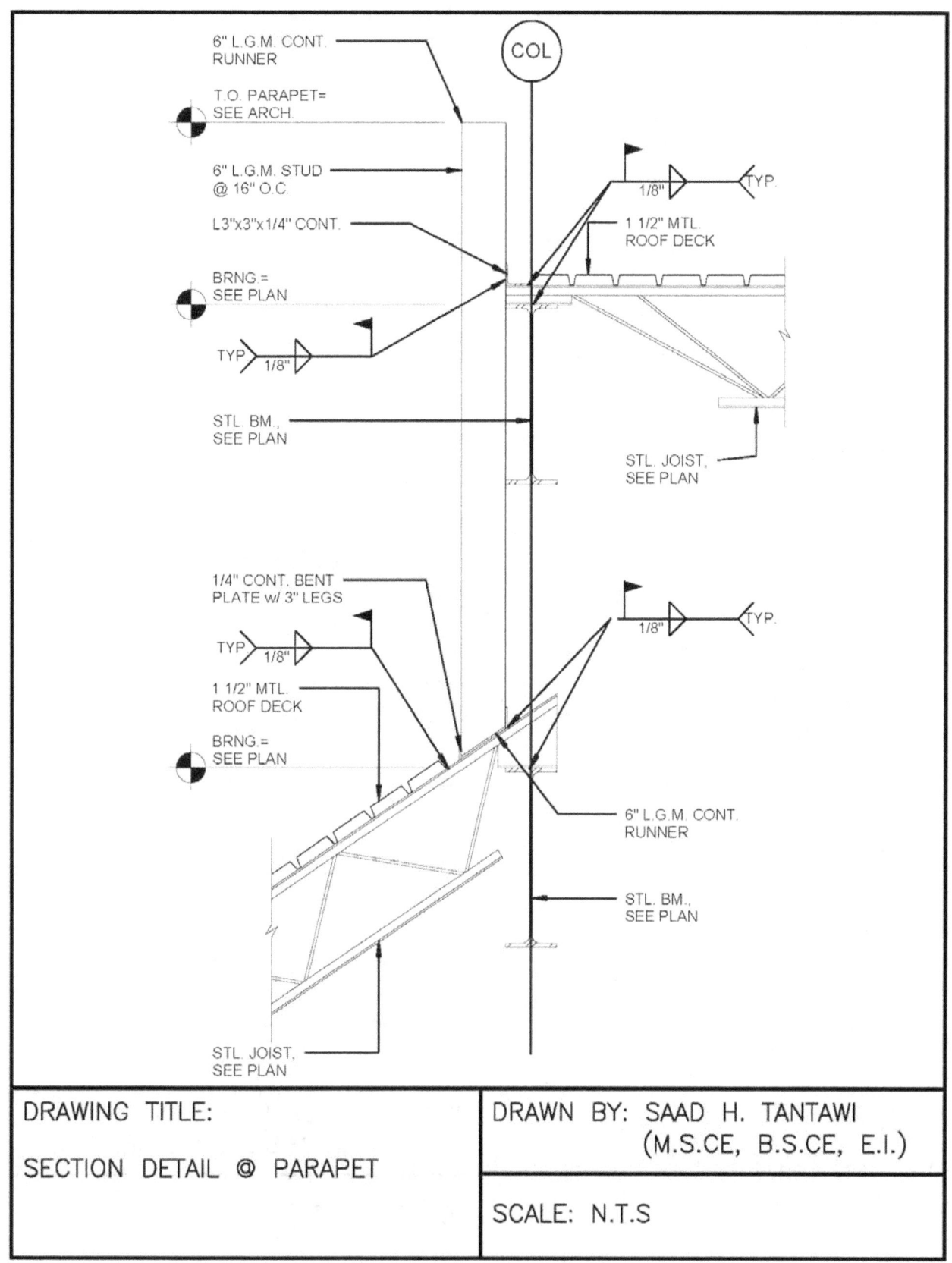

6" L.G.M. CONT.
RUNNER

COL

T.O. PARAPET=
SEE ARCH.

TYP.

6" L.G.M. STUD
@ 16" O.C.

1/8"

1 1/2" MTL.
ROOF DECK

L3"x3"x1/4" CONT.

BRNG.=
SEE PLAN

TYP
1/8"

STL. BM.,
SEE PLAN

STL. JOIST,
SEE PLAN

1/4" CONT. BENT
PLATE w/ 3" LEGS

TYP.

TYP
1/8"

1/8"

1 1/2" MTL.
ROOF DECK

BRNG.=
SEE PLAN

6" L.G.M. CONT.
RUNNER

STL. BM.,
SEE PLAN

STL. JOIST,
SEE PLAN

DRAWING TITLE:

SECTION DETAIL @ PARAPET

DRAWN BY: SAAD H. TANTAWI
(M.S.CE, B.S.CE, E.I.)

SCALE: N.T.S

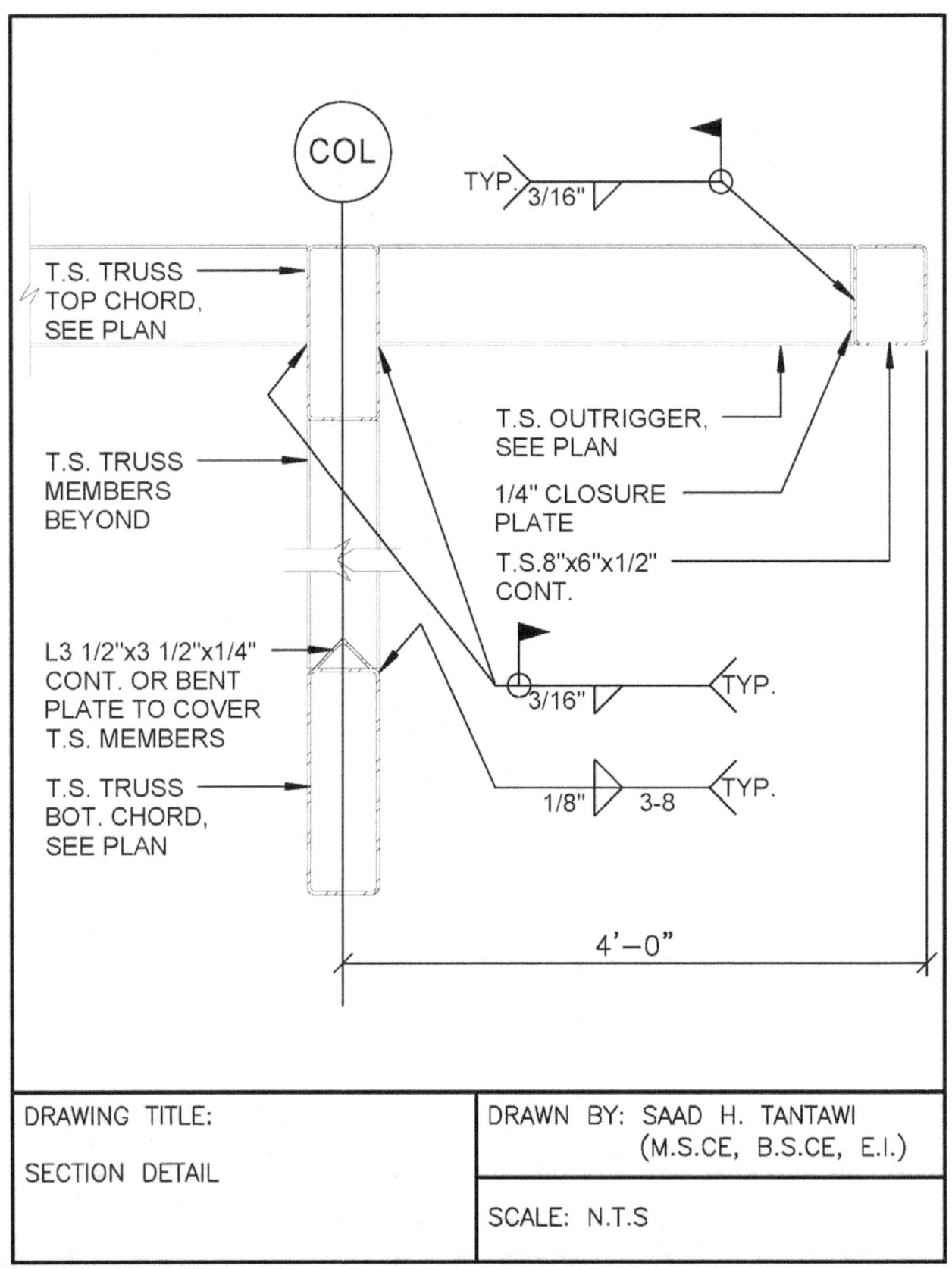

COL

TYP. 3/16"

T.S. TRUSS
TOP CHORD,
SEE PLAN

T.S. TRUSS
MEMBERS
BEYOND

T.S. OUTRIGGER,
SEE PLAN

1/4" CLOSURE
PLATE

T.S.8"x6"x1/2"
CONT.

L3 1/2"x3 1/2"x1/4"
CONT. OR BENT
PLATE TO COVER
T.S. MEMBERS

T.S. TRUSS
BOT. CHORD,
SEE PLAN

3/16"   TYP.

1/8"   3-8   TYP.

4'–0"

DRAWING TITLE:

SECTION DETAIL

DRAWN BY: SAAD H. TANTAWI
(M.S.CE, B.S.CE, E.I.)

SCALE: N.T.S

140

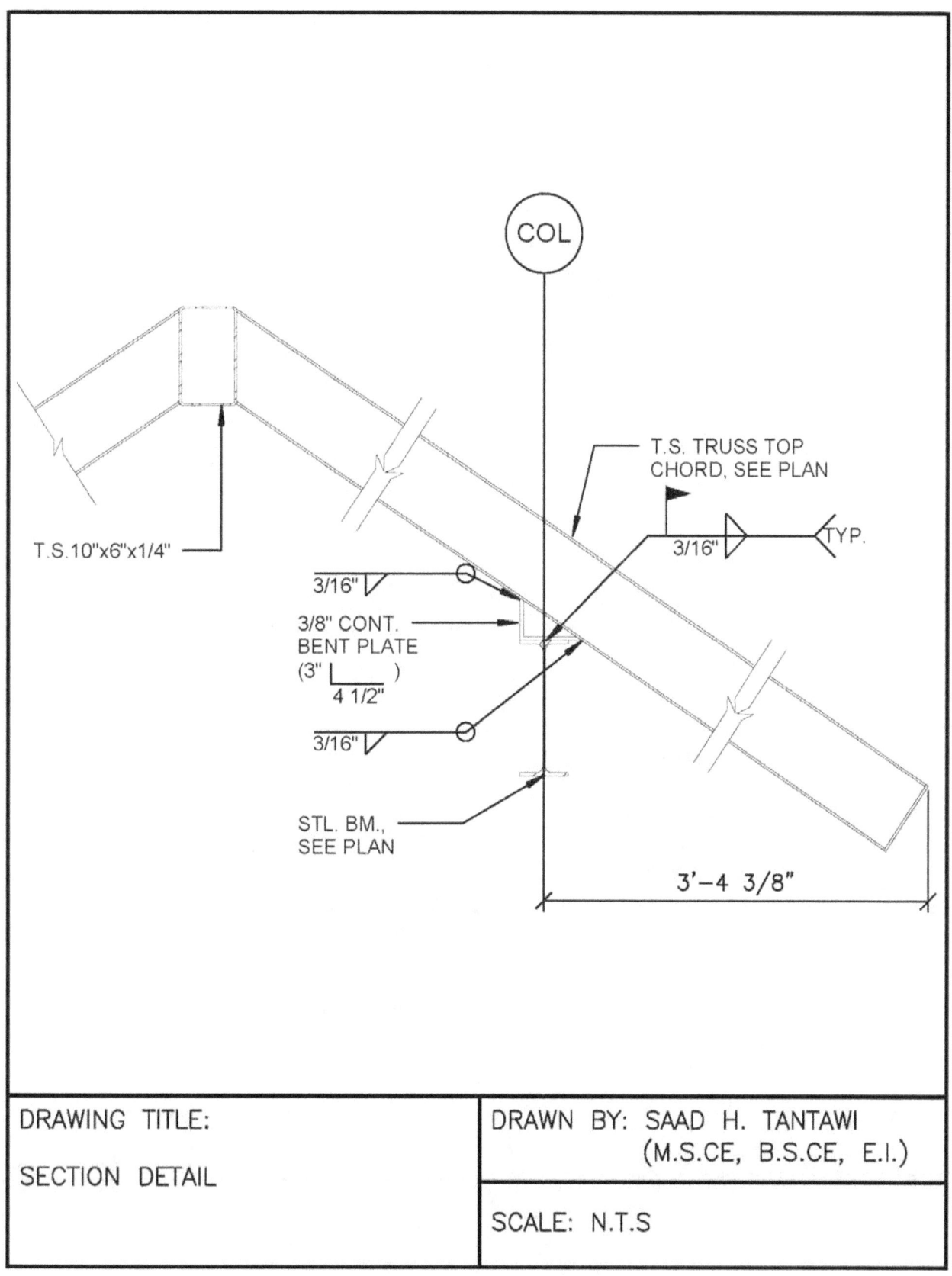

COL

T.S. TRUSS TOP
CHORD, SEE PLAN

3/16" ⊿ ◁ TYP.

T.S.10"x6"x1/4"

3/16" ⊿ ◯

3/8" CONT.
BENT PLATE
(3" ∟)
4 1/2"

3/16" ⊿ ◯

STL. BM.,
SEE PLAN

3'-4 3/8"

DRAWING TITLE: SECTION DETAIL	DRAWN BY: SAAD H. TANTAWI (M.S.CE, B.S.CE, E.I.)
	SCALE: N.T.S

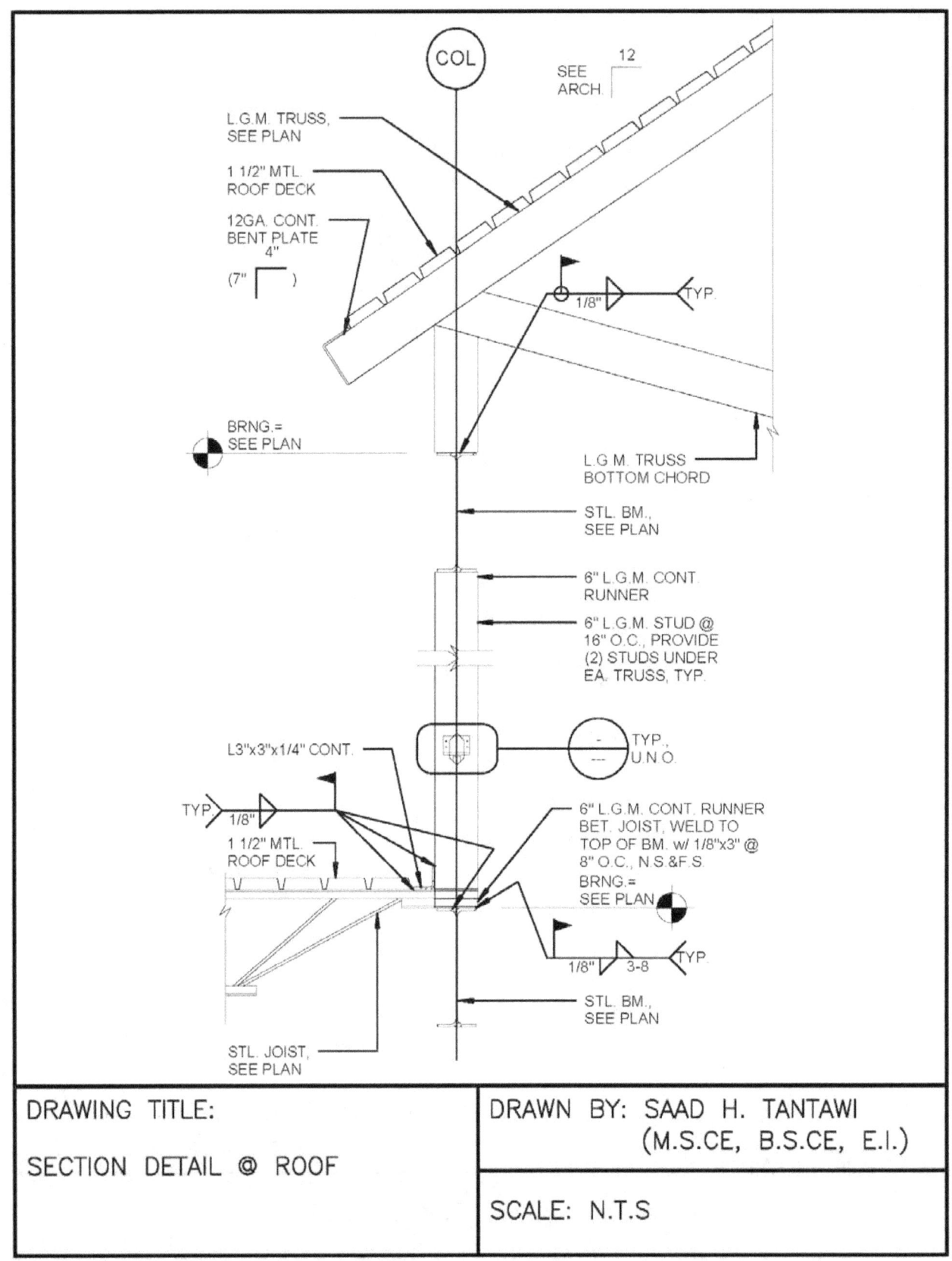

DRAWING TITLE:	DRAWN BY: SAAD H. TANTAWI
SECTION DETAIL @ ROOF	(M.S.CE, B.S.CE, E.I.)
	SCALE: N.T.S

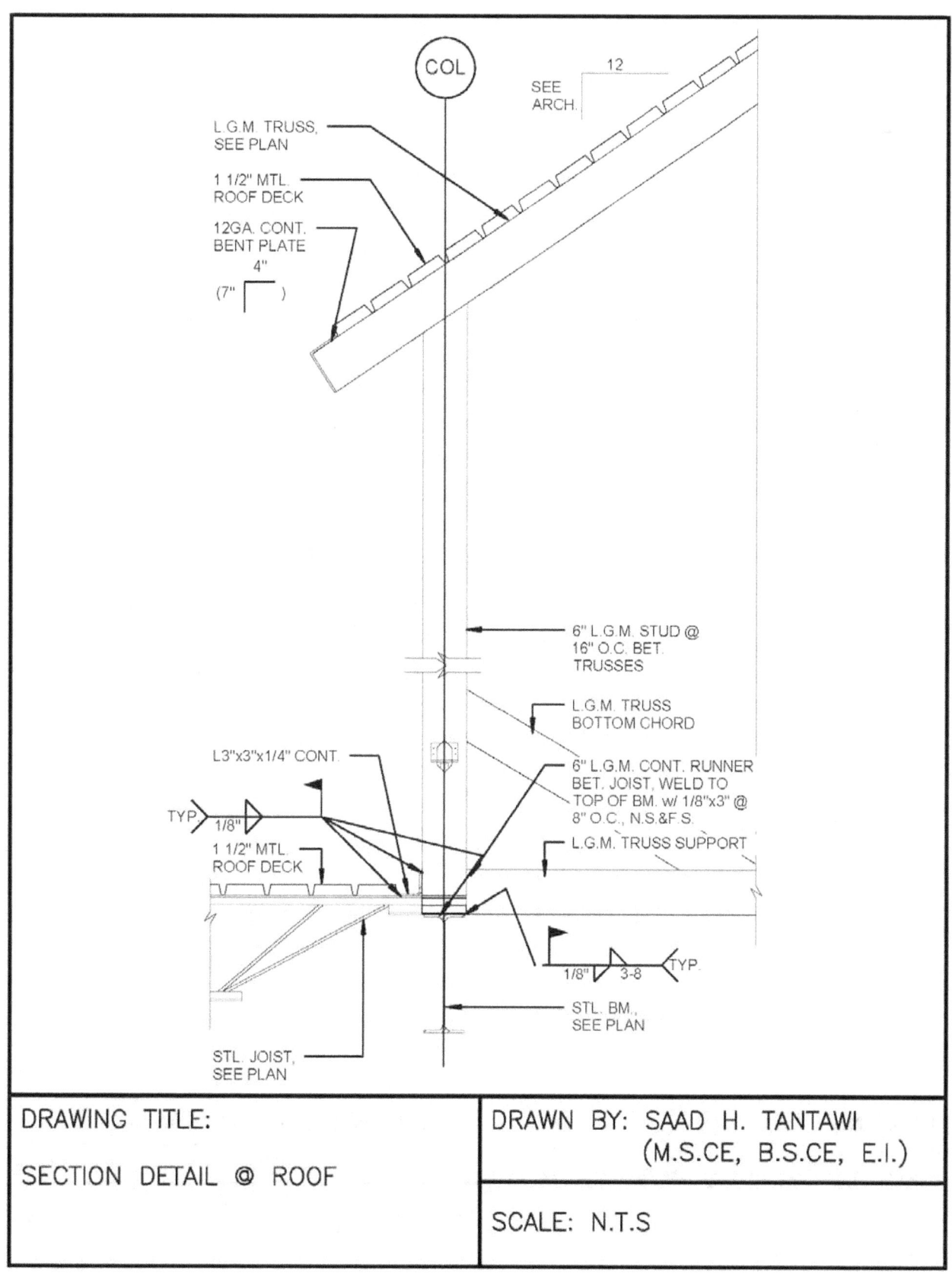

COL

SEE
ARCH.

12

L.G.M. TRUSS,
SEE PLAN

1 1/2" MTL.
ROOF DECK

12GA. CONT.
BENT PLATE
4"

(7" ⌐)

6" L.G.M. STUD @
16" O.C. BET.
TRUSSES

L.G.M. TRUSS
BOTTOM CHORD

L3"x3"x1/4" CONT.

6" L.G.M. CONT. RUNNER
BET. JOIST, WELD TO
TOP OF BM. w/ 1/8"x3" @
8" O.C., N.S.&F.S.

TYP.
1/8"

1 1/2" MTL.
ROOF DECK

L.G.M. TRUSS SUPPORT

1/8"   3-8   TYP.

STL. BM.,
SEE PLAN

STL. JOIST,
SEE PLAN

DRAWING TITLE:	DRAWN BY: SAAD H. TANTAWI
	(M.S.CE, B.S.CE, E.I.)
SECTION DETAIL @ ROOF	
	SCALE: N.T.S

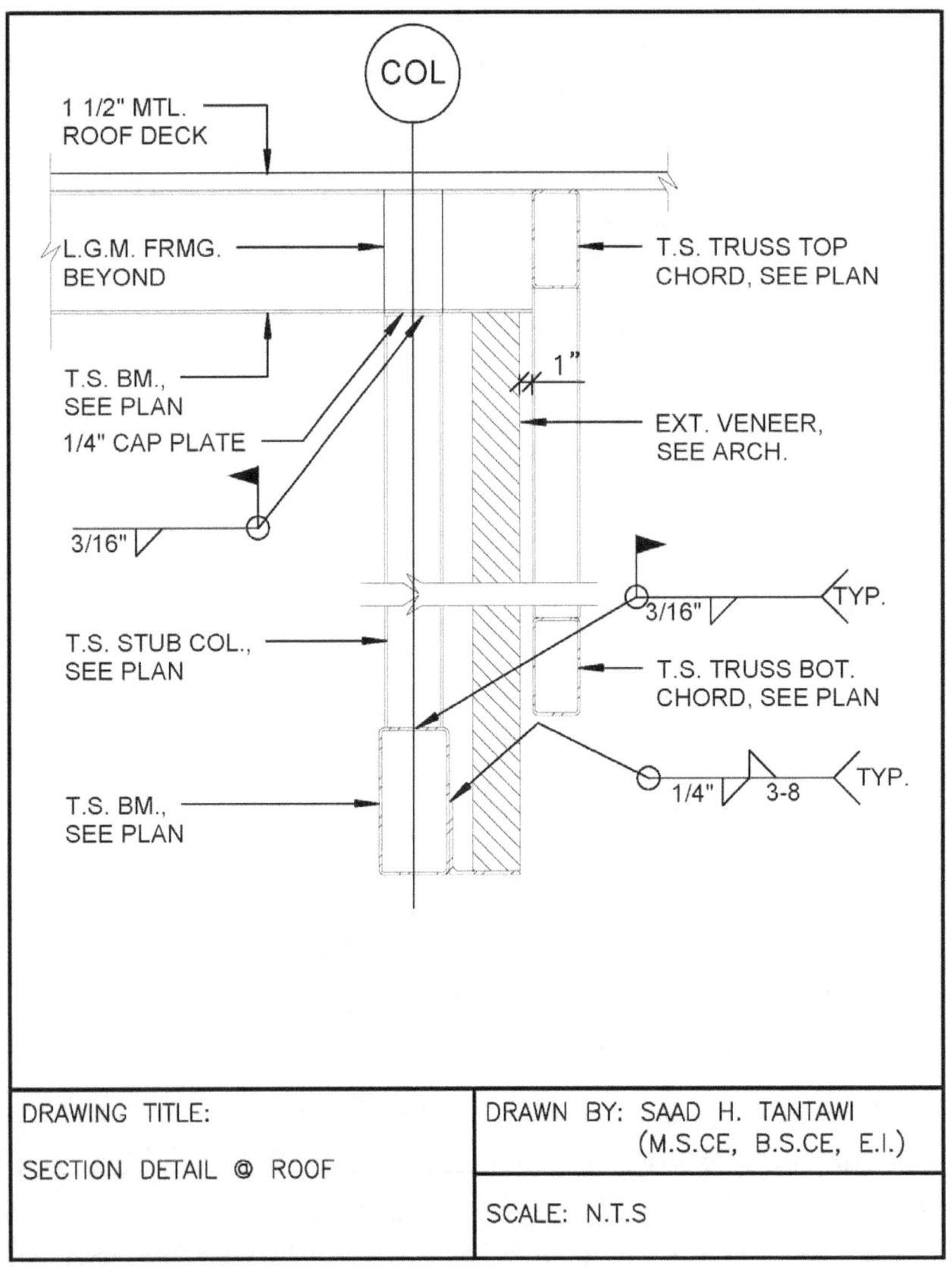

1 1/2" MTL.
ROOF DECK

L.G.M. FRMG.
BEYOND

T.S. TRUSS TOP
CHORD, SEE PLAN

T.S. BM.,
SEE PLAN

1/4" CAP PLATE

1"

EXT. VENEER,
SEE ARCH.

3/16"

3/16"

T.S. STUB COL.,
SEE PLAN

TYP.

T.S. TRUSS BOT.
CHORD, SEE PLAN

1/4"   3-8   TYP.

T.S. BM.,
SEE PLAN

DRAWING TITLE:	DRAWN BY: SAAD H. TANTAWI
	(M.S.CE, B.S.CE, E.I.)
SECTION DETAIL @ ROOF	
	SCALE: N.T.S

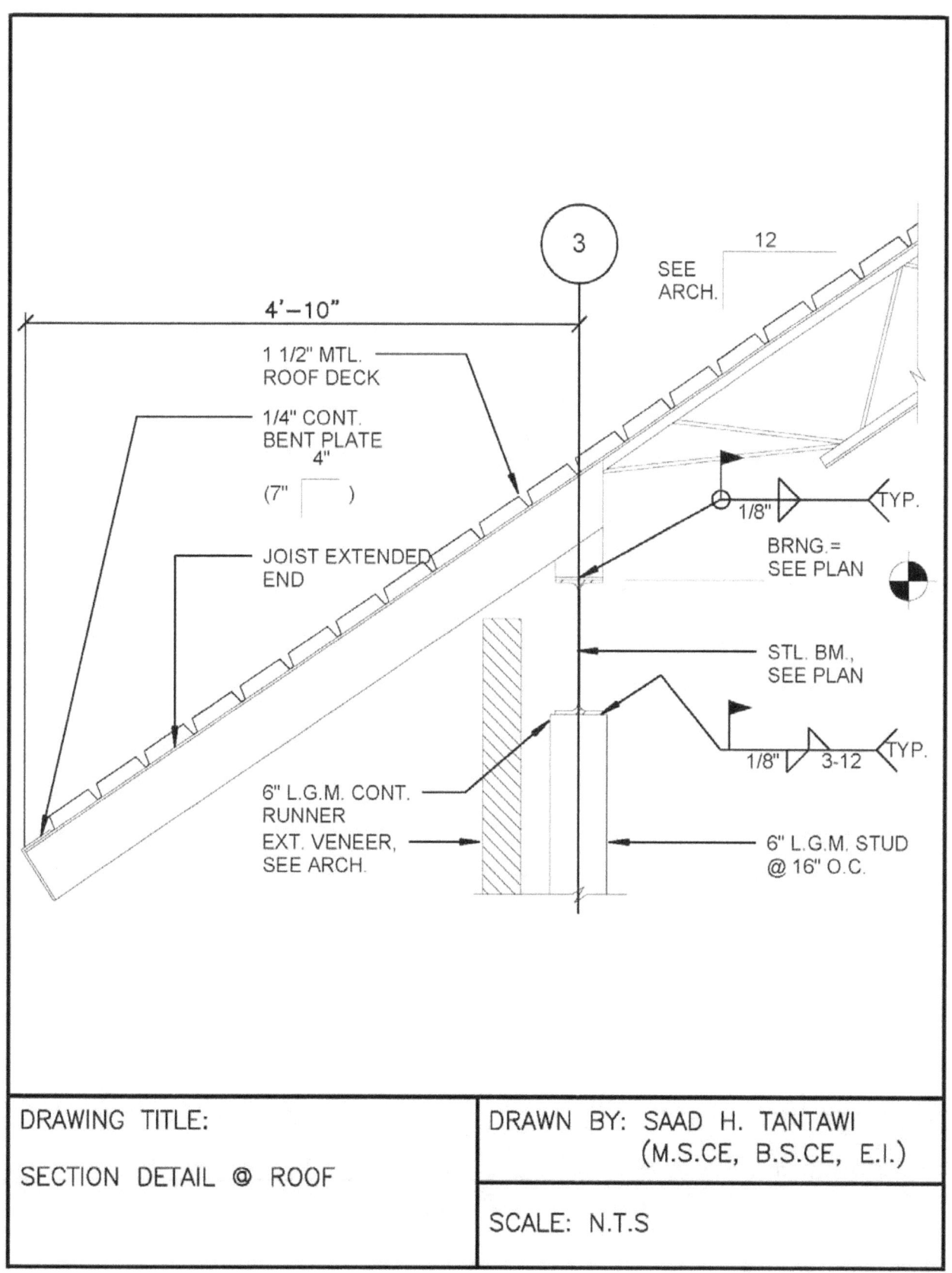

4'–10"

③

12

SEE
ARCH.

1 1/2" MTL.
ROOF DECK

1/4" CONT.
BENT PLATE
4"

(7"   )

JOIST EXTENDED
END

1/8"

BRNG.=
SEE PLAN

TYP.

STL. BM.,
SEE PLAN

1/8"   3-12   TYP.

6" L.G.M. CONT.
RUNNER

EXT. VENEER,
SEE ARCH.

6" L.G.M. STUD
@ 16" O.C.

DRAWING TITLE:	DRAWN BY: SAAD H. TANTAWI
	(M.S.CE, B.S.CE, E.I.)
SECTION DETAIL @ ROOF	
	SCALE: N.T.S

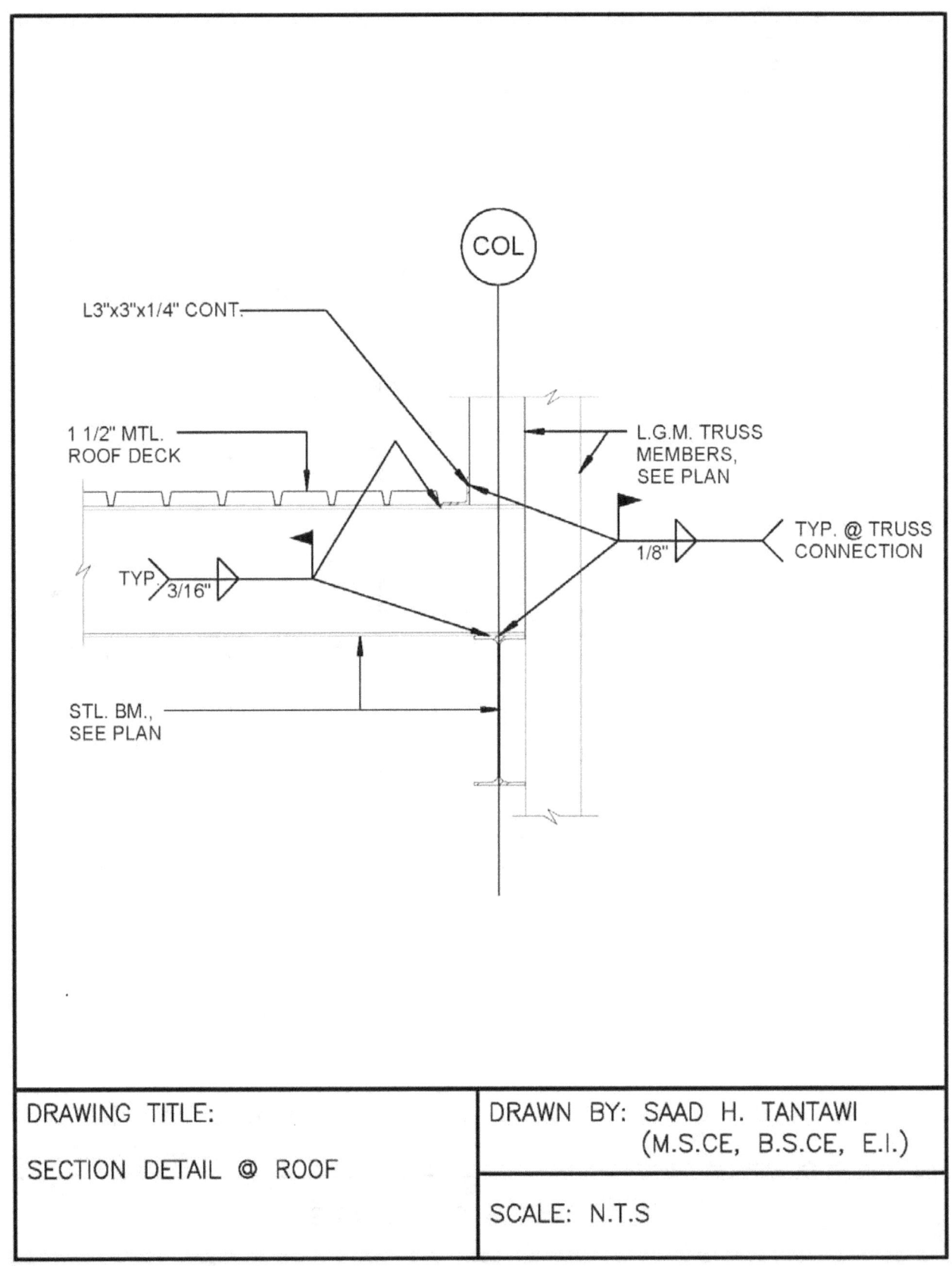

DRAWING TITLE:	DRAWN BY: SAAD H. TANTAWI
SECTION DETAIL @ ROOF	(M.S.CE, B.S.CE, E.I.)
	SCALE: N.T.S

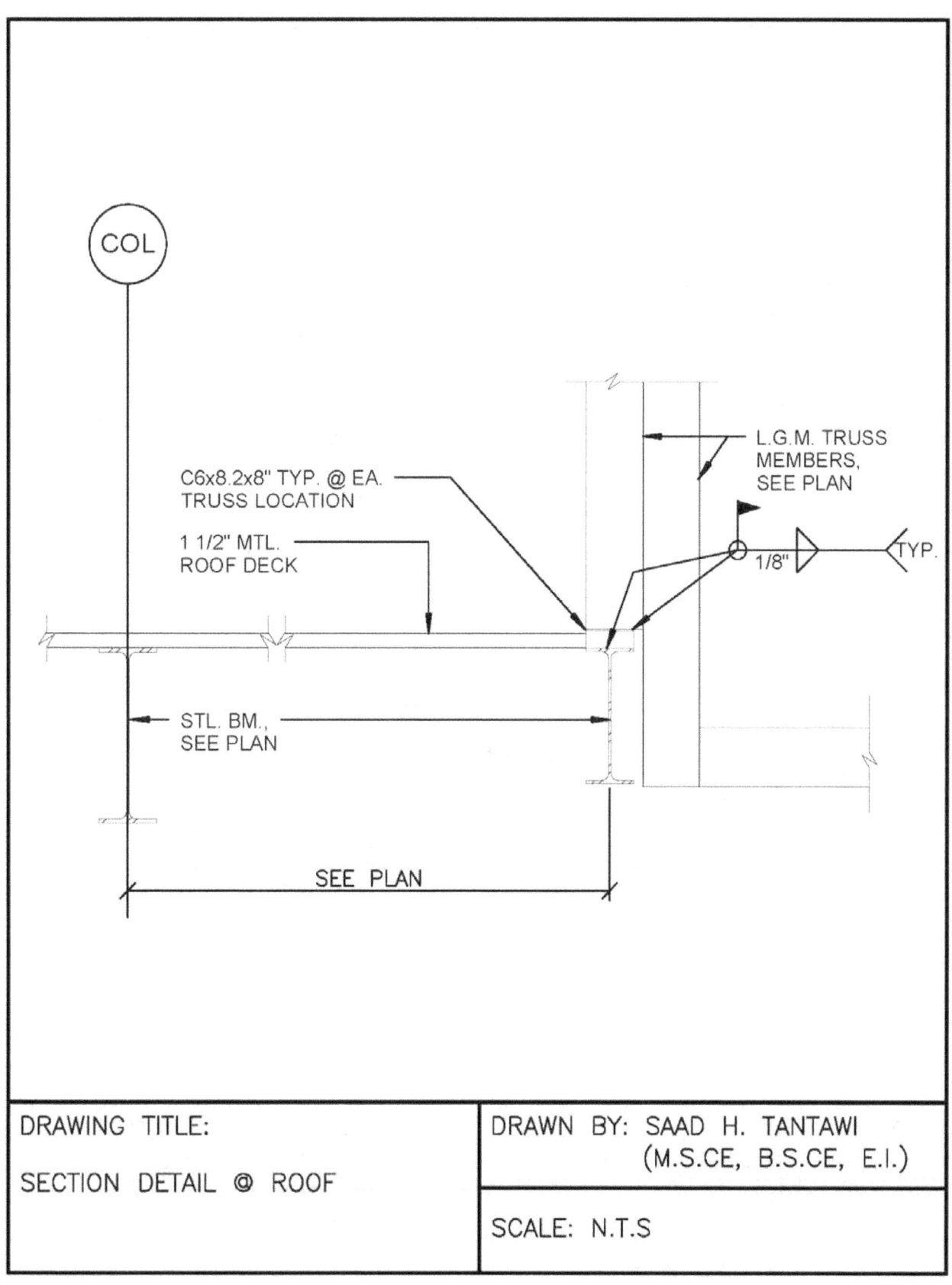

COL

C6x8.2x8" TYP. @ EA.
TRUSS LOCATION

1 1/2" MTL.
ROOF DECK

L.G.M. TRUSS
MEMBERS,
SEE PLAN

1/8"

TYP.

STL. BM.,
SEE PLAN

SEE PLAN

DRAWING TITLE:	DRAWN BY: SAAD H. TANTAWI
	(M.S.CE, B.S.CE, E.I.)
SECTION DETAIL @ ROOF	
	SCALE: N.T.S

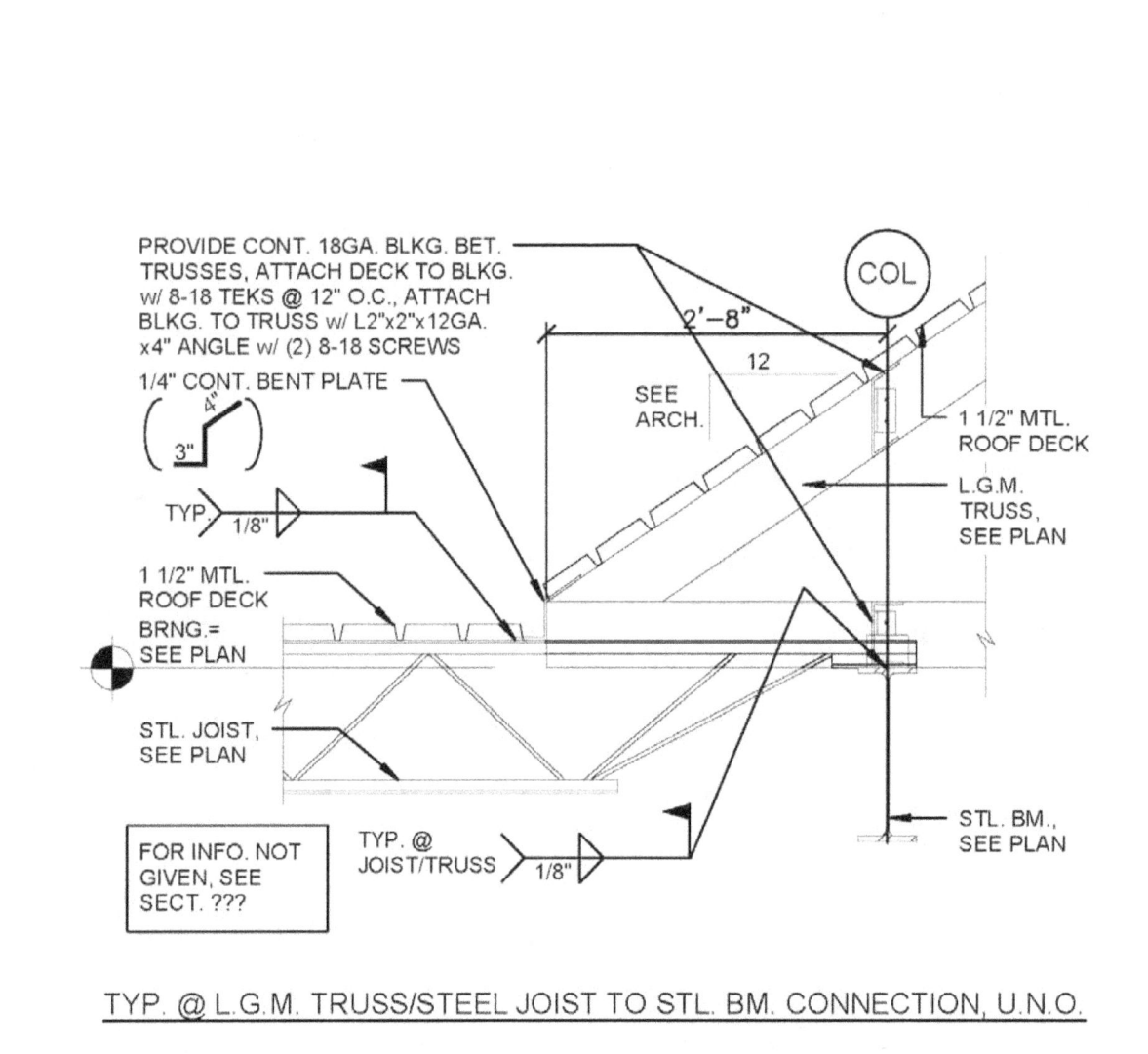

PROVIDE CONT. 18GA. BLKG. BET.
TRUSSES, ATTACH DECK TO BLKG.
w/ 8-18 TEKS @ 12" O.C., ATTACH
BLKG. TO TRUSS w/ L2"x2"x12GA.
x4" ANGLE w/ (2) 8-18 SCREWS

1/4" CONT. BENT PLATE

4"

3"

TYP.

1/8"

1 1/2" MTL.
ROOF DECK

BRNG.=
SEE PLAN

STL. JOIST,
SEE PLAN

FOR INFO. NOT
GIVEN, SEE
SECT. ???

TYP. @
JOIST/TRUSS

1/8"

2'-8"

12

SEE
ARCH.

COL

1 1/2" MTL.
ROOF DECK

L.G.M.
TRUSS,
SEE PLAN

STL. BM.,
SEE PLAN

TYP. @ L.G.M. TRUSS/STEEL JOIST TO STL. BM. CONNECTION, U.N.O.

DRAWING TITLE:	DRAWN BY: SAAD H. TANTAWI
	(M.S.CE, B.S.CE, E.I.)
SECTION DETAIL @ ROOF	
	SCALE: N.T.S

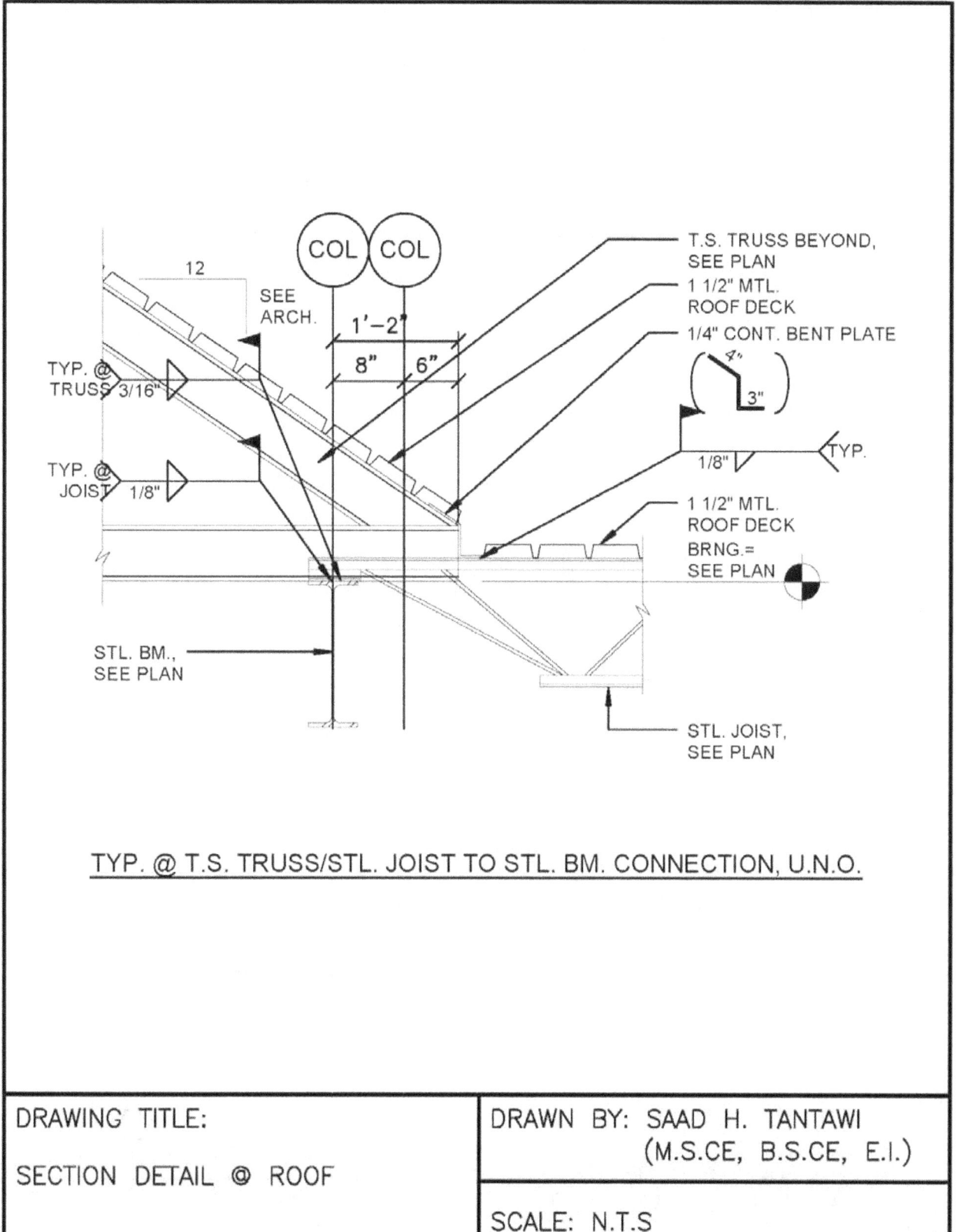

TYP. @ T.S. TRUSS/STL. JOIST TO STL. BM. CONNECTION, U.N.O.

DRAWING TITLE:	DRAWN BY: SAAD H. TANTAWI
	(M.S.CE, B.S.CE, E.I.)
SECTION DETAIL @ ROOF	
	SCALE: N.T.S

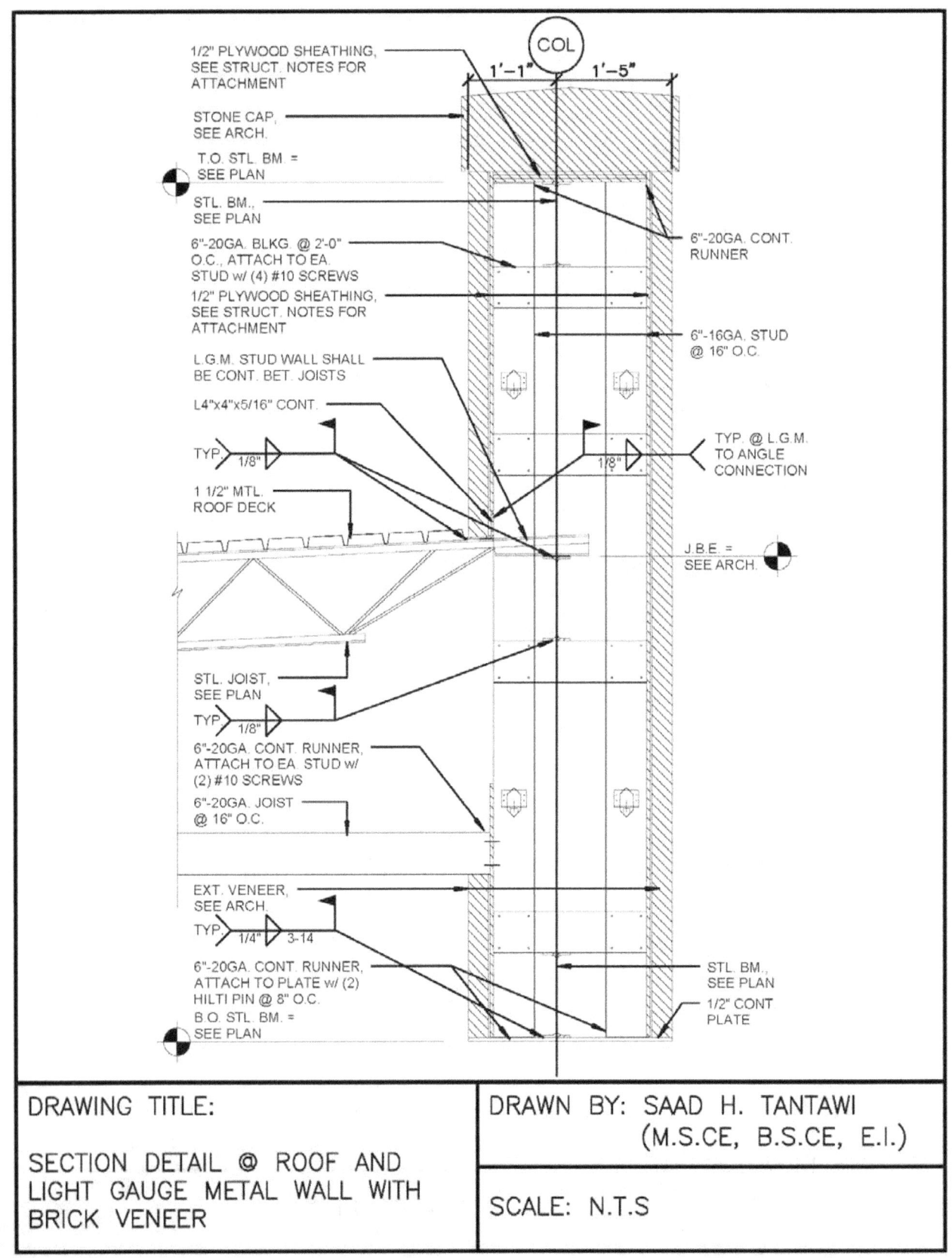

1/2" PLYWOOD SHEATHING, SEE STRUCT. NOTES FOR ATTACHMENT

STONE CAP, SEE ARCH.

T.O. STL. BM. = SEE PLAN

STL. BM., SEE PLAN

6"-20GA. BLKG. @ 2'-0" O.C., ATTACH TO EA. STUD w/ (4) #10 SCREWS

1/2" PLYWOOD SHEATHING, SEE STRUCT. NOTES FOR ATTACHMENT

L.G.M. STUD WALL SHALL BE CONT. BET. JOISTS

L4"x4"x5/16" CONT.

TYP. 1/8"

1 1/2" MTL. ROOF DECK

STL. JOIST, SEE PLAN

TYP. 1/8"

6"-20GA. CONT. RUNNER, ATTACH TO EA. STUD w/ (2) #10 SCREWS

6"-20GA. JOIST @ 16" O.C.

EXT. VENEER, SEE ARCH.

TYP. 1/4" 3-14

6"-20GA. CONT. RUNNER, ATTACH TO PLATE w/ (2) HILTI PIN @ 8" O.C.

B.O. STL. BM. = SEE PLAN

COL

1'-1"   1'-5"

6"-20GA. CONT. RUNNER

6"-16GA. STUD @ 16" O.C.

TYP. @ L.G.M. TO ANGLE CONNECTION

1/8"

J.B.E. = SEE ARCH.

STL. BM., SEE PLAN

1/2" CONT PLATE

DRAWING TITLE:	DRAWN BY: SAAD H. TANTAWI (M.S.CE, B.S.CE, E.I.)
SECTION DETAIL @ ROOF AND LIGHT GAUGE METAL WALL WITH BRICK VENEER	SCALE: N.T.S

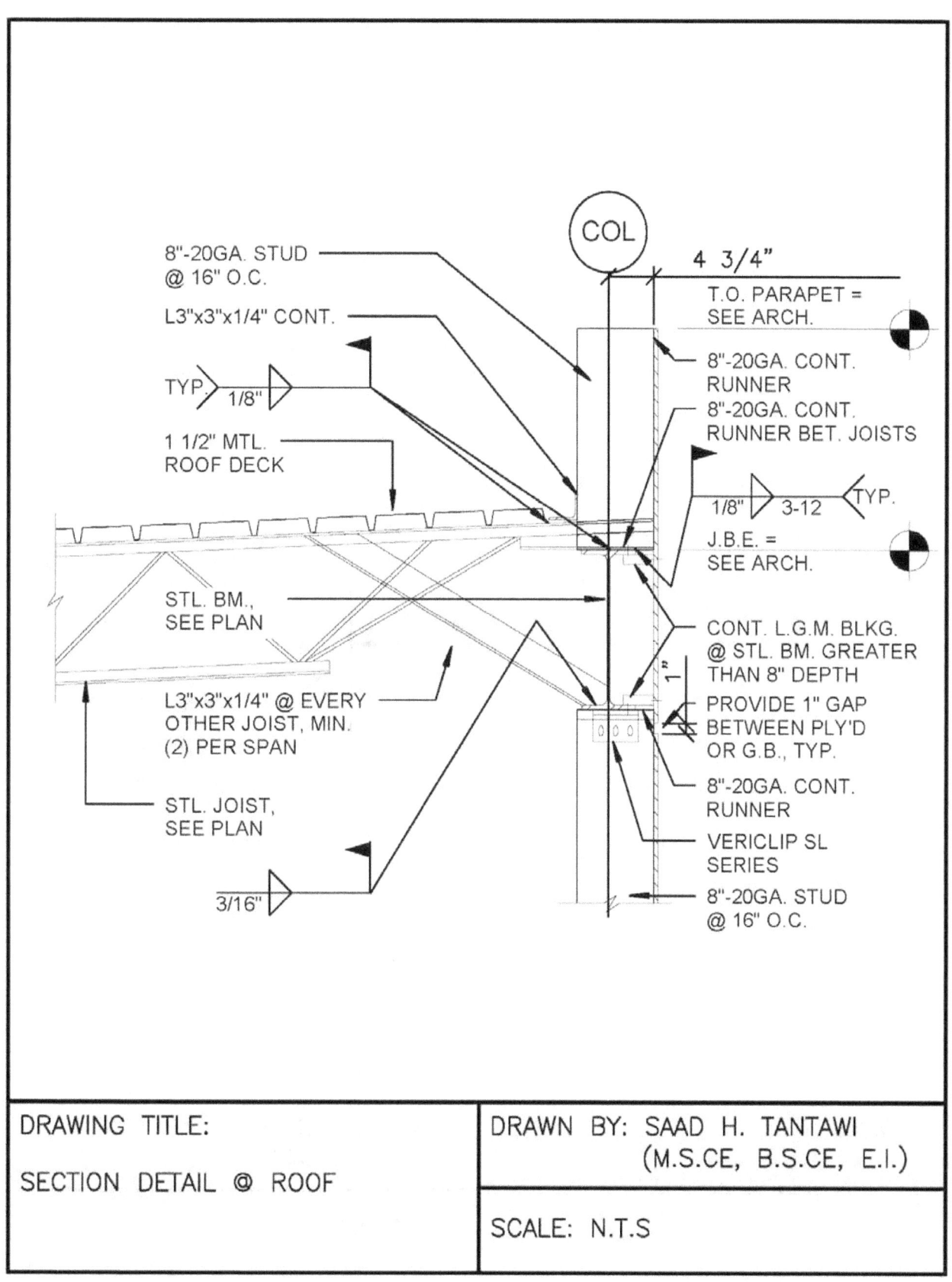

8"-20GA. STUD
@ 16" O.C.

L3"x3"x1/4" CONT.

TYP. 1/8"

1 1/2" MTL.
ROOF DECK

STL. BM.,
SEE PLAN

L3"x3"x1/4" @ EVERY
OTHER JOIST, MIN.
(2) PER SPAN

STL. JOIST,
SEE PLAN

3/16"

COL

4 3/4"

T.O. PARAPET =
SEE ARCH.

8"-20GA. CONT.
RUNNER

8"-20GA. CONT.
RUNNER BET. JOISTS

1/8" 3-12 TYP.

J.B.E. =
SEE ARCH.

CONT. L.G.M. BLKG.
@ STL. BM. GREATER
THAN 8" DEPTH

1"

PROVIDE 1" GAP
BETWEEN PLY'D
OR G.B., TYP.

8"-20GA. CONT.
RUNNER

VERICLIP SL
SERIES

8"-20GA. STUD
@ 16" O.C.

DRAWING TITLE:	DRAWN BY: SAAD H. TANTAWI
	(M.S.CE, B.S.CE, E.I.)
SECTION DETAIL @ ROOF	
	SCALE: N.T.S

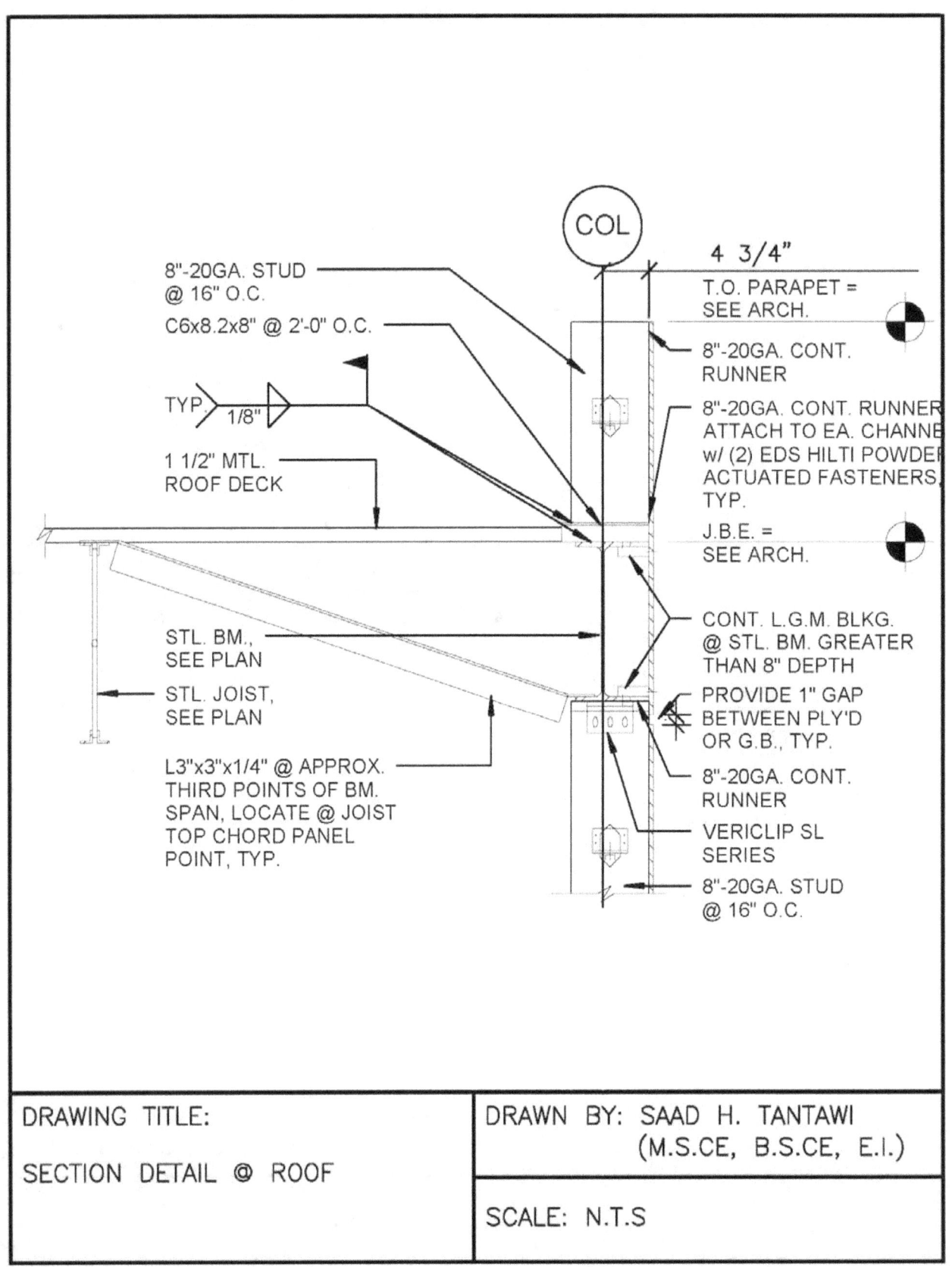

8"-20GA. STUD
@ 16" O.C.

C6x8.2x8" @ 2'-0" O.C.

TYP.
1/8"

1 1/2" MTL.
ROOF DECK

STL. BM.,
SEE PLAN

STL. JOIST,
SEE PLAN

L3"x3"x1/4" @ APPROX.
THIRD POINTS OF BM.
SPAN, LOCATE @ JOIST
TOP CHORD PANEL
POINT, TYP.

COL

4 3/4"

T.O. PARAPET =
SEE ARCH.

8"-20GA. CONT.
RUNNER

8"-20GA. CONT. RUNNER
ATTACH TO EA. CHANNE
w/ (2) EDS HILTI POWDER
ACTUATED FASTENERS,
TYP.

J.B.E. =
SEE ARCH.

CONT. L.G.M. BLKG.
@ STL. BM. GREATER
THAN 8" DEPTH

PROVIDE 1" GAP
BETWEEN PLY'D
OR G.B., TYP.

8"-20GA. CONT.
RUNNER

VERICLIP SL
SERIES

8"-20GA. STUD
@ 16" O.C.

DRAWING TITLE:	DRAWN BY: SAAD H. TANTAWI
	(M.S.CE, B.S.CE, E.I.)
SECTION DETAIL @ ROOF	
	SCALE: N.T.S

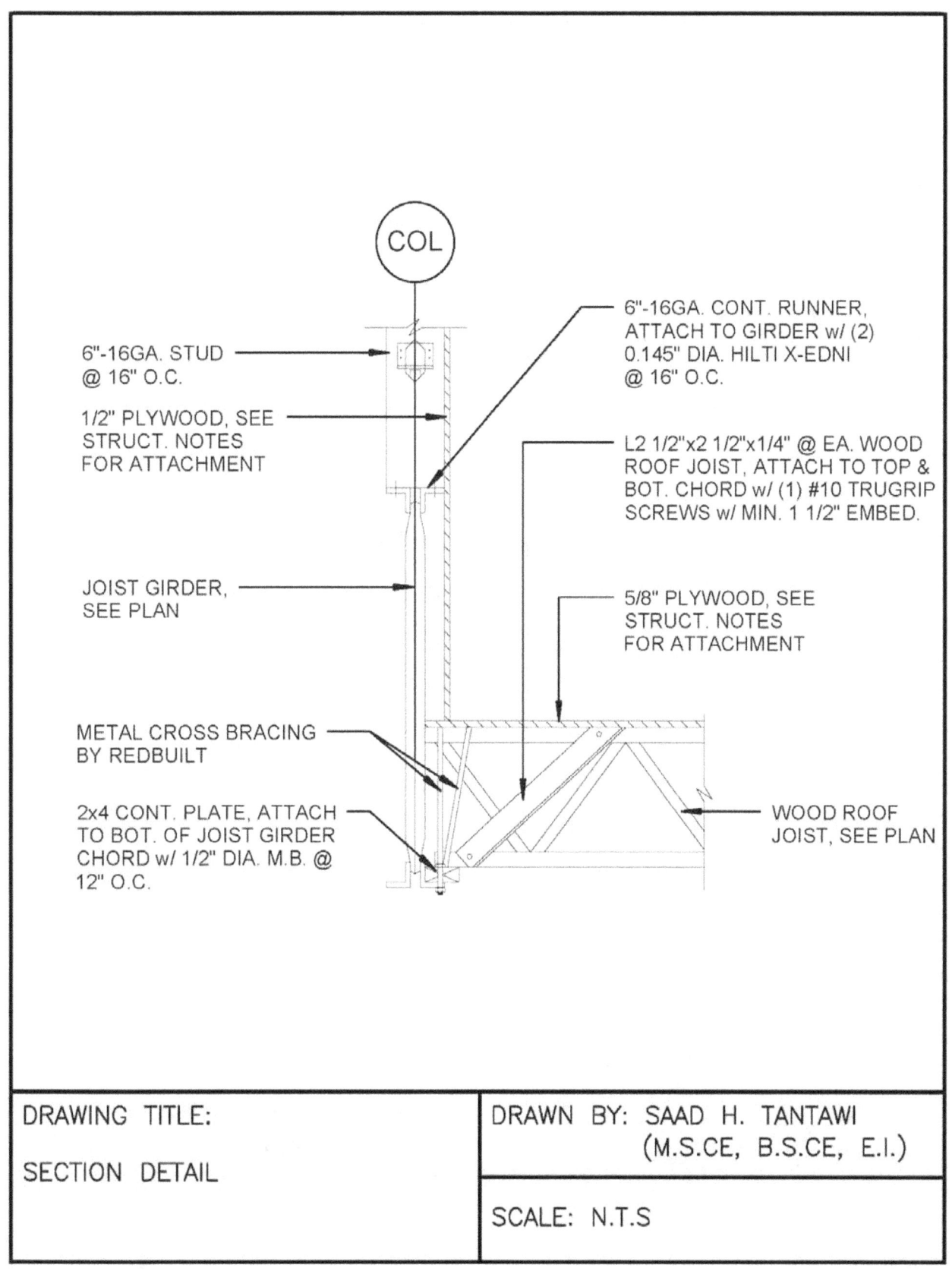

6"-16GA. STUD
@ 16" O.C.

1/2" PLYWOOD, SEE
STRUCT. NOTES
FOR ATTACHMENT

JOIST GIRDER,
SEE PLAN

METAL CROSS BRACING
BY REDBUILT

2x4 CONT. PLATE, ATTACH
TO BOT. OF JOIST GIRDER
CHORD w/ 1/2" DIA. M.B. @
12" O.C.

COL

6"-16GA. CONT. RUNNER,
ATTACH TO GIRDER w/ (2)
0.145" DIA. HILTI X-EDNI
@ 16" O.C.

L2 1/2"x2 1/2"x1/4" @ EA. WOOD
ROOF JOIST, ATTACH TO TOP &
BOT. CHORD w/ (1) #10 TRUGRIP
SCREWS w/ MIN. 1 1/2" EMBED.

5/8" PLYWOOD, SEE
STRUCT. NOTES
FOR ATTACHMENT

WOOD ROOF
JOIST, SEE PLAN

DRAWING TITLE: SECTION DETAIL	DRAWN BY: SAAD H. TANTAWI (M.S.CE, B.S.CE, E.I.)
	SCALE: N.T.S

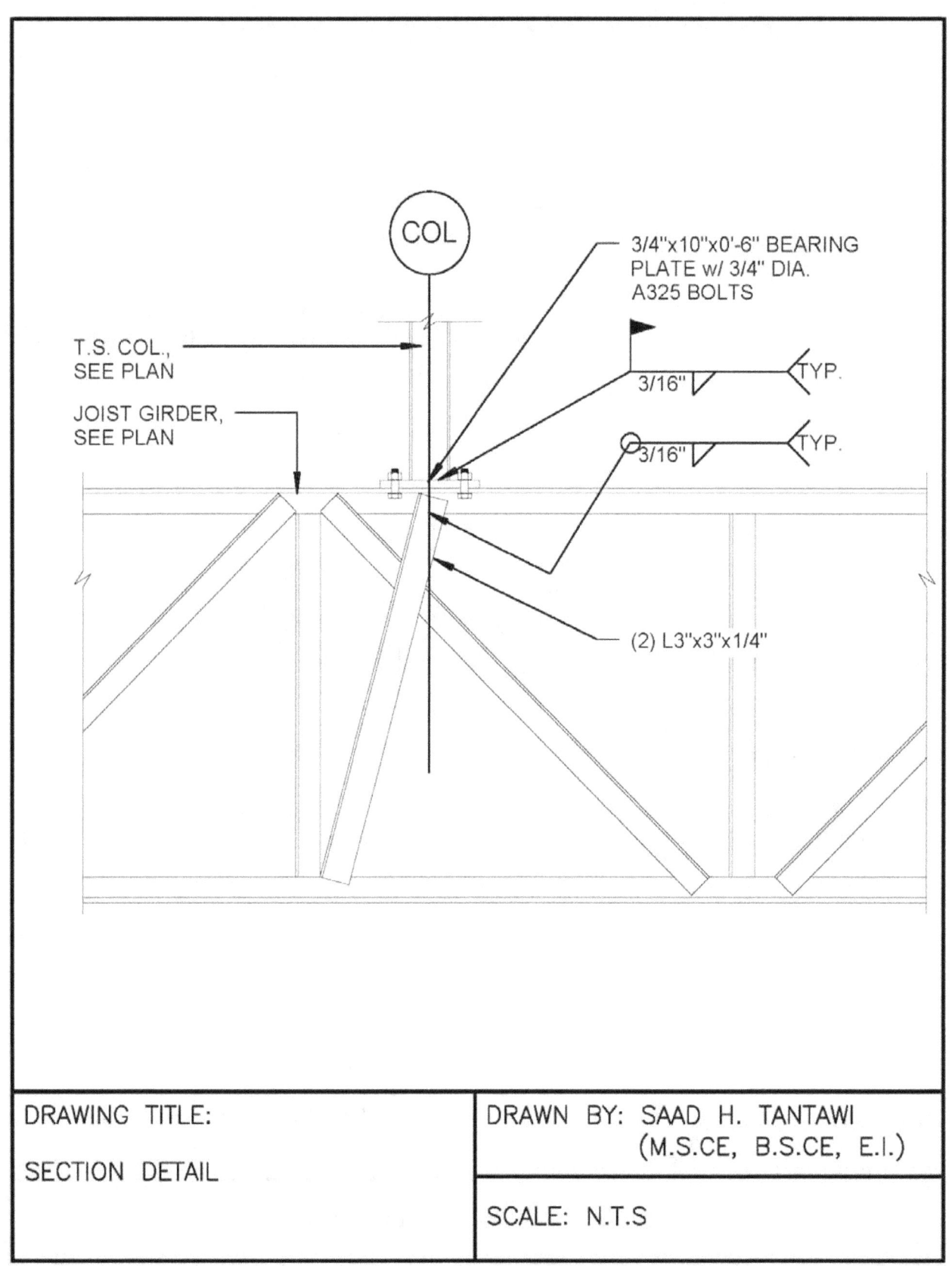

COL

3/4"x10"x0'-6" BEARING
PLATE w/ 3/4" DIA.
A325 BOLTS

3/16" TYP.

3/16" TYP.

T.S. COL.,
SEE PLAN

JOIST GIRDER,
SEE PLAN

(2) L3"x3"x1/4"

DRAWING TITLE:	DRAWN BY: SAAD H. TANTAWI
	(M.S.CE, B.S.CE, E.I.)
SECTION DETAIL	
	SCALE: N.T.S

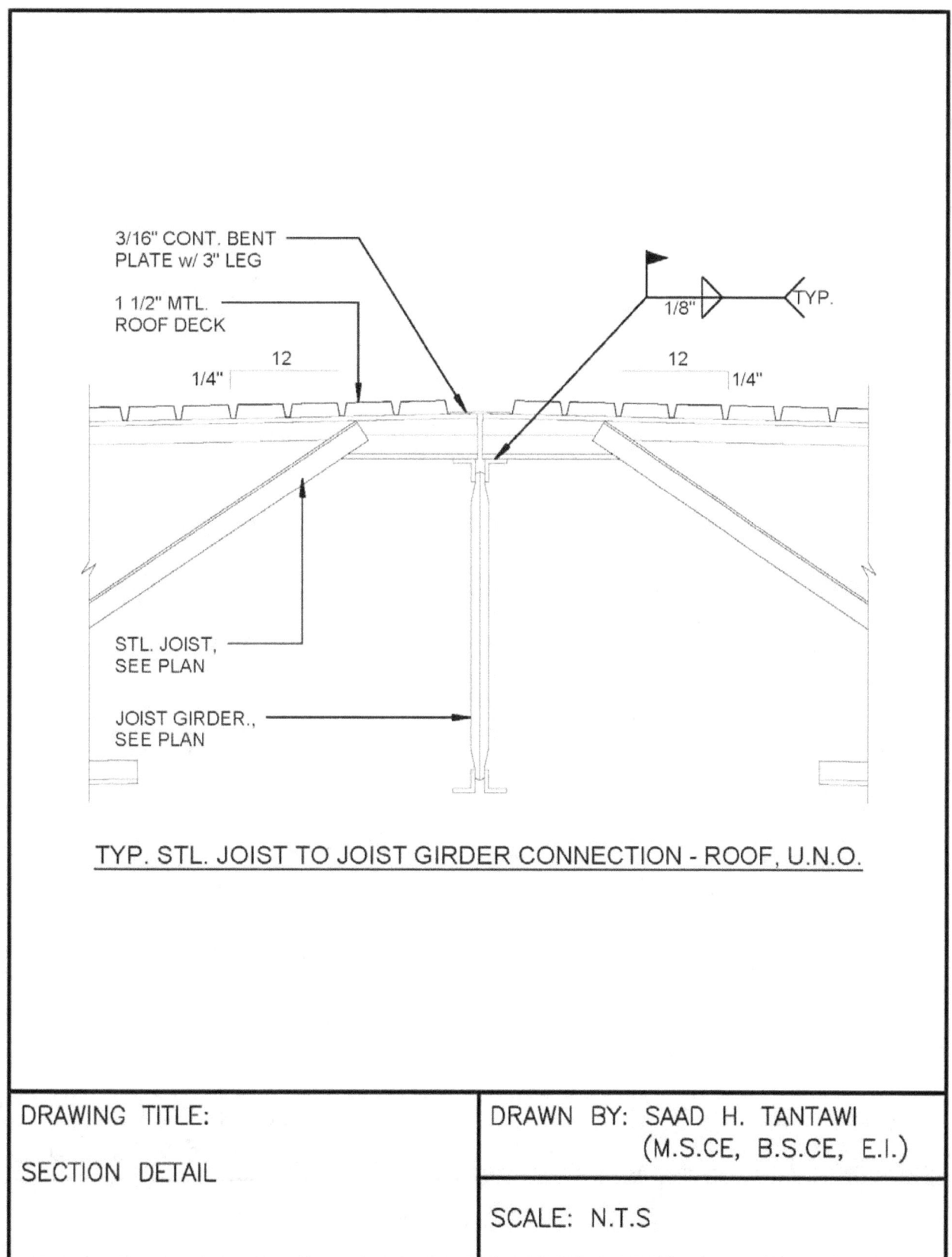

TYP. STL. JOIST TO JOIST GIRDER CONNECTION - ROOF, U.N.O.

DRAWING TITLE:	DRAWN BY: SAAD H. TANTAWI
	(M.S.CE, B.S.CE, E.I.)
SECTION DETAIL	
	SCALE: N.T.S

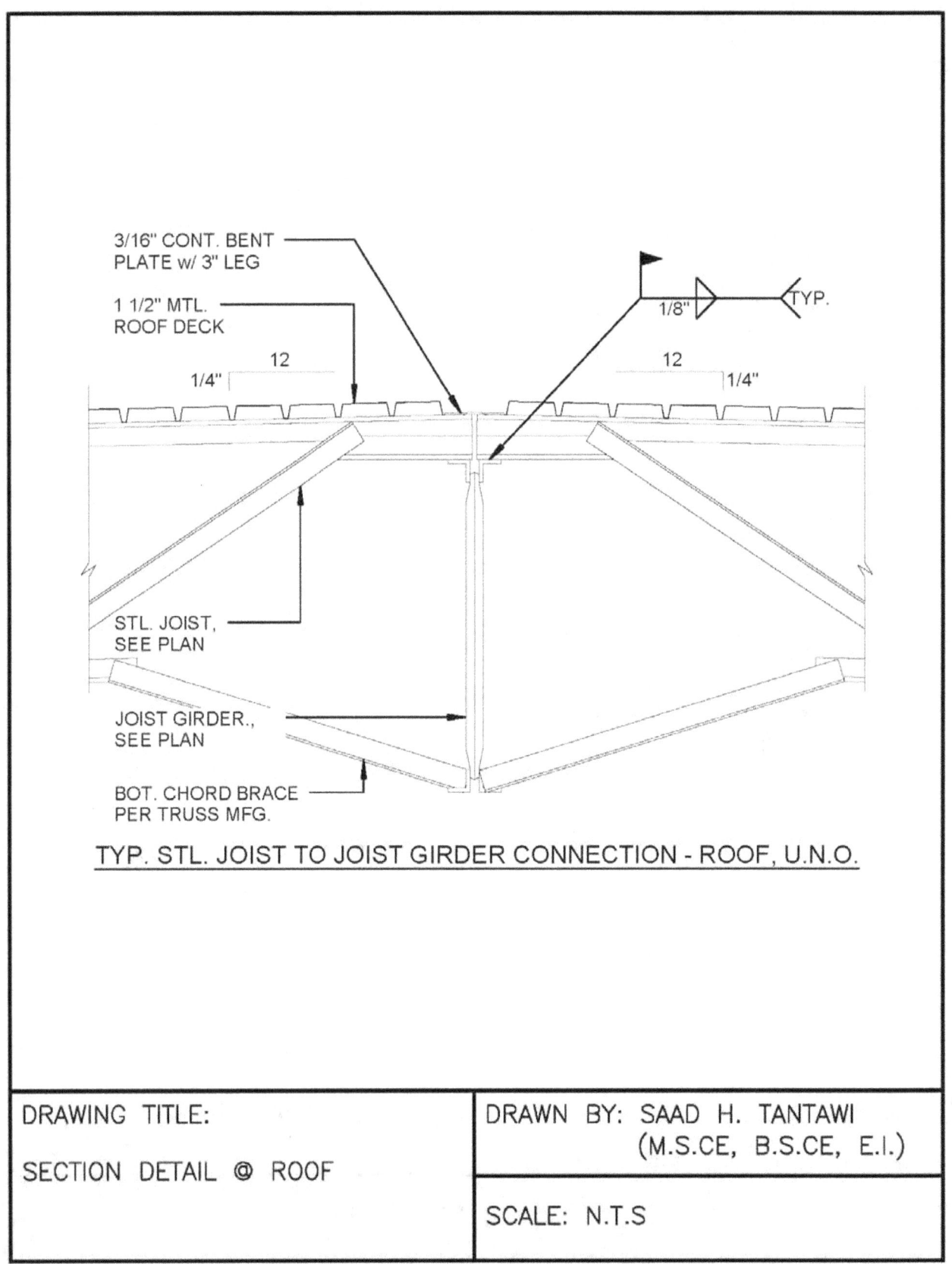

3/16" CONT. BENT
PLATE w/ 3" LEG

1 1/2" MTL.
ROOF DECK

12

1/4"

1/8"

TYP.

12

1/4"

STL. JOIST,
SEE PLAN

JOIST GIRDER.,
SEE PLAN

BOT. CHORD BRACE
PER TRUSS MFG.

TYP. STL. JOIST TO JOIST GIRDER CONNECTION - ROOF, U.N.O.

DRAWING TITLE: SECTION DETAIL @ ROOF	DRAWN BY: SAAD H. TANTAWI (M.S.CE, B.S.CE, E.I.)
	SCALE: N.T.S

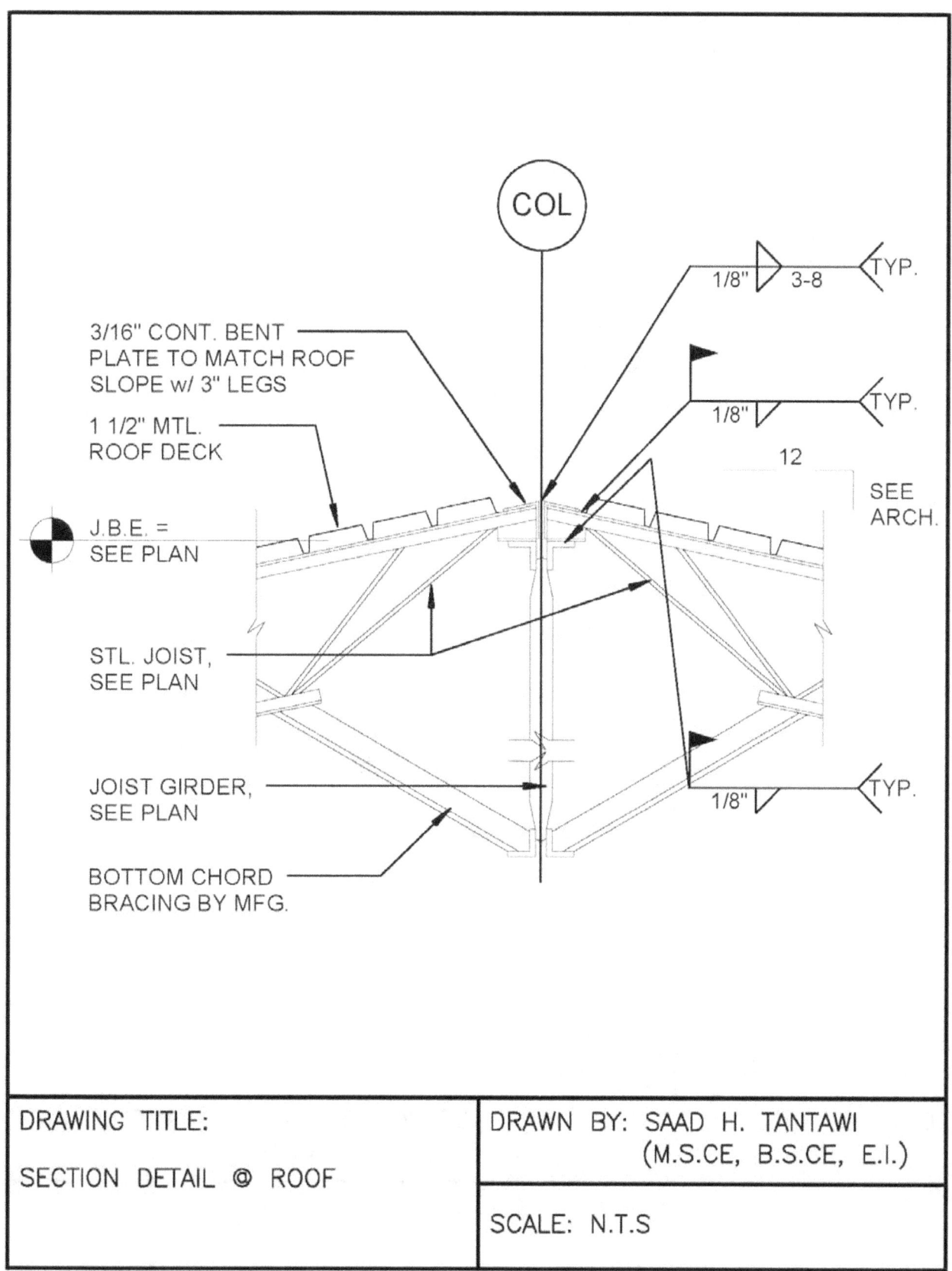

COL

3/16" CONT. BENT PLATE TO MATCH ROOF SLOPE w/ 3" LEGS

1 1/2" MTL. ROOF DECK

J.B.E. = SEE PLAN

STL. JOIST, SEE PLAN

JOIST GIRDER, SEE PLAN

BOTTOM CHORD BRACING BY MFG.

1/8" 3-8 TYP.

1/8" TYP.

12

SEE ARCH.

1/8" TYP.

DRAWING TITLE:	DRAWN BY: SAAD H. TANTAWI
	(M.S.CE, B.S.CE, E.I.)
SECTION DETAIL @ ROOF	
	SCALE: N.T.S

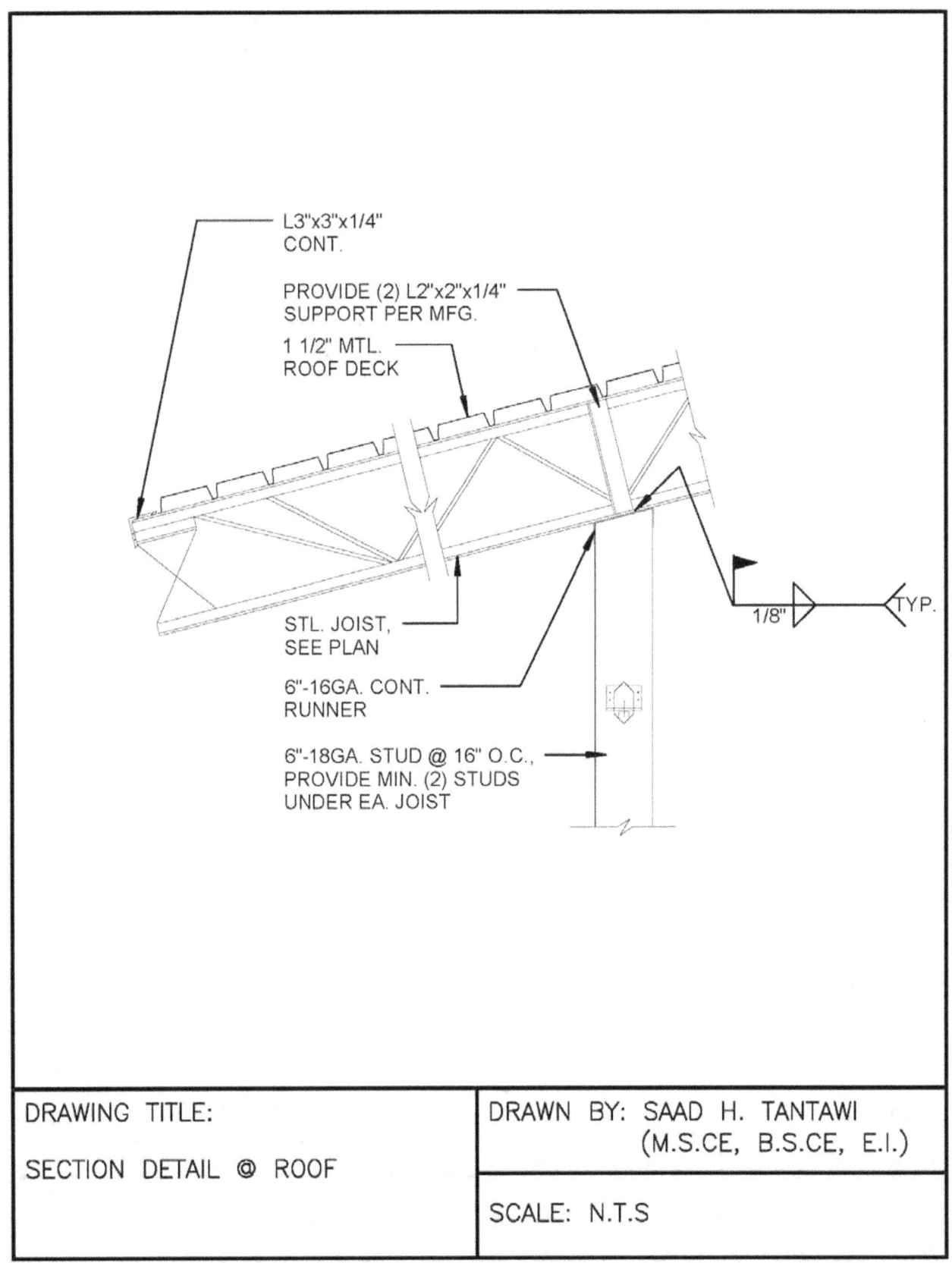

L3"x3"x1/4"
CONT.

PROVIDE (2) L2"x2"x1/4"
SUPPORT PER MFG.

1 1/2" MTL.
ROOF DECK

STL. JOIST,
SEE PLAN

6"-16GA. CONT.
RUNNER

6"-18GA. STUD @ 16" O.C.,
PROVIDE MIN. (2) STUDS
UNDER EA. JOIST

1/8"   TYP.

DRAWING TITLE:	DRAWN BY: SAAD H. TANTAWI
	(M.S.CE, B.S.CE, E.I.)
SECTION DETAIL @ ROOF	
	SCALE: N.T.S

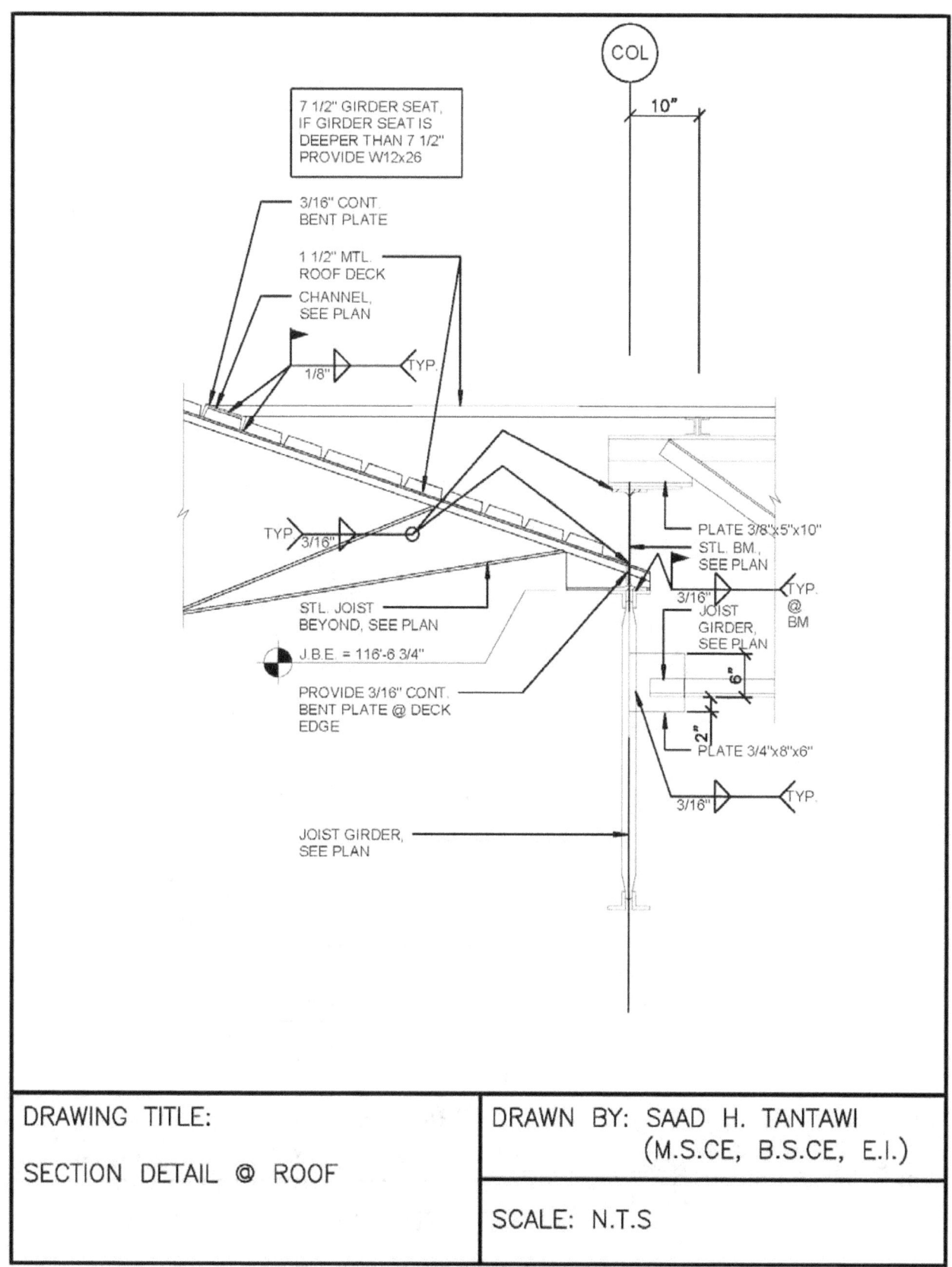

7 1/2" GIRDER SEAT,
IF GIRDER SEAT IS
DEEPER THAN 7 1/2"
PROVIDE W12x26

COL

10"

3/16" CONT.
BENT PLATE

1 1/2" MTL.
ROOF DECK

CHANNEL,
SEE PLAN

1/8"   TYP.

TYP. 3/16"

STL. JOIST
BEYOND, SEE PLAN

J.B.E. = 116'-6 3/4"

PROVIDE 3/16" CONT.
BENT PLATE @ DECK
EDGE

PLATE 3/8"x5"x10"

STL. BM.,
SEE PLAN

3/16"   TYP.
           @
JOIST   BM
GIRDER,
SEE PLAN

6"

2"

PLATE 3/4"x8"x6"

3/16"   TYP.

JOIST GIRDER,
SEE PLAN

DRAWING TITLE:	DRAWN BY: SAAD H. TANTAWI
	(M.S.CE, B.S.CE, E.I.)
SECTION DETAIL @ ROOF	
	SCALE: N.T.S

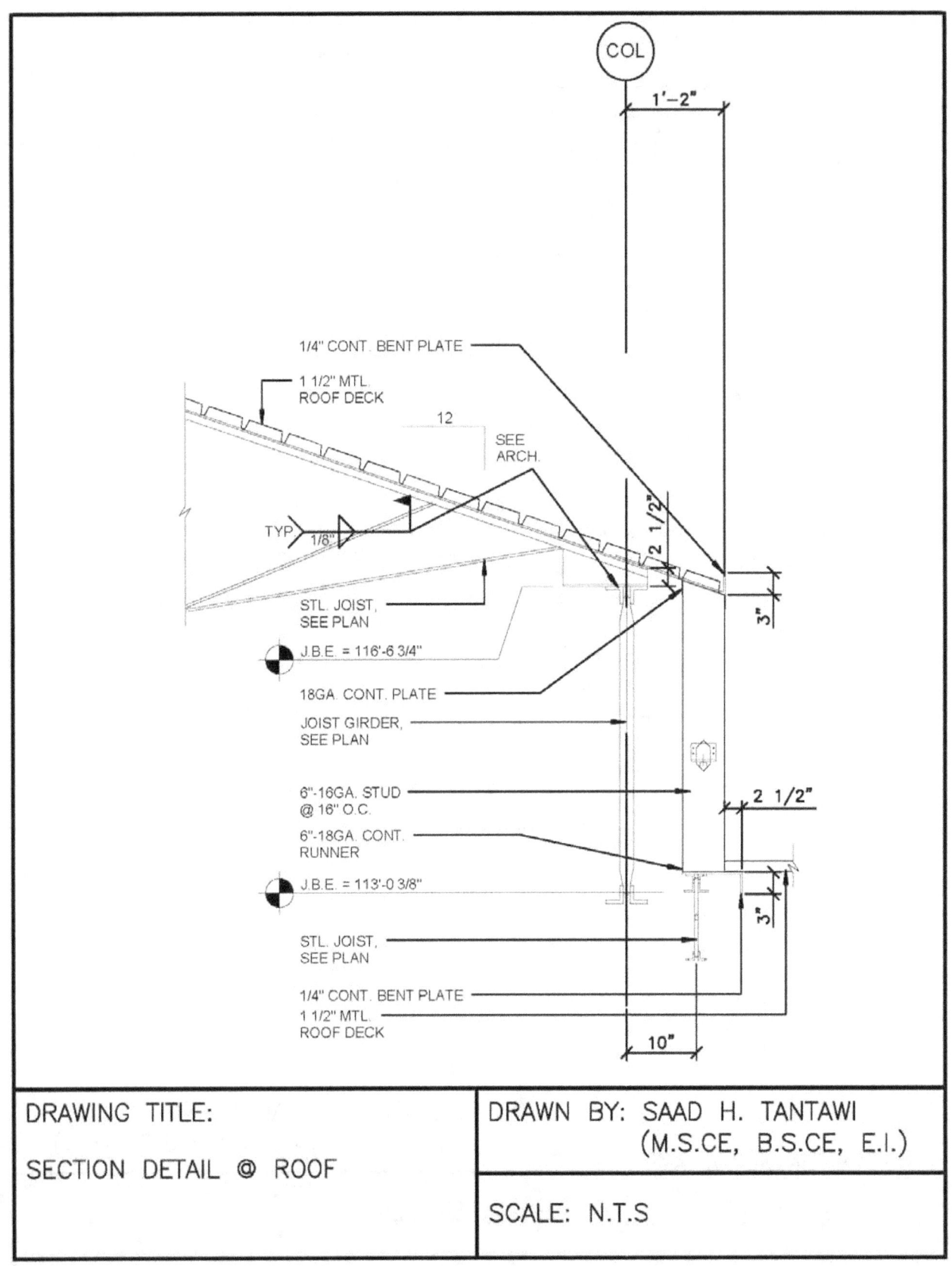

DRAWING TITLE: SECTION DETAIL @ ROOF	DRAWN BY: SAAD H. TANTAWI (M.S.CE, B.S.CE, E.I.)
	SCALE: N.T.S

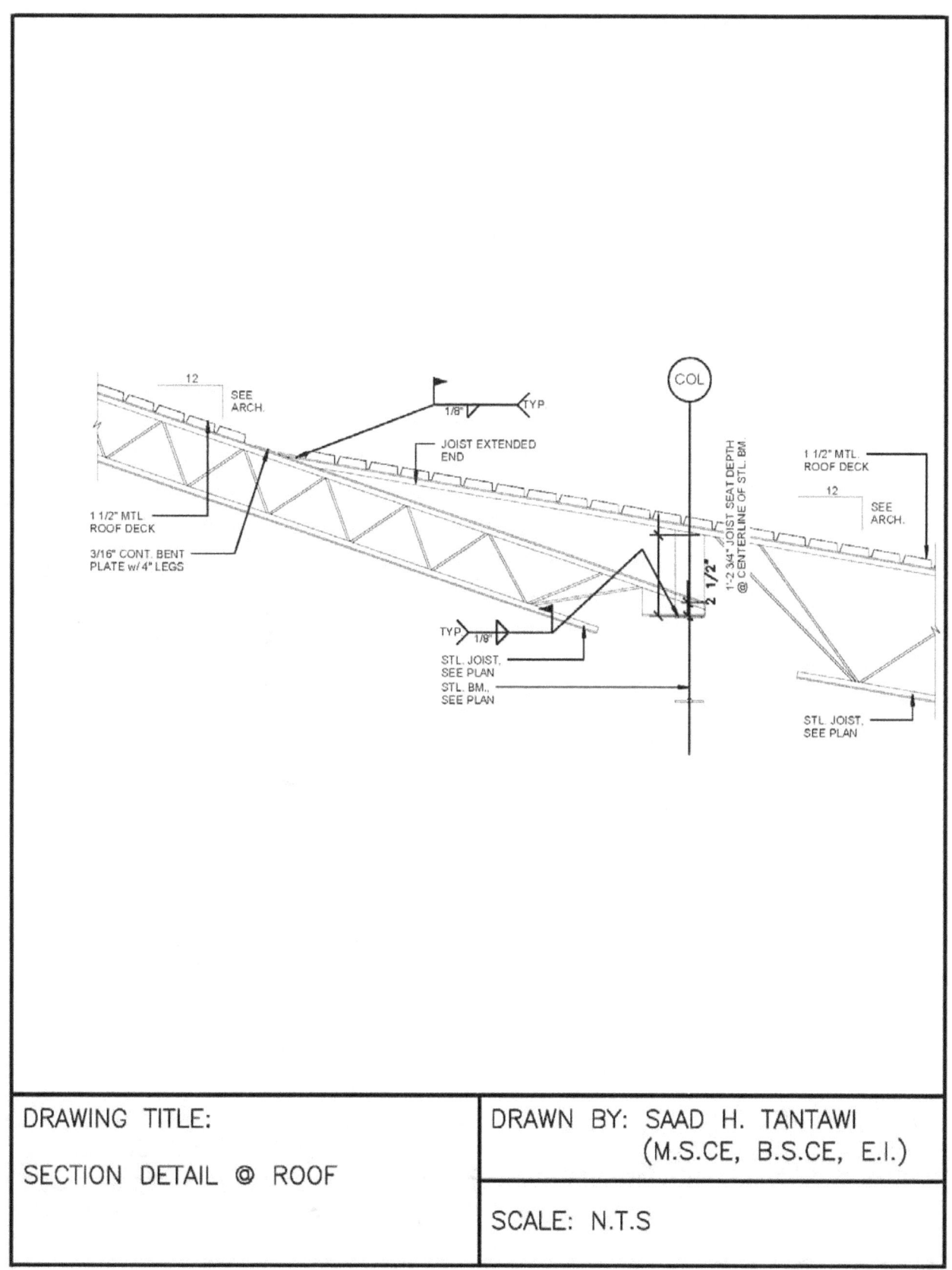

DRAWING TITLE:

SECTION DETAIL @ ROOF

DRAWN BY: SAAD H. TANTAWI
(M.S.CE, B.S.CE, E.I.)

SCALE: N.T.S

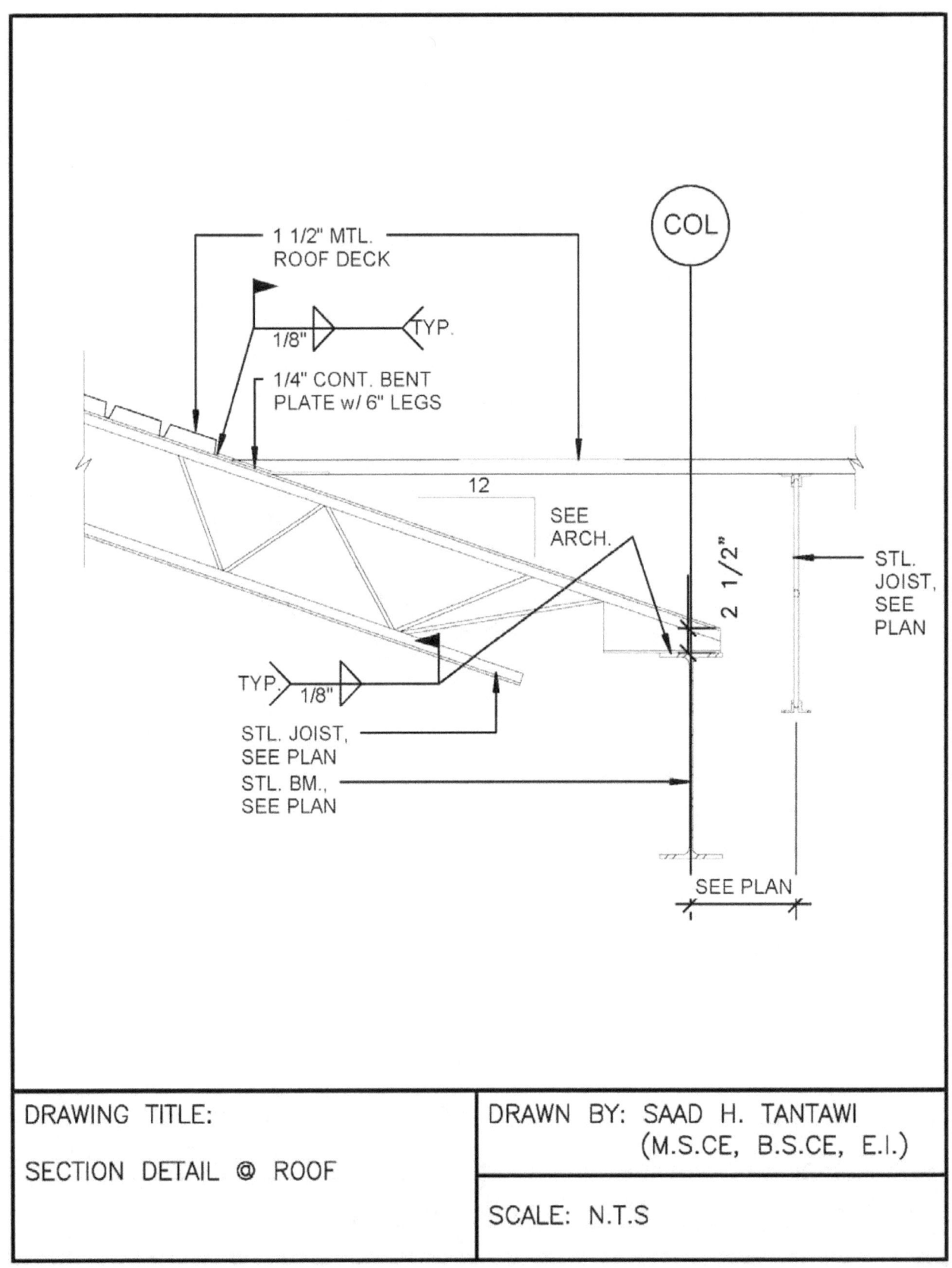

DRAWING TITLE:	DRAWN BY: SAAD H. TANTAWI
	(M.S.CE, B.S.CE, E.I.)
SECTION DETAIL @ ROOF	SCALE: N.T.S

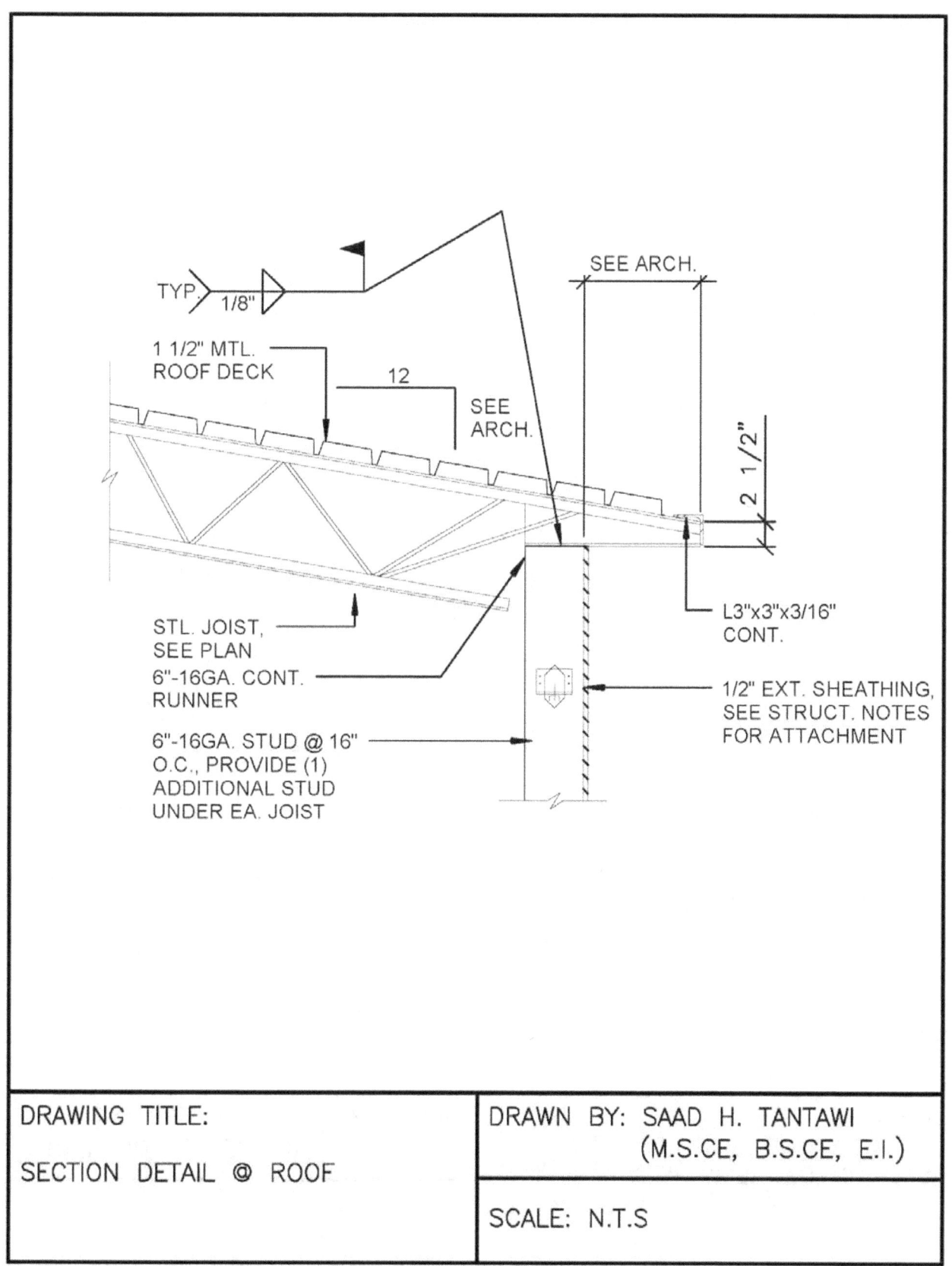

TYP. 1/8"

1 1/2" MTL.
ROOF DECK

12

SEE
ARCH.

SEE ARCH.

2 1/2"

STL. JOIST,
SEE PLAN

6"-16GA. CONT.
RUNNER

6"-16GA. STUD @ 16"
O.C., PROVIDE (1)
ADDITIONAL STUD
UNDER EA. JOIST

L3"x3"x3/16"
CONT.

1/2" EXT. SHEATHING,
SEE STRUCT. NOTES
FOR ATTACHMENT

DRAWING TITLE:	DRAWN BY: SAAD H. TANTAWI
SECTION DETAIL @ ROOF	(M.S.CE, B.S.CE, E.I.)
	SCALE: N.T.S

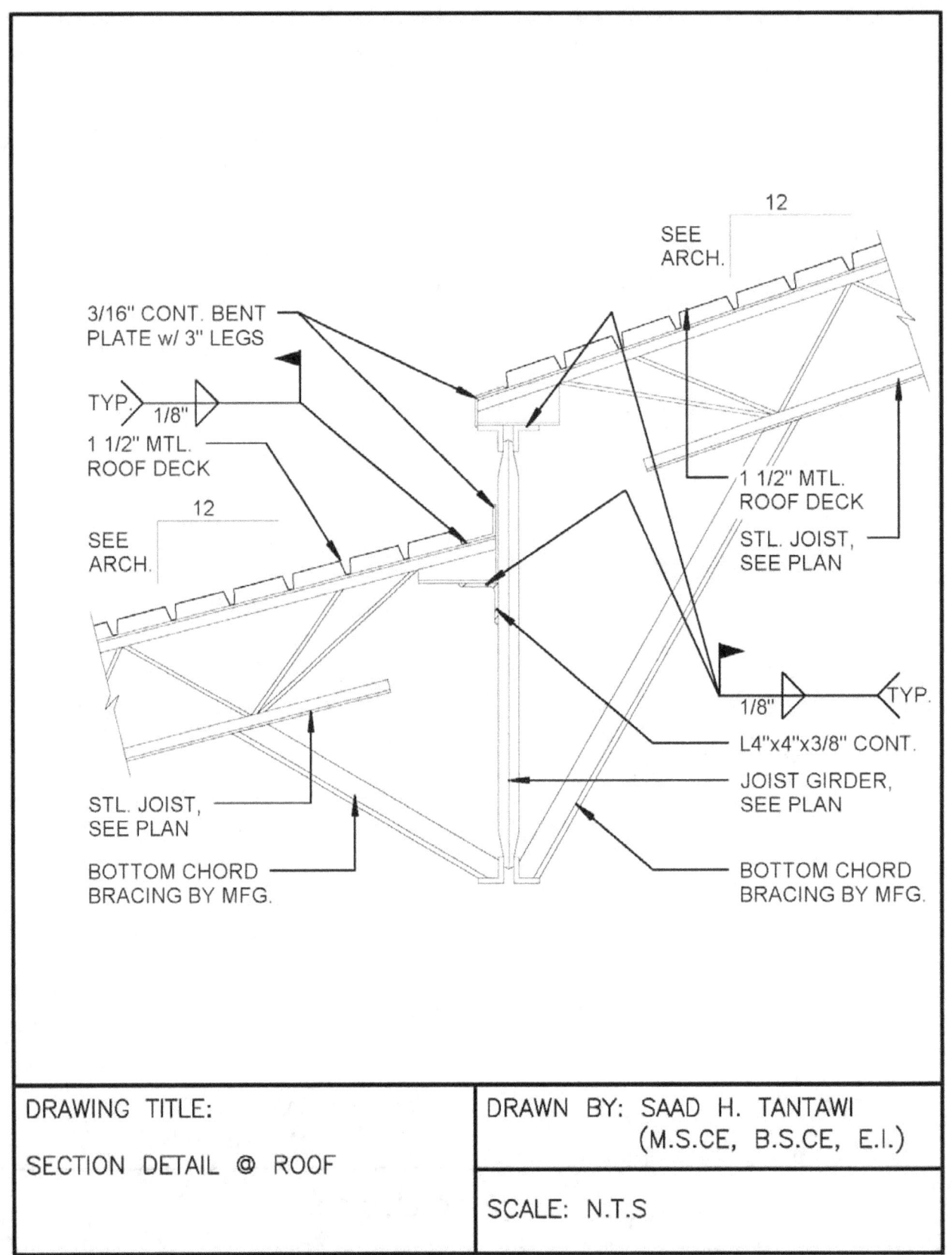

12

SEE
ARCH.

3/16" CONT. BENT
PLATE w/ 3" LEGS

TYP.
1/8"

1 1/2" MTL.
ROOF DECK

12

SEE
ARCH.

1 1/2" MTL.
ROOF DECK

STL. JOIST,
SEE PLAN

1/8"

TYP.

L4"x4"x3/8" CONT.

JOIST GIRDER,
SEE PLAN

STL. JOIST,
SEE PLAN

BOTTOM CHORD
BRACING BY MFG.

BOTTOM CHORD
BRACING BY MFG.

DRAWING TITLE:	DRAWN BY: SAAD H. TANTAWI
	(M.S.CE, B.S.CE, E.I.)
SECTION DETAIL @ ROOF	
	SCALE: N.T.S

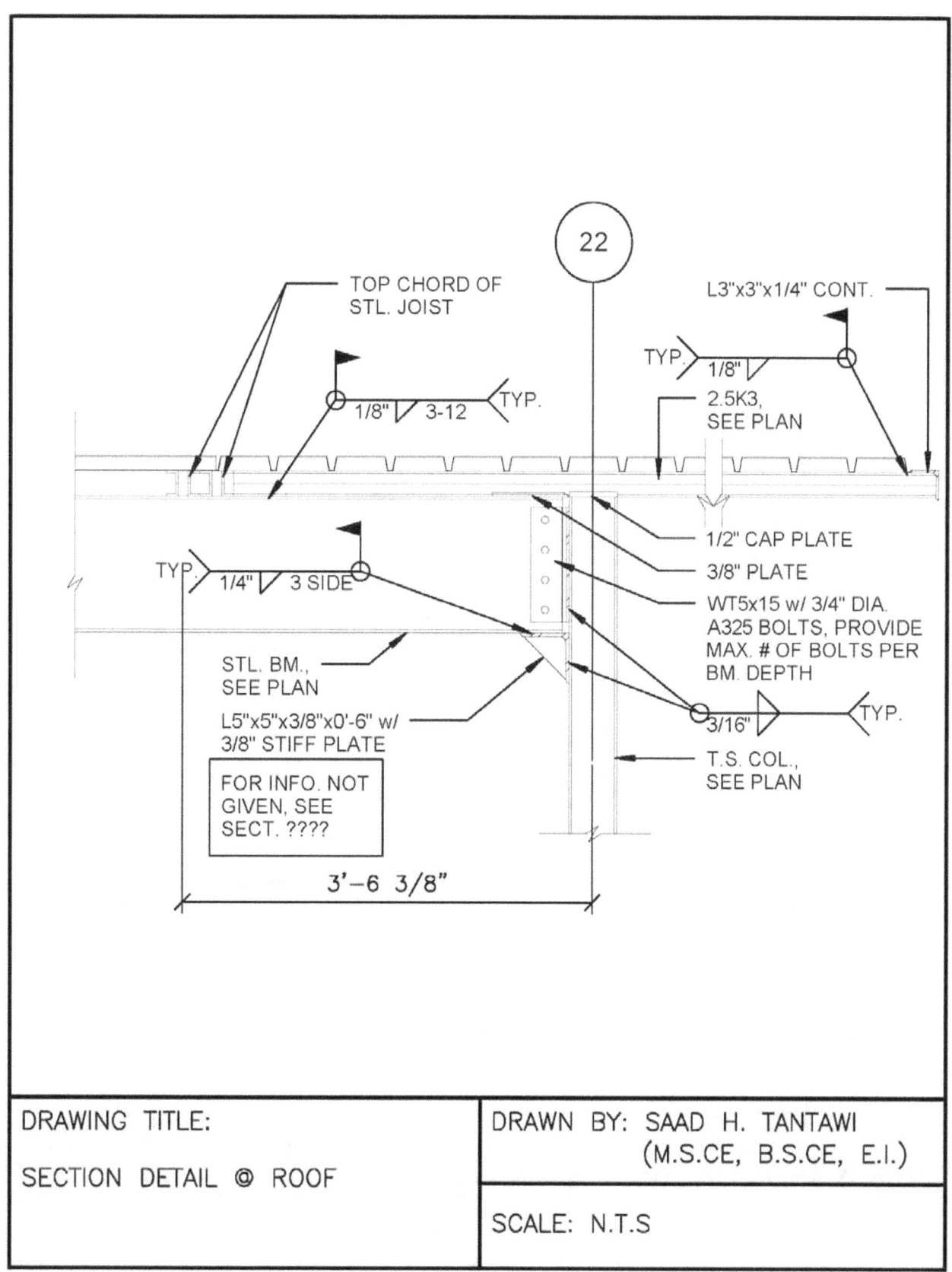

DRAWING TITLE:	DRAWN BY: SAAD H. TANTAWI
SECTION DETAIL @ ROOF	(M.S.CE, B.S.CE, E.I.)
	SCALE: N.T.S

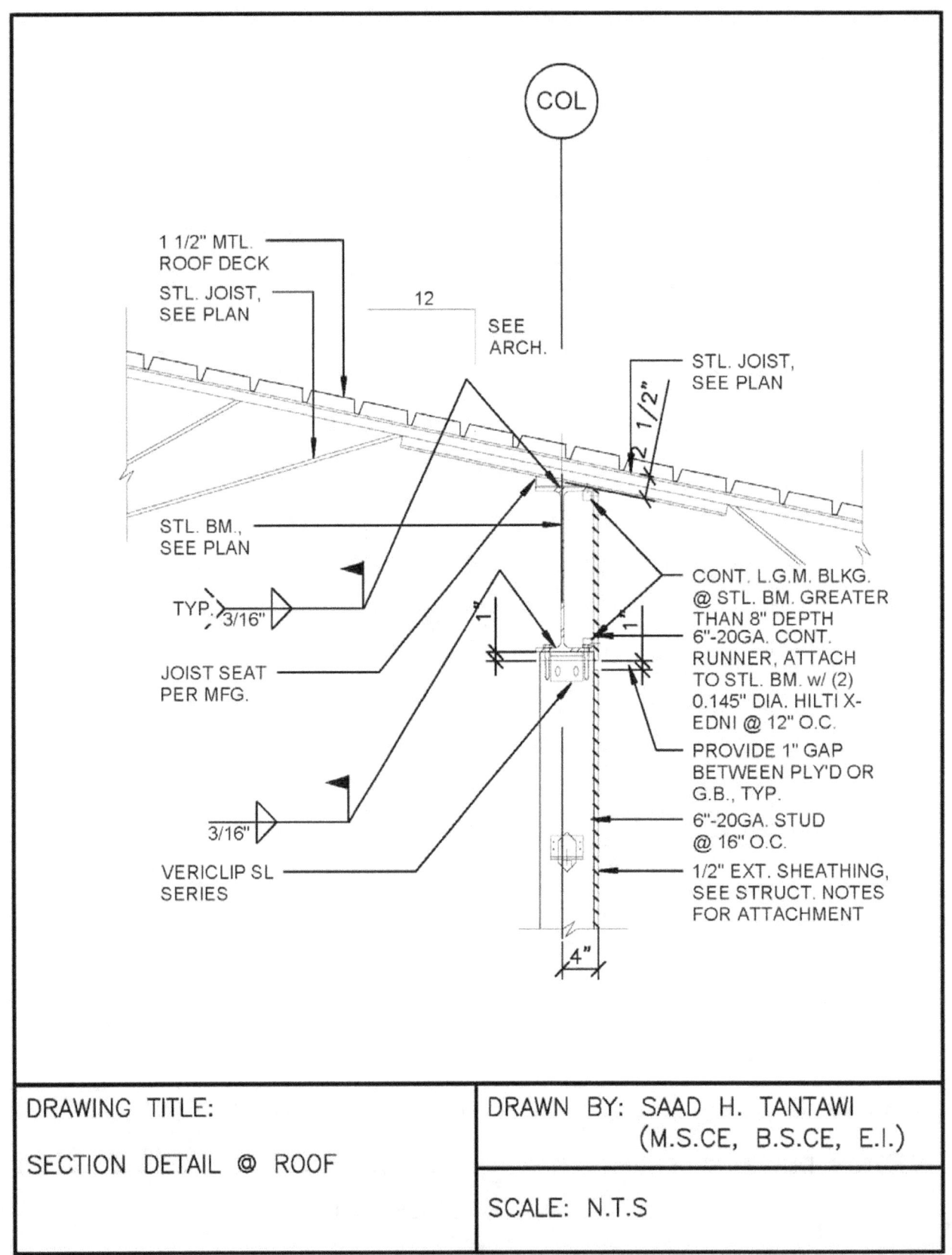

1 1/2" MTL.
ROOF DECK

STL. JOIST,
SEE PLAN

12

SEE
ARCH.

COL

STL. JOIST,
SEE PLAN

2 1/2"

STL. BM.,
SEE PLAN

TYP. 3/16"

JOIST SEAT
PER MFG.

1"

1"

CONT. L.G.M. BLKG.
@ STL. BM. GREATER
THAN 8" DEPTH
6"-20GA. CONT.
RUNNER, ATTACH
TO STL. BM. w/ (2)
0.145" DIA. HILTI X-
EDNI @ 12" O.C.

PROVIDE 1" GAP
BETWEEN PLY'D OR
G.B., TYP.

6"-20GA. STUD
@ 16" O.C.

1/2" EXT. SHEATHING,
SEE STRUCT. NOTES
FOR ATTACHMENT

3/16"

VERICLIP SL
SERIES

4"

DRAWING TITLE:	DRAWN BY: SAAD H. TANTAWI
	(M.S.CE, B.S.CE, E.I.)
SECTION DETAIL @ ROOF	SCALE: N.T.S

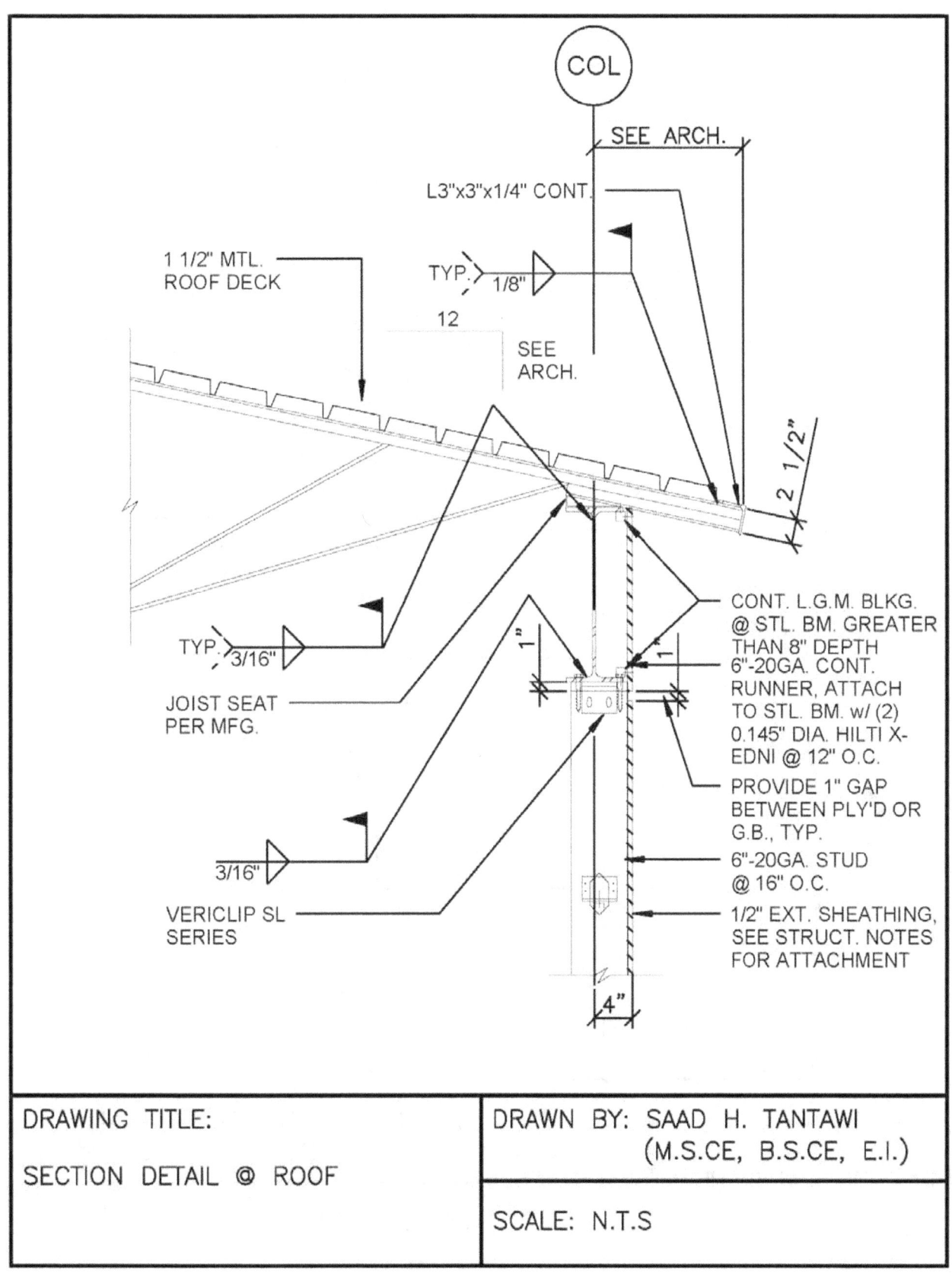

COL

SEE ARCH.

L3"x3"x1/4" CONT.

1 1/2" MTL.
ROOF DECK

TYP. 1/8"

12

SEE
ARCH.

2 1/2"

TYP. 3/16"

JOIST SEAT
PER MFG.

1"

1"

CONT. L.G.M. BLKG.
@ STL. BM. GREATER
THAN 8" DEPTH
6"-20GA. CONT.
RUNNER, ATTACH
TO STL. BM. w/ (2)
0.145" DIA. HILTI X-
EDNI @ 12" O.C.

PROVIDE 1" GAP
BETWEEN PLY'D OR
G.B., TYP.

6"-20GA. STUD
@ 16" O.C.

3/16"

VERICLIP SL
SERIES

1/2" EXT. SHEATHING,
SEE STRUCT. NOTES
FOR ATTACHMENT

4"

DRAWING TITLE:	DRAWN BY: SAAD H. TANTAWI
SECTION DETAIL @ ROOF	(M.S.CE, B.S.CE, E.I.)
	SCALE: N.T.S

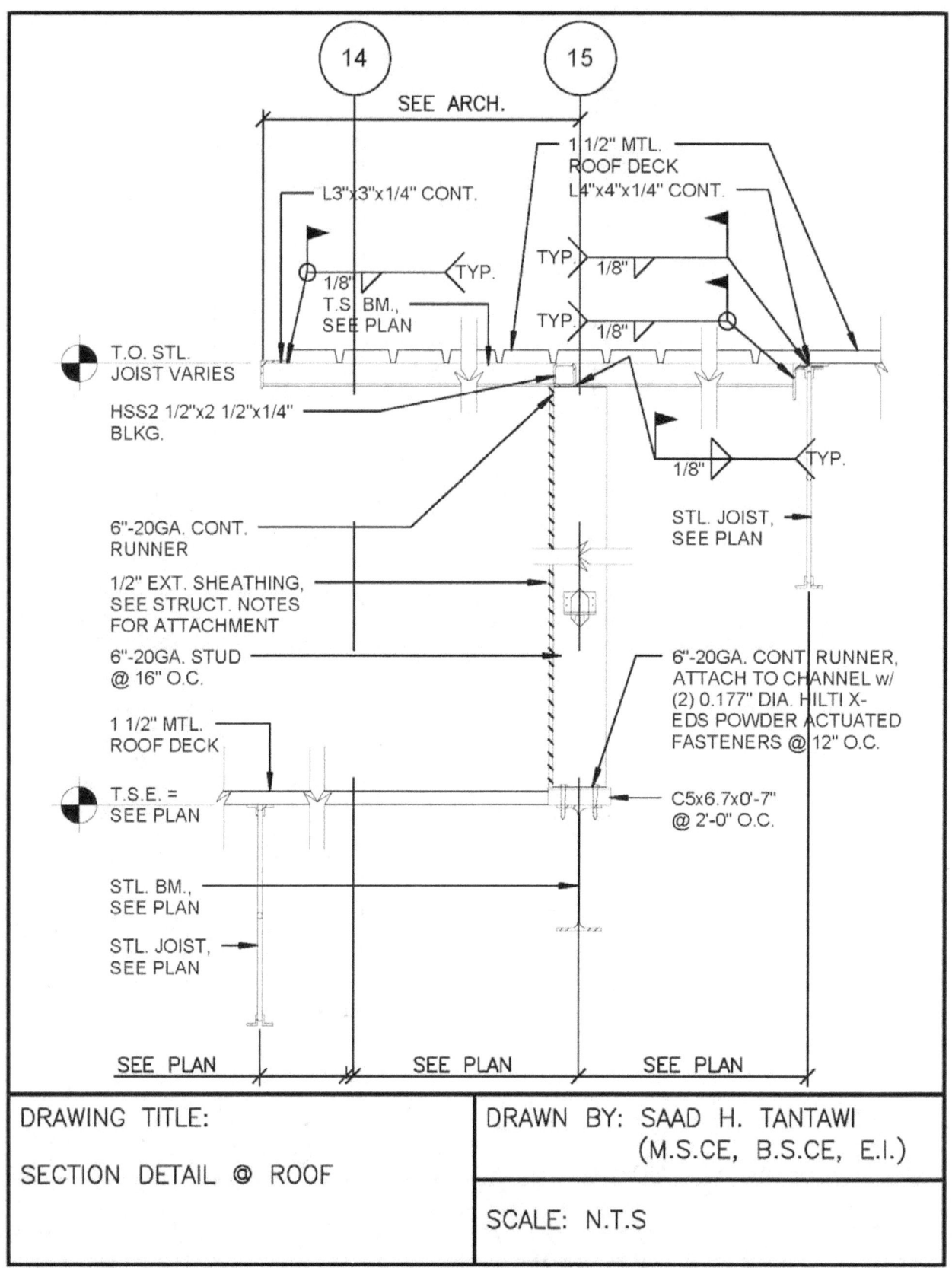

DRAWING TITLE:

SECTION DETAIL @ ROOF

DRAWN BY: SAAD H. TANTAWI
(M.S.CE, B.S.CE, E.I.)

SCALE: N.T.S

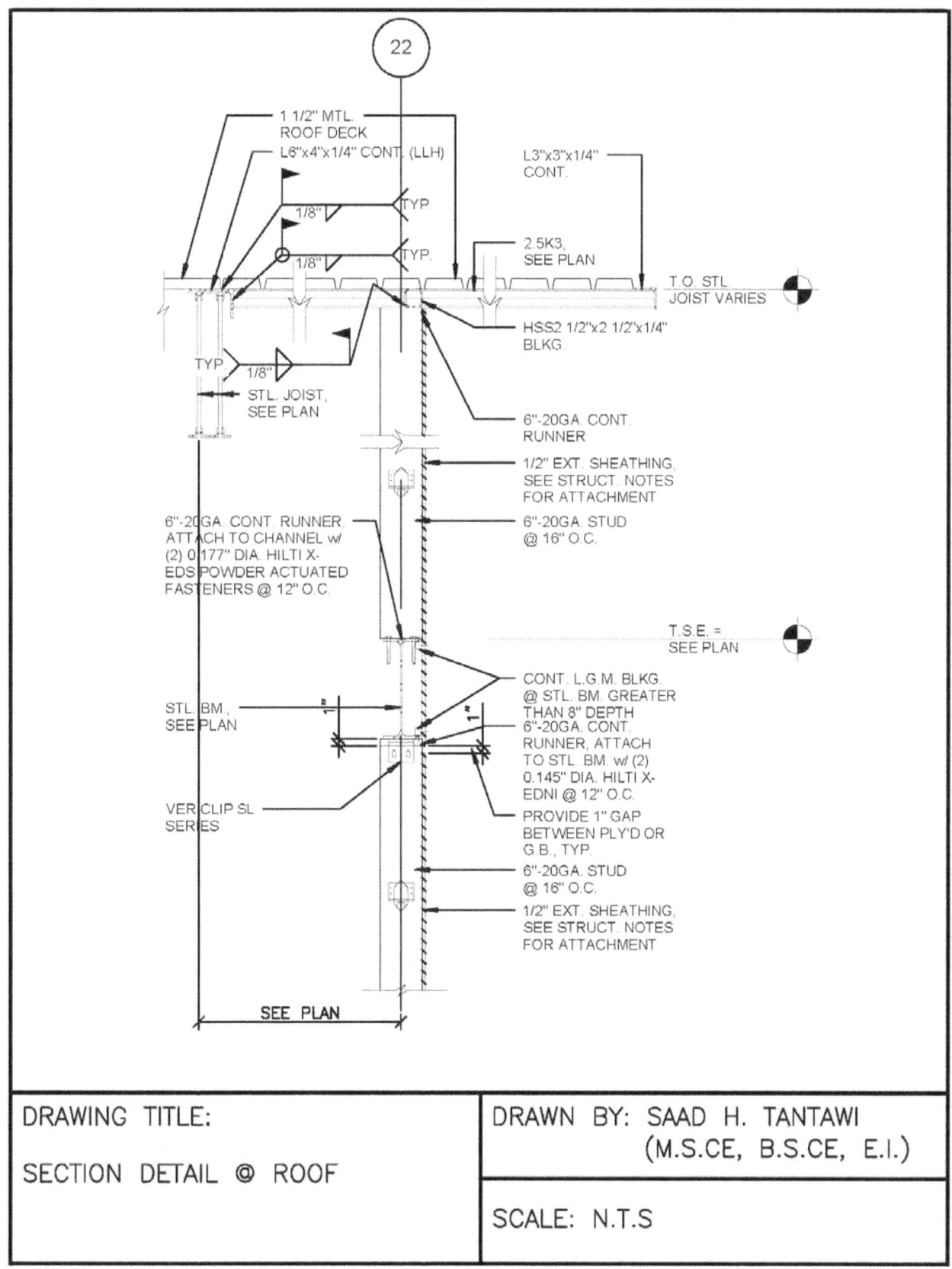

22

1 1/2" MTL.
ROOF DECK

L6"x4"x1/4" CONT. (LLH)

L3"x3"x1/4"
CONT.

1/8"   TYP.

1/8"   TYP.

2.5K3,
SEE PLAN

T.O. STL.
JOIST VARIES

TYP.   1/8"

HSS2 1/2"x2 1/2"x1/4"
BLKG.

STL. JOIST,
SEE PLAN

6"-20GA. CONT.
RUNNER

1/2" EXT. SHEATHING,
SEE STRUCT. NOTES
FOR ATTACHMENT

6"-20GA. CONT. RUNNER,
ATTACH TO CHANNEL w/
(2) 0.177" DIA. HILTI X-
EDS POWDER ACTUATED
FASTENERS @ 12" O.C.

6"-20GA. STUD
@ 16" O.C.

T.S.E. =
SEE PLAN

STL. BM.,
SEE PLAN

CONT. L.G.M. BLKG.
@ STL. BM. GREATER
THAN 8" DEPTH
6"-20GA. CONT.
RUNNER, ATTACH
TO STL. BM. w/ (2)
0.145" DIA. HILTI X-
EDNI @ 12" O.C.

VER CLIP SL
SERIES

PROVIDE 1" GAP
BETWEEN PLY'D OR
G.B., TYP.

6"-20GA. STUD
@ 16" O.C.

1/2" EXT. SHEATHING,
SEE STRUCT. NOTES
FOR ATTACHMENT

SEE PLAN

DRAWING TITLE:	DRAWN BY: SAAD H. TANTAWI
	(M.S.CE, B.S.CE, E.I.)
SECTION DETAIL @ ROOF	
	SCALE: N.T.S

169

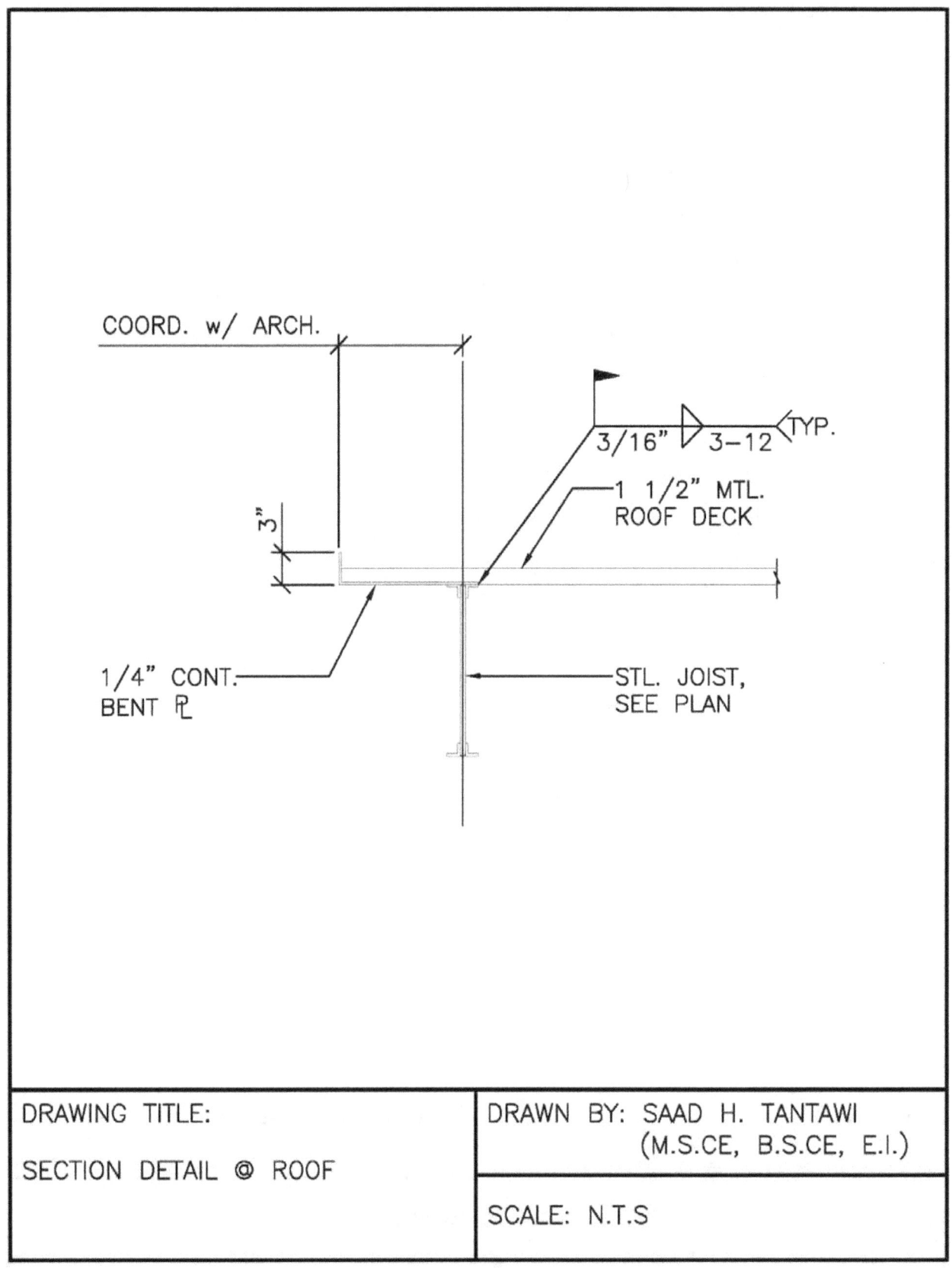

COORD. w/ ARCH.

3/16" 3-12 TYP.

1 1/2" MTL.
ROOF DECK

3"

1/4" CONT.
BENT ℙℒ

STL. JOIST,
SEE PLAN

DRAWING TITLE:	DRAWN BY: SAAD H. TANTAWI
	(M.S.CE, B.S.CE, E.I.)
SECTION DETAIL @ ROOF	
	SCALE: N.T.S

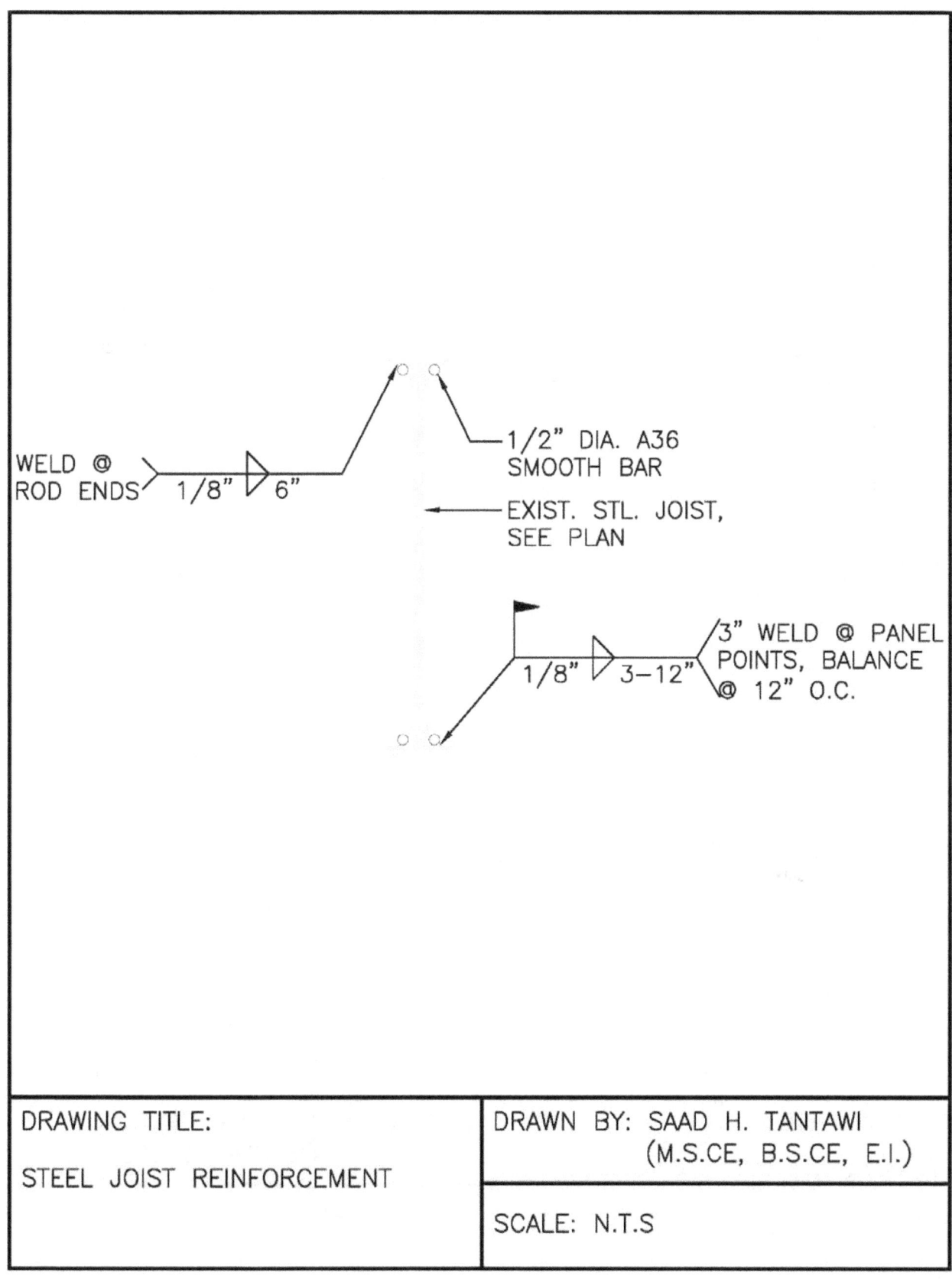

DRAWING TITLE:	DRAWN BY: SAAD H. TANTAWI
STEEL JOIST REINFORCEMENT	(M.S.CE, B.S.CE, E.I.)
	SCALE: N.T.S

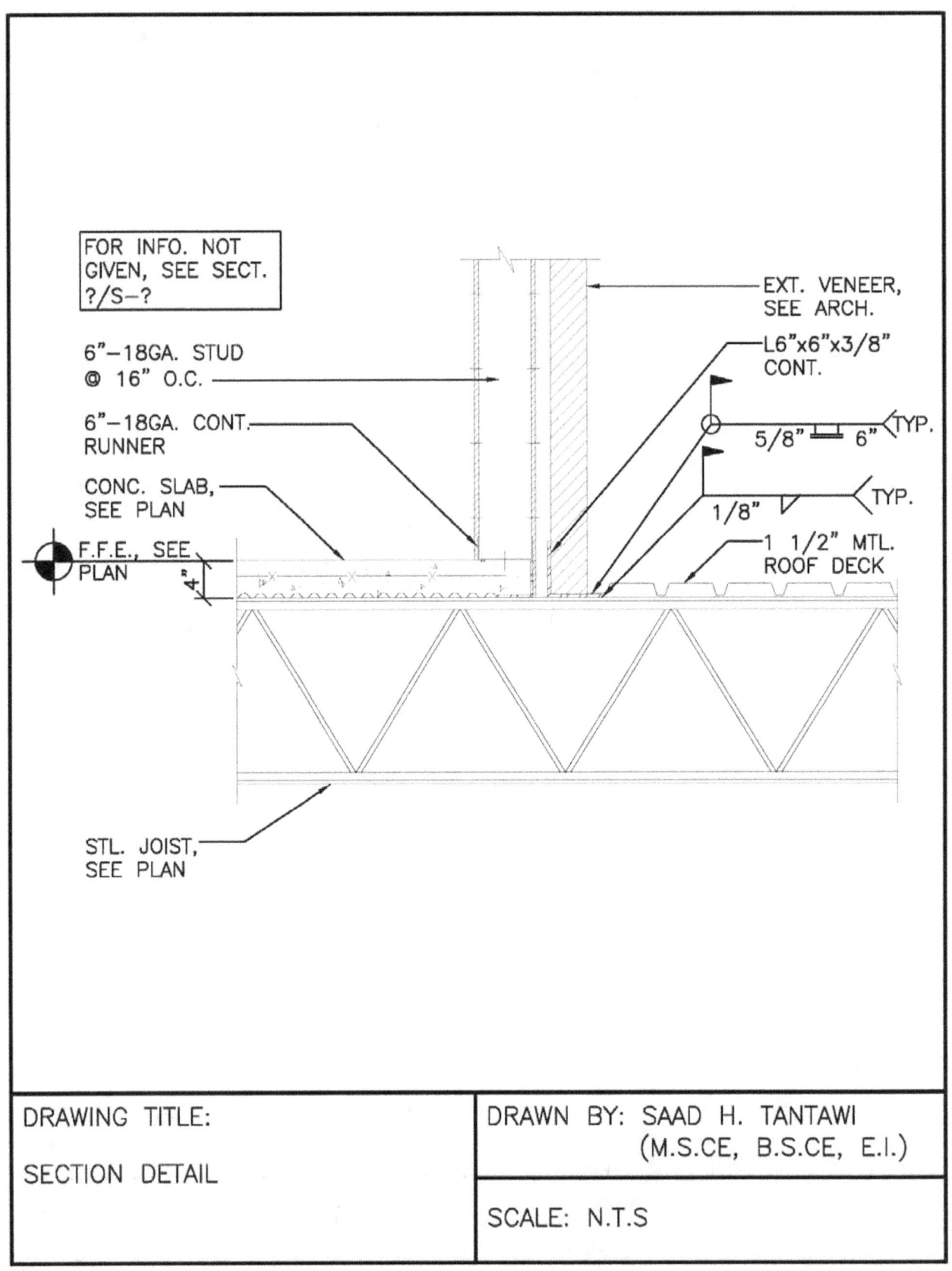

FOR INFO. NOT GIVEN, SEE SECT. ?/S-?

6"-18GA. STUD @ 16" O.C.

6"-18GA. CONT. RUNNER

CONC. SLAB, SEE PLAN

F.F.E., SEE PLAN

4"

STL. JOIST, SEE PLAN

EXT. VENEER, SEE ARCH.

L6"x6"x3/8" CONT.

5/8" 6" TYP.

1/8" TYP.

1 1/2" MTL. ROOF DECK

DRAWING TITLE:	DRAWN BY: SAAD H. TANTAWI (M.S.CE, B.S.CE, E.I.)
SECTION DETAIL	SCALE: N.T.S

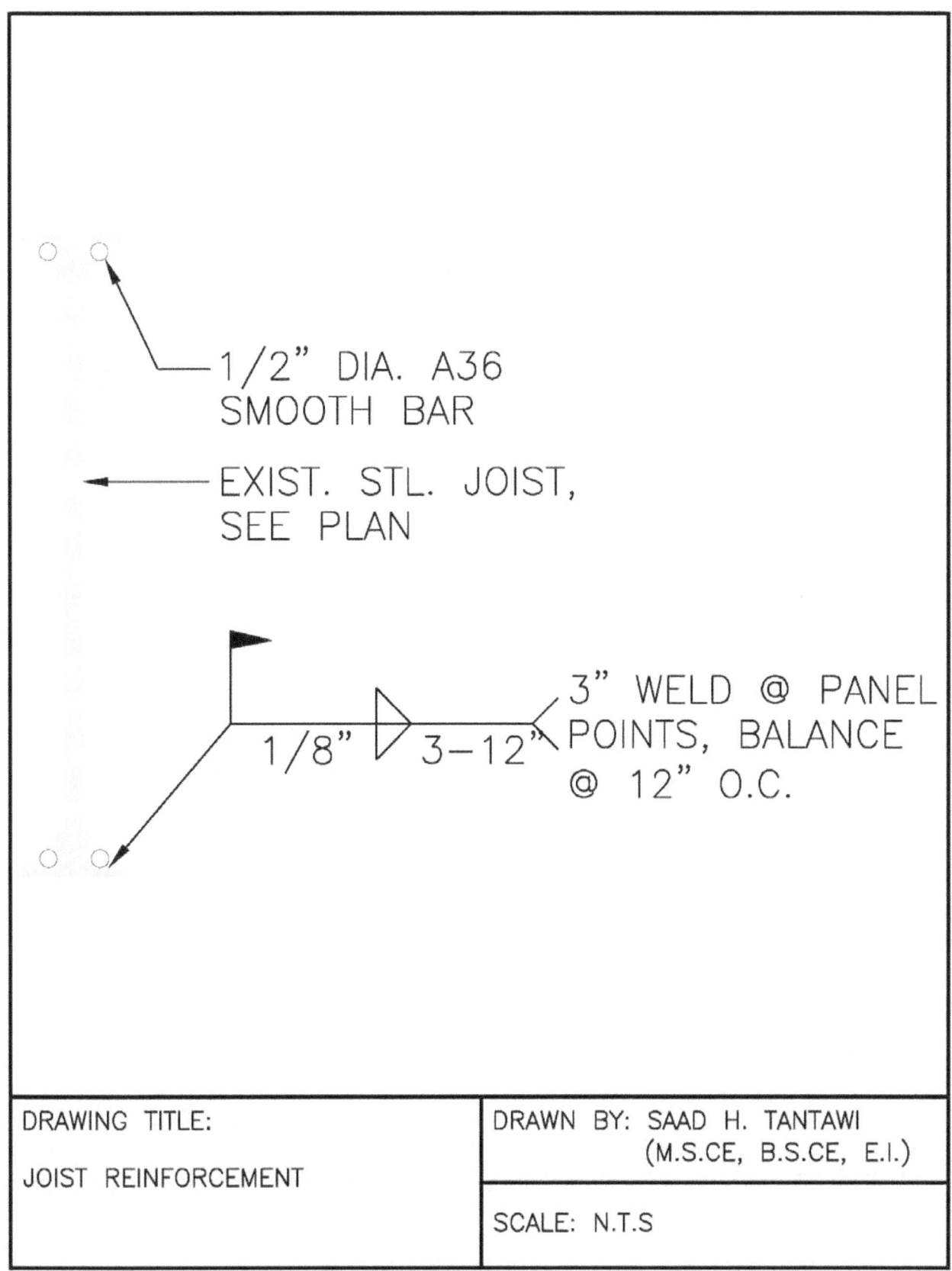

DRAWING TITLE:	DRAWN BY: SAAD H. TANTAWI
	(M.S.CE, B.S.CE, E.I.)
JOIST REINFORCEMENT	SCALE: N.T.S

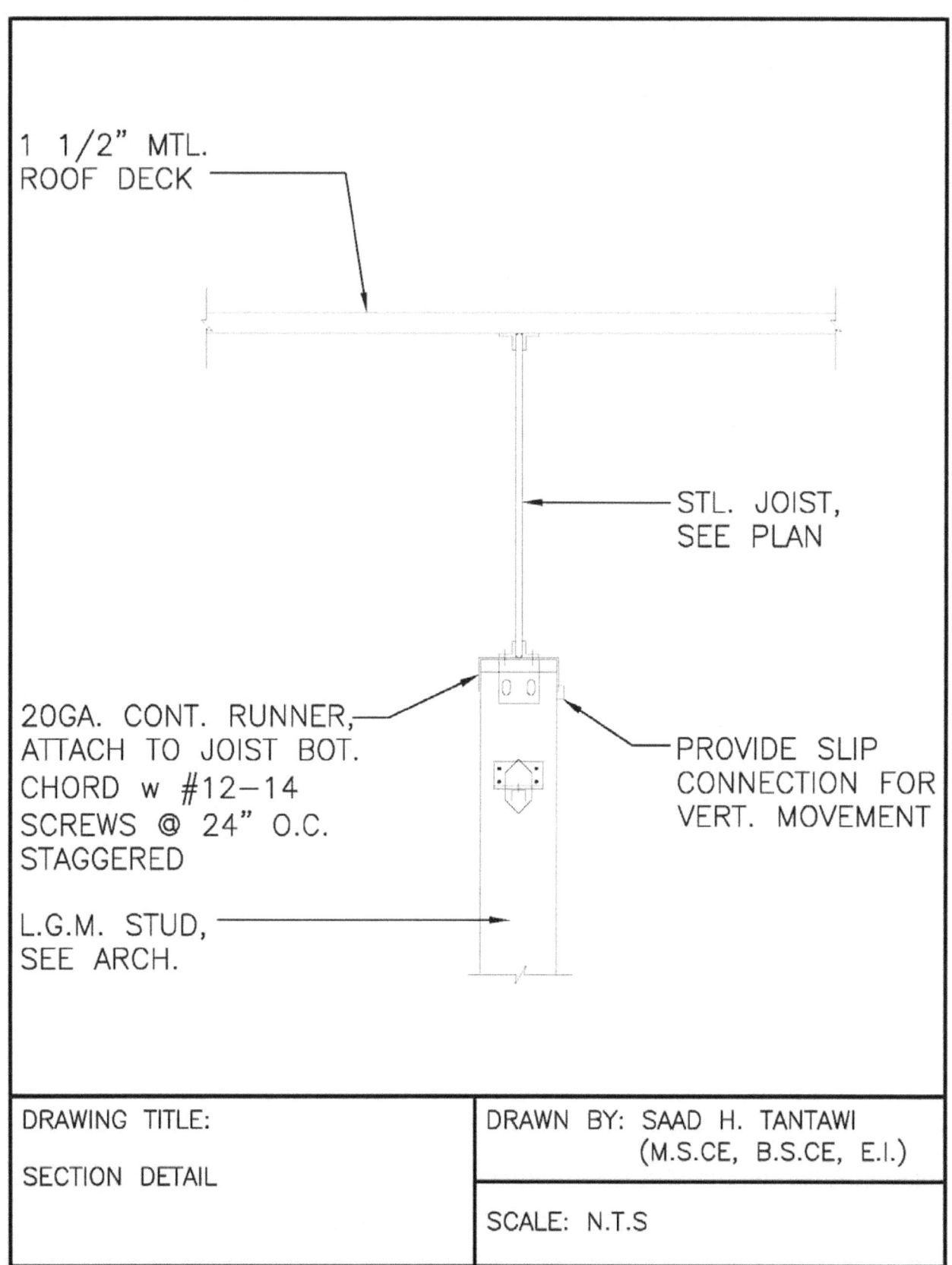

1 1/2" MTL.
ROOF DECK

STL. JOIST,
SEE PLAN

20GA. CONT. RUNNER,
ATTACH TO JOIST BOT.
CHORD w #12-14
SCREWS @ 24" O.C.
STAGGERED

PROVIDE SLIP
CONNECTION FOR
VERT. MOVEMENT

L.G.M. STUD,
SEE ARCH.

DRAWING TITLE:	DRAWN BY: SAAD H. TANTAWI
	(M.S.CE, B.S.CE, E.I.)
SECTION DETAIL	
	SCALE: N.T.S

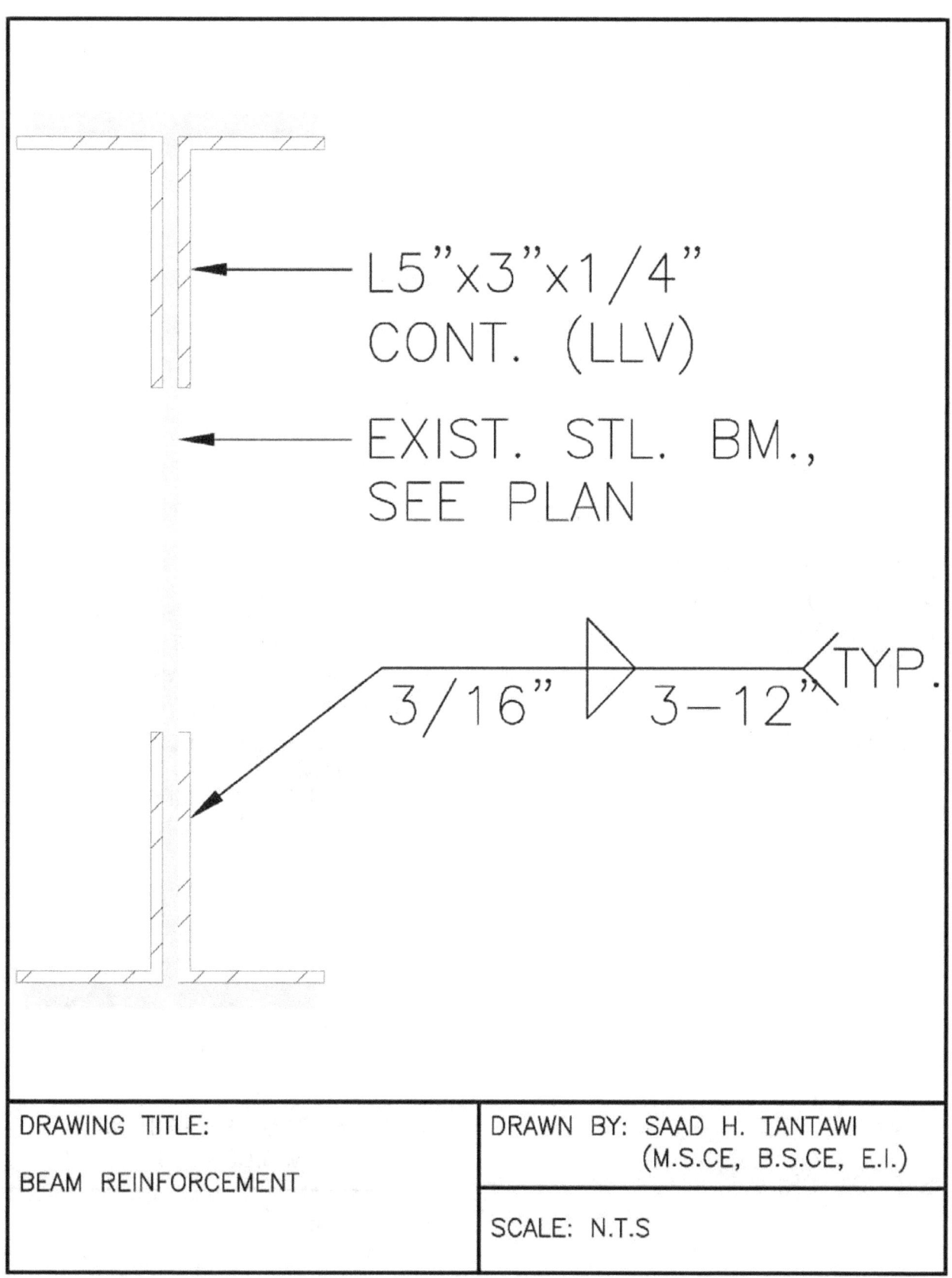

L5"x3"x1/4" CONT. (LLV)

EXIST. STL. BM., SEE PLAN

3/16" 3−12" TYP.

DRAWING TITLE:	DRAWN BY: SAAD H. TANTAWI
BEAM REINFORCEMENT	(M.S.CE, B.S.CE, E.I.)
	SCALE: N.T.S

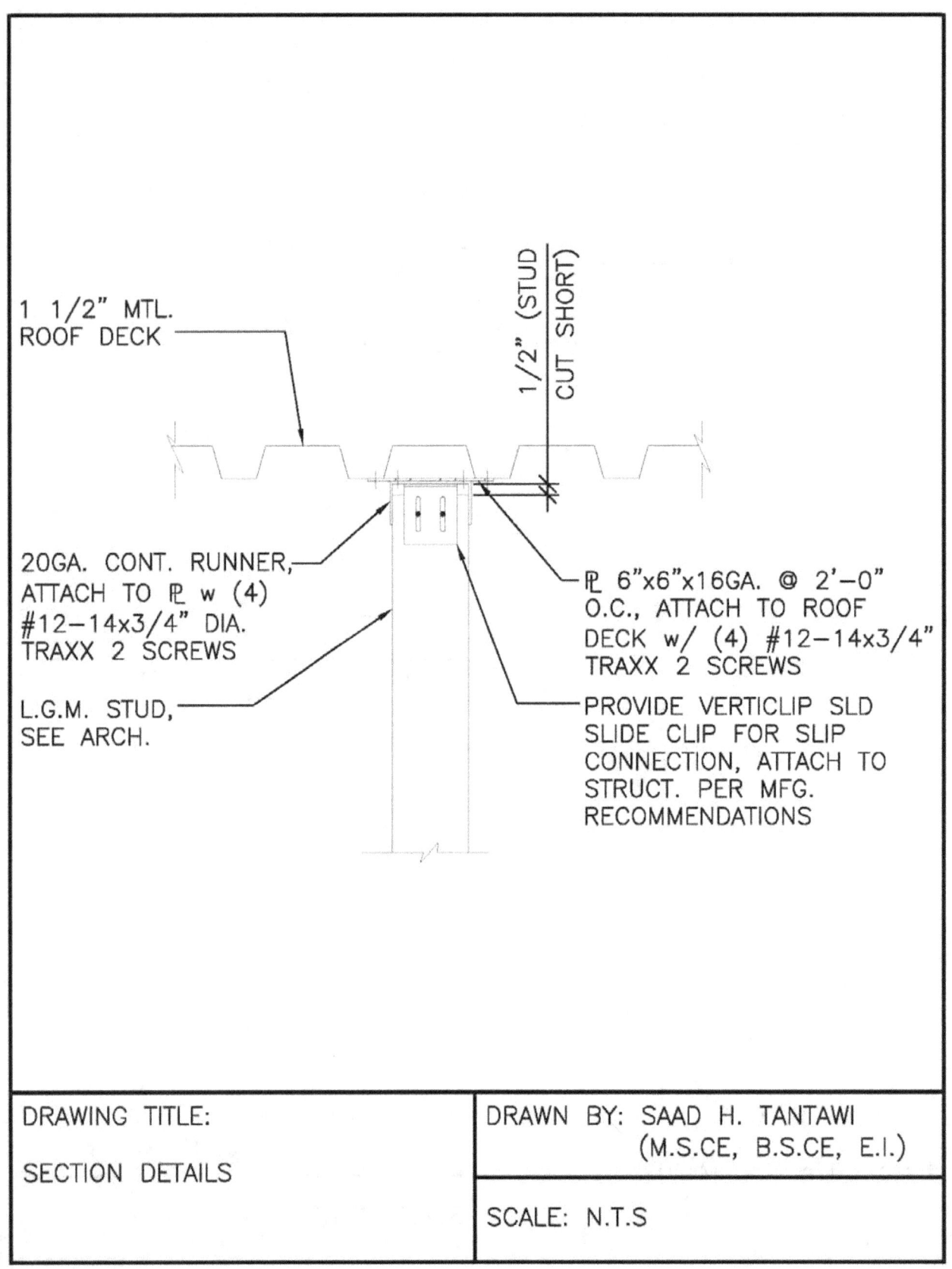

1 1/2" MTL.
ROOF DECK

1/2" (STUD
CUT SHORT)

20GA. CONT. RUNNER,
ATTACH TO ℙ w (4)
#12–14x3/4" DIA.
TRAXX 2 SCREWS

L.G.M. STUD,
SEE ARCH.

ℙ 6"x6"x16GA. @ 2'–0"
O.C., ATTACH TO ROOF
DECK w/ (4) #12–14x3/4"
TRAXX 2 SCREWS

PROVIDE VERTICLIP SLD
SLIDE CLIP FOR SLIP
CONNECTION, ATTACH TO
STRUCT. PER MFG.
RECOMMENDATIONS

DRAWING TITLE:	DRAWN BY: SAAD H. TANTAWI
	(M.S.CE, B.S.CE, E.I.)
SECTION DETAILS	
	SCALE: N.T.S

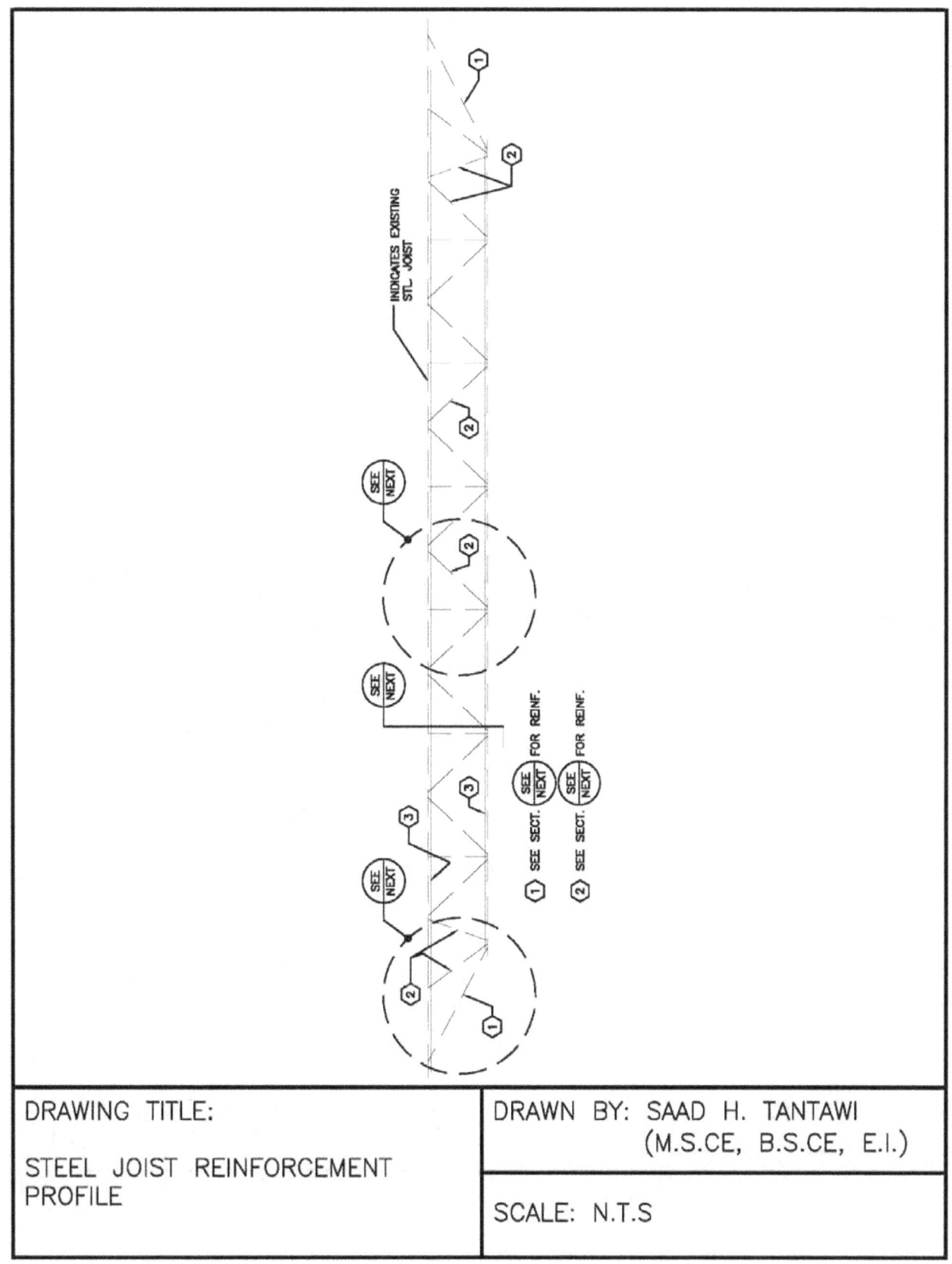

DRAWING TITLE:	DRAWN BY: SAAD H. TANTAWI
	(M.S.CE, B.S.CE, E.I.)
STEEL JOIST REINFORCEMENT PROFILE	SCALE: N.T.S

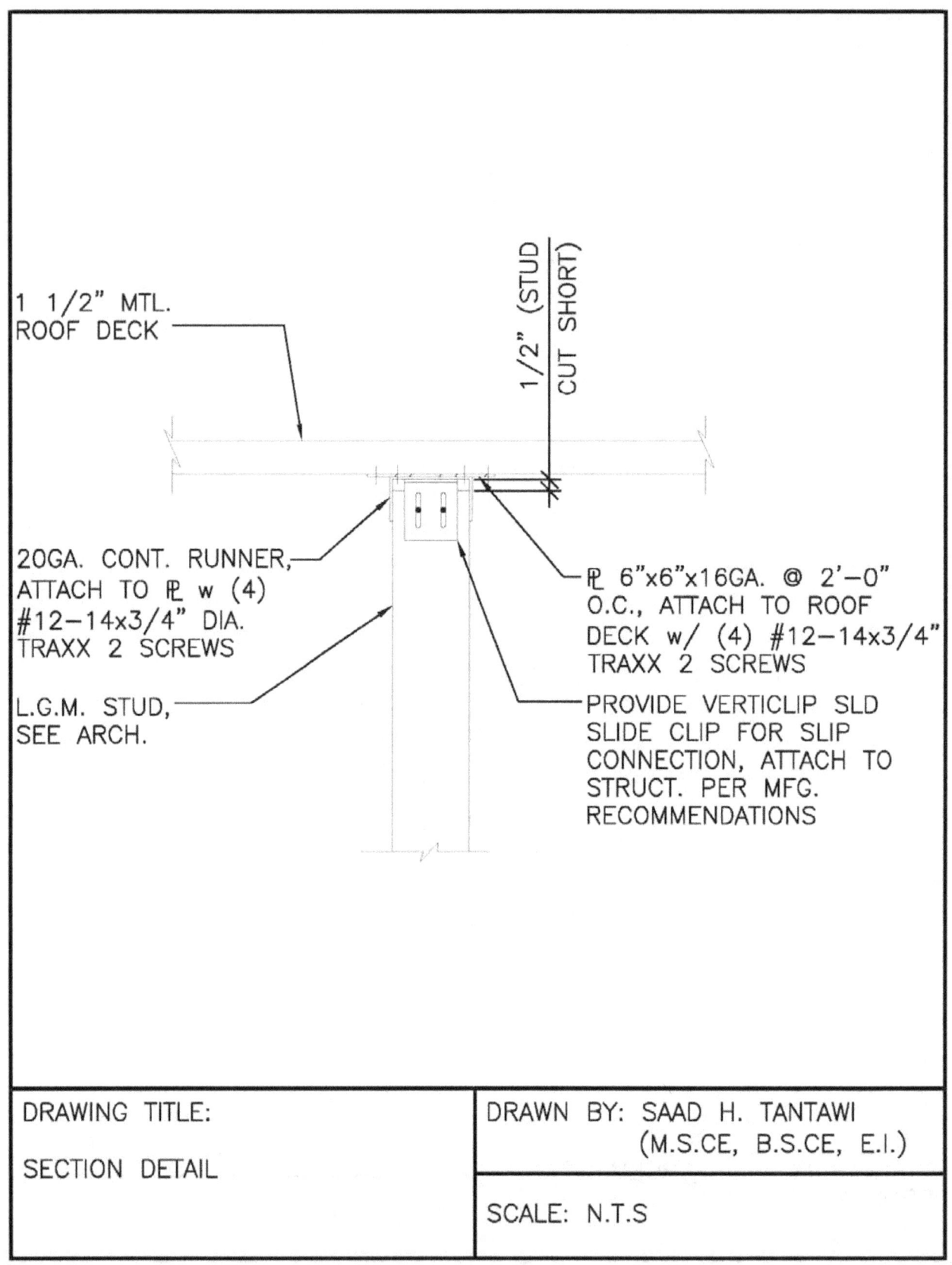

1 1/2" MTL.
ROOF DECK

1/2" (STUD CUT SHORT)

20GA. CONT. RUNNER,
ATTACH TO ℄ w (4)
#12−14x3/4" DIA.
TRAXX 2 SCREWS

L.G.M. STUD,
SEE ARCH.

℄ 6"x6"x16GA. @ 2'−0"
O.C., ATTACH TO ROOF
DECK w/ (4) #12−14x3/4"
TRAXX 2 SCREWS

PROVIDE VERTICLIP SLD
SLIDE CLIP FOR SLIP
CONNECTION, ATTACH TO
STRUCT. PER MFG.
RECOMMENDATIONS

DRAWING TITLE:	DRAWN BY: SAAD H. TANTAWI
	(M.S.CE, B.S.CE, E.I.)
SECTION DETAIL	
	SCALE: N.T.S

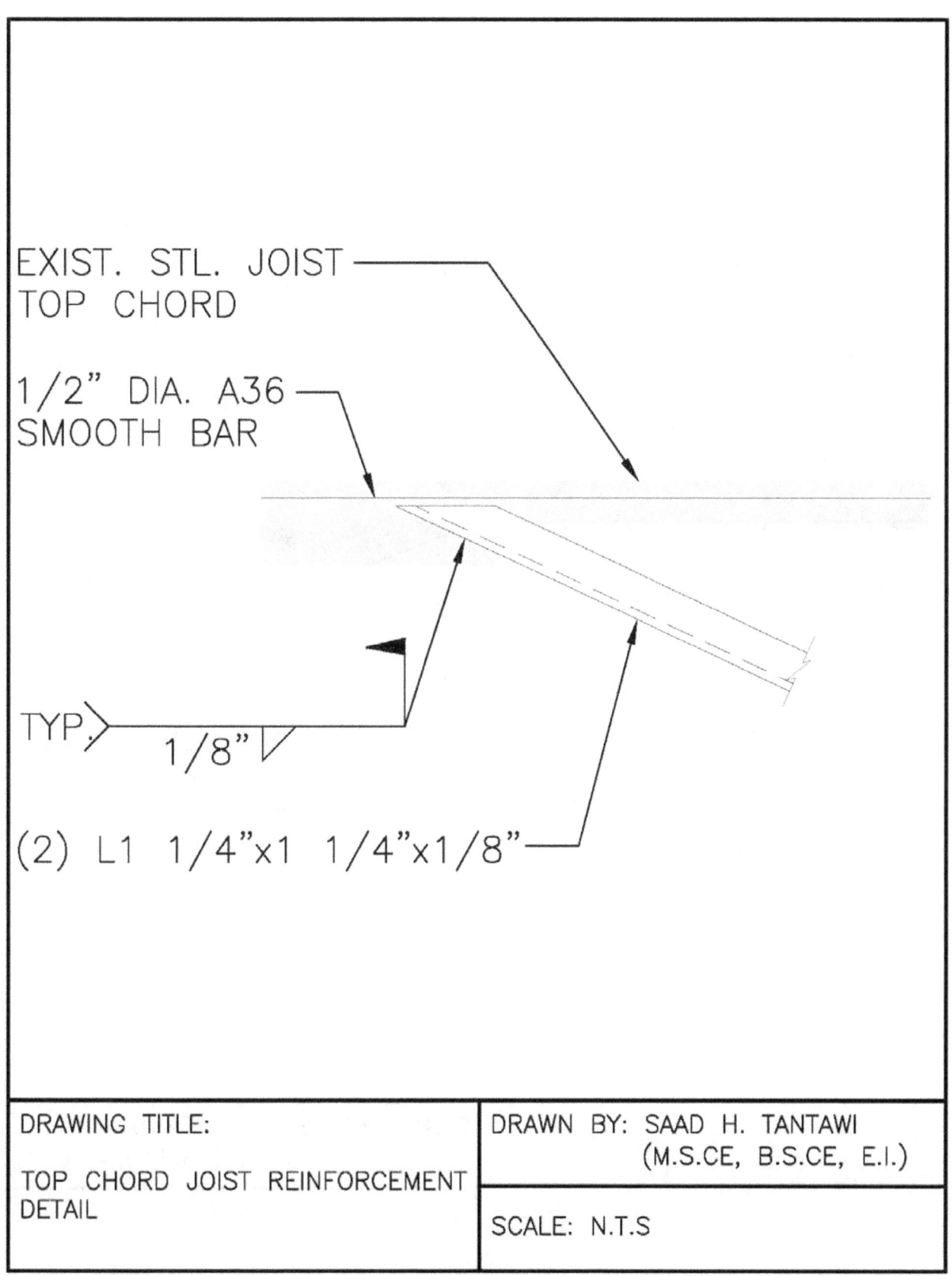

EXIST. STL. JOIST
TOP CHORD

1/2" DIA. A36
SMOOTH BAR

TYP.

1/8"

(2) L1 1/4"x1 1/4"x1/8"

DRAWING TITLE:	DRAWN BY: SAAD H. TANTAWI
TOP CHORD JOIST REINFORCEMENT DETAIL	(M.S.CE, B.S.CE, E.I.)
	SCALE: N.T.S

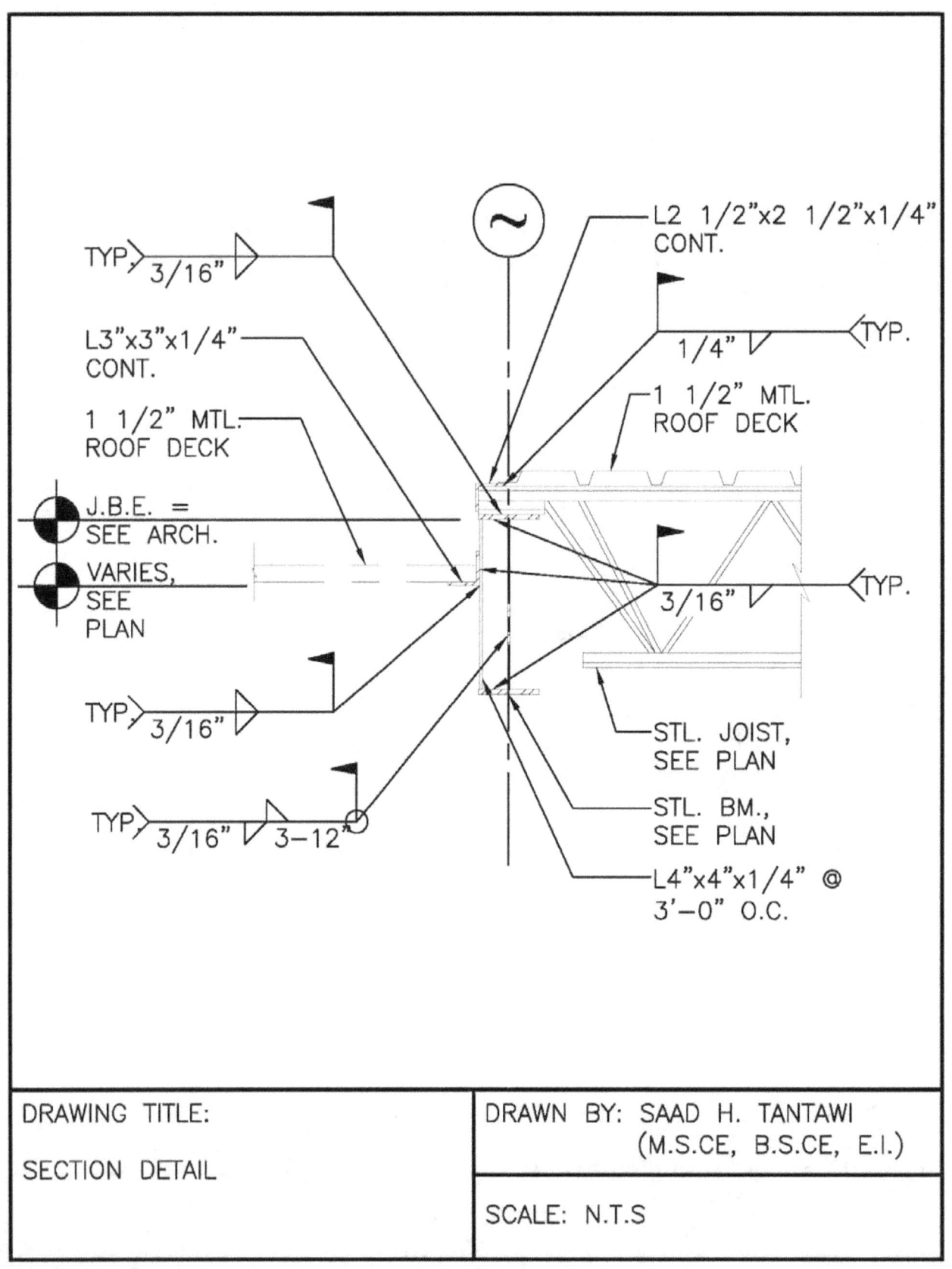

DRAWING TITLE:	DRAWN BY: SAAD H. TANTAWI
SECTION DETAIL	(M.S.CE, B.S.CE, E.I.)
	SCALE: N.T.S

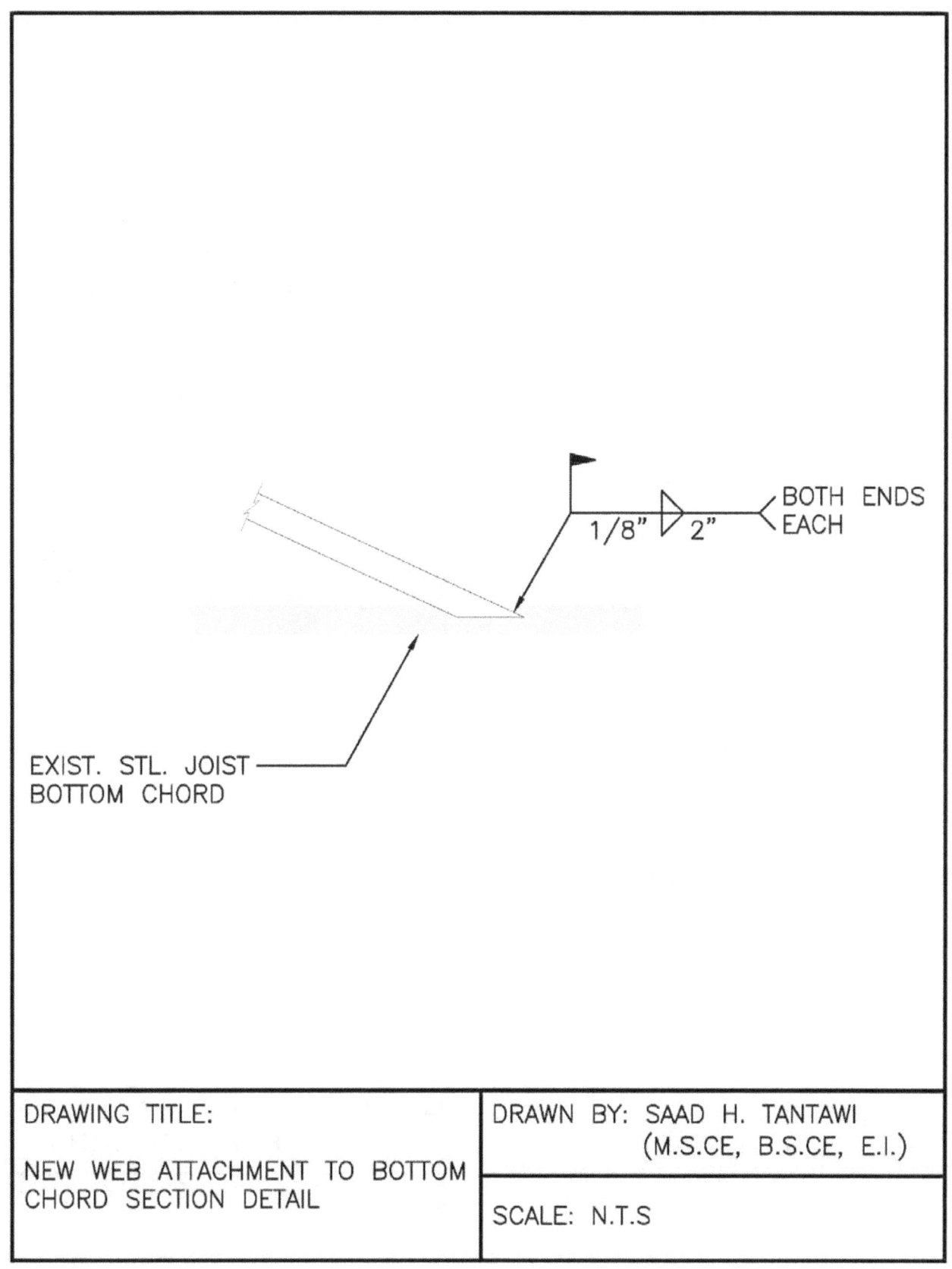

EXIST. STL. JOIST
BOTTOM CHORD

1/8" 2" BOTH ENDS
EACH

DRAWING TITLE:	DRAWN BY: SAAD H. TANTAWI
NEW WEB ATTACHMENT TO BOTTOM CHORD SECTION DETAIL	(M.S.CE, B.S.CE, E.I.)
	SCALE: N.T.S

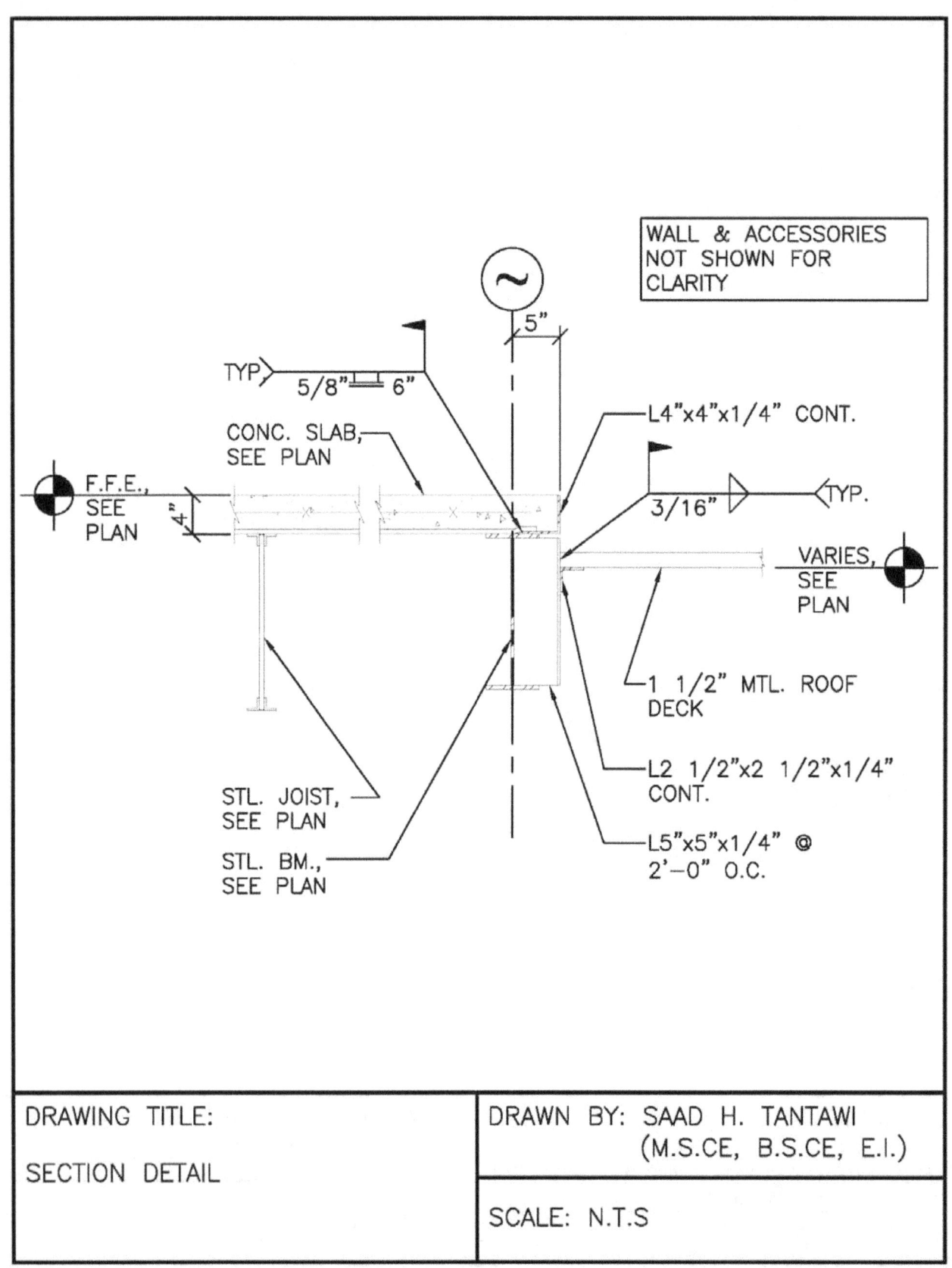

WALL & ACCESSORIES
NOT SHOWN FOR
CLARITY

5"

TYP
5/8" 6"

CONC. SLAB,
SEE PLAN

L4"x4"x1/4" CONT.

3/16" TYP.

F.F.E.,
SEE
PLAN

4"

VARIES,
SEE
PLAN

1 1/2" MTL. ROOF
DECK

STL. JOIST,
SEE PLAN

STL. BM.,
SEE PLAN

L2 1/2"x2 1/2"x1/4"
CONT.

L5"x5"x1/4" @
2'-0" O.C.

DRAWING TITLE:	DRAWN BY: SAAD H. TANTAWI
	(M.S.CE, B.S.CE, E.I.)
SECTION DETAIL	
	SCALE: N.T.S

L 1 1/4"x1 1/4"x1/8" ────◁──── ←──── EXIST. 1/2" DIA. BAR

DRAWING TITLE:	DRAWN BY: SAAD H. TANTAWI
STEEL JOISTS CHORD REINFORCEMENT SECTION DETAIL	(M.S.CE, B.S.CE, E.I.)
	SCALE: N.T.S

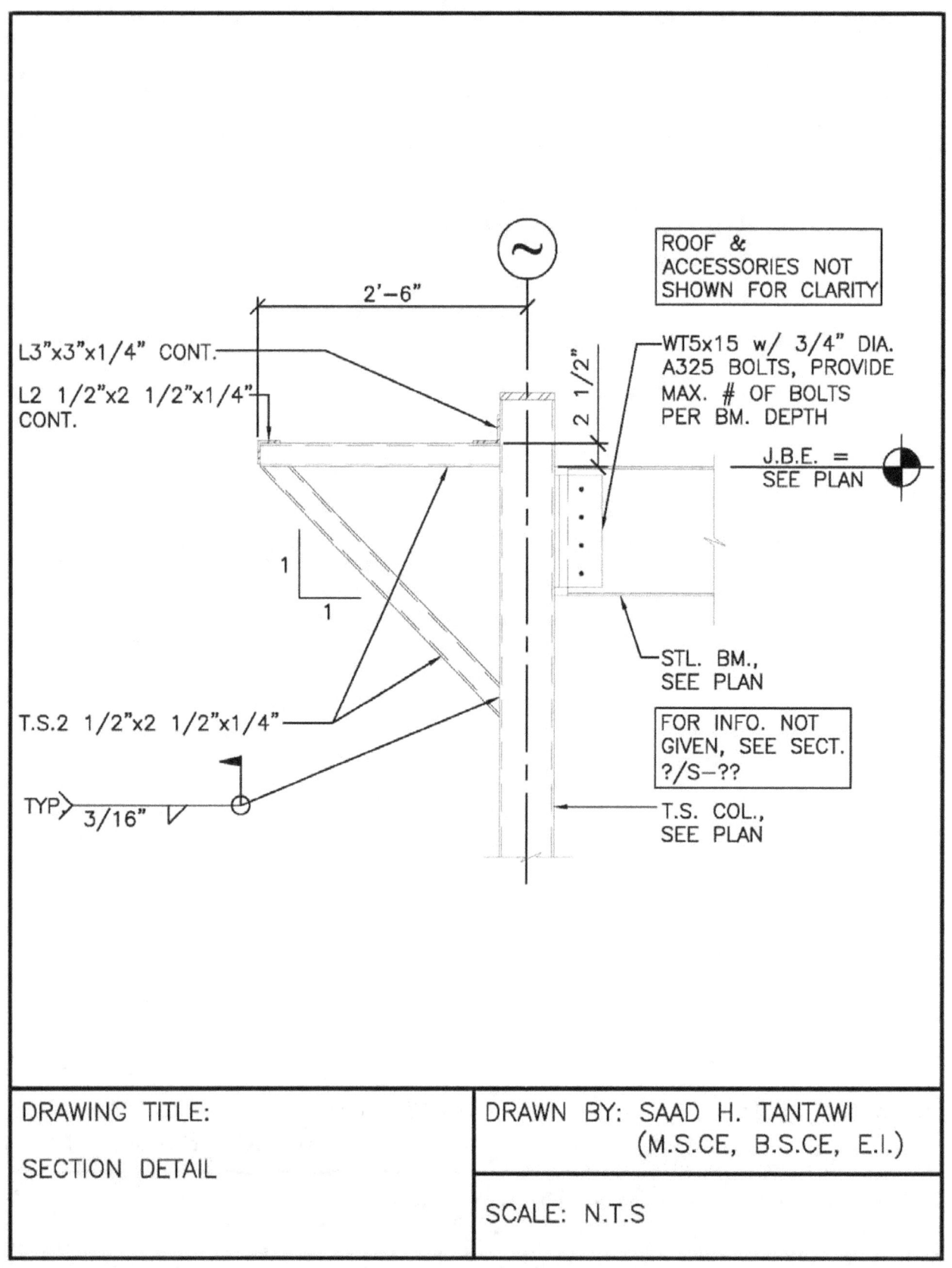

DRAWING TITLE: SECTION DETAIL	DRAWN BY: SAAD H. TANTAWI (M.S.CE, B.S.CE, E.I.)
	SCALE: N.T.S

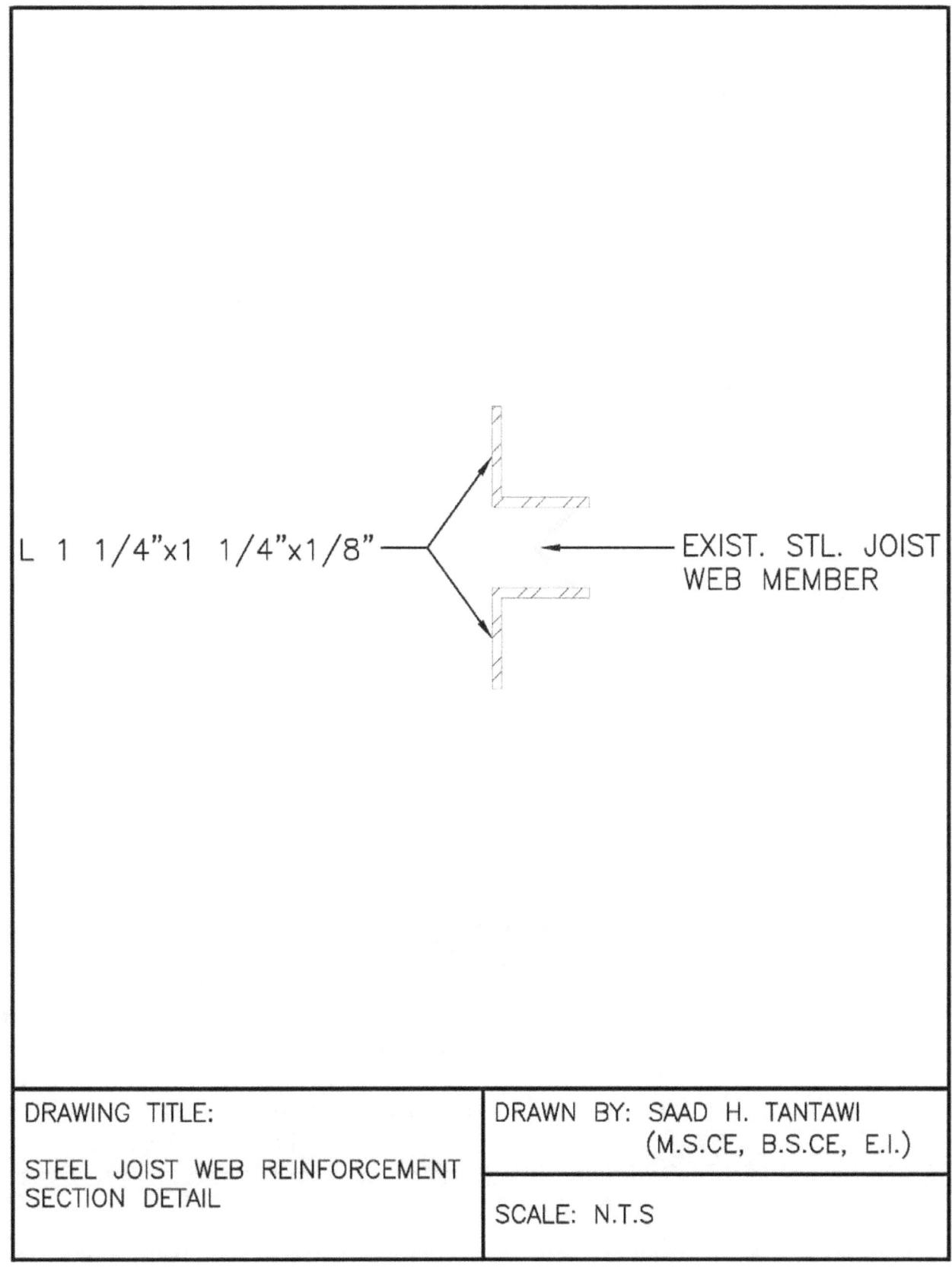

L 1 1/4"x1 1/4"x1/8" ——

EXIST. STL. JOIST
WEB MEMBER

DRAWING TITLE:	DRAWN BY: SAAD H. TANTAWI
	(M.S.CE, B.S.CE, E.I.)
STEEL JOIST WEB REINFORCEMENT SECTION DETAIL	SCALE: N.T.S

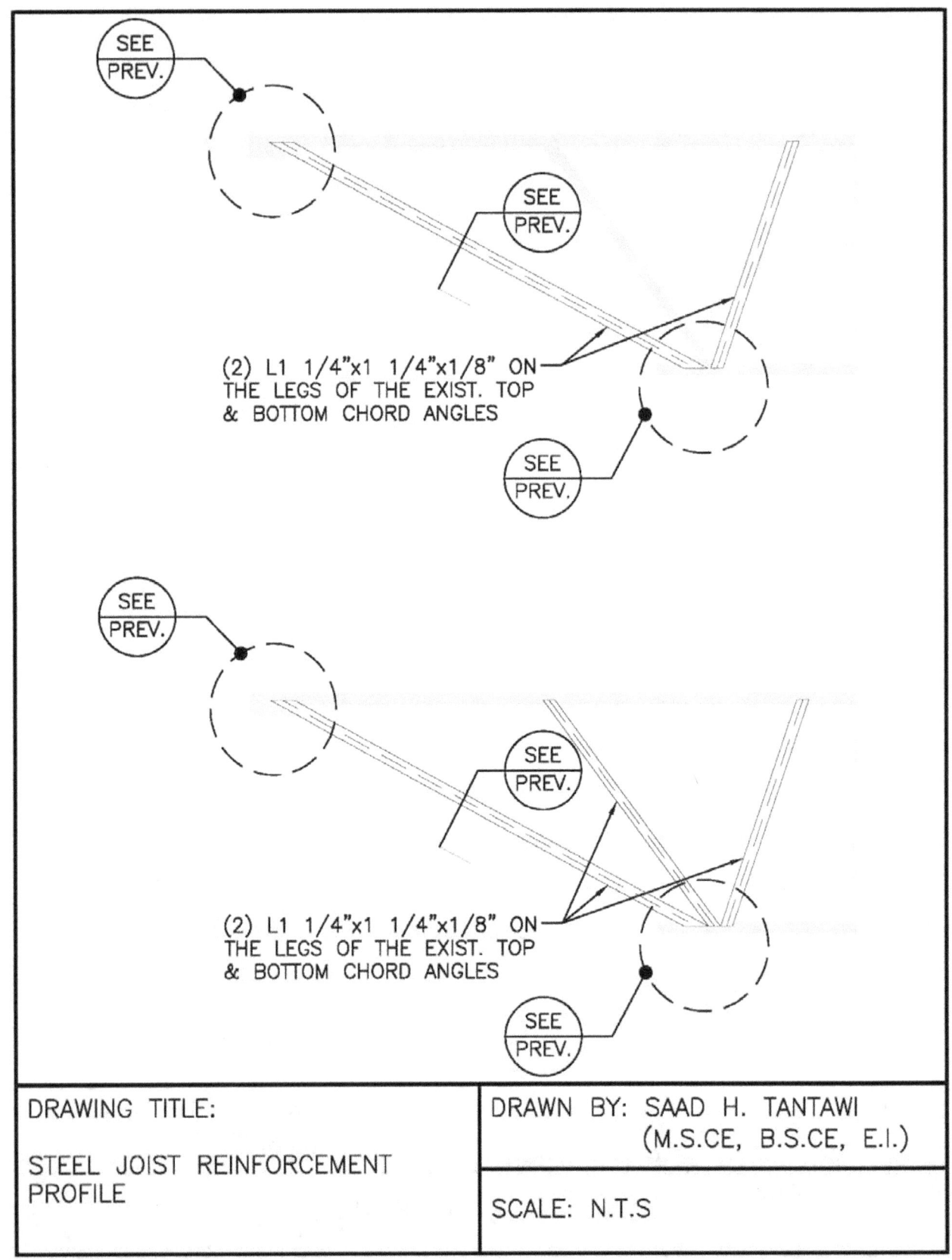

SEE
PREV.

SEE
PREV.

(2) L1 1/4"x1 1/4"x1/8" ON
THE LEGS OF THE EXIST. TOP
& BOTTOM CHORD ANGLES

SEE
PREV.

SEE
PREV.

SEE
PREV.

(2) L1 1/4"x1 1/4"x1/8" ON
THE LEGS OF THE EXIST. TOP
& BOTTOM CHORD ANGLES

SEE
PREV.

DRAWING TITLE:	DRAWN BY: SAAD H. TANTAWI
	(M.S.CE, B.S.CE, E.I.)
STEEL JOIST REINFORCEMENT PROFILE	SCALE: N.T.S

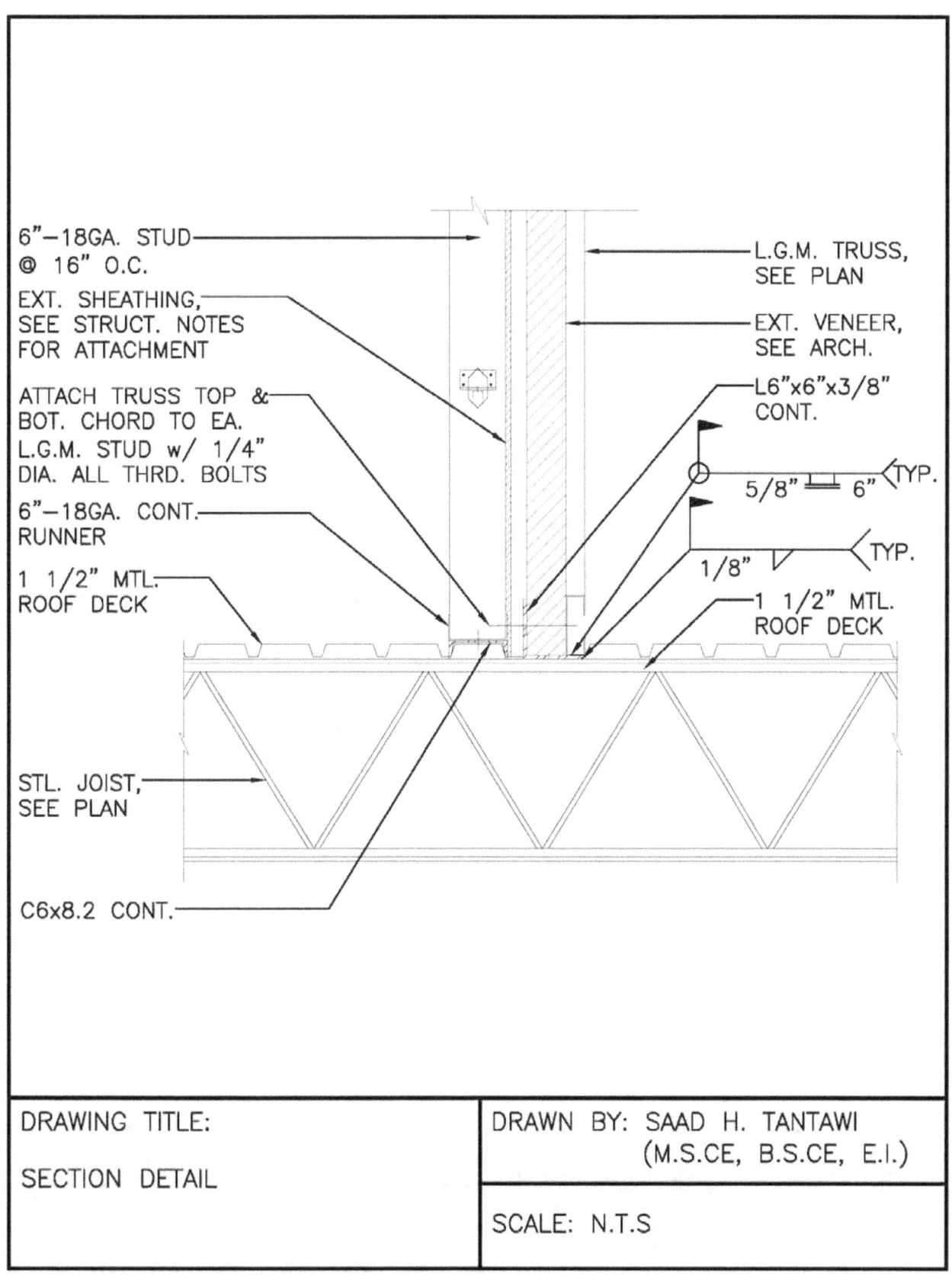

6"–18GA. STUD
@ 16" O.C.

EXT. SHEATHING,
SEE STRUCT. NOTES
FOR ATTACHMENT

ATTACH TRUSS TOP &
BOT. CHORD TO EA.
L.G.M. STUD w/ 1/4"
DIA. ALL THRD. BOLTS

6"–18GA. CONT.
RUNNER

1 1/2" MTL.
ROOF DECK

STL. JOIST,
SEE PLAN

C6x8.2 CONT.

L.G.M. TRUSS,
SEE PLAN

EXT. VENEER,
SEE ARCH.

L6"x6"x3/8"
CONT.

5/8"   6"   TYP.

1/8"   TYP.

1 1/2" MTL.
ROOF DECK

DRAWING TITLE:	DRAWN BY: SAAD H. TANTAWI
	(M.S.CE, B.S.CE, E.I.)
SECTION DETAIL	
	SCALE: N.T.S

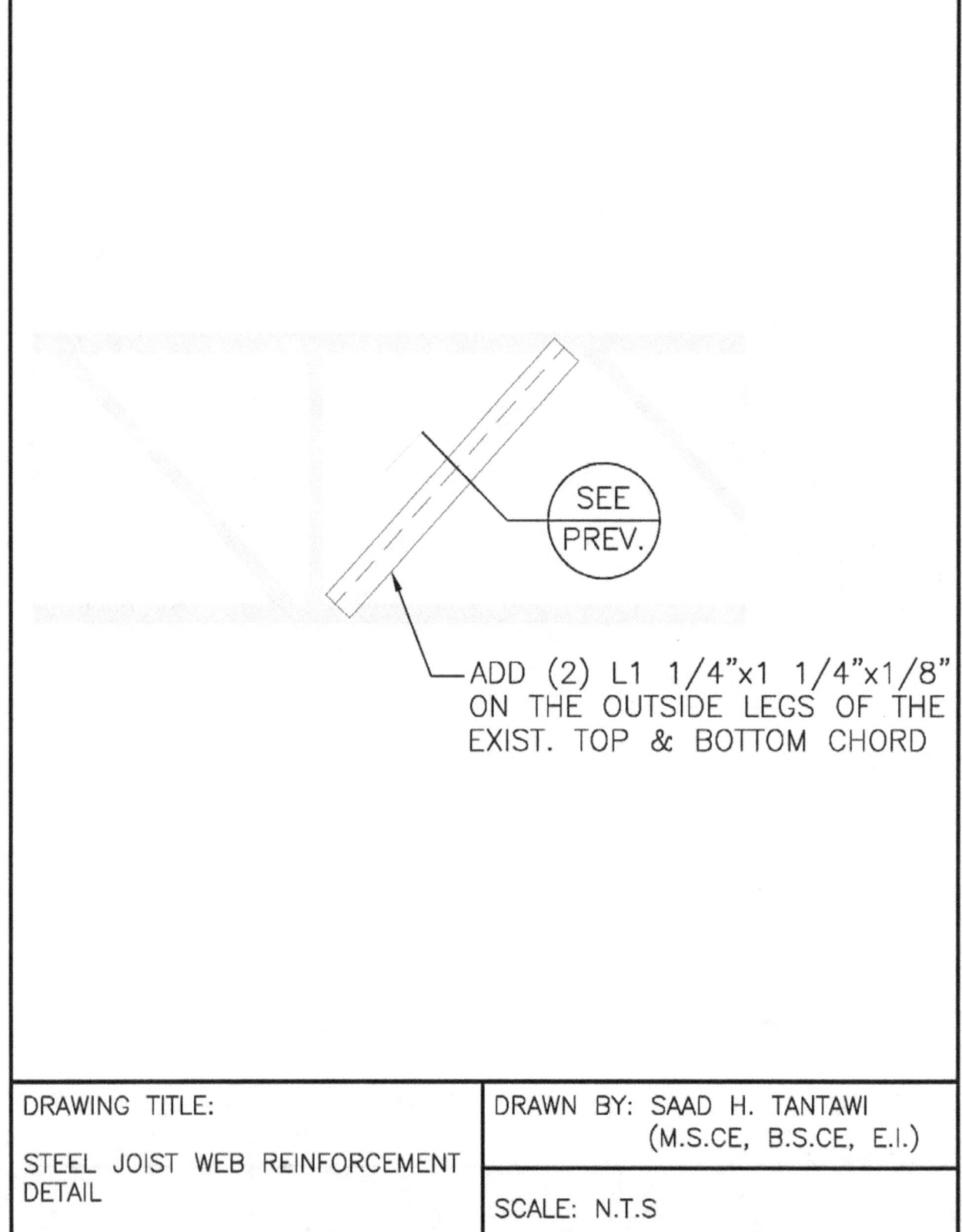

SEE
PREV.

ADD (2) L1 1/4"x1 1/4"x1/8"
ON THE OUTSIDE LEGS OF THE
EXIST. TOP & BOTTOM CHORD

DRAWING TITLE:	DRAWN BY: SAAD H. TANTAWI
STEEL JOIST WEB REINFORCEMENT DETAIL	(M.S.CE, B.S.CE, E.I.)
	SCALE: N.T.S

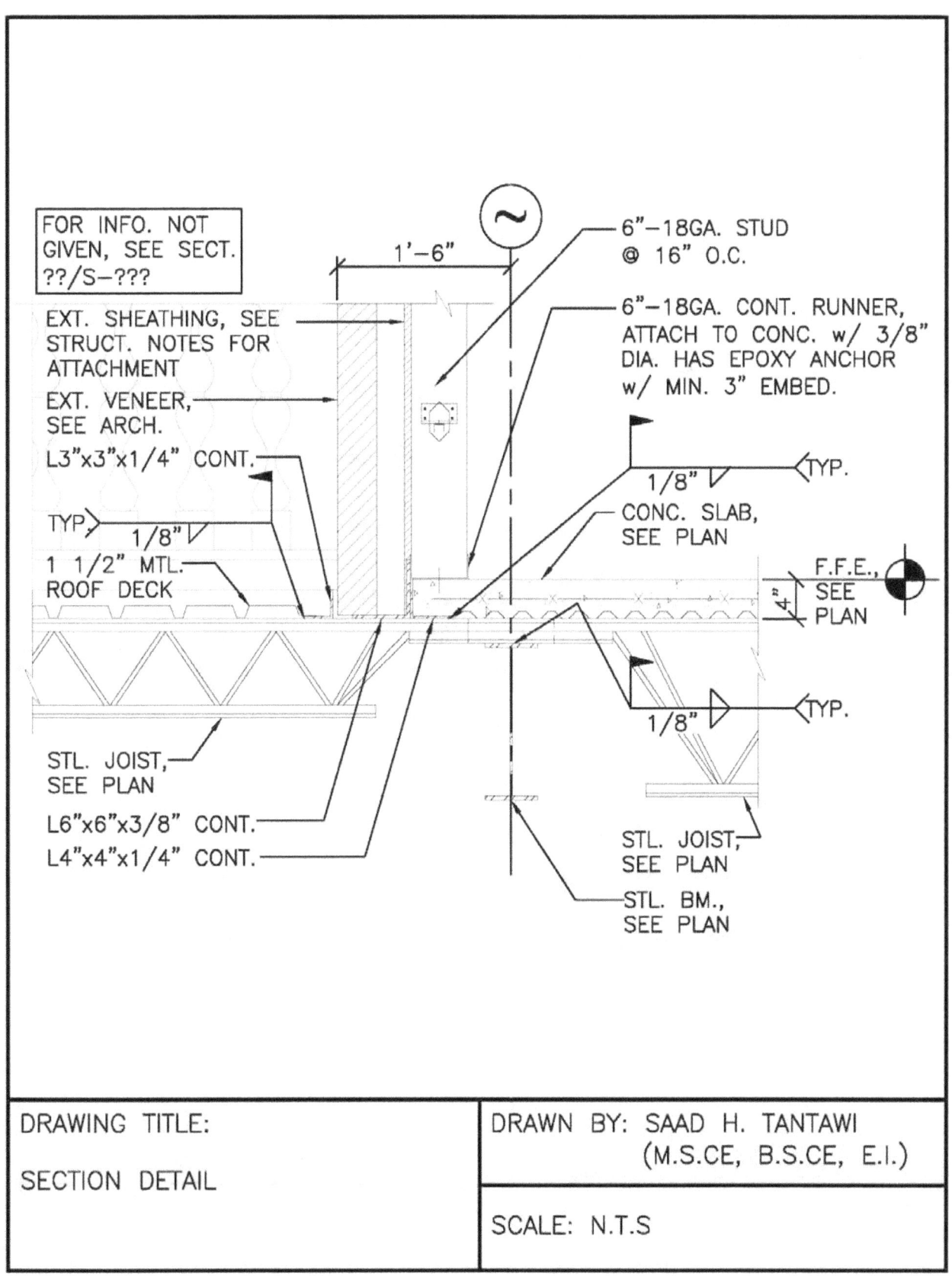

FOR INFO. NOT GIVEN, SEE SECT. ??/S-???

1'-6"

6"-18GA. STUD @ 16" O.C.

6"-18GA. CONT. RUNNER, ATTACH TO CONC. w/ 3/8" DIA. HAS EPOXY ANCHOR w/ MIN. 3" EMBED.

EXT. SHEATHING, SEE STRUCT. NOTES FOR ATTACHMENT

EXT. VENEER, SEE ARCH.

L3"x3"x1/4" CONT.

TYP. 1/8"

1 1/2" MTL. ROOF DECK

1/8" TYP.

CONC. SLAB, SEE PLAN

F.F.E., SEE PLAN

4"

1/8" TYP.

STL. JOIST, SEE PLAN

L6"x6"x3/8" CONT.

L4"x4"x1/4" CONT.

STL. JOIST, SEE PLAN

STL. BM., SEE PLAN

DRAWING TITLE:	DRAWN BY: SAAD H. TANTAWI
	(M.S.CE, B.S.CE, E.I.)
SECTION DETAIL	
	SCALE: N.T.S

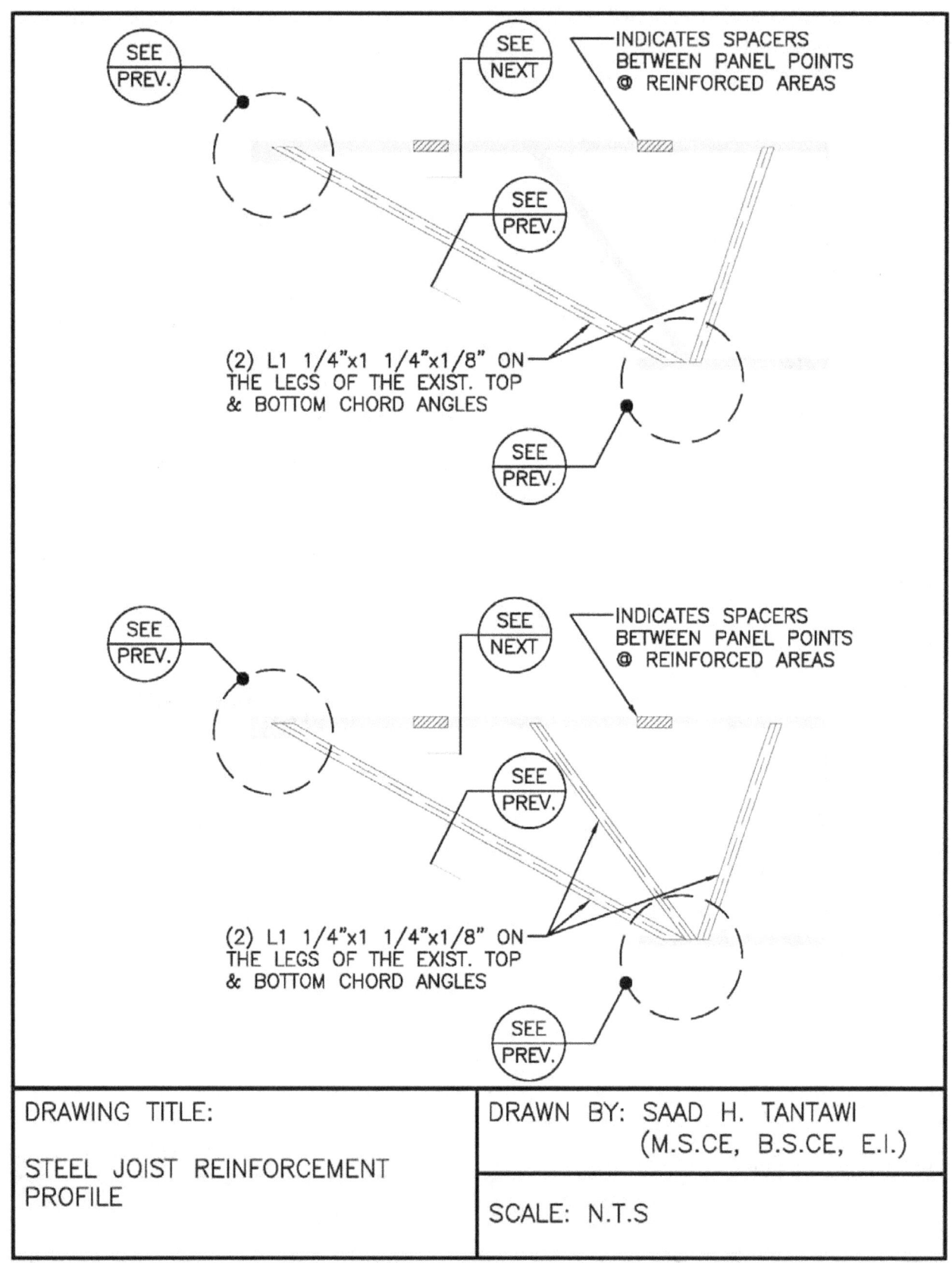

INDICATES SPACERS
BETWEEN PANEL POINTS
@ REINFORCED AREAS

(2) L1 1/4"x1 1/4"x1/8" ON
THE LEGS OF THE EXIST. TOP
& BOTTOM CHORD ANGLES

SEE PREV.

SEE NEXT

SEE PREV.

INDICATES SPACERS
BETWEEN PANEL POINTS
@ REINFORCED AREAS

(2) L1 1/4"x1 1/4"x1/8" ON
THE LEGS OF THE EXIST. TOP
& BOTTOM CHORD ANGLES

DRAWING TITLE:	DRAWN BY: SAAD H. TANTAWI
	(M.S.CE, B.S.CE, E.I.)
STEEL JOIST REINFORCEMENT PROFILE	
	SCALE: N.T.S

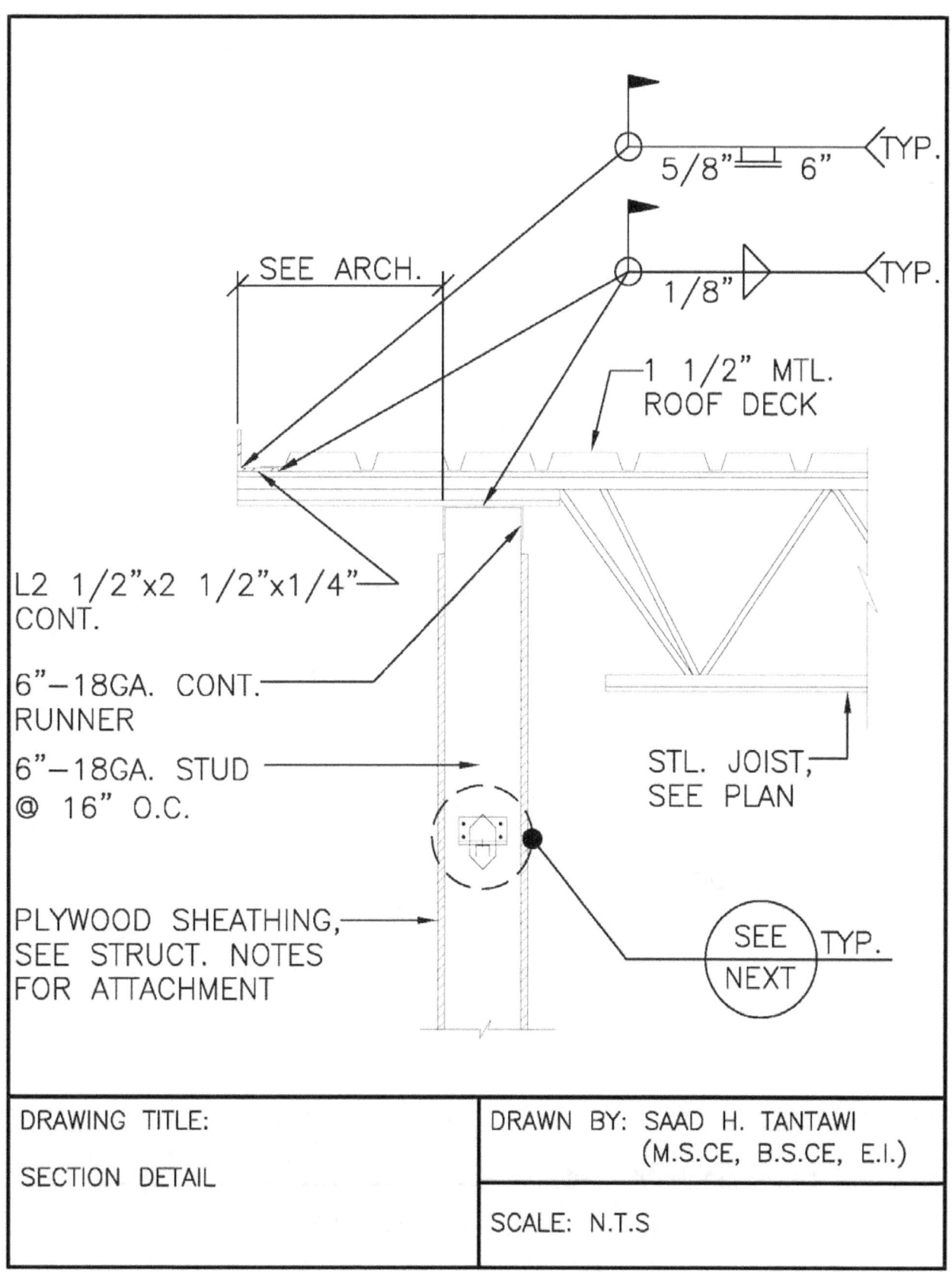

5/8"⌐⊓ 6" ⟨TYP.

1/8" ⟨TYP.

1 1/2" MTL. ROOF DECK

SEE ARCH.

L2 1/2"x2 1/2"x1/4" CONT.

6"-18GA. CONT. RUNNER

6"-18GA. STUD @ 16" O.C.

STL. JOIST, SEE PLAN

PLYWOOD SHEATHING, SEE STRUCT. NOTES FOR ATTACHMENT

SEE NEXT TYP.

DRAWING TITLE:	DRAWN BY: SAAD H. TANTAWI
SECTION DETAIL	(M.S.CE, B.S.CE, E.I.)
	SCALE: N.T.S

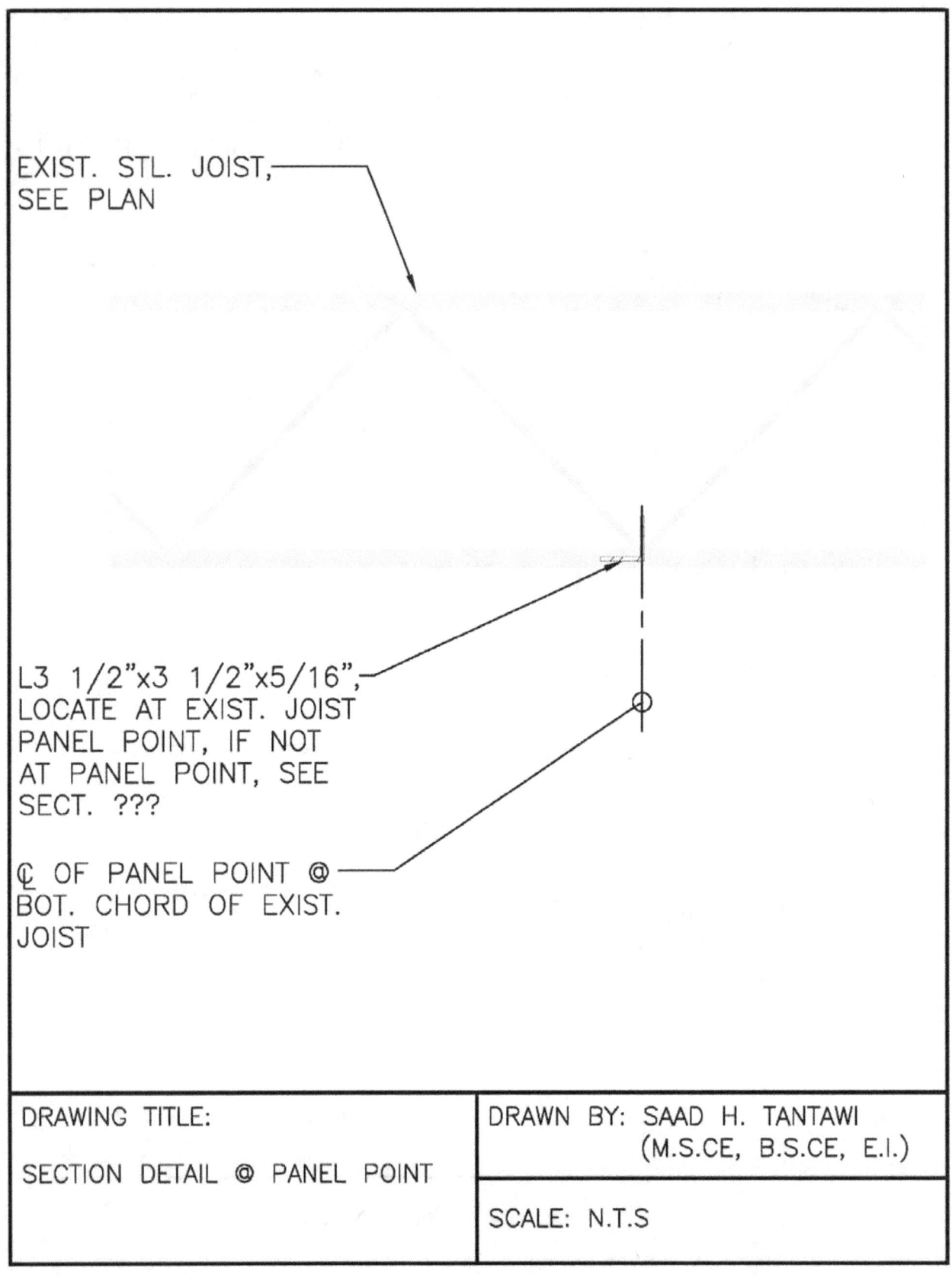

EXIST. STL. JOIST,
SEE PLAN

L3 1/2"x3 1/2"x5/16",
LOCATE AT EXIST. JOIST
PANEL POINT, IF NOT
AT PANEL POINT, SEE
SECT. ???

℄ OF PANEL POINT @
BOT. CHORD OF EXIST.
JOIST

DRAWING TITLE:	DRAWN BY: SAAD H. TANTAWI
	(M.S.CE, B.S.CE, E.I.)
SECTION DETAIL @ PANEL POINT	
	SCALE: N.T.S

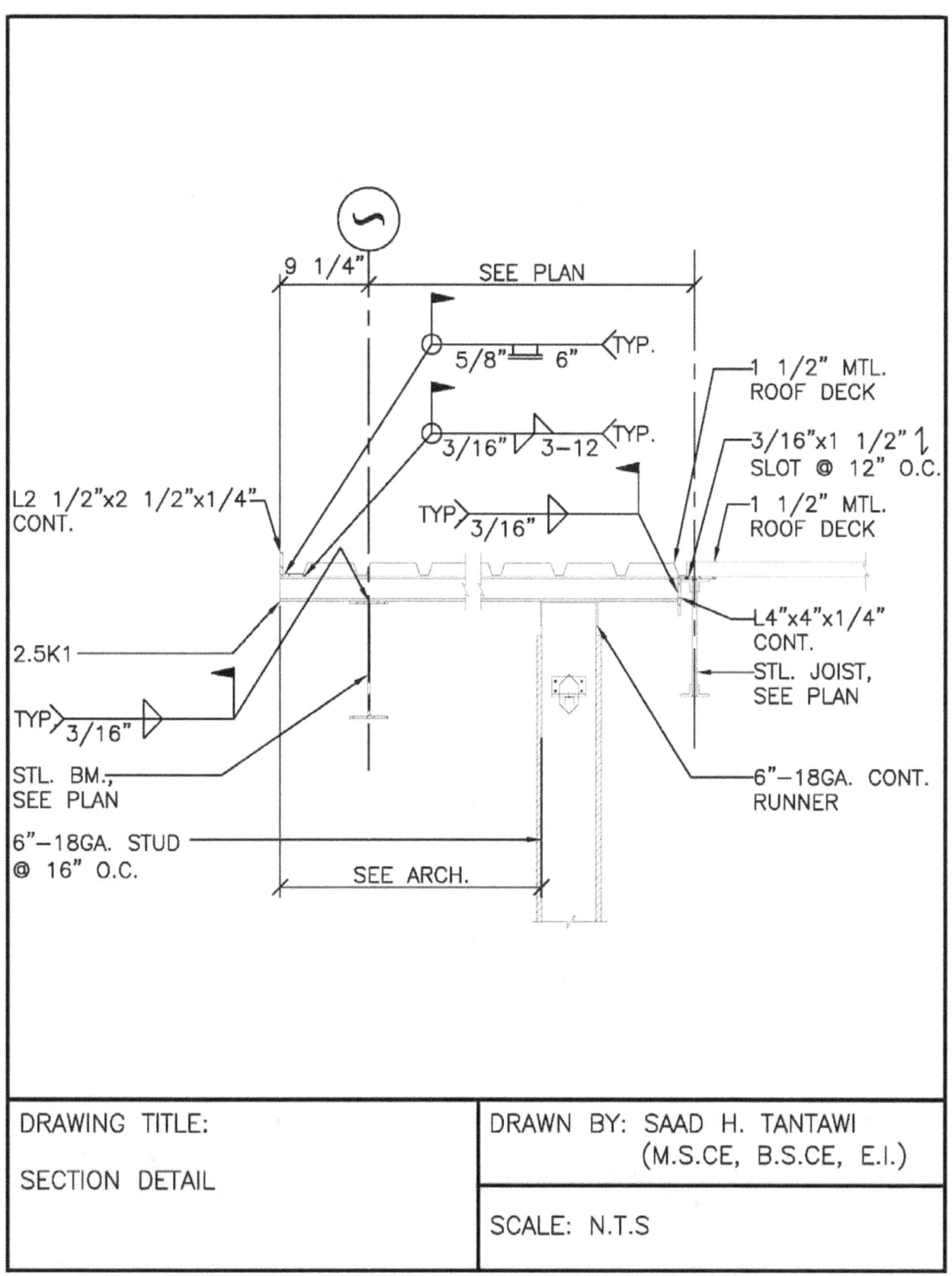

DRAWING TITLE:	DRAWN BY: SAAD H. TANTAWI
	(M.S.CE, B.S.CE, E.I.)
SECTION DETAIL	
	SCALE: N.T.S

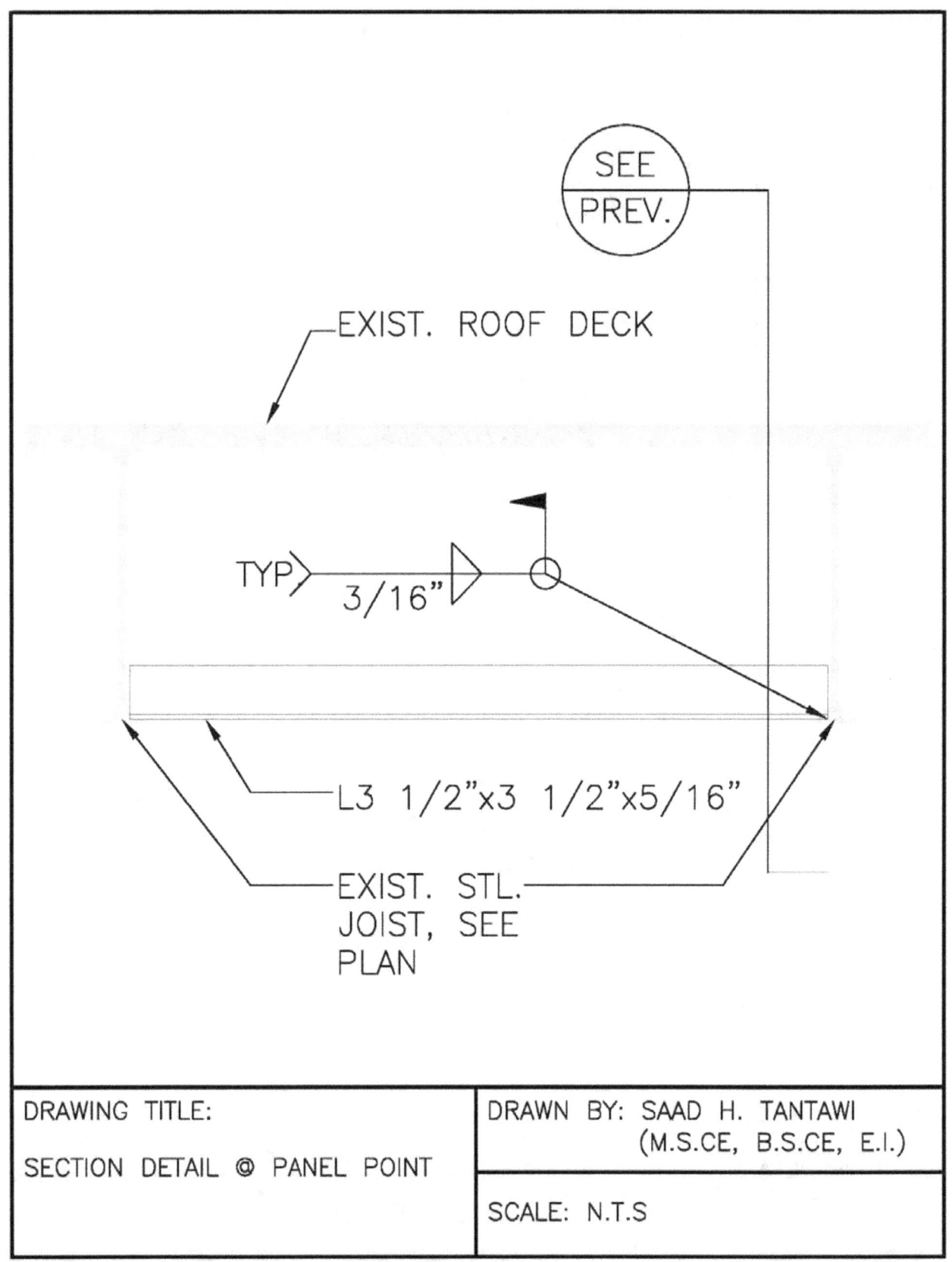

SEE PREV.

EXIST. ROOF DECK

TYP

3/16"

L3 1/2"x3 1/2"x5/16"

EXIST. STL.
JOIST, SEE
PLAN

DRAWING TITLE:	DRAWN BY: SAAD H. TANTAWI
	(M.S.CE, B.S.CE, E.I.)
SECTION DETAIL @ PANEL POINT	
	SCALE: N.T.S

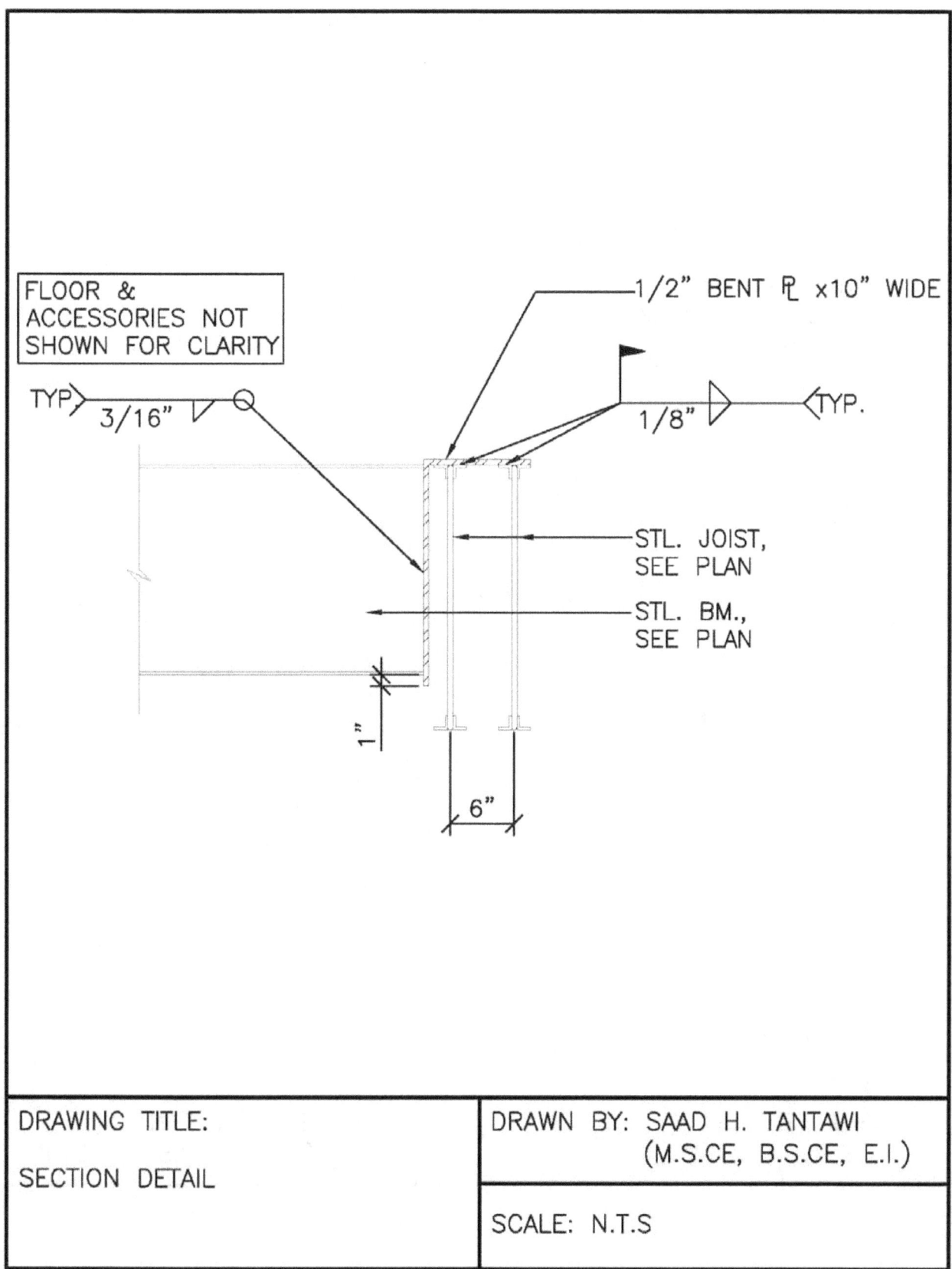

FLOOR & ACCESSORIES NOT SHOWN FOR CLARITY

1/2" BENT ℞ x10" WIDE

TYP

3/16"

1/8"

TYP.

STL. JOIST, SEE PLAN

STL. BM., SEE PLAN

1"

6"

DRAWING TITLE:	DRAWN BY: SAAD H. TANTAWI
SECTION DETAIL	(M.S.CE, B.S.CE, E.I.)
	SCALE: N.T.S

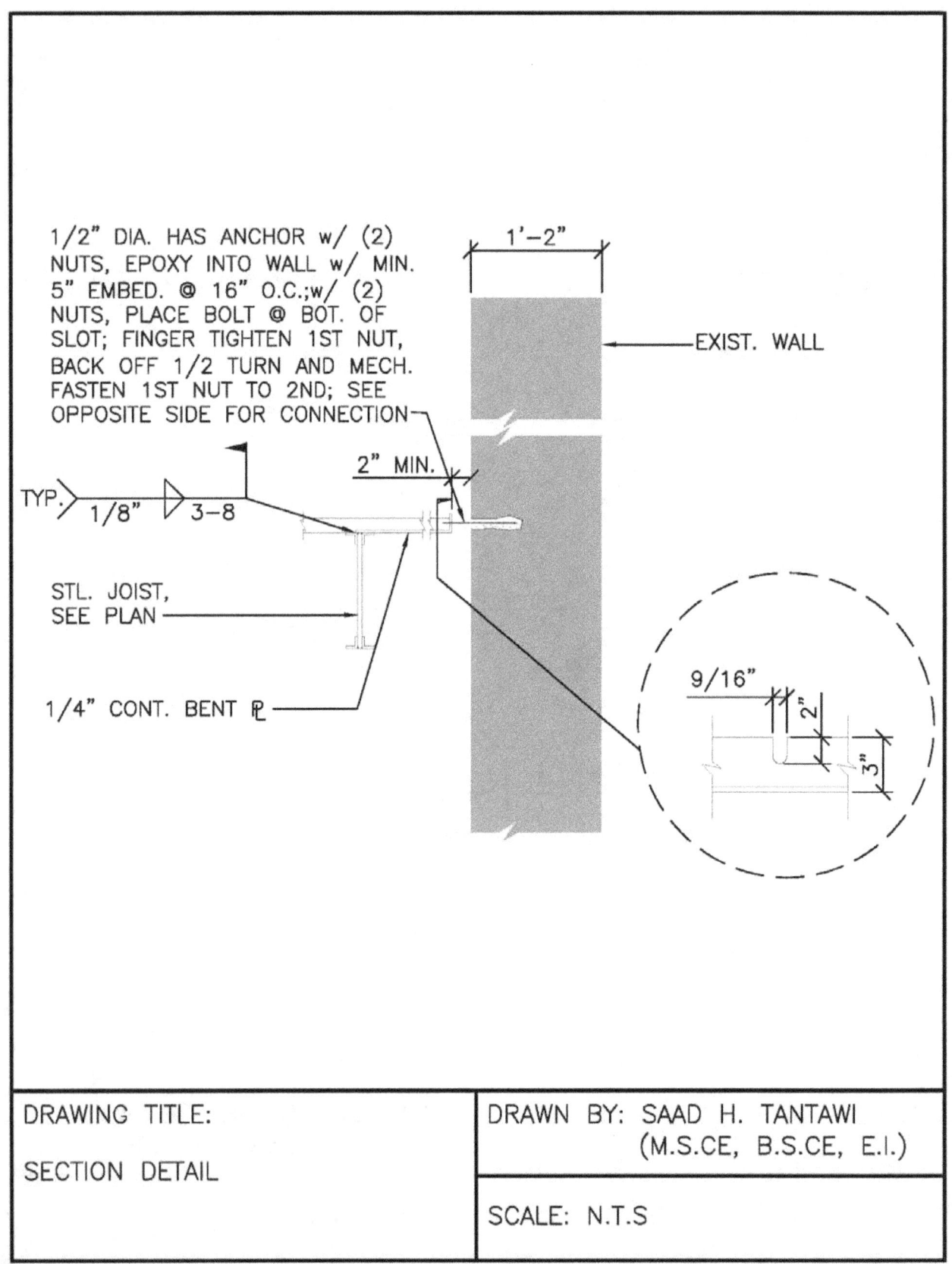

1/2" DIA. HAS ANCHOR w/ (2) NUTS, EPOXY INTO WALL w/ MIN. 5" EMBED. @ 16" O.C.;w/ (2) NUTS, PLACE BOLT @ BOT. OF SLOT; FINGER TIGHTEN 1ST NUT, BACK OFF 1/2 TURN AND MECH. FASTEN 1ST NUT TO 2ND; SEE OPPOSITE SIDE FOR CONNECTION

1'-2"

EXIST. WALL

2" MIN.

TYP. 1/8" 3-8

STL. JOIST, SEE PLAN

1/4" CONT. BENT ℙ

9/16"

2"

3"

DRAWING TITLE:	DRAWN BY: SAAD H. TANTAWI
	(M.S.CE, B.S.CE, E.I.)
SECTION DETAIL	
	SCALE: N.T.S

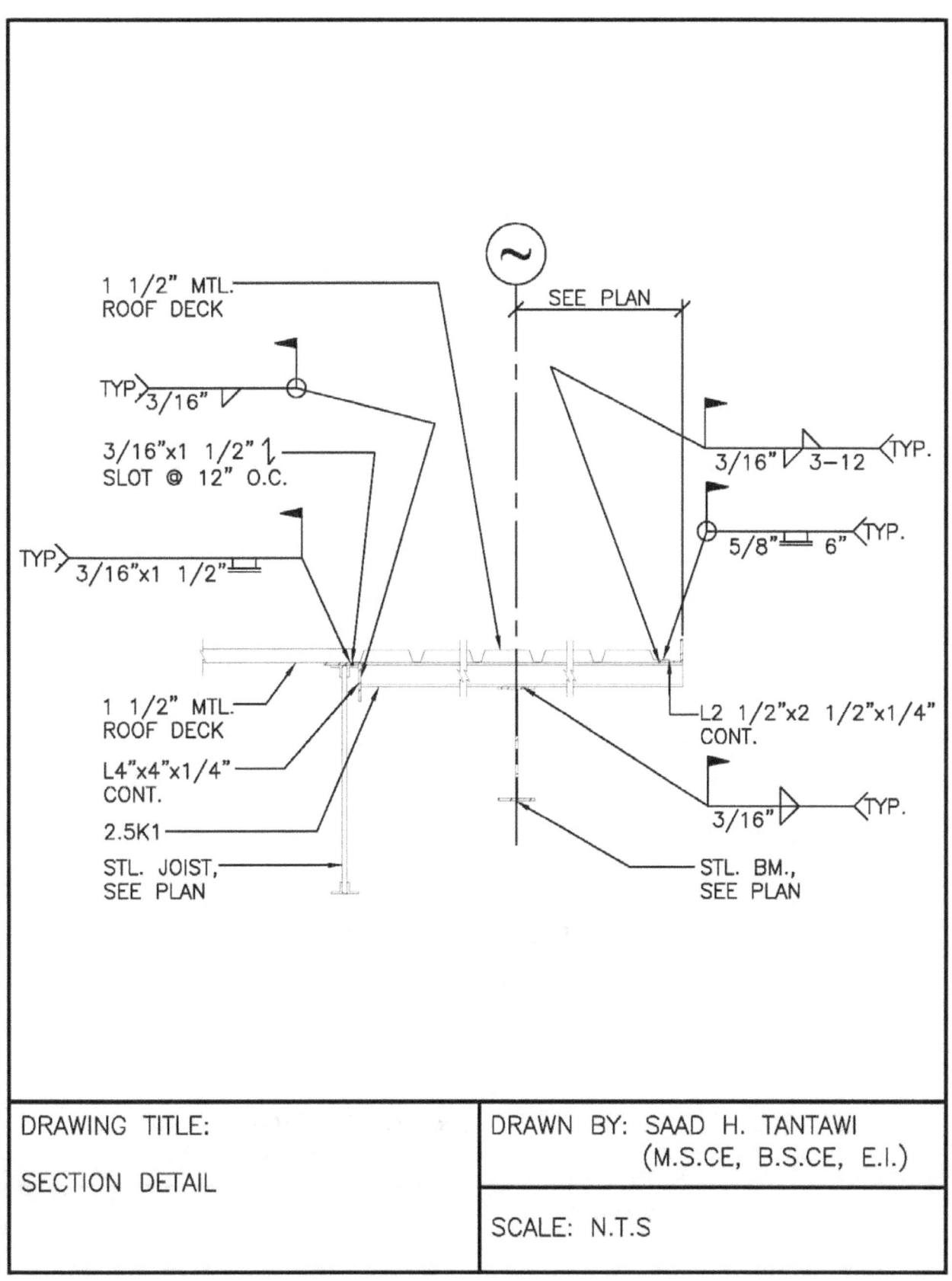

1 1/2" MTL. ROOF DECK

TYP 3/16"

3/16"x1 1/2" SLOT @ 12" O.C.

TYP 3/16"x1 1/2"

SEE PLAN

3/16" 3-12 TYP.

5/8" 6" TYP.

1 1/2" MTL. ROOF DECK

L4"x4"x1/4" CONT.

2.5K1

STL. JOIST, SEE PLAN

L2 1/2"x2 1/2"x1/4" CONT.

3/16" TYP.

STL. BM., SEE PLAN

DRAWING TITLE:	DRAWN BY: SAAD H. TANTAWI
	(M.S.CE, B.S.CE, E.I.)
SECTION DETAIL	
	SCALE: N.T.S

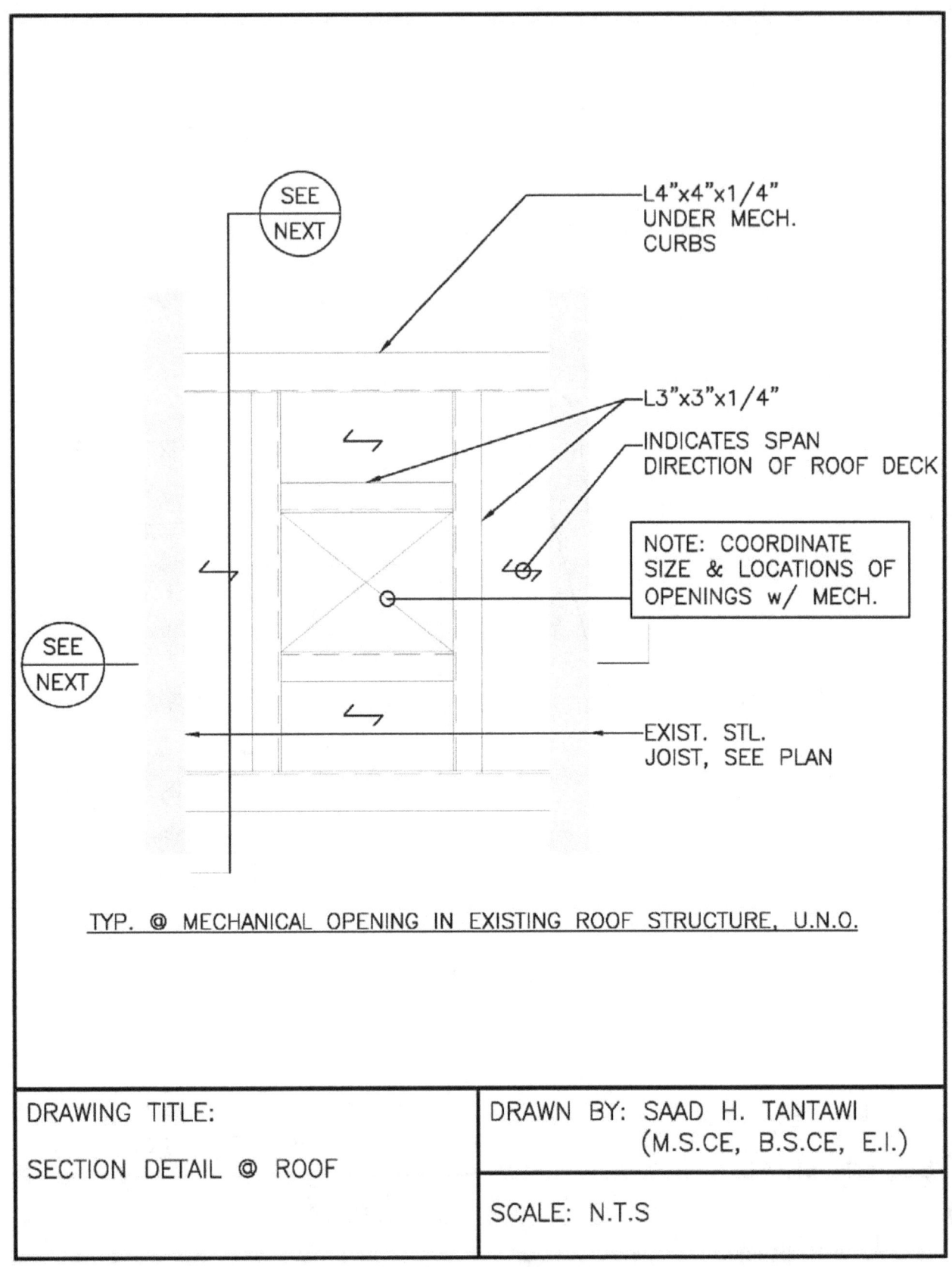

TYP. @ MECHANICAL OPENING IN EXISTING ROOF STRUCTURE, U.N.O.

DRAWING TITLE: SECTION DETAIL @ ROOF	DRAWN BY: SAAD H. TANTAWI (M.S.CE, B.S.CE, E.I.)
	SCALE: N.T.S

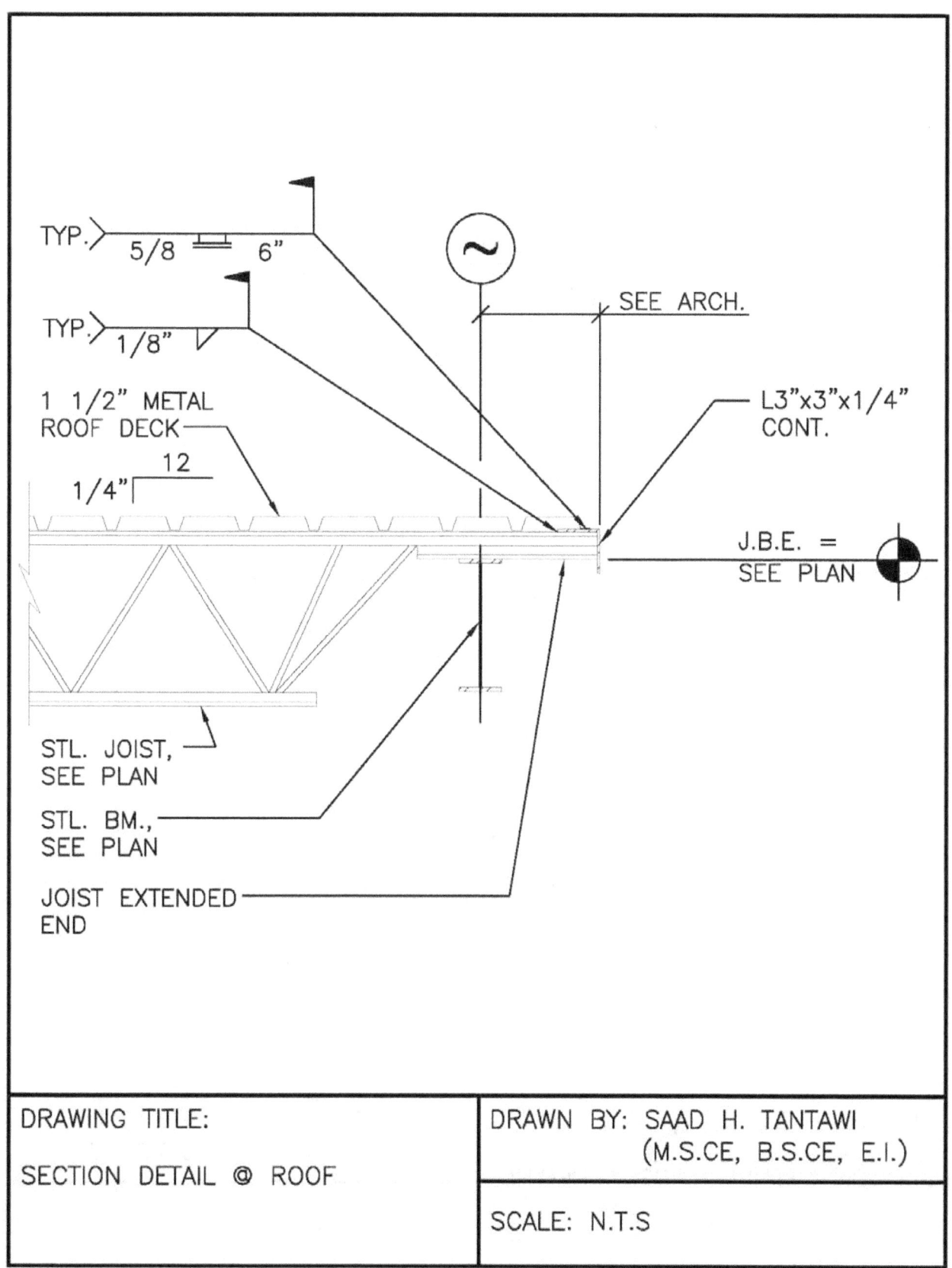

TYP. 5/8 6"

TYP. 1/8"

SEE ARCH.

1 1/2" METAL
ROOF DECK

12

1/4"

L3"x3"x1/4"
CONT.

J.B.E. =
SEE PLAN

STL. JOIST,
SEE PLAN

STL. BM.,
SEE PLAN

JOIST EXTENDED
END

DRAWING TITLE:	DRAWN BY: SAAD H. TANTAWI
	(M.S.CE, B.S.CE, E.I.)
SECTION DETAIL @ ROOF	
	SCALE: N.T.S

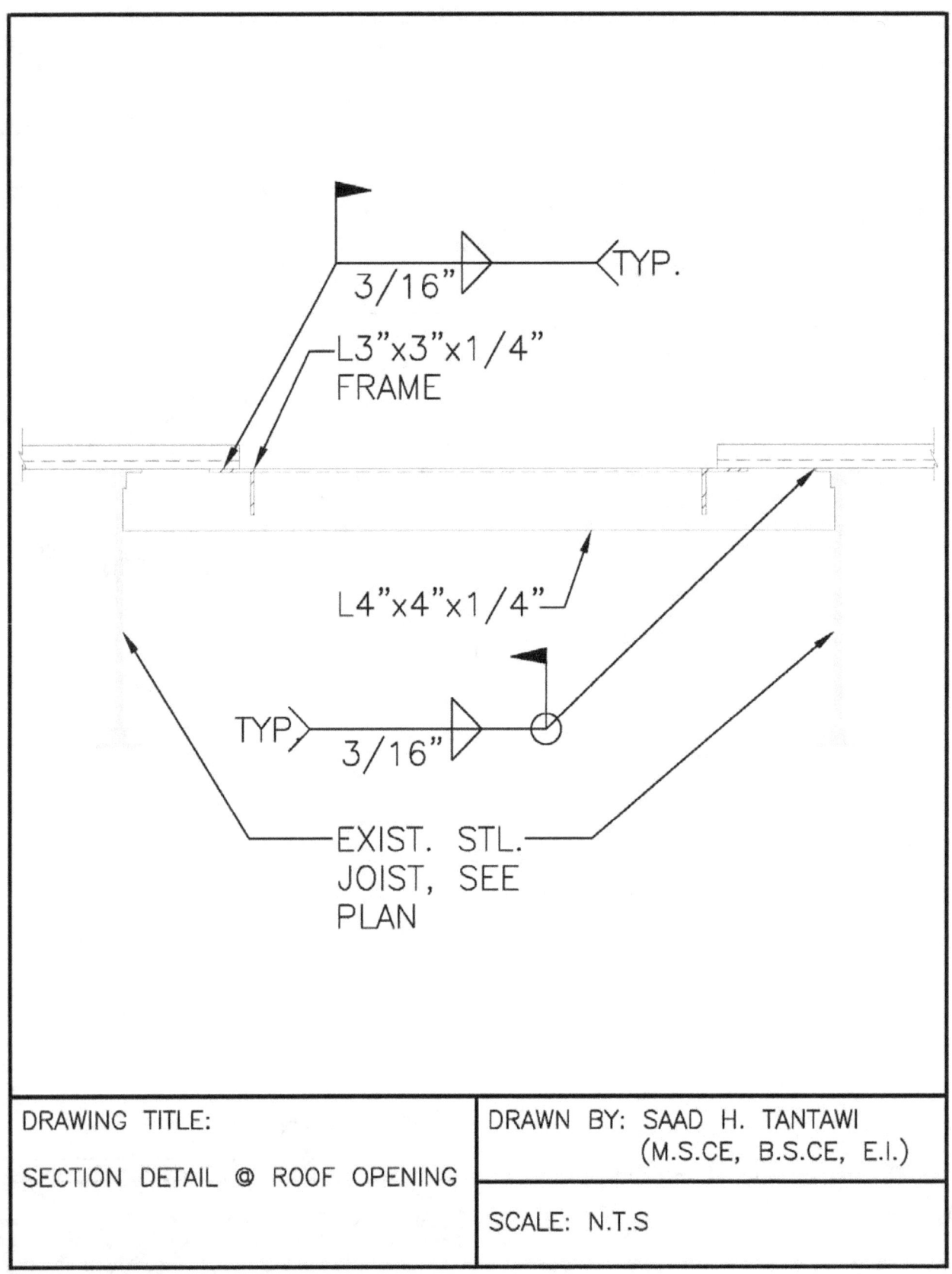

3/16"   TYP.

L3"x3"x1/4"
FRAME

L4"x4"x1/4"

TYP   3/16"

EXIST. STL.
JOIST, SEE
PLAN

DRAWING TITLE:	DRAWN BY: SAAD H. TANTAWI
	(M.S.CE, B.S.CE, E.I.)
SECTION DETAIL @ ROOF OPENING	
	SCALE: N.T.S

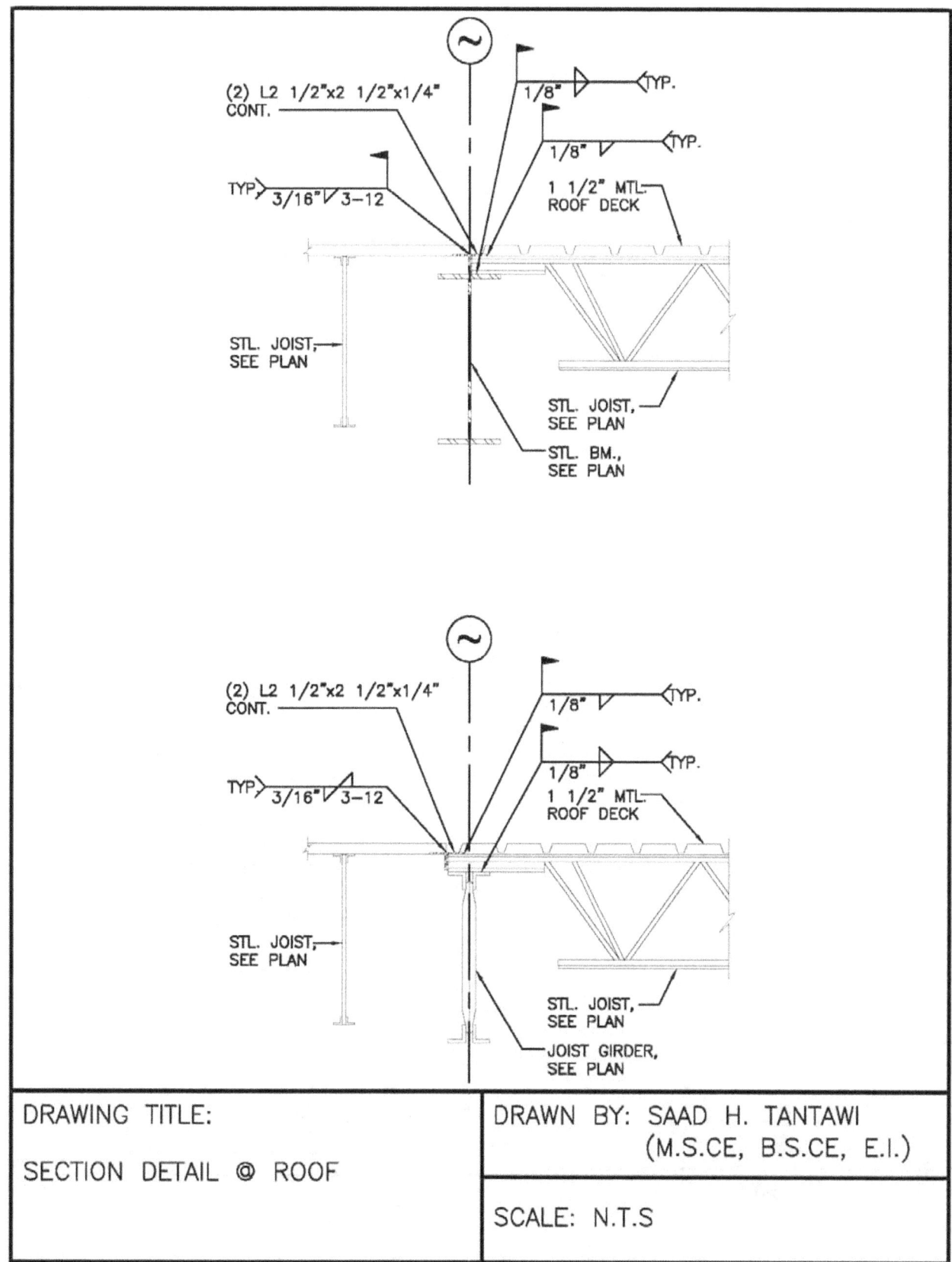

DRAWING TITLE:	DRAWN BY: SAAD H. TANTAWI
SECTION DETAIL @ ROOF	(M.S.CE, B.S.CE, E.I.)
	SCALE: N.T.S

Saad Hasan Tantawi (M.S.CE, B.S.CE, E.I., A.M.ASCE)

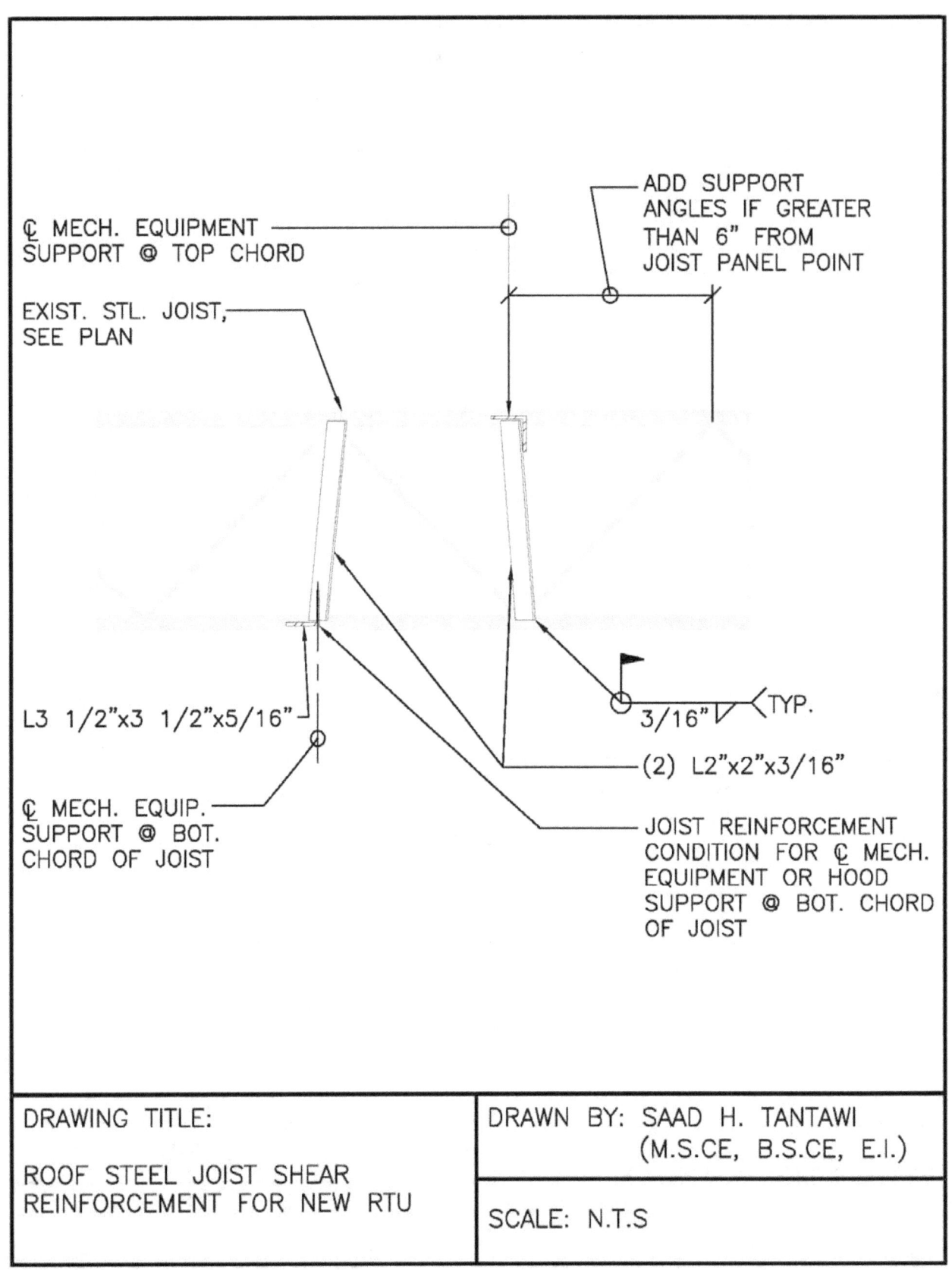

DRAWING TITLE:

ROOF STEEL JOIST SHEAR
REINFORCEMENT FOR NEW RTU

DRAWN BY: SAAD H. TANTAWI
(M.S.CE, B.S.CE, E.I.)

SCALE: N.T.S

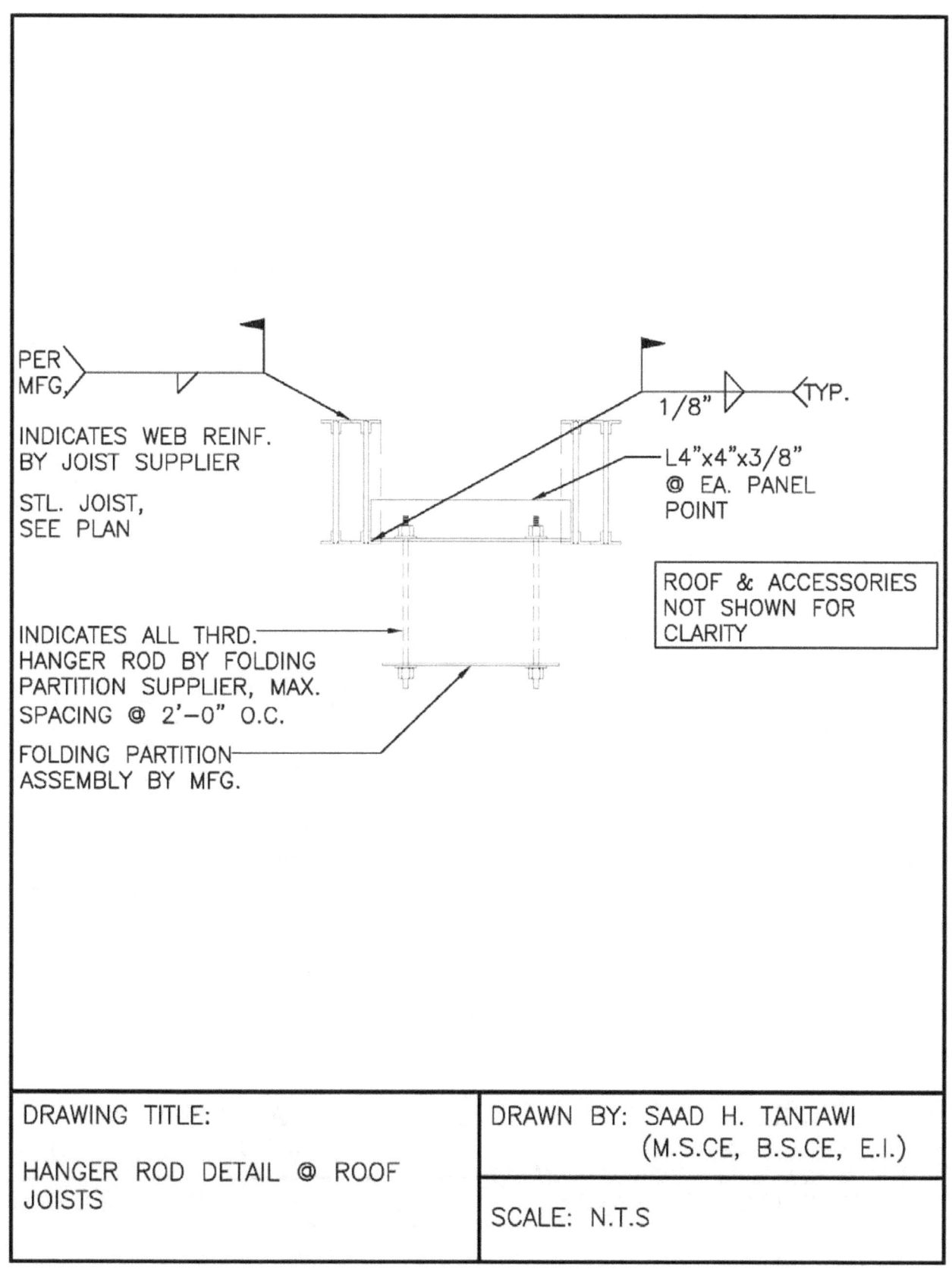

PER
MFG,

INDICATES WEB REINF.
BY JOIST SUPPLIER

STL. JOIST,
SEE PLAN

1/8"

TYP.

L4"x4"x3/8"
@ EA. PANEL
POINT

ROOF & ACCESSORIES
NOT SHOWN FOR
CLARITY

INDICATES ALL THRD.
HANGER ROD BY FOLDING
PARTITION SUPPLIER, MAX.
SPACING @ 2'-0" O.C.

FOLDING PARTITION
ASSEMBLY BY MFG.

DRAWING TITLE:	DRAWN BY: SAAD H. TANTAWI
	(M.S.CE, B.S.CE, E.I.)
HANGER ROD DETAIL @ ROOF JOISTS	SCALE: N.T.S

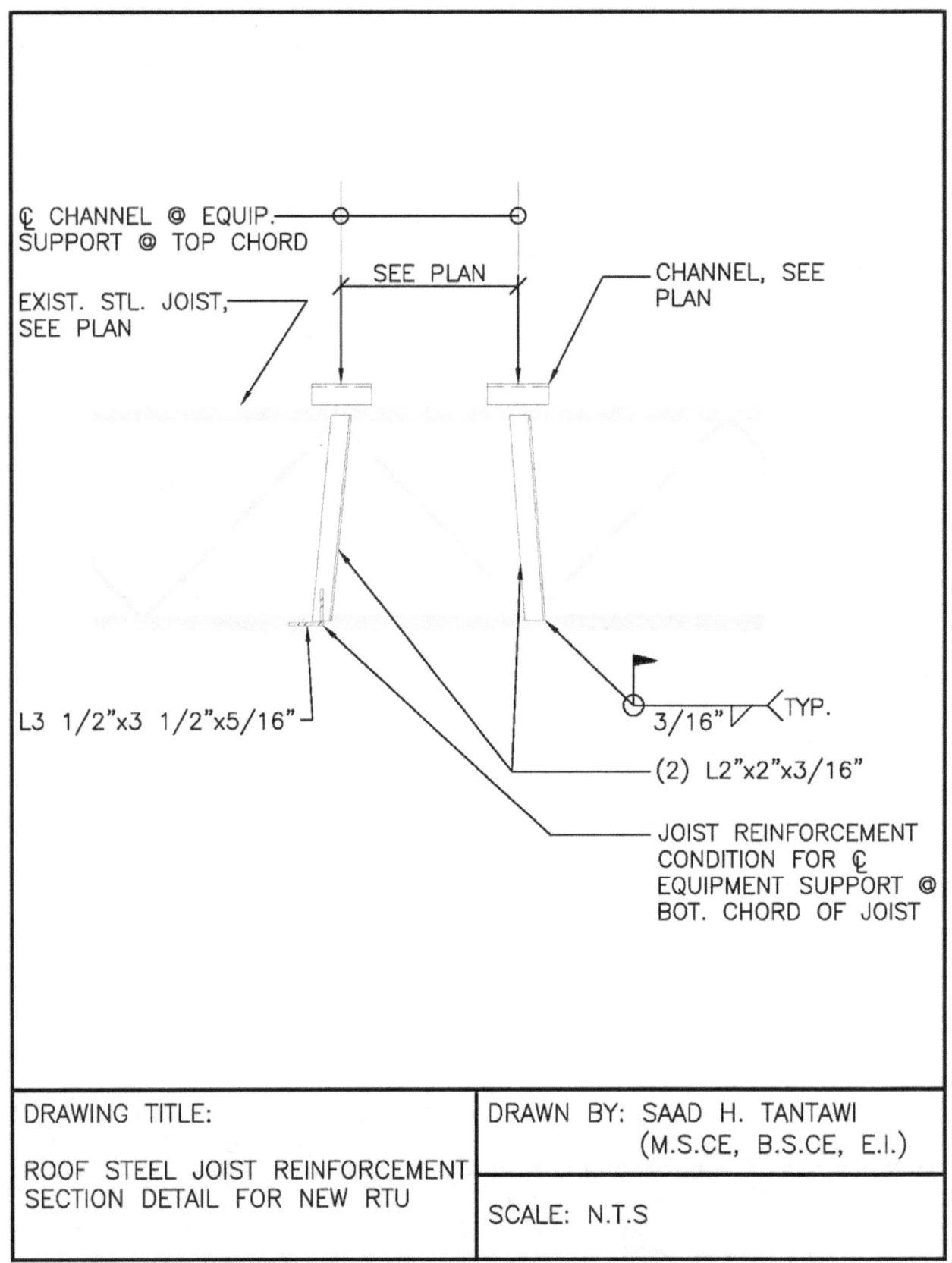

₵ CHANNEL @ EQUIP.
SUPPORT @ TOP CHORD

EXIST. STL. JOIST,
SEE PLAN

SEE PLAN

CHANNEL, SEE
PLAN

L3 1/2"x3 1/2"x5/16"

3/16" TYP.

(2) L2"x2"x3/16"

JOIST REINFORCEMENT
CONDITION FOR ₵
EQUIPMENT SUPPORT @
BOT. CHORD OF JOIST

DRAWING TITLE:	DRAWN BY: SAAD H. TANTAWI (M.S.CE, B.S.CE, E.I.)
ROOF STEEL JOIST REINFORCEMENT SECTION DETAIL FOR NEW RTU	SCALE: N.T.S

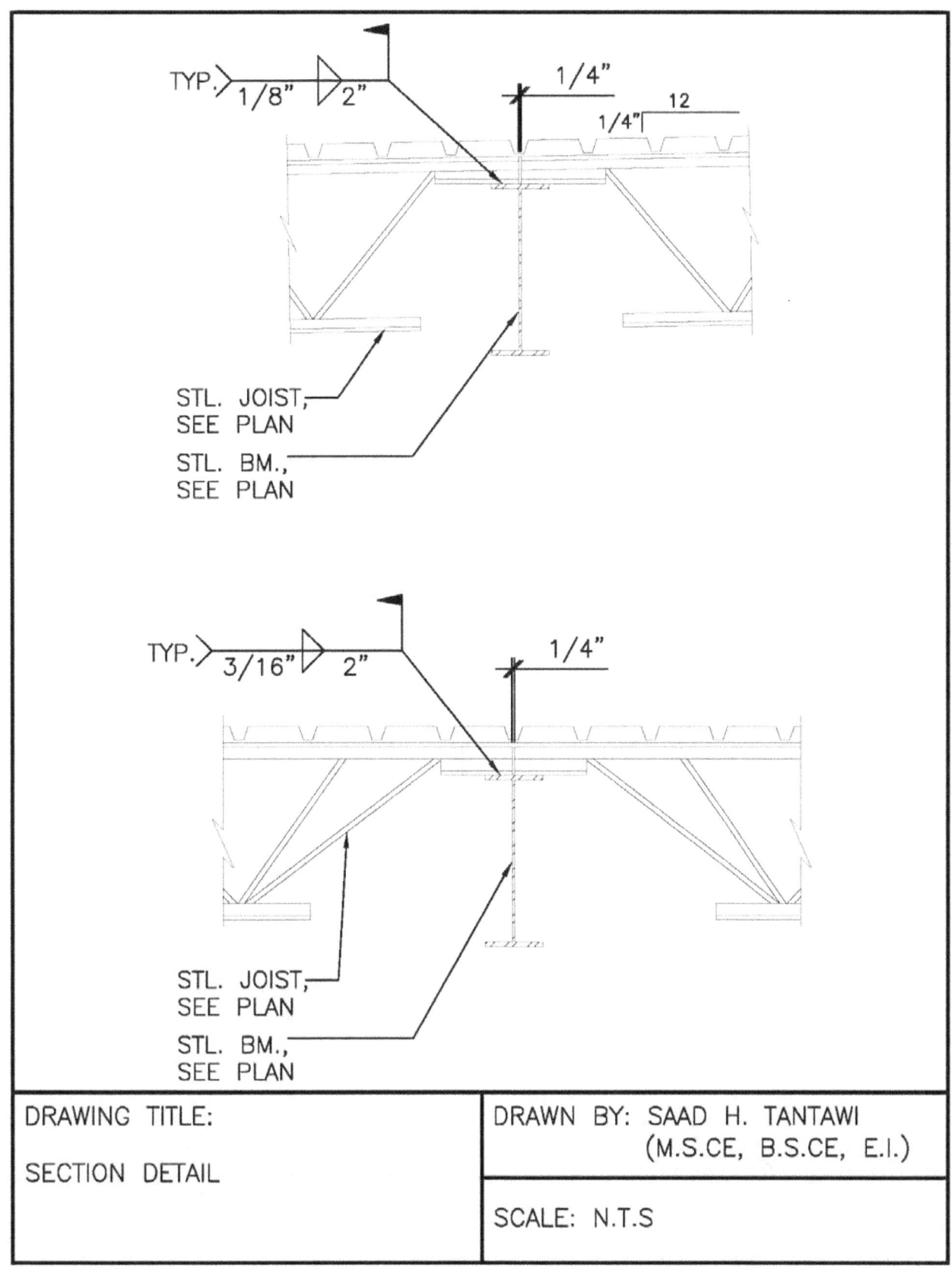

DRAWING TITLE:	DRAWN BY: SAAD H. TANTAWI
	(M.S.CE, B.S.CE, E.I.)
SECTION DETAIL	SCALE: N.T.S

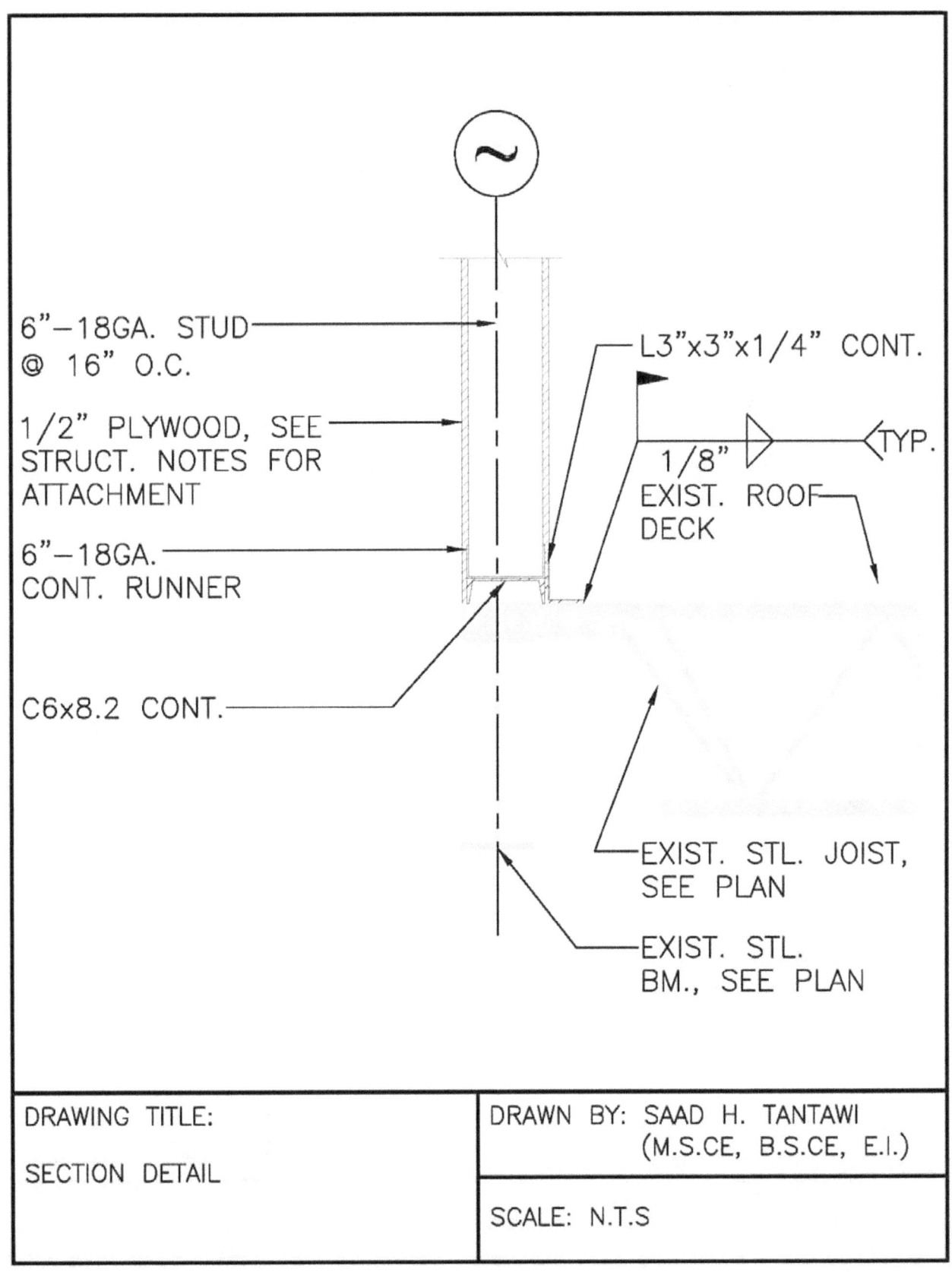

6"−18GA. STUD @ 16" O.C.

1/2" PLYWOOD, SEE STRUCT. NOTES FOR ATTACHMENT

6"−18GA. CONT. RUNNER

C6x8.2 CONT.

L3"x3"x1/4" CONT.

1/8"

EXIST. ROOF DECK

TYP.

EXIST. STL. JOIST, SEE PLAN

EXIST. STL. BM., SEE PLAN

DRAWING TITLE:	DRAWN BY: SAAD H. TANTAWI (M.S.CE, B.S.CE, E.I.)
SECTION DETAIL	SCALE: N.T.S

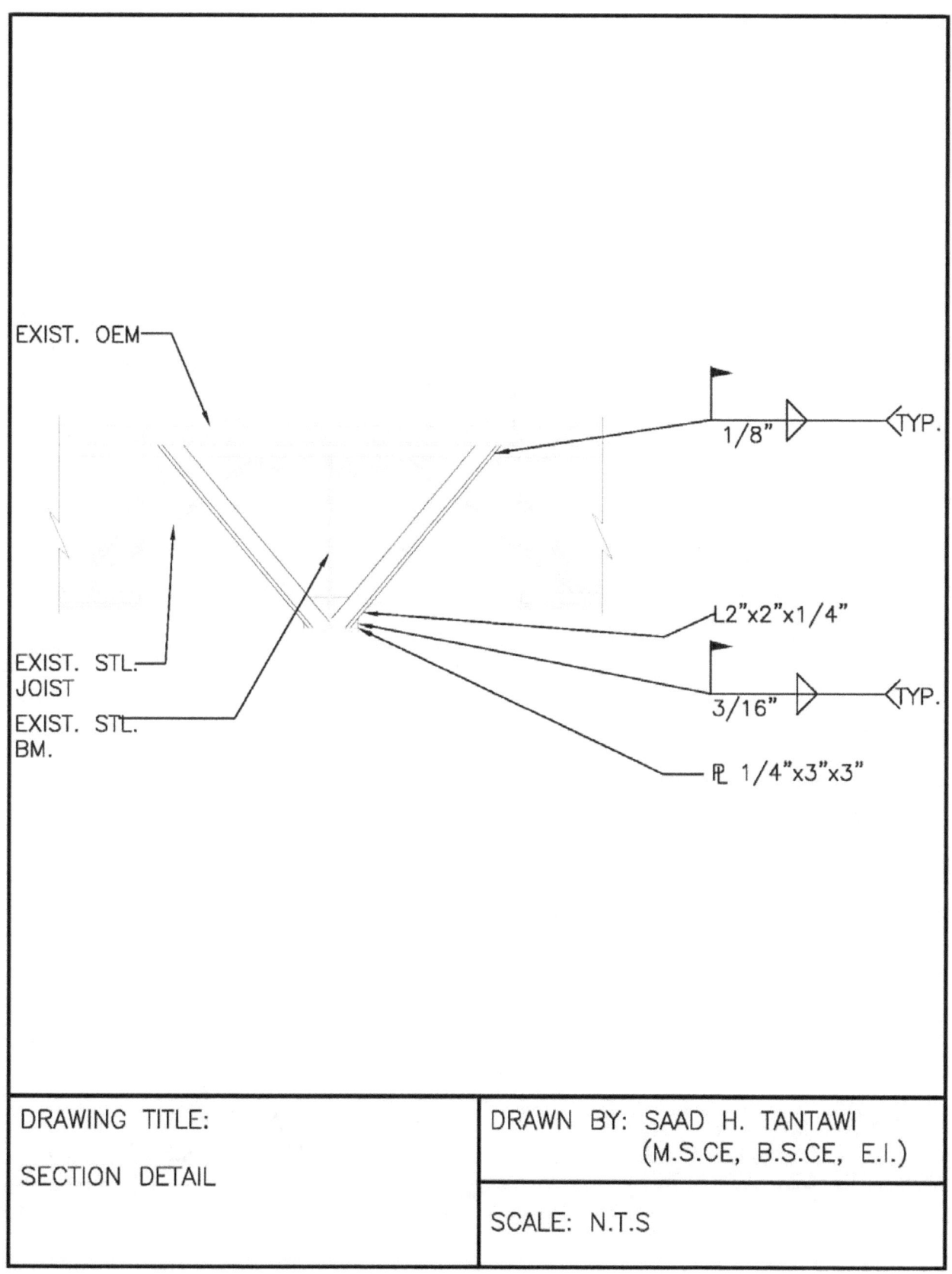

EXIST. OEM

1/8"   TYP.

EXIST. STL.
JOIST

EXIST. STL.
BM.

L2"x2"x1/4"

3/16"   TYP.

PL 1/4"x3"x3"

DRAWING TITLE:	DRAWN BY: SAAD H. TANTAWI
SECTION DETAIL	(M.S.CE, B.S.CE, E.I.)
	SCALE: N.T.S

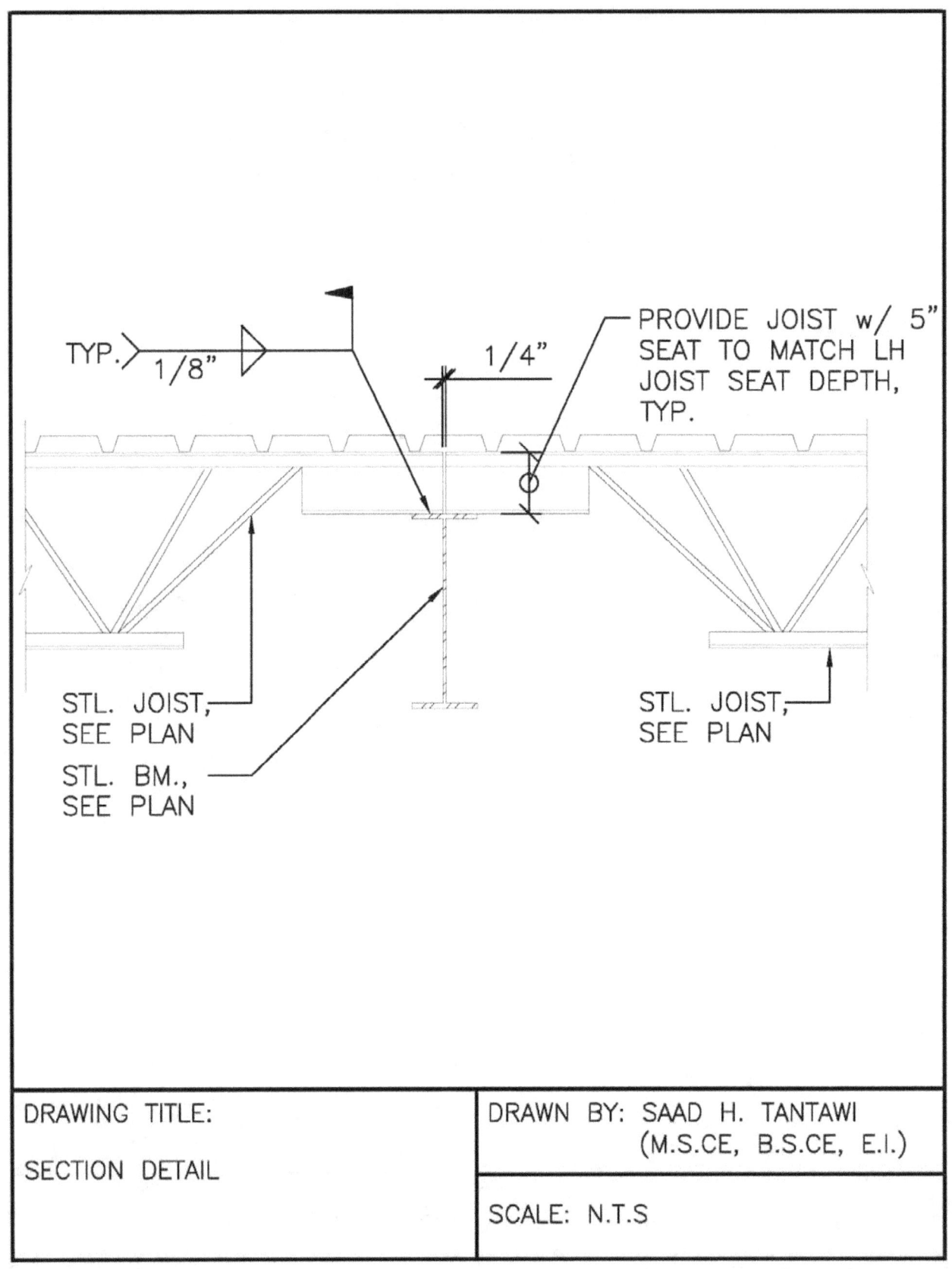

TYP. 1/8"

1/4"

PROVIDE JOIST w/ 5"
SEAT TO MATCH LH
JOIST SEAT DEPTH,
TYP.

STL. JOIST,
SEE PLAN

STL. BM.,
SEE PLAN

STL. JOIST,
SEE PLAN

DRAWING TITLE:	DRAWN BY: SAAD H. TANTAWI
	(M.S.CE, B.S.CE, E.I.)
SECTION DETAIL	
	SCALE: N.T.S

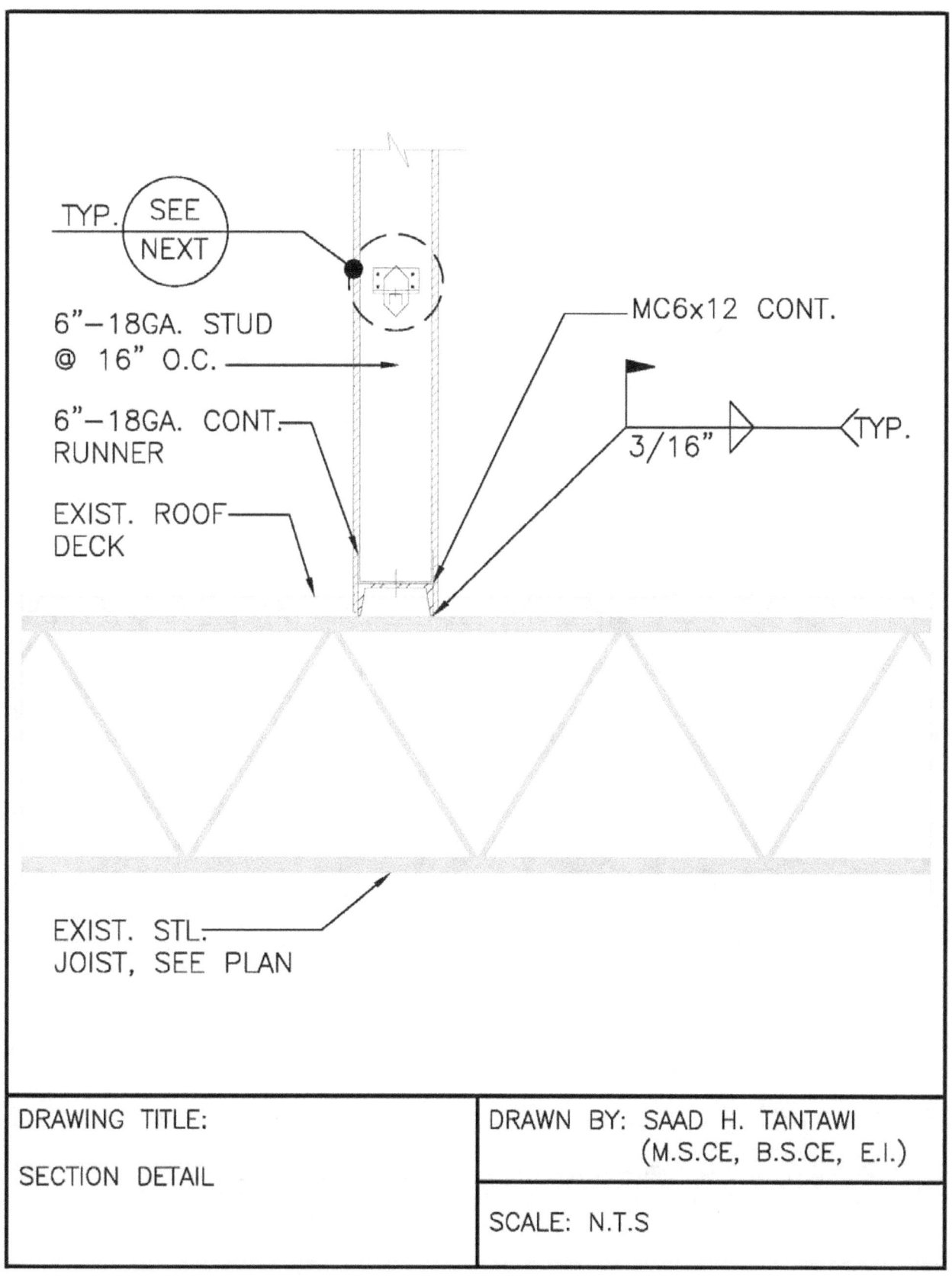

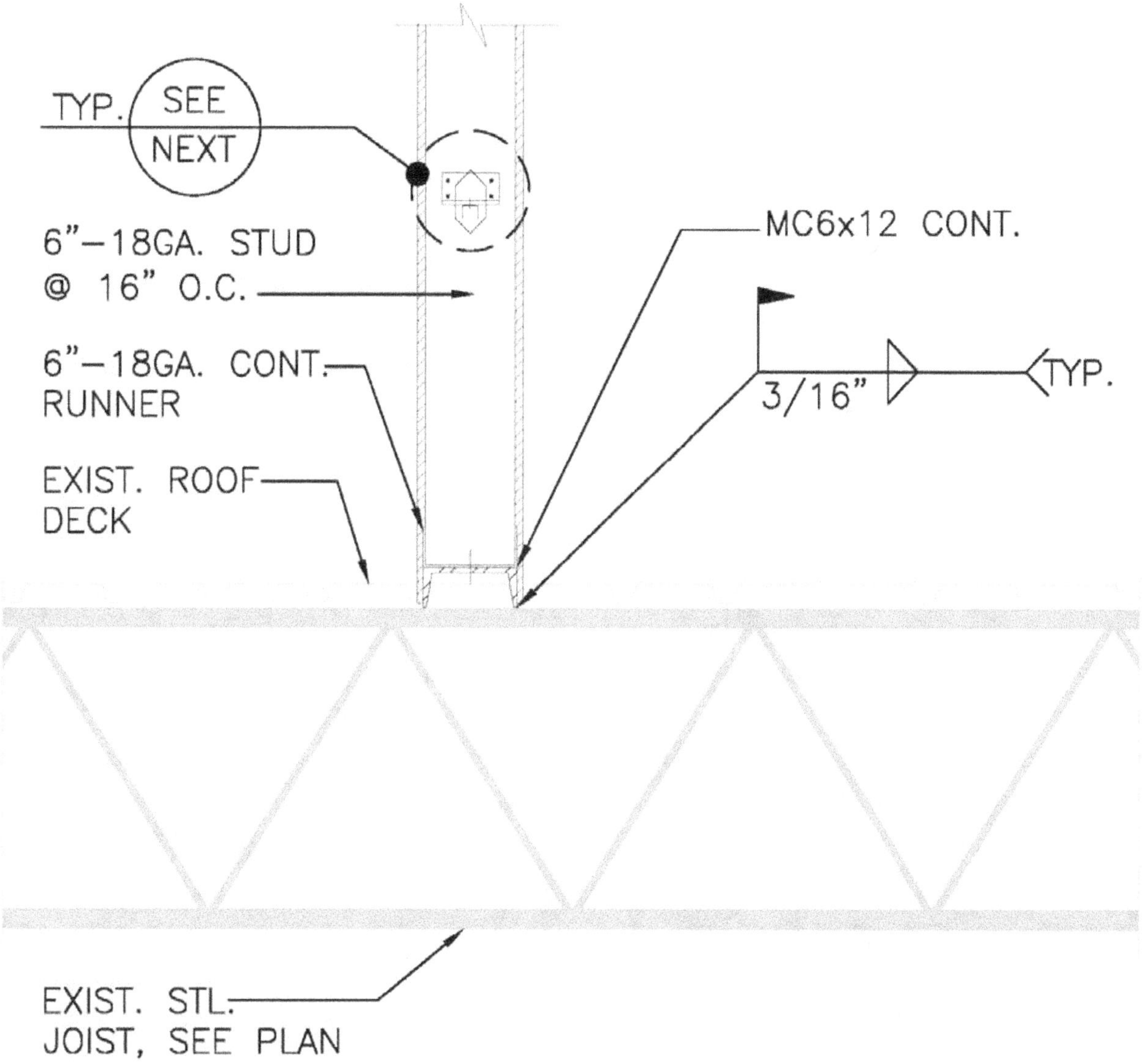
TYP.
SEE
NEXT
6"–18GA. STUD
@ 16" O.C.
6"–18GA. CONT.
RUNNER
EXIST. ROOF
DECK
MC6x12 CONT.
3/16"
TYP.
EXIST. STL.
JOIST, SEE PLAN

DRAWING TITLE: SECTION DETAIL	DRAWN BY: SAAD H. TANTAWI (M.S.CE, B.S.CE, E.I.)
	SCALE: N.T.S

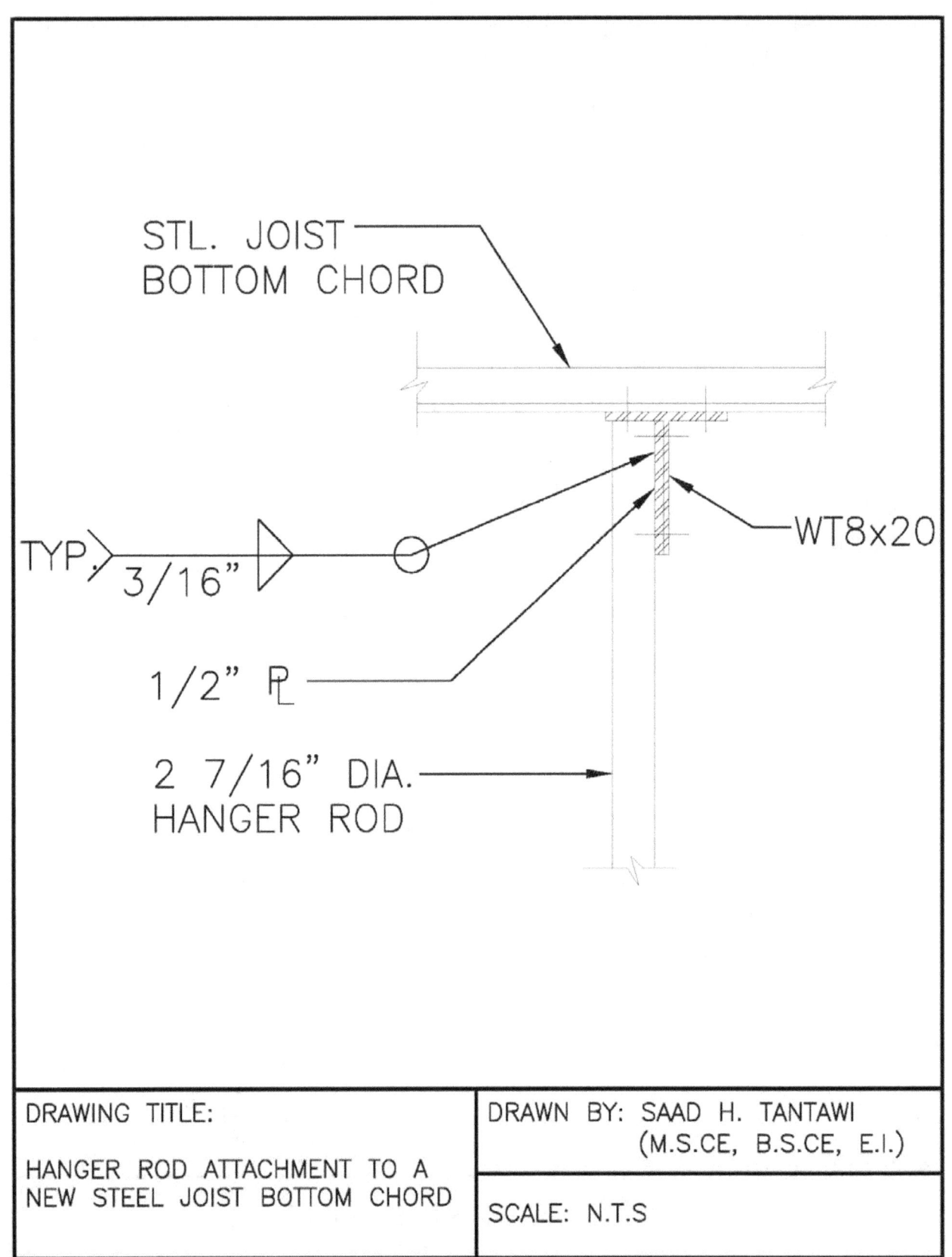

STL. JOIST
BOTTOM CHORD

TYP. 3/16"

1/2" PL

2 7/16" DIA.
HANGER ROD

WT8x20

DRAWING TITLE:	DRAWN BY: SAAD H. TANTAWI (M.S.CE, B.S.CE, E.I.)
HANGER ROD ATTACHMENT TO A NEW STEEL JOIST BOTTOM CHORD	SCALE: N.T.S

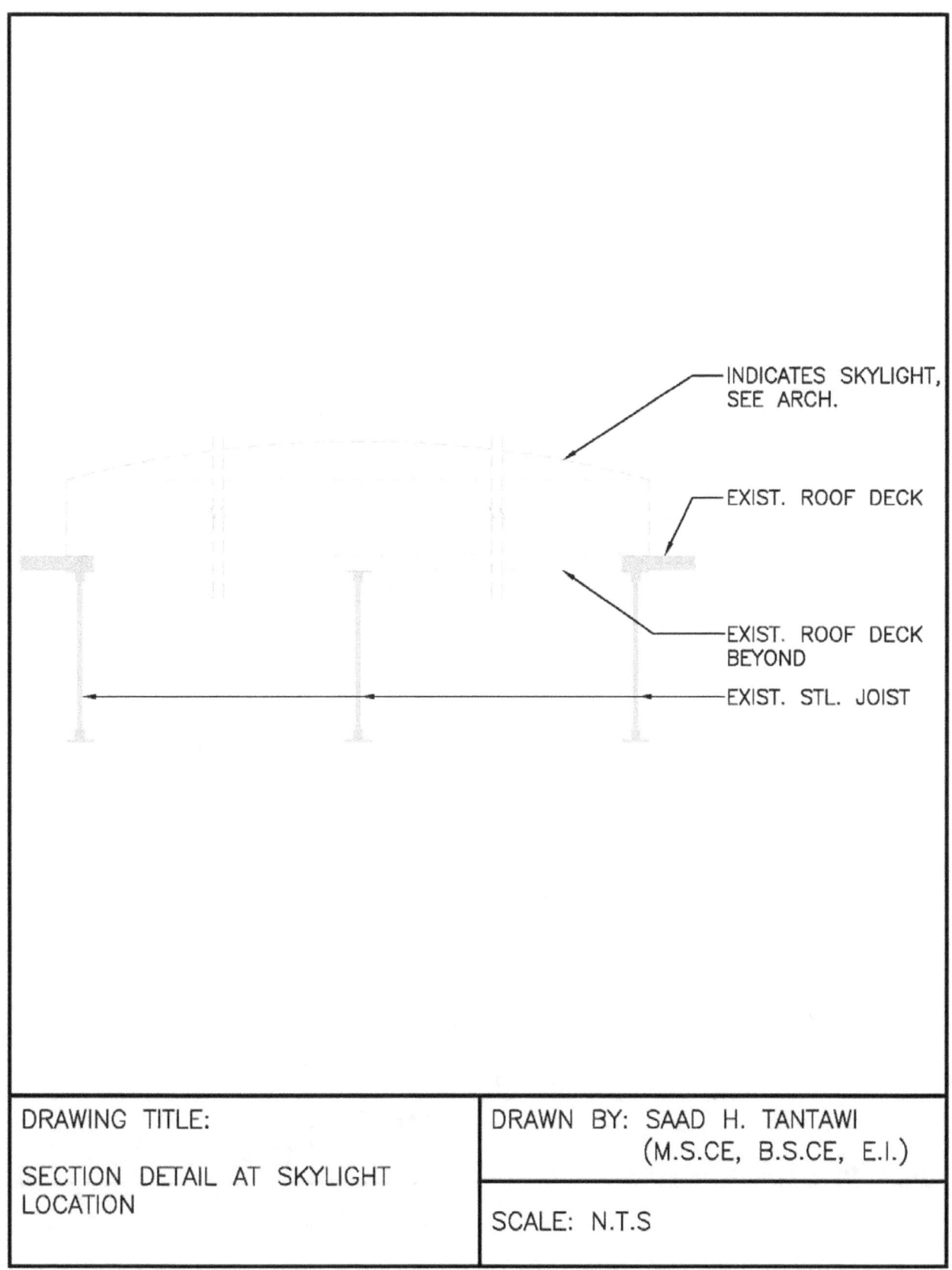

INDICATES SKYLIGHT, SEE ARCH.

EXIST. ROOF DECK

EXIST. ROOF DECK BEYOND

EXIST. STL. JOIST

DRAWING TITLE: SECTION DETAIL AT SKYLIGHT LOCATION	DRAWN BY: SAAD H. TANTAWI (M.S.CE, B.S.CE, E.I.)
	SCALE: N.T.S

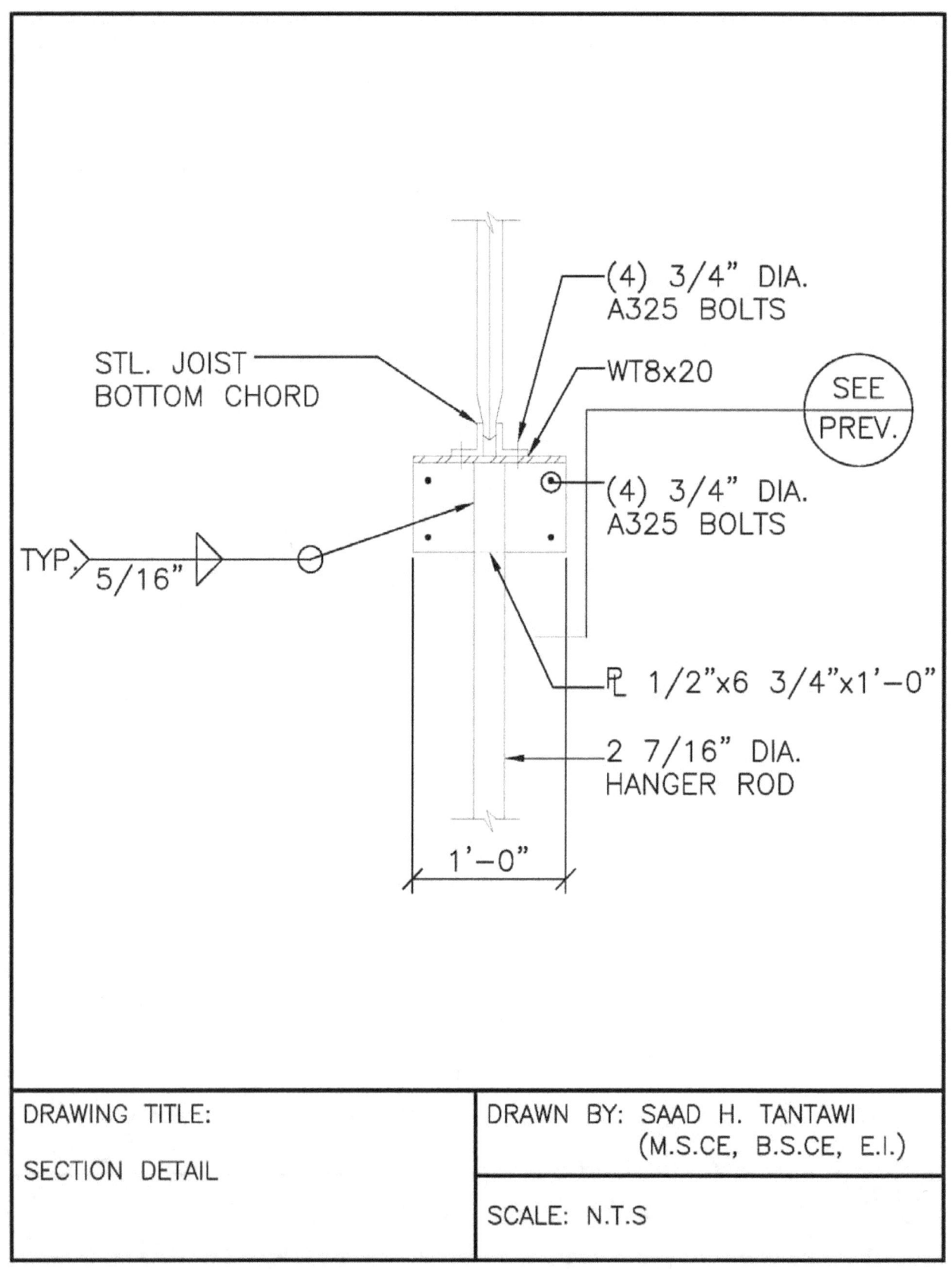

STL. JOIST
BOTTOM CHORD

(4) 3/4" DIA.
A325 BOLTS

WT8x20

SEE
PREV.

(4) 3/4" DIA.
A325 BOLTS

TYP
5/16"

℔ 1/2"x6 3/4"x1'-0"

2 7/16" DIA.
HANGER ROD

1'-0"

DRAWING TITLE:	DRAWN BY: SAAD H. TANTAWI
	(M.S.CE, B.S.CE, E.I.)
SECTION DETAIL	
	SCALE: N.T.S

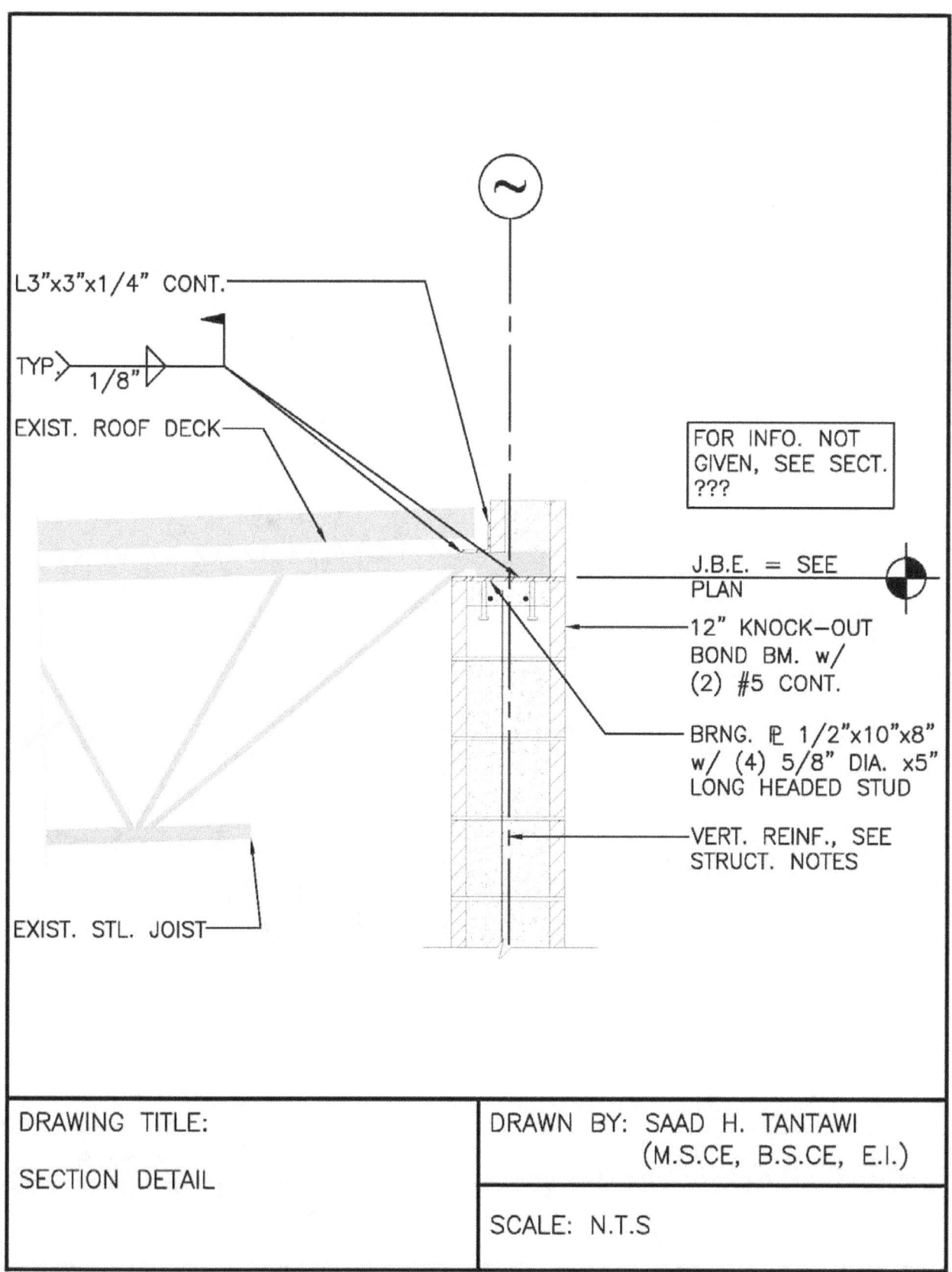

L3"x3"x1/4" CONT.

TYP. 1/8"

EXIST. ROOF DECK

EXIST. STL. JOIST

FOR INFO. NOT GIVEN, SEE SECT. ???

J.B.E. = SEE PLAN

12" KNOCK—OUT BOND BM. w/ (2) #5 CONT.

BRNG. ℔ 1/2"x10"x8" w/ (4) 5/8" DIA. x5" LONG HEADED STUD

VERT. REINF., SEE STRUCT. NOTES

DRAWING TITLE:	DRAWN BY: SAAD H. TANTAWI
SECTION DETAIL	(M.S.CE, B.S.CE, E.I.)
	SCALE: N.T.S

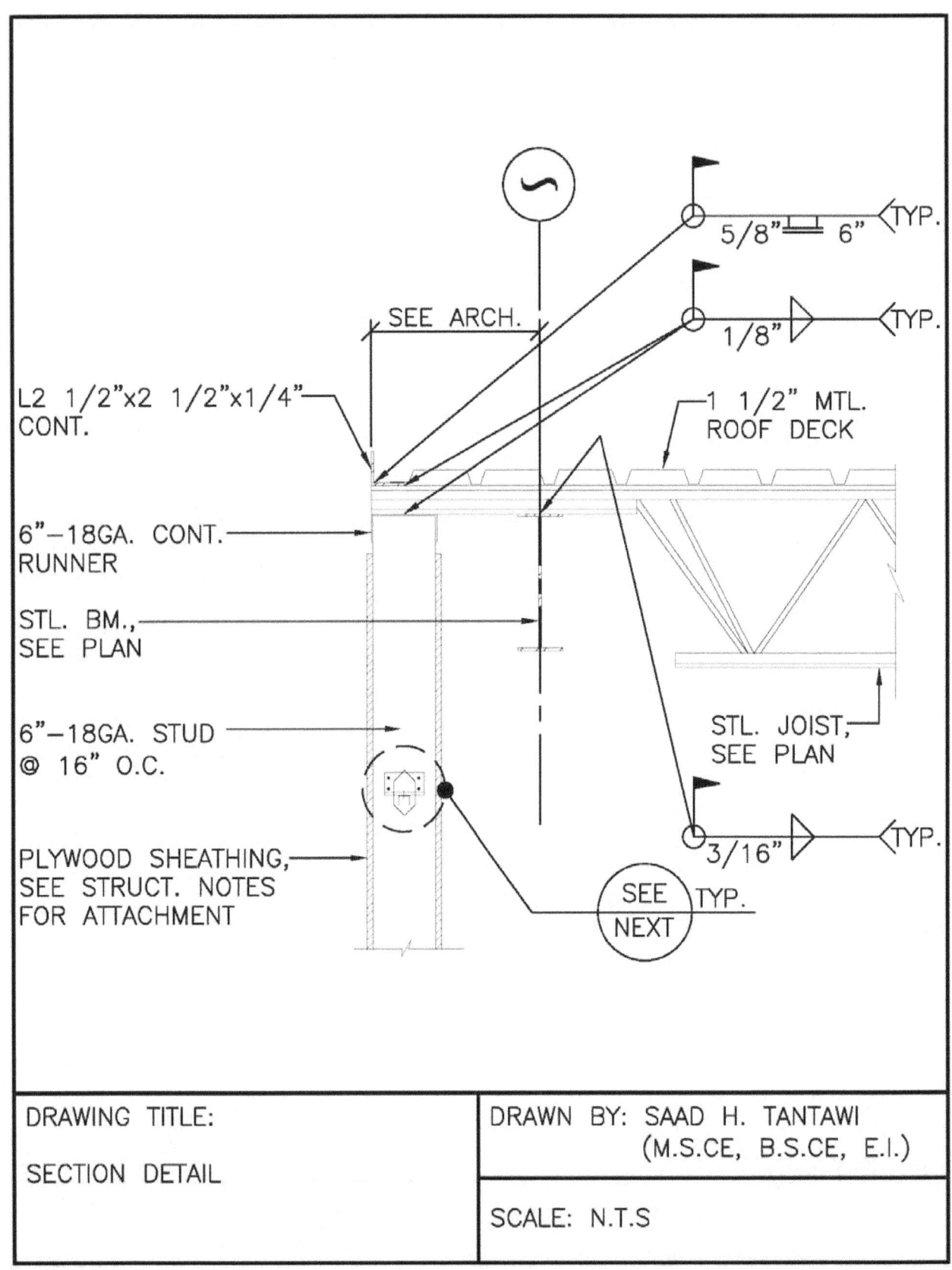

SEE ARCH.

TYP.

5/8" ☐ 6"

TYP.

1/8"

L2 1/2"x2 1/2"x1/4" CONT.

1 1/2" MTL. ROOF DECK

6"—18GA. CONT. RUNNER

STL. BM., SEE PLAN

6"—18GA. STUD @ 16" O.C.

STL. JOIST, SEE PLAN

PLYWOOD SHEATHING, SEE STRUCT. NOTES FOR ATTACHMENT

3/16"

TYP.

SEE NEXT

TYP.

DRAWING TITLE: SECTION DETAIL	DRAWN BY: SAAD H. TANTAWI (M.S.CE, B.S.CE, E.I.)
	SCALE: N.T.S

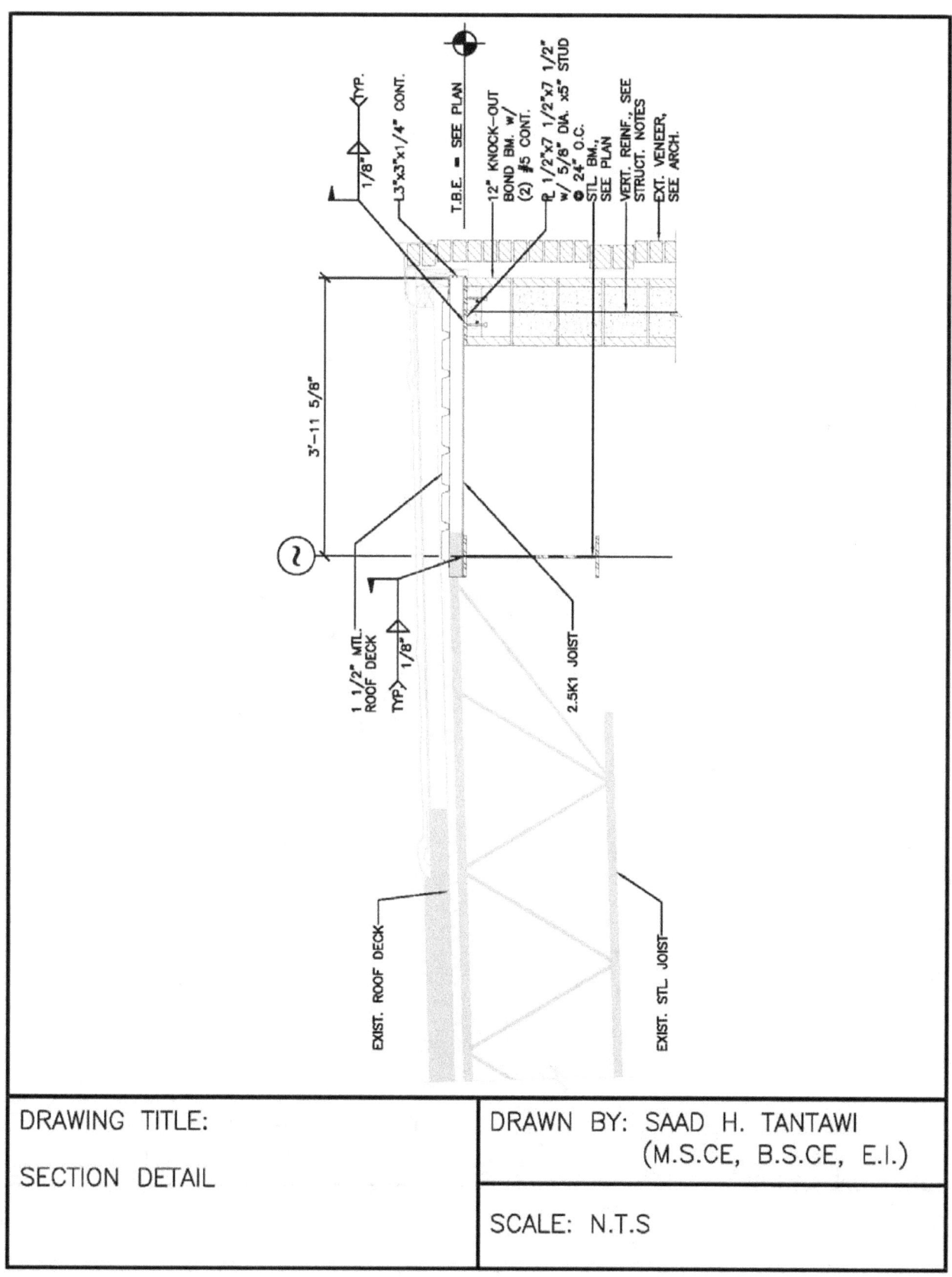

DRAWING TITLE:	DRAWN BY: SAAD H. TANTAWI
SECTION DETAIL	(M.S.CE, B.S.CE, E.I.)
	SCALE: N.T.S

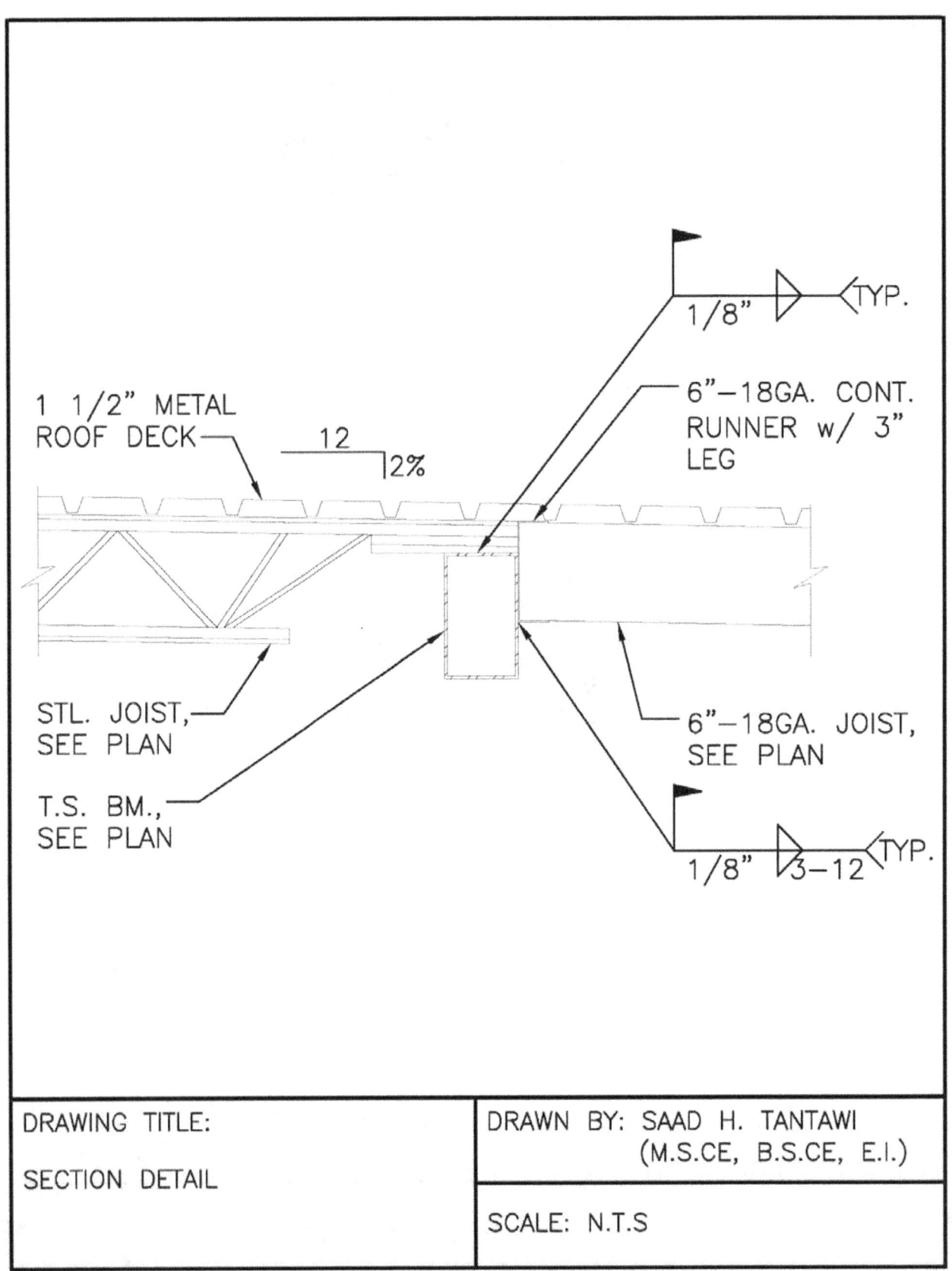

1 1/2" METAL
ROOF DECK

12
2%

1/8" TYP.

6"−18GA. CONT.
RUNNER w/ 3"
LEG

STL. JOIST,
SEE PLAN

T.S. BM.,
SEE PLAN

6"−18GA. JOIST,
SEE PLAN

1/8" 3−12 TYP.

DRAWING TITLE:	DRAWN BY: SAAD H. TANTAWI (M.S.CE, B.S.CE, E.I.)
SECTION DETAIL	SCALE: N.T.S

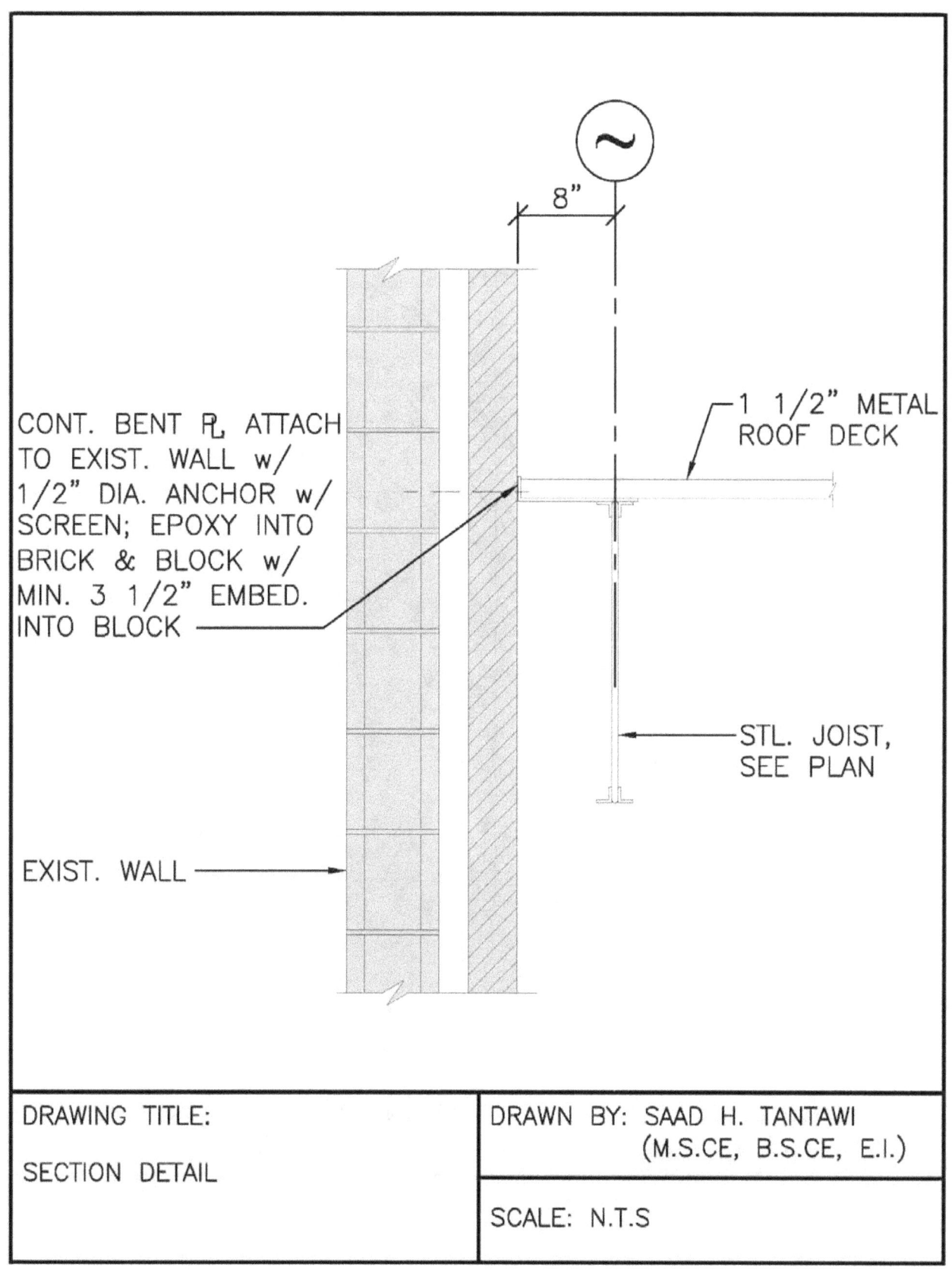

CONT. BENT P̶L̶ ATTACH
TO EXIST. WALL w/
1/2" DIA. ANCHOR w/
SCREEN; EPOXY INTO
BRICK & BLOCK w/
MIN. 3 1/2" EMBED.
INTO BLOCK

8"

1 1/2" METAL
ROOF DECK

STL. JOIST,
SEE PLAN

EXIST. WALL

DRAWING TITLE:	DRAWN BY: SAAD H. TANTAWI
	(M.S.CE, B.S.CE, E.I.)
SECTION DETAIL	
	SCALE: N.T.S

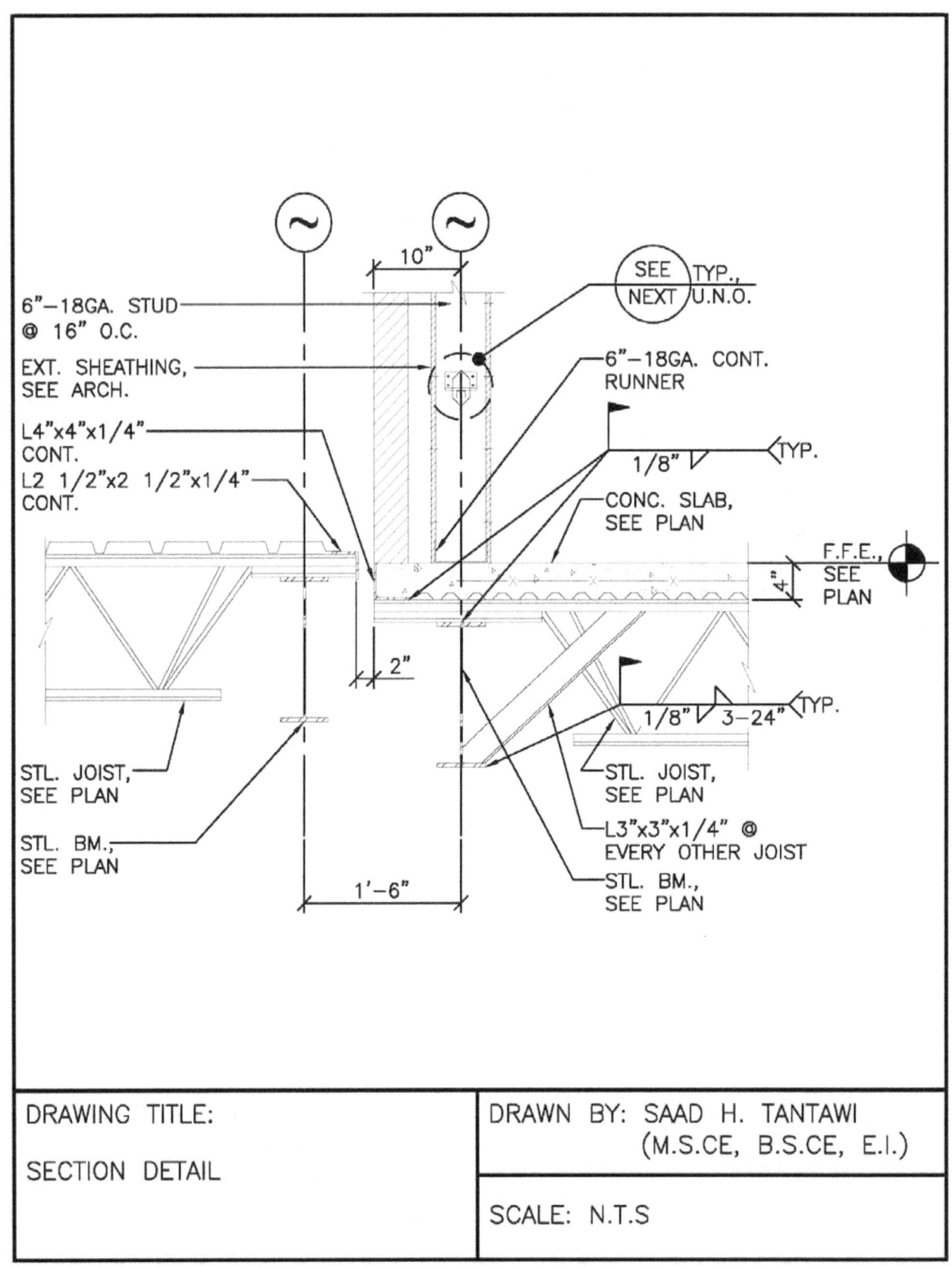

DRAWING TITLE:	DRAWN BY: SAAD H. TANTAWI
SECTION DETAIL	(M.S.CE, B.S.CE, E.I.)
	SCALE: N.T.S

Within the drawing:

6"-18GA. STUD @ 16" O.C.

EXT. SHEATHING, SEE ARCH.

L4"x4"x1/4" CONT.

L2 1/2"x2 1/2"x1/4" CONT.

10"

SEE NEXT TYP., U.N.O.

6"-18GA. CONT. RUNNER

1/8" TYP.

CONC. SLAB, SEE PLAN

F.F.E., SEE PLAN

4"

2"

1/8" 3-24" TYP.

STL. JOIST, SEE PLAN

STL. JOIST, SEE PLAN

STL. BM., SEE PLAN

L3"x3"x1/4" @ EVERY OTHER JOIST

STL. BM., SEE PLAN

1'-6"

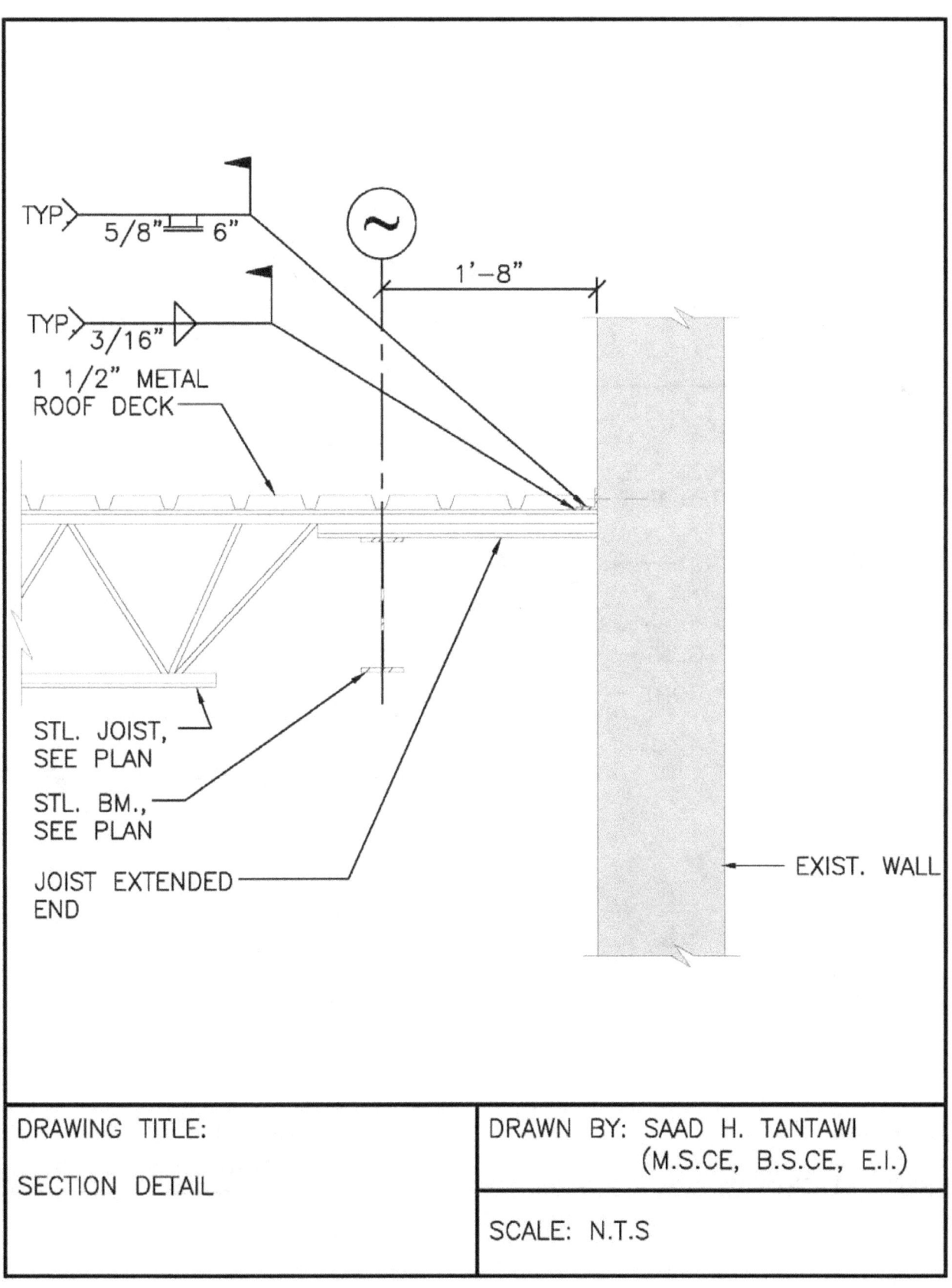

TYP⟩ 5/8" ⊢ 6"

TYP⟩ 3/16"

∼

1'–8"

1 1/2" METAL
ROOF DECK

STL. JOIST,
SEE PLAN

STL. BM.,
SEE PLAN

JOIST EXTENDED
END

EXIST. WALL

DRAWING TITLE:	DRAWN BY: SAAD H. TANTAWI
	(M.S.CE, B.S.CE, E.I.)
SECTION DETAIL	
	SCALE: N.T.S

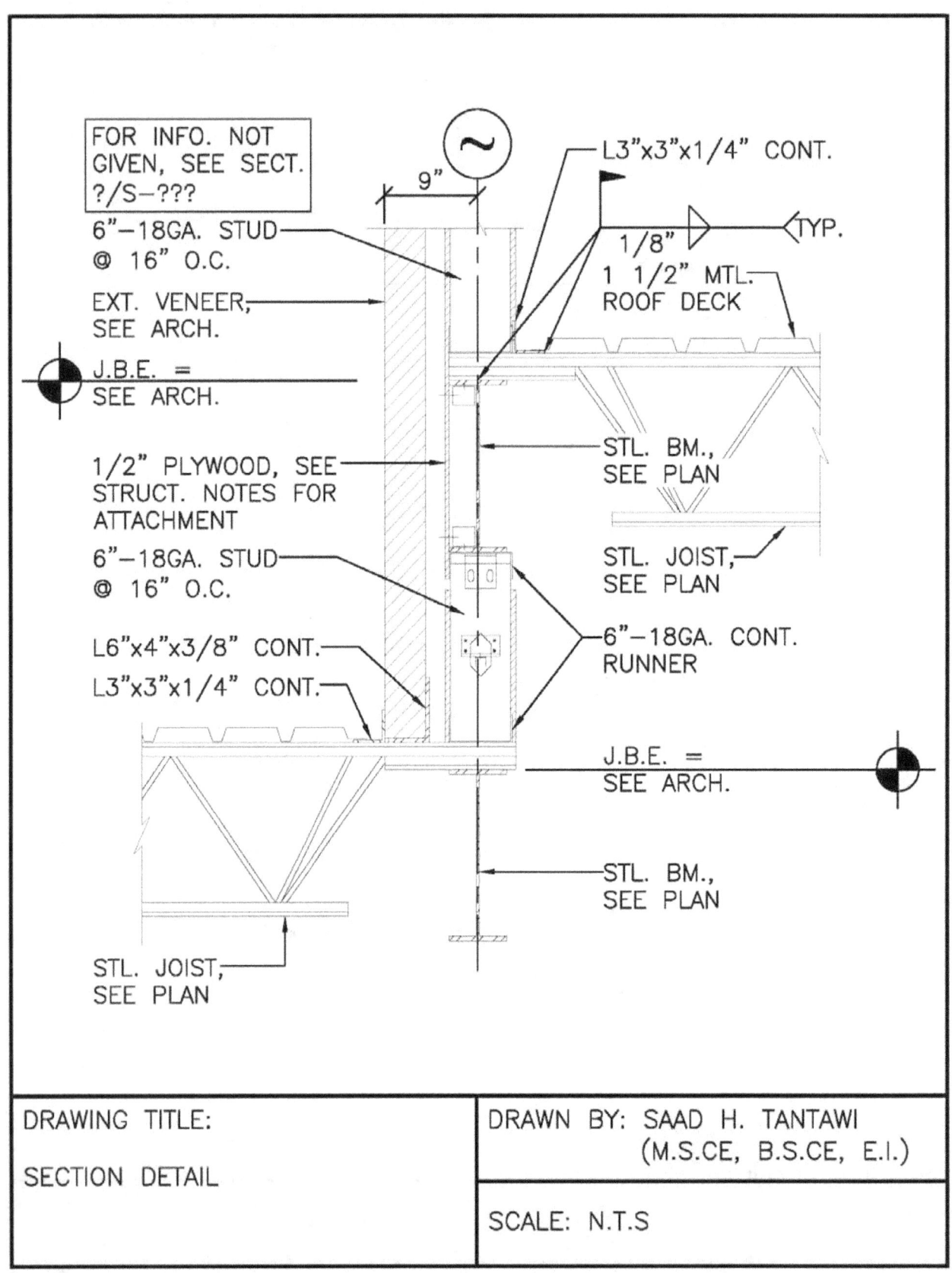

FOR INFO. NOT
GIVEN, SEE SECT.
?/S-???

6"-18GA. STUD
@ 16" O.C.

EXT. VENEER,
SEE ARCH.

J.B.E. =
SEE ARCH.

1/2" PLYWOOD, SEE
STRUCT. NOTES FOR
ATTACHMENT

6"-18GA. STUD
@ 16" O.C.

L6"x4"x3/8" CONT.
L3"x3"x1/4" CONT.

STL. JOIST,
SEE PLAN

9"

L3"x3"x1/4" CONT.

1/8"   TYP.

1 1/2" MTL.
ROOF DECK

STL. BM.,
SEE PLAN

STL. JOIST,
SEE PLAN

6"-18GA. CONT.
RUNNER

J.B.E. =
SEE ARCH.

STL. BM.,
SEE PLAN

DRAWING TITLE:	DRAWN BY: SAAD H. TANTAWI
	(M.S.CE, B.S.CE, E.I.)
SECTION DETAIL	
	SCALE: N.T.S

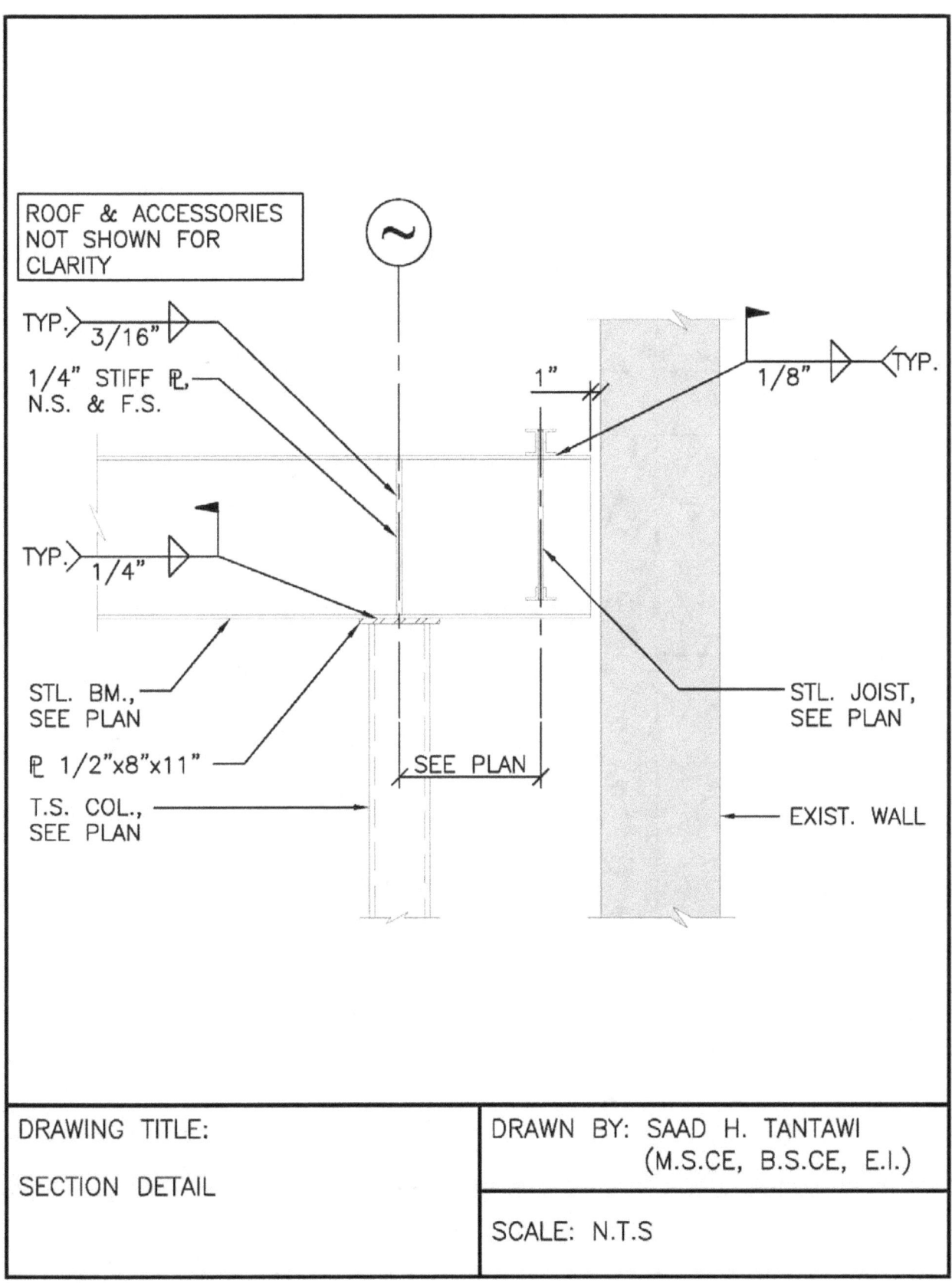

ROOF & ACCESSORIES
NOT SHOWN FOR
CLARITY

TYP. 3/16"

1/4" STIFF PL,
N.S. & F.S.

TYP. 1/4"

1"

1/8"   TYP.

STL. BM.,
SEE PLAN

PL 1/2"x8"x11"

T.S. COL.,
SEE PLAN

SEE PLAN

STL. JOIST,
SEE PLAN

EXIST. WALL

DRAWING TITLE:	DRAWN BY: SAAD H. TANTAWI (M.S.CE, B.S.CE, E.I.)
SECTION DETAIL	
	SCALE: N.T.S

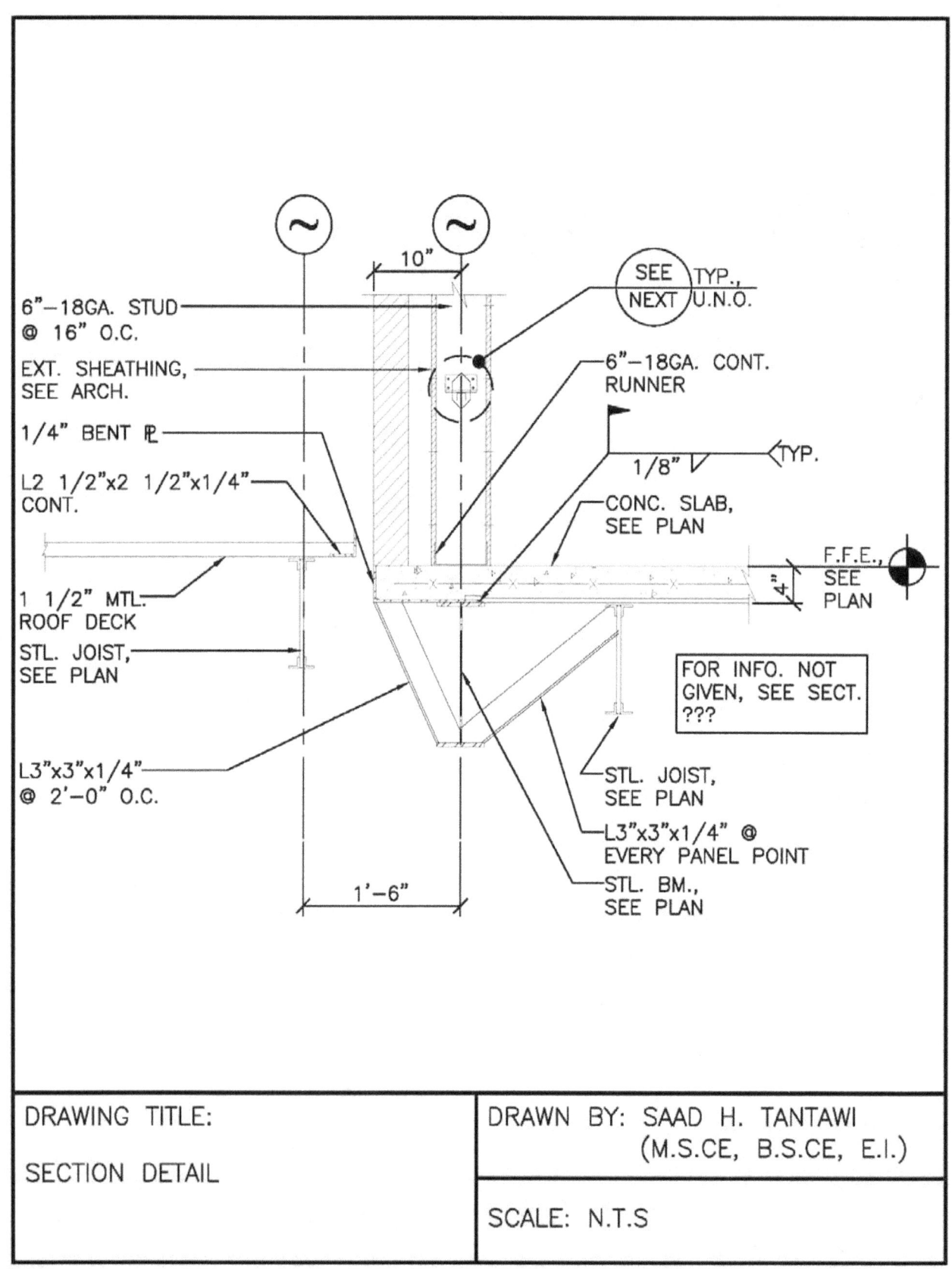

6"–18GA. STUD
@ 16" O.C.

EXT. SHEATHING,
SEE ARCH.

1/4" BENT PL

L2 1/2"x2 1/2"x1/4"
CONT.

1 1/2" MTL.
ROOF DECK

STL. JOIST,
SEE PLAN

L3"x3"x1/4"
@ 2'-0" O.C.

10"

SEE TYP.,
NEXT U.N.O.

6"–18GA. CONT.
RUNNER

1/8" TYP.

CONC. SLAB,
SEE PLAN

F.F.E.,
SEE
PLAN

4"

FOR INFO. NOT
GIVEN, SEE SECT.
???

STL. JOIST,
SEE PLAN

L3"x3"x1/4" @
EVERY PANEL POINT

STL. BM.,
SEE PLAN

1'-6"

DRAWING TITLE:	DRAWN BY: SAAD H. TANTAWI
	(M.S.CE, B.S.CE, E.I.)
SECTION DETAIL	
	SCALE: N.T.S

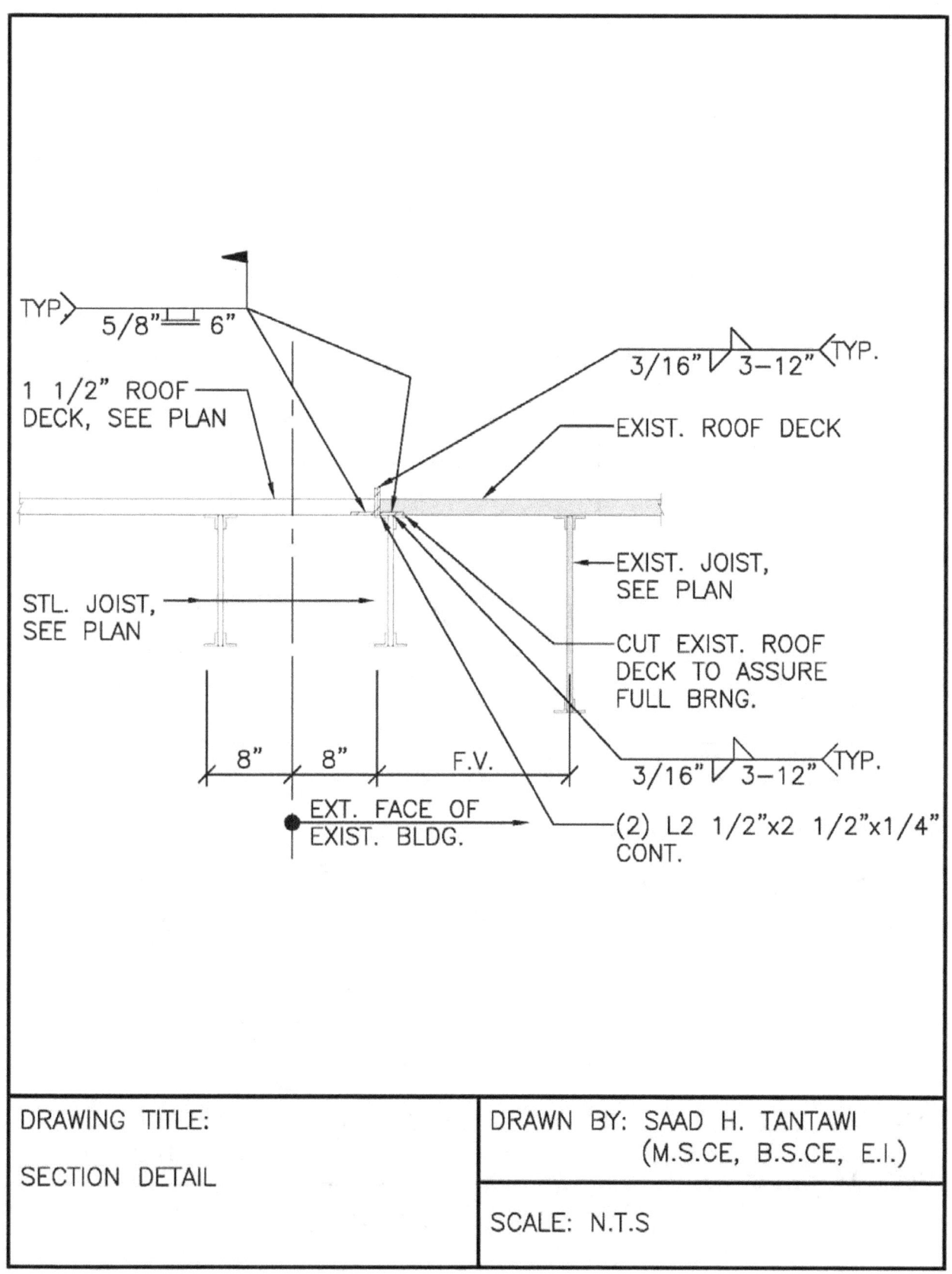

TYP
5/8" ⌑ 6"

3/16" ╲ 3-12" TYP.

1 1/2" ROOF
DECK, SEE PLAN

EXIST. ROOF DECK

STL. JOIST,
SEE PLAN

EXIST. JOIST,
SEE PLAN

CUT EXIST. ROOF
DECK TO ASSURE
FULL BRNG.

8"   8"   F.V.

3/16" ╲ 3-12" TYP.

EXT. FACE OF
EXIST. BLDG.

(2) L2 1/2"x2 1/2"x1/4"
CONT.

DRAWING TITLE:	DRAWN BY: SAAD H. TANTAWI
	(M.S.CE, B.S.CE, E.I.)
SECTION DETAIL	
	SCALE: N.T.S

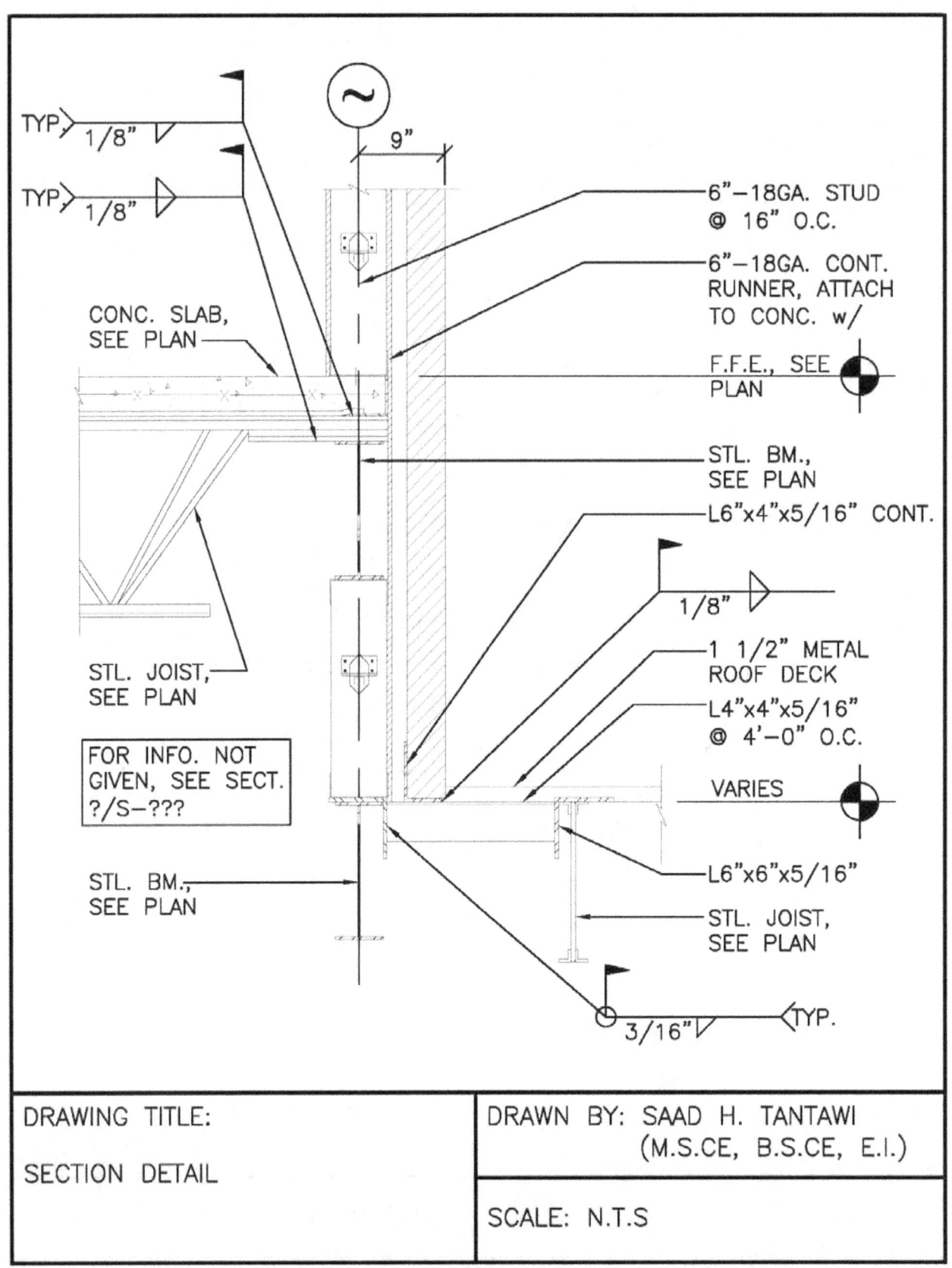

TYP 1/8"

TYP 1/8"

9"

6"—18GA. STUD @ 16" O.C.

6"—18GA. CONT. RUNNER, ATTACH TO CONC. w/

F.F.E., SEE PLAN

CONC. SLAB, SEE PLAN

STL. BM., SEE PLAN

L6"x4"x5/16" CONT.

1/8"

STL. JOIST, SEE PLAN

1 1/2" METAL ROOF DECK

L4"x4"x5/16" @ 4'-0" O.C.

VARIES

FOR INFO. NOT GIVEN, SEE SECT. ?/S-???

L6"x6"x5/16"

STL. BM., SEE PLAN

STL. JOIST, SEE PLAN

3/16" TYP.

DRAWING TITLE:	DRAWN BY: SAAD H. TANTAWI
SECTION DETAIL	(M.S.CE, B.S.CE, E.I.)
	SCALE: N.T.S

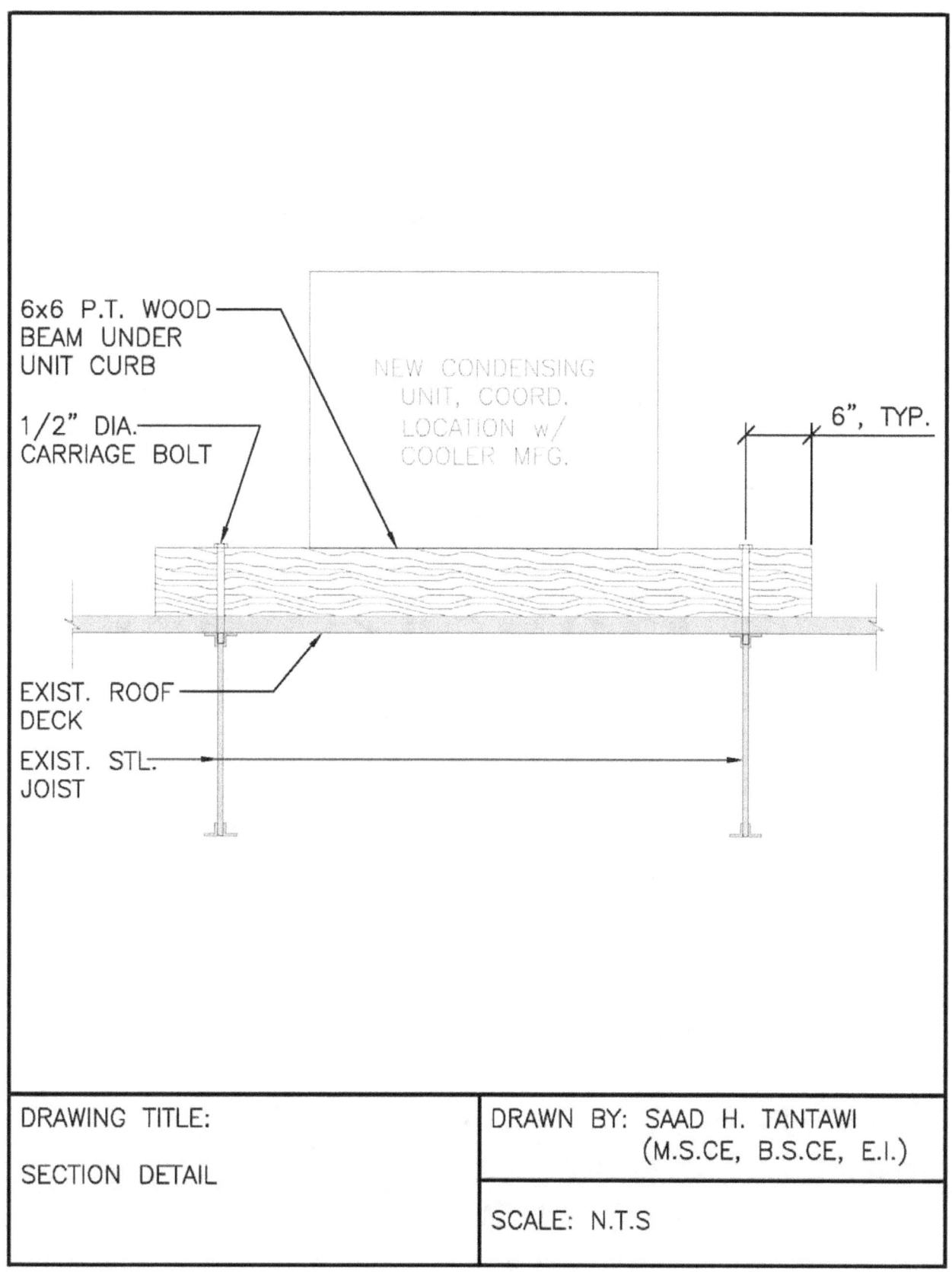

6x6 P.T. WOOD BEAM UNDER UNIT CURB

1/2" DIA. CARRIAGE BOLT

NEW CONDENSING UNIT, COORD. LOCATION w/ COOLER MFG.

6", TYP.

EXIST. ROOF DECK

EXIST. STL. JOIST

DRAWING TITLE:	DRAWN BY: SAAD H. TANTAWI
SECTION DETAIL	(M.S.CE, B.S.CE, E.I.)
	SCALE: N.T.S

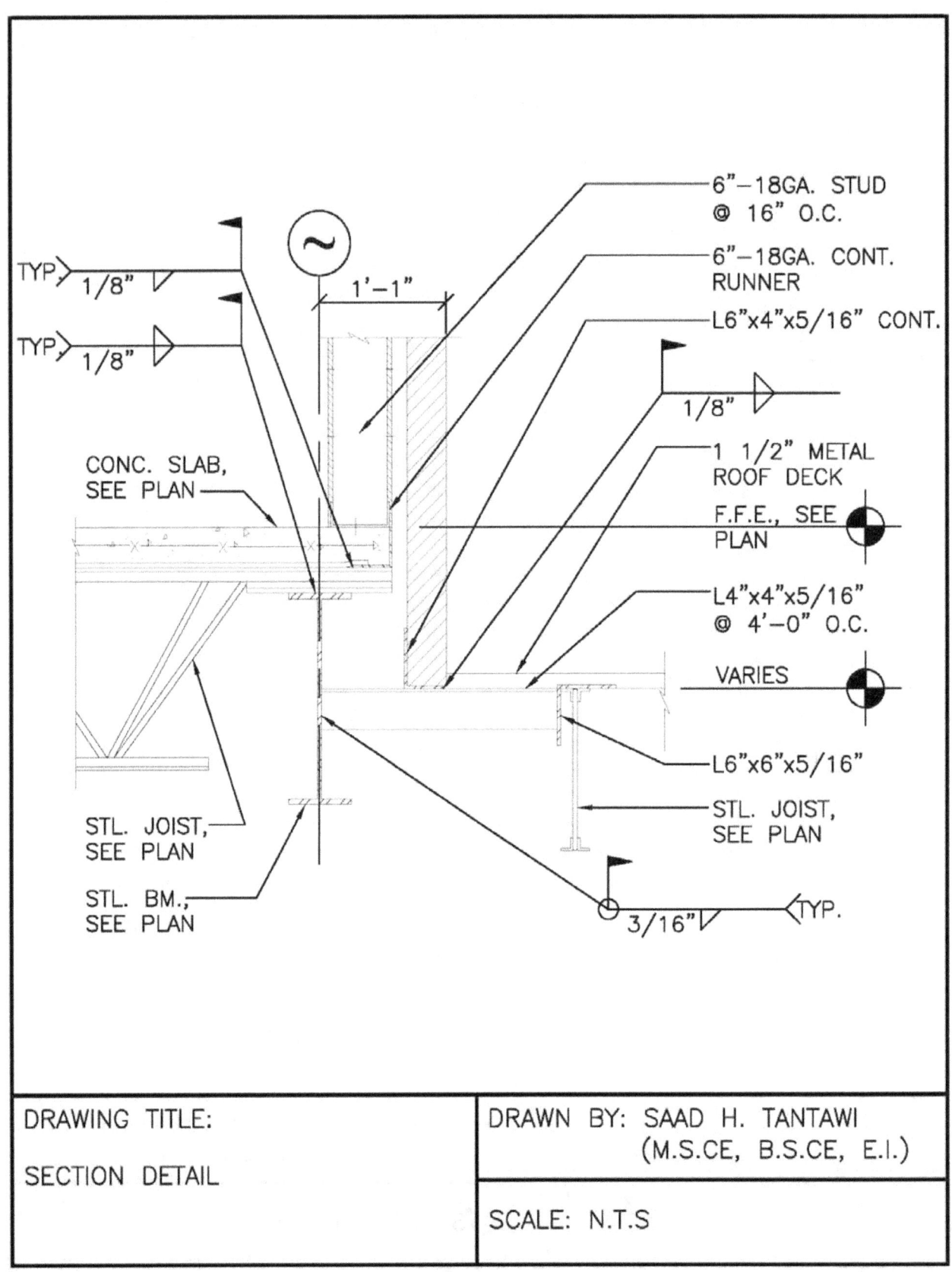

DRAWING TITLE: SECTION DETAIL	DRAWN BY: SAAD H. TANTAWI (M.S.CE, B.S.CE, E.I.)
	SCALE: N.T.S

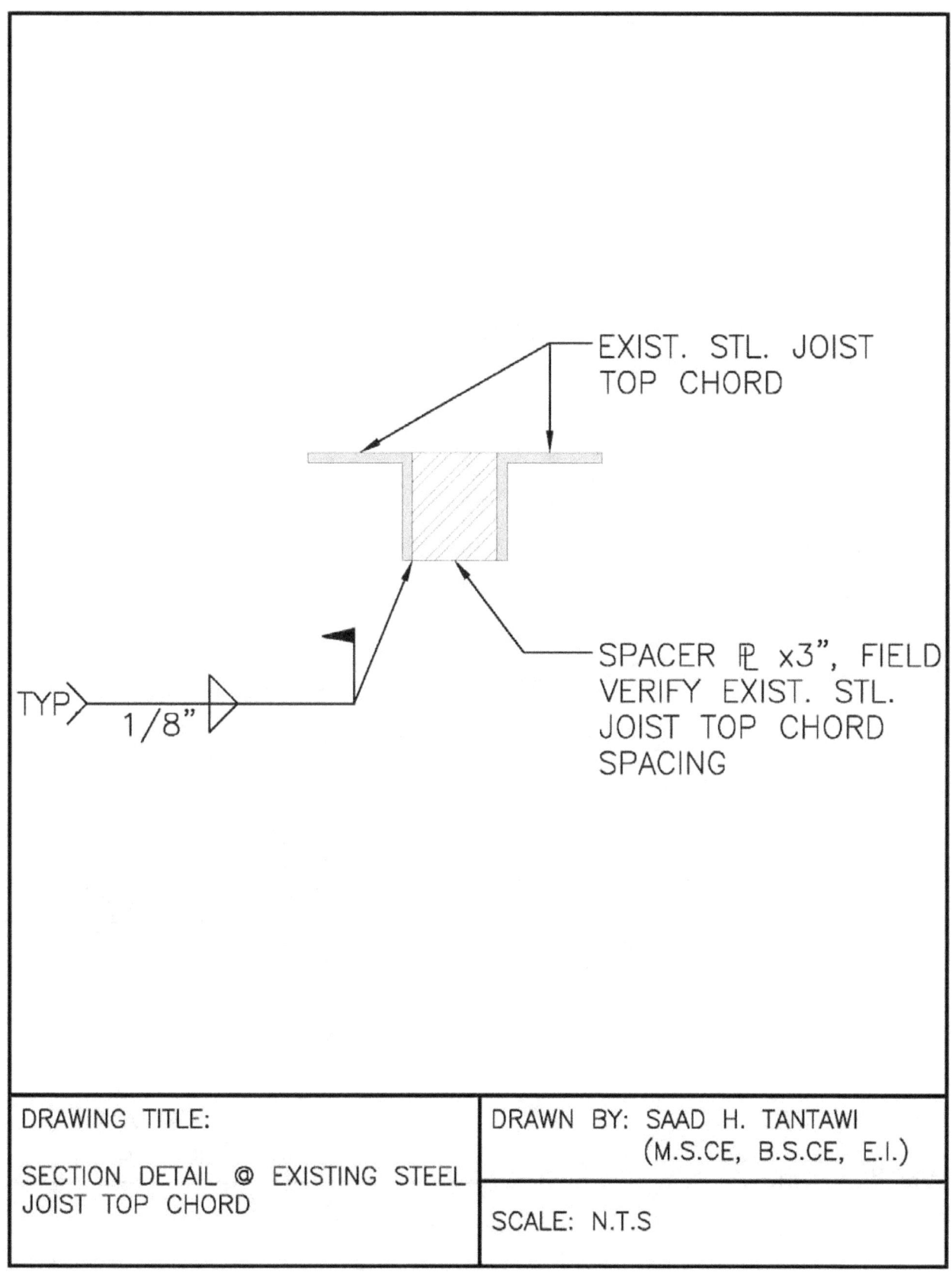

DRAWING TITLE: SECTION DETAIL @ EXISTING STEEL JOIST TOP CHORD	DRAWN BY: SAAD H. TANTAWI (M.S.CE, B.S.CE, E.I.)
	SCALE: N.T.S

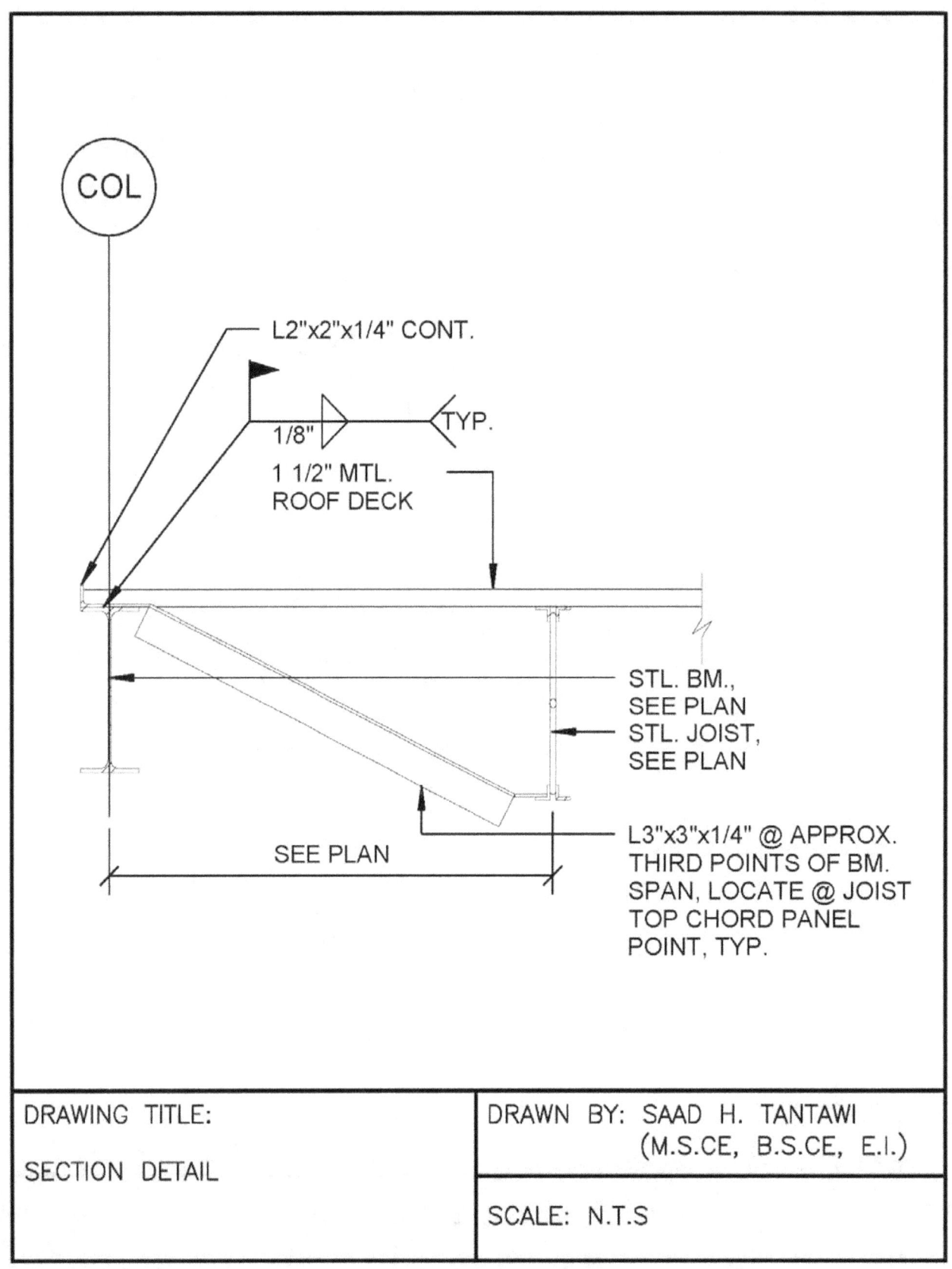

COL

L2"x2"x1/4" CONT.

1/8"

TYP.

1 1/2" MTL.
ROOF DECK

STL. BM.,
SEE PLAN
STL. JOIST,
SEE PLAN

L3"x3"x1/4" @ APPROX.
THIRD POINTS OF BM.
SPAN, LOCATE @ JOIST
TOP CHORD PANEL
POINT, TYP.

SEE PLAN

DRAWING TITLE:	DRAWN BY: SAAD H. TANTAWI
	(M.S.CE, B.S.CE, E.I.)
SECTION DETAIL	
	SCALE: N.T.S

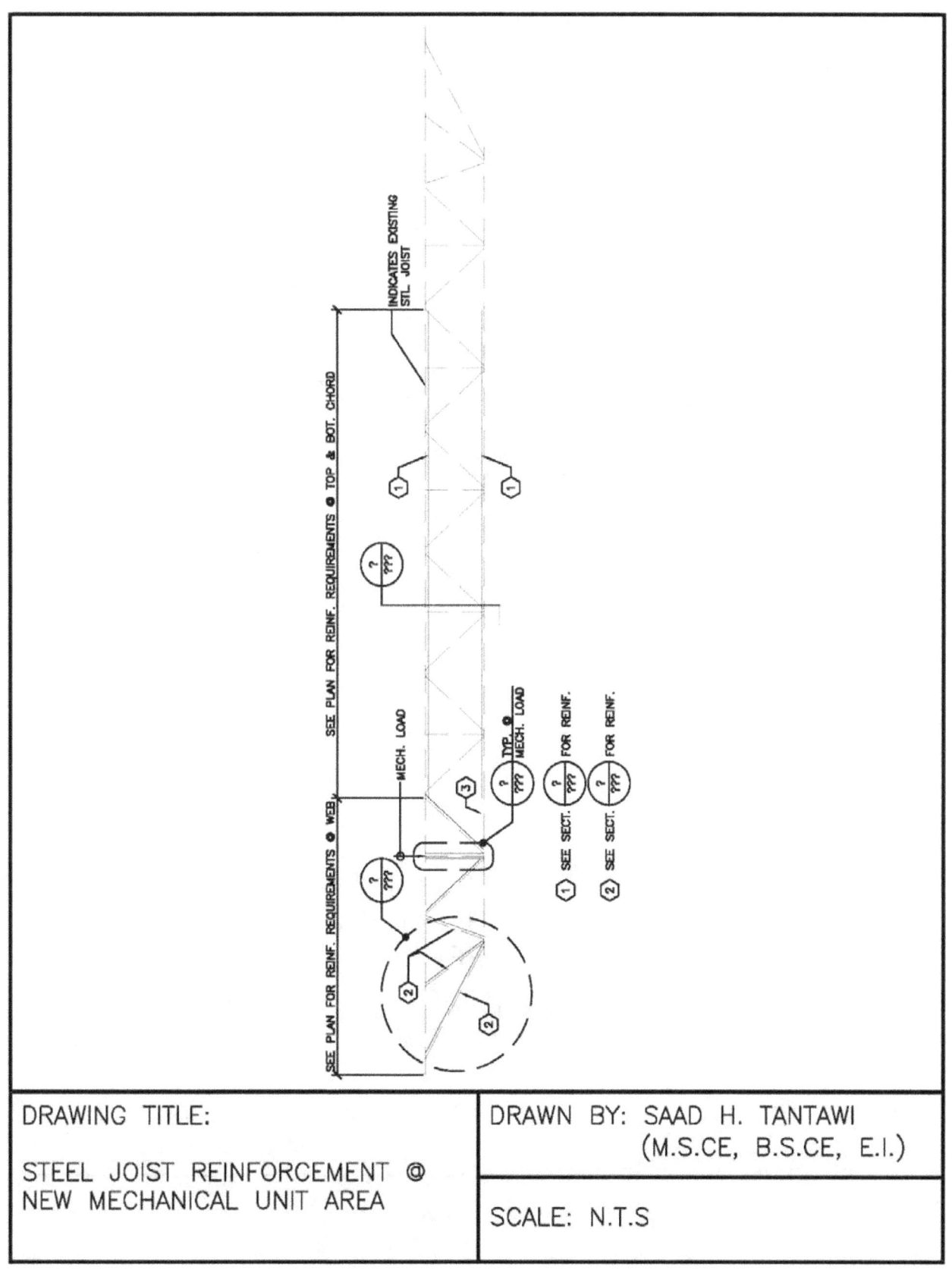

DRAWING TITLE:	DRAWN BY: SAAD H. TANTAWI
STEEL JOIST REINFORCEMENT @ NEW MECHANICAL UNIT AREA	(M.S.CE, B.S.CE, E.I.)
	SCALE: N.T.S

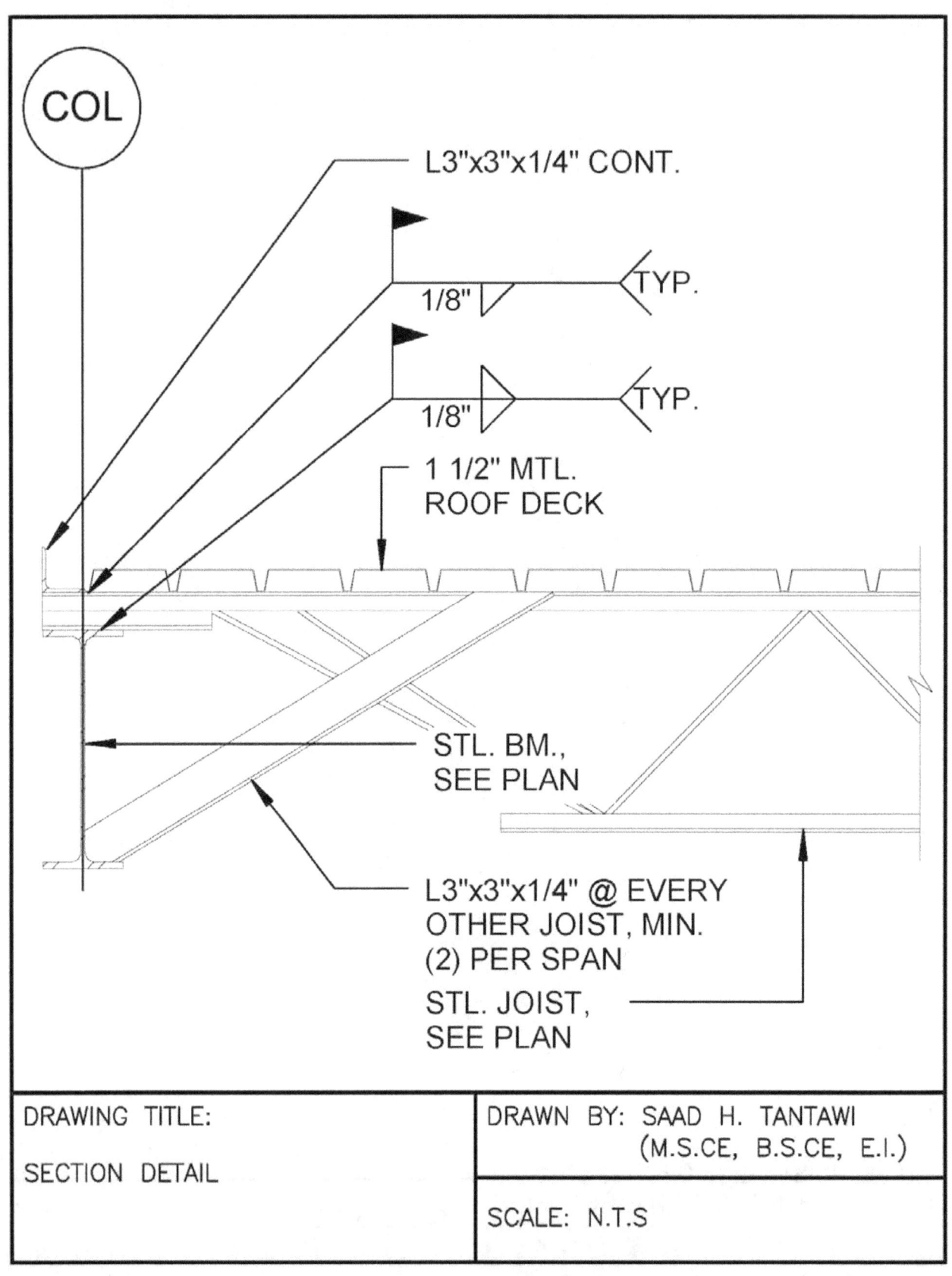

COL

L3"x3"x1/4" CONT.

1/8" TYP.

1/8" TYP.

1 1/2" MTL.
ROOF DECK

STL. BM.,
SEE PLAN

L3"x3"x1/4" @ EVERY
OTHER JOIST, MIN.
(2) PER SPAN

STL. JOIST,
SEE PLAN

DRAWING TITLE:	DRAWN BY: SAAD H. TANTAWI
	(M.S.CE, B.S.CE, E.I.)
SECTION DETAIL	
	SCALE: N.T.S

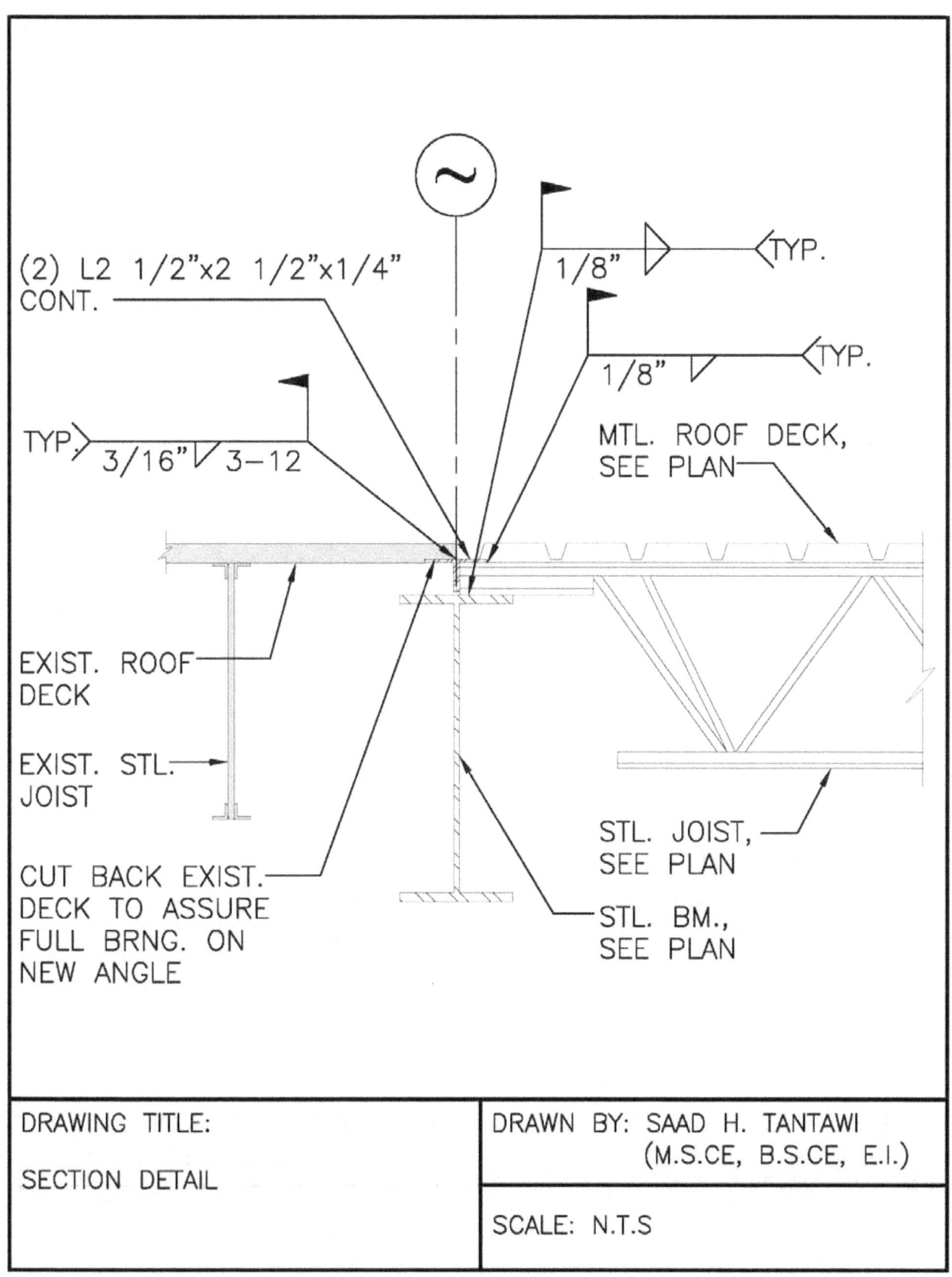

(2) L2 1/2"x2 1/2"x1/4"
CONT.

1/8"   TYP.

1/8"   TYP.

TYP.
3/16" 3-12

MTL. ROOF DECK,
SEE PLAN

EXIST. ROOF
DECK

EXIST. STL.
JOIST

STL. JOIST,
SEE PLAN

CUT BACK EXIST.
DECK TO ASSURE
FULL BRNG. ON
NEW ANGLE

STL. BM.,
SEE PLAN

DRAWING TITLE:	DRAWN BY: SAAD H. TANTAWI
	(M.S.CE, B.S.CE, E.I.)
SECTION DETAIL	
	SCALE: N.T.S

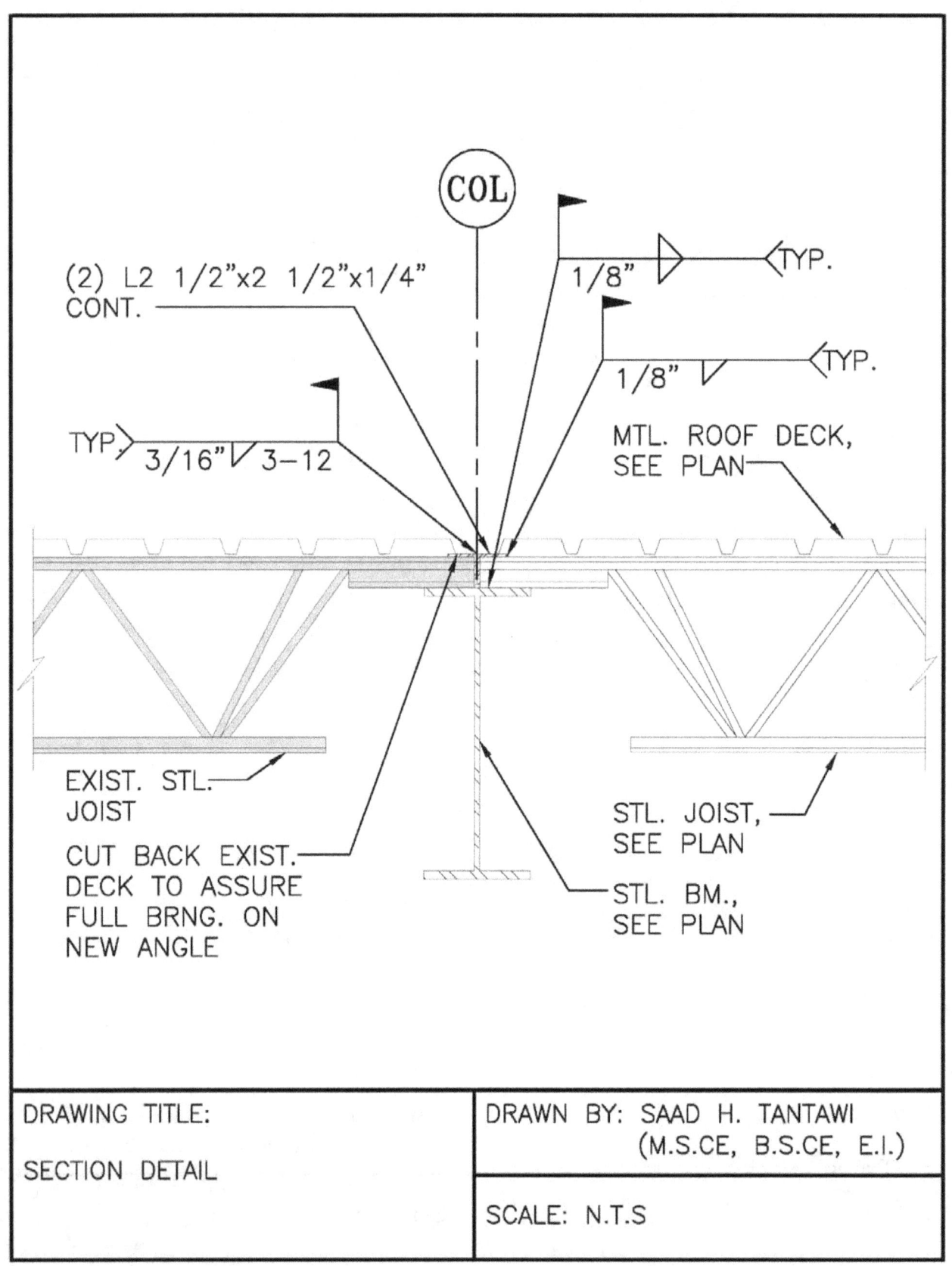

DRAWING TITLE:	DRAWN BY: SAAD H. TANTAWI
	(M.S.CE, B.S.CE, E.I.)
SECTION DETAIL	
	SCALE: N.T.S

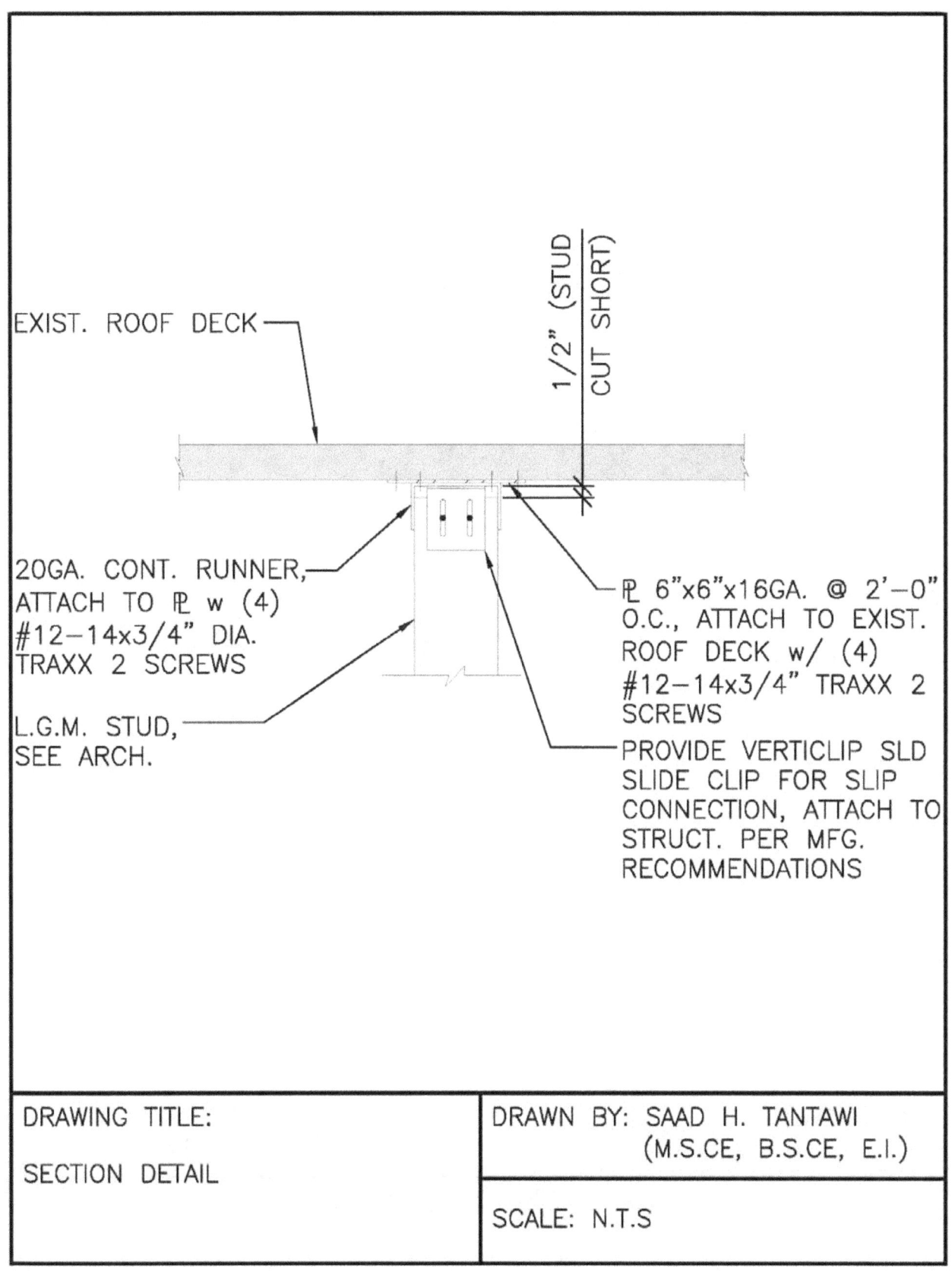

EXIST. ROOF DECK

1/2" (STUD CUT SHORT)

20GA. CONT. RUNNER, ATTACH TO ℙ w (4) #12-14x3/4" DIA. TRAXX 2 SCREWS

L.G.M. STUD, SEE ARCH.

ℙ 6"x6"x16GA. @ 2'-0" O.C., ATTACH TO EXIST. ROOF DECK w/ (4) #12-14x3/4" TRAXX 2 SCREWS

PROVIDE VERTICLIP SLD SLIDE CLIP FOR SLIP CONNECTION, ATTACH TO STRUCT. PER MFG. RECOMMENDATIONS

DRAWING TITLE: SECTION DETAIL	DRAWN BY: SAAD H. TANTAWI (M.S.CE, B.S.CE, E.I.)
	SCALE: N.T.S

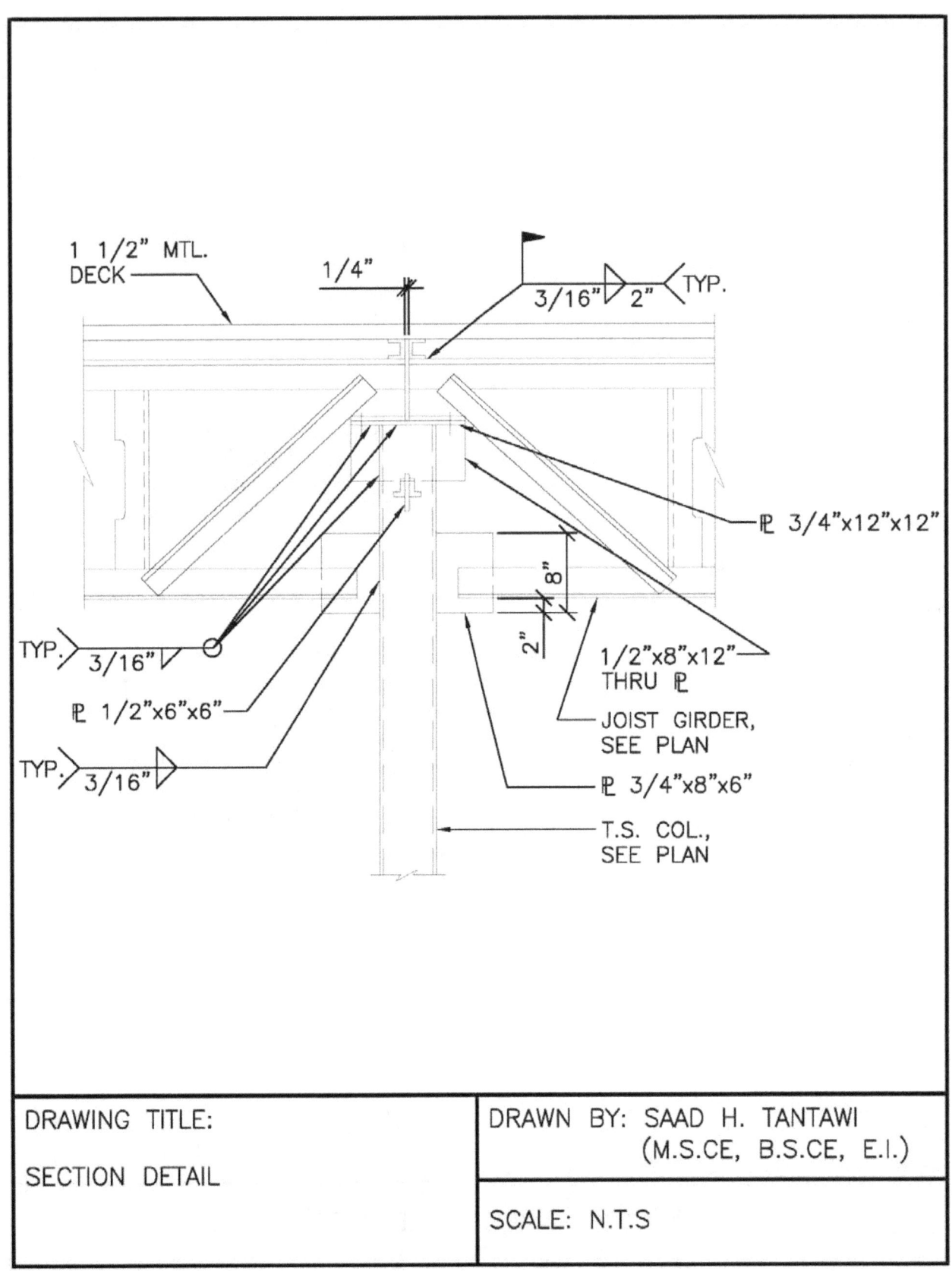

1 1/2" MTL. DECK

1/4"

3/16" 2" TYP.

⊾ 3/4"x12"x12"

8"

2"

1/2"x8"x12" THRU ⊾

JOIST GIRDER, SEE PLAN

⊾ 3/4"x8"x6"

T.S. COL., SEE PLAN

TYP. 3/16"

⊾ 1/2"x6"x6"

TYP. 3/16"

DRAWING TITLE:	DRAWN BY: SAAD H. TANTAWI
	(M.S.CE, B.S.CE, E.I.)
SECTION DETAIL	SCALE: N.T.S

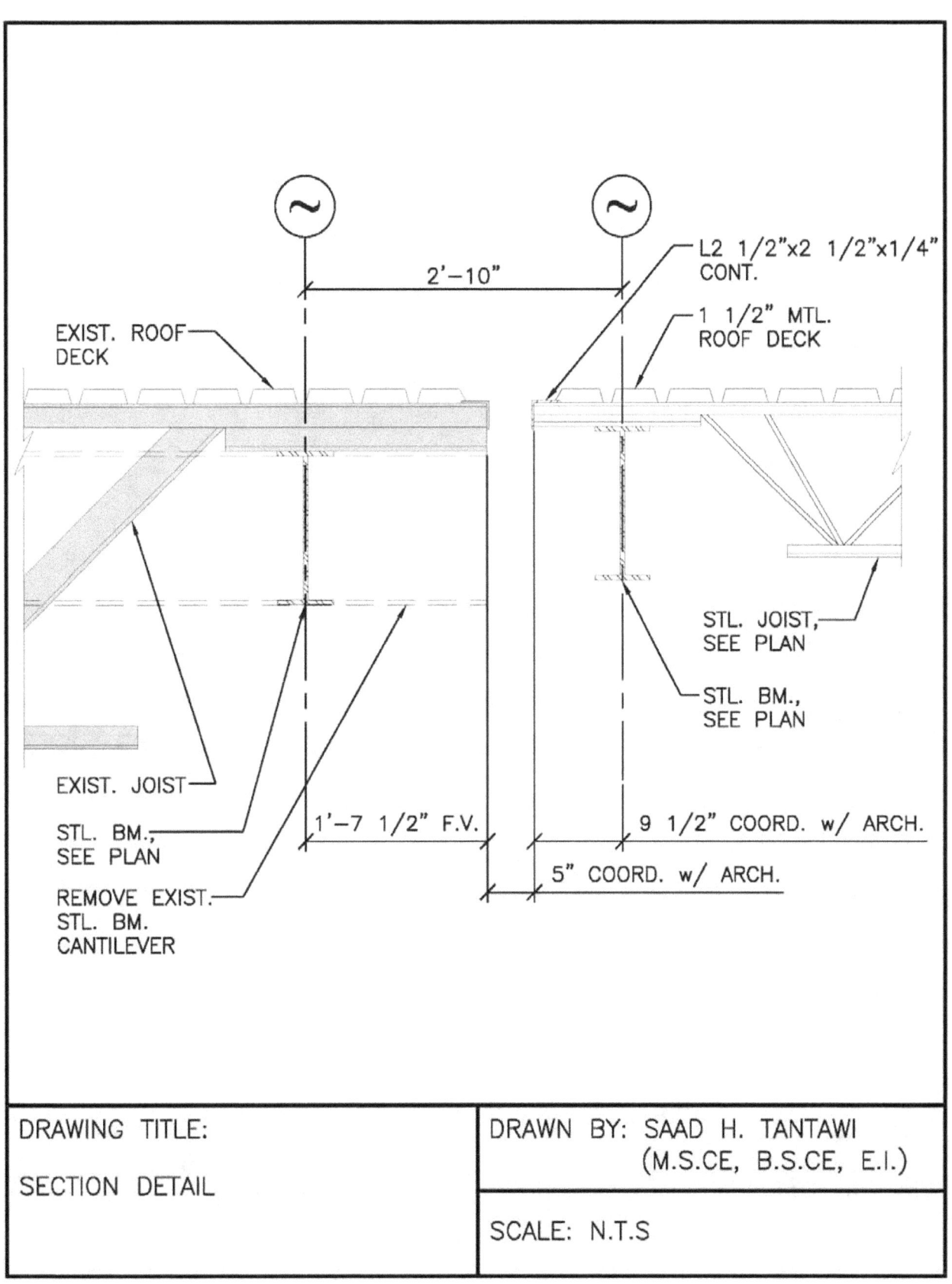

EXIST. ROOF DECK

L2 1/2"x2 1/2"x1/4" CONT.

1 1/2" MTL. ROOF DECK

2'-10"

EXIST. JOIST

STL. BM., SEE PLAN

REMOVE EXIST. STL. BM. CANTILEVER

STL. JOIST, SEE PLAN

STL. BM., SEE PLAN

1'-7 1/2" F.V.

9 1/2" COORD. w/ ARCH.

5" COORD. w/ ARCH.

DRAWING TITLE: SECTION DETAIL	DRAWN BY: SAAD H. TANTAWI (M.S.CE, B.S.CE, E.I.)
	SCALE: N.T.S

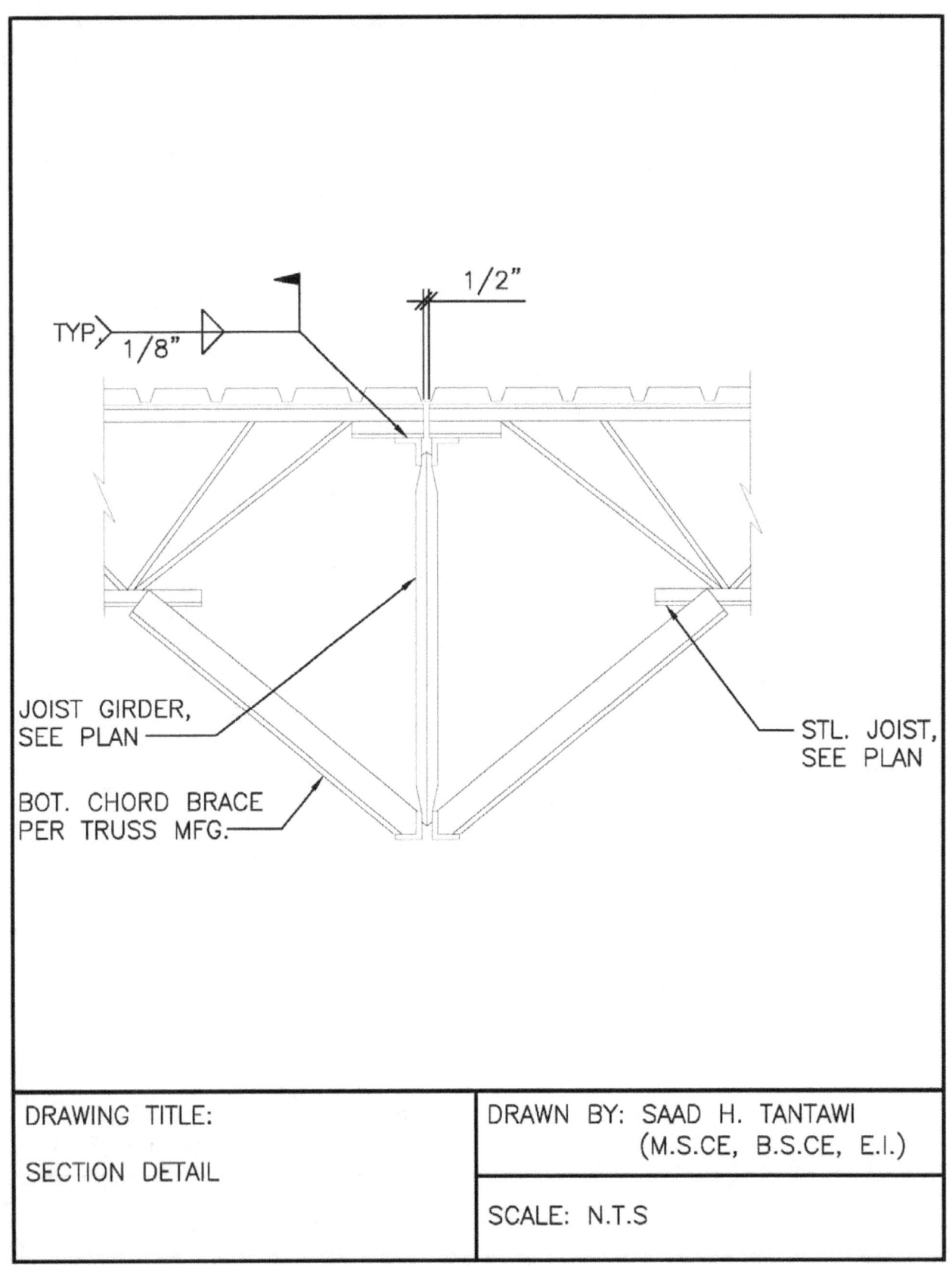

TYP
1/8"

1/2"

JOIST GIRDER,
SEE PLAN

BOT. CHORD BRACE
PER TRUSS MFG.

STL. JOIST,
SEE PLAN

DRAWING TITLE: SECTION DETAIL	DRAWN BY: SAAD H. TANTAWI (M.S.CE, B.S.CE, E.I.)
	SCALE: N.T.S

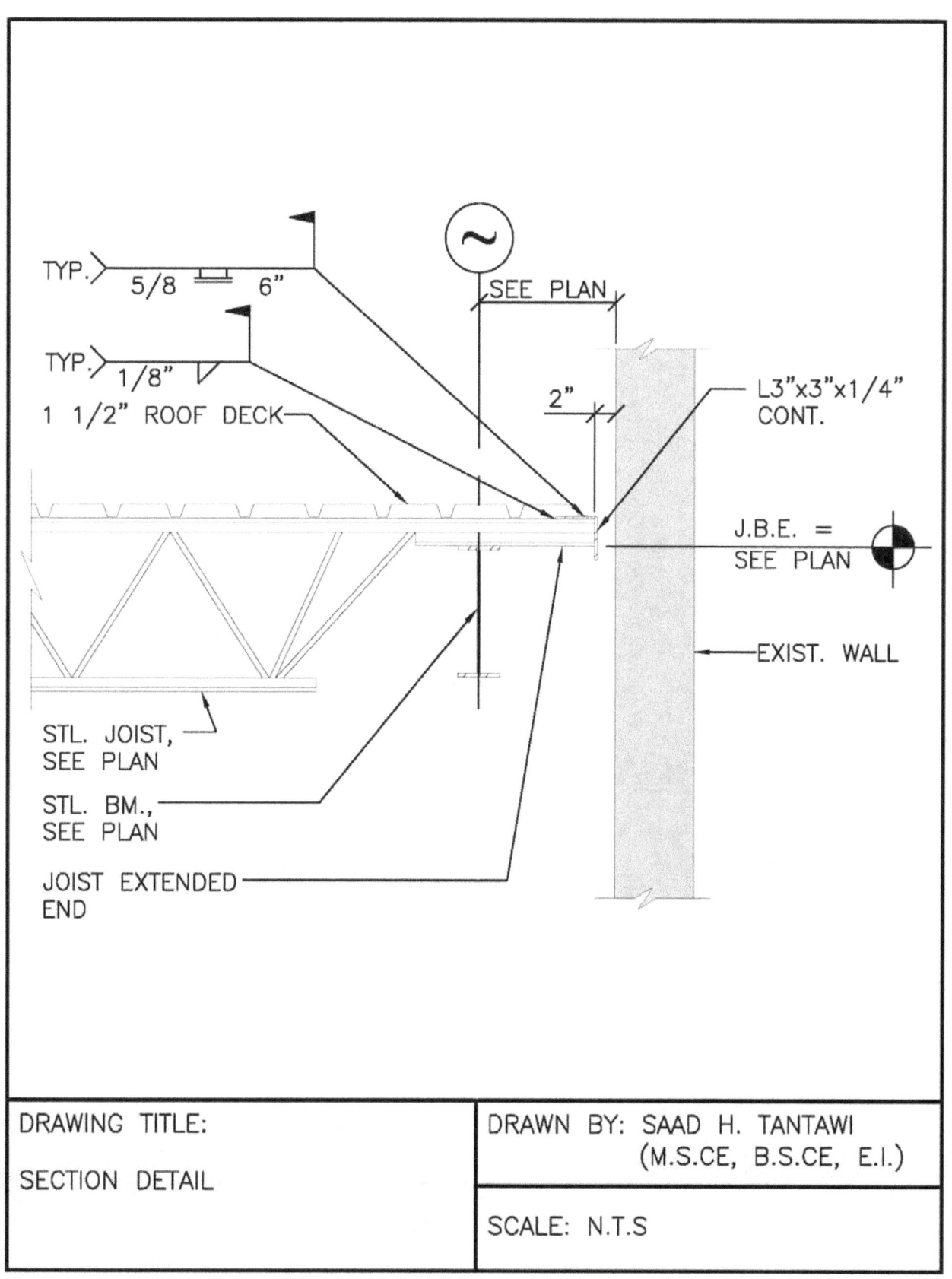

TYP.
5/8   6"

TYP.
1/8"
1 1/2" ROOF DECK

SEE PLAN

2"

L3"x3"x1/4"
CONT.

J.B.E. =
SEE PLAN

EXIST. WALL

STL. JOIST,
SEE PLAN

STL. BM.,
SEE PLAN

JOIST EXTENDED
END

DRAWING TITLE:	DRAWN BY: SAAD H. TANTAWI
	(M.S.CE, B.S.CE, E.I.)
SECTION DETAIL	
	SCALE: N.T.S

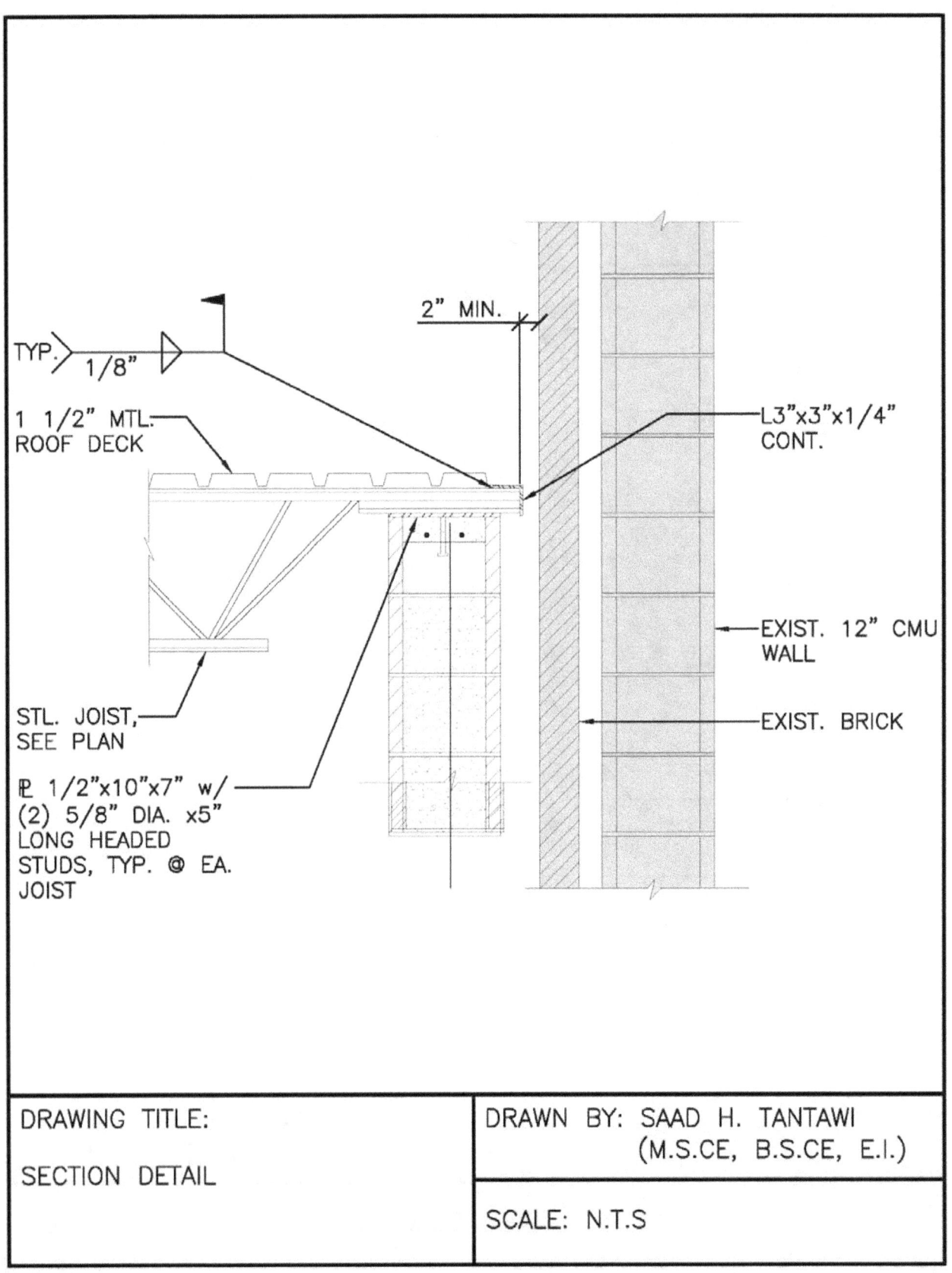

TYP. 1/8"

2" MIN.

1 1/2" MTL. ROOF DECK

L3"x3"x1/4" CONT.

EXIST. 12" CMU WALL

EXIST. BRICK

STL. JOIST, SEE PLAN

℗ 1/2"x10"x7" w/ (2) 5/8" DIA. x5" LONG HEADED STUDS, TYP. @ EA. JOIST

DRAWING TITLE:	DRAWN BY: SAAD H. TANTAWI
SECTION DETAIL	(M.S.CE, B.S.CE, E.I.)
	SCALE: N.T.S

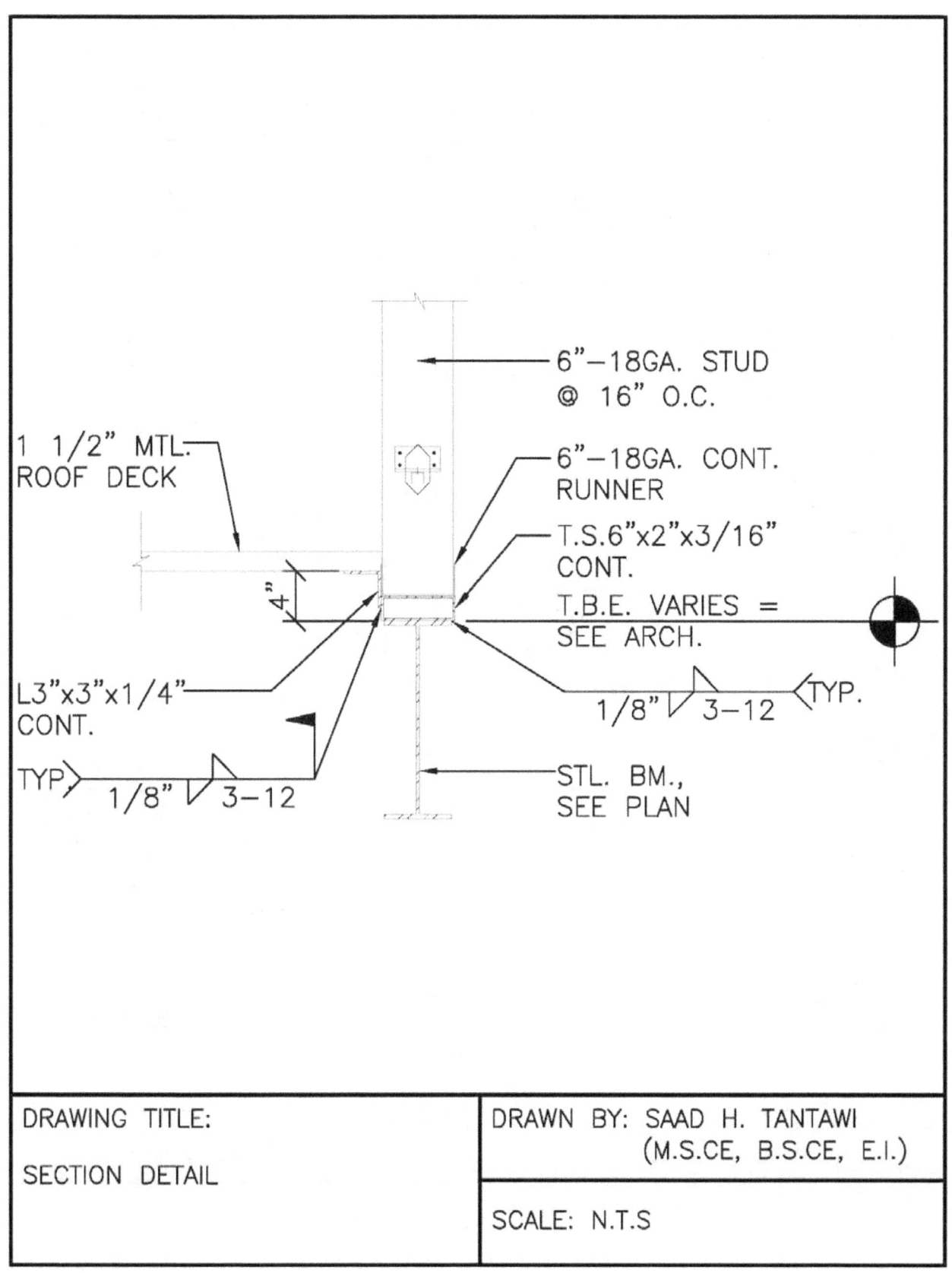

6"–18GA. STUD
@ 16" O.C.

1 1/2" MTL.
ROOF DECK

6"–18GA. CONT.
RUNNER

T.S.6"x2"x3/16"
CONT.

T.B.E. VARIES =
SEE ARCH.

L3"x3"x1/4"
CONT.

1/8" 3–12   TYP.

TYP   1/8" 3–12

STL. BM.,
SEE PLAN

4"

DRAWING TITLE: SECTION DETAIL	DRAWN BY: SAAD H. TANTAWI (M.S.CE, B.S.CE, E.I.)
	SCALE: N.T.S

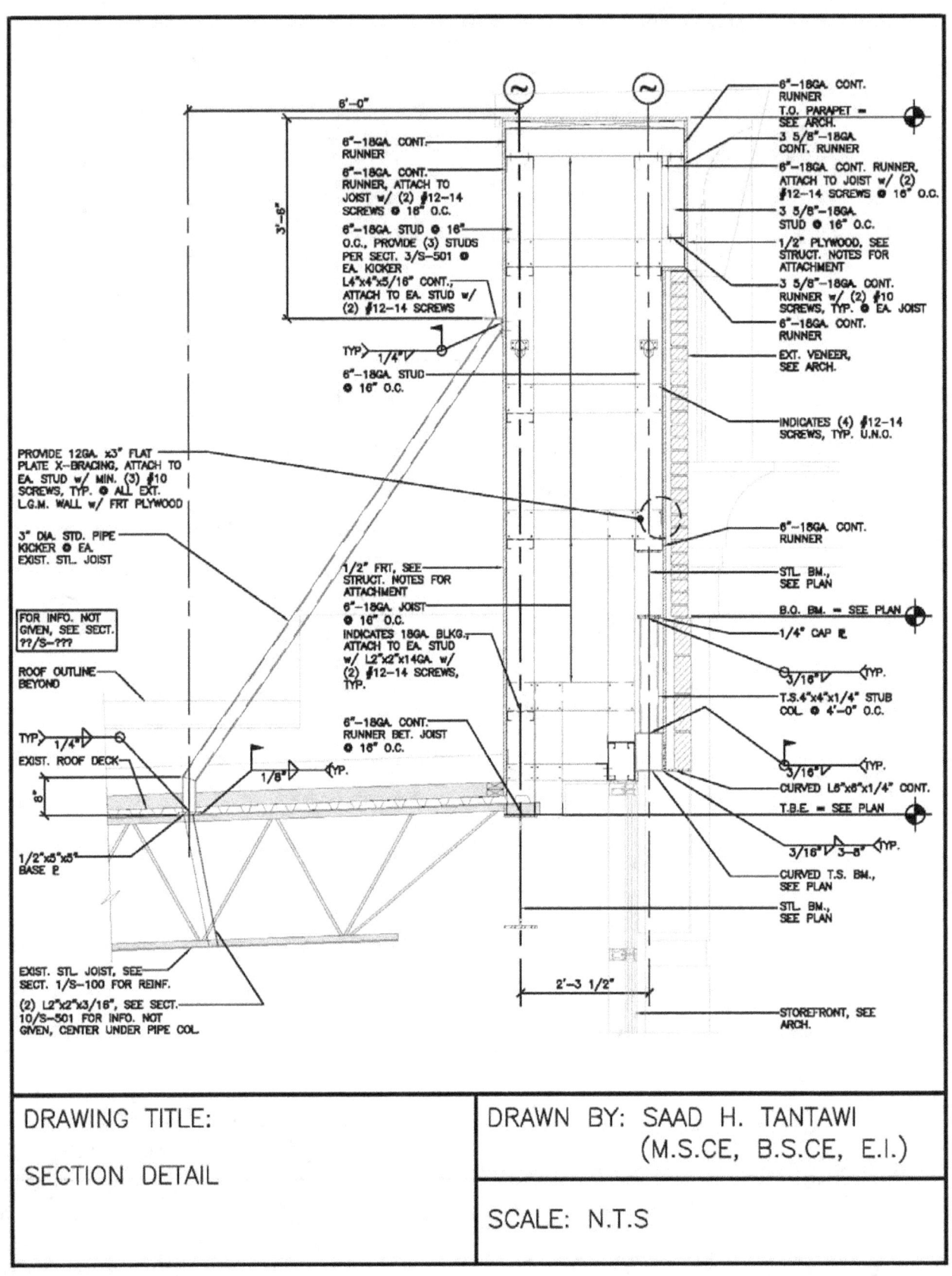

6'-0"

3'-6"

6"-18GA. CONT.
RUNNER

6"-18GA. CONT.
RUNNER, ATTACH TO
JOIST w/ (2) #12-14
SCREWS @ 16" O.C.

6"-18GA. STUD @ 16"
O.C., PROVIDE (3) STUDS
PER SECT. 3/S-501 @
EA. KICKER

L4"x4"x5/16" CONT.,
ATTACH TO EA. STUD w/
(2) #12-14 SCREWS

TYP 1/4"

6"-18GA. STUD
@ 16" O.C.

6"-18GA. CONT.
RUNNER
T.O. PARAPET =
SEE ARCH.

3 5/8"-18GA.
CONT. RUNNER

6"-18GA. CONT. RUNNER,
ATTACH TO JOIST w/ (2)
#12-14 SCREWS @ 16" O.C.

3 5/8"-18GA.
STUD @ 16" O.C.

1/2" PLYWOOD, SEE
STRUCT. NOTES FOR
ATTACHMENT

3 5/8"-18GA. CONT.
RUNNER w/ (2) #10
SCREWS, TYP. @ EA. JOIST

6"-18GA. CONT.
RUNNER

EXT. VENEER,
SEE ARCH.

INDICATES (4) #12-14
SCREWS, TYP. U.N.O.

PROVIDE 12GA. x3" FLAT
PLATE X-BRACING, ATTACH TO
EA. STUD w/ MIN. (3) #10
SCREWS, TYP. @ ALL EXT.
L.G.M. WALL w/ FRT PLYWOOD

3" DIA. STD. PIPE
KICKER @ EA.
EXIST. STL. JOIST

FOR INFO. NOT
GIVEN, SEE SECT.
??/S-???

ROOF OUTLINE
BEYOND

TYP 1/4"

EXIST. ROOF DECK

1/2"x5"x5"
BASE P

EXIST. STL. JOIST, SEE
SECT. 1/S-100 FOR REINF.

(2) L2"x2"x3/16", SEE SECT.
10/S-501 FOR INFO. NOT
GIVEN, CENTER UNDER PIPE COL.

1/2" FRT, SEE
STRUCT. NOTES FOR
ATTACHMENT

6"-18GA. JOIST
@ 16" O.C.

INDICATES 18GA. BLKG.
ATTACH TO EA. STUD
w/ L2"x2"x14GA. w/
(2) #12-14 SCREWS,
TYP.

6"-18GA. CONT.
RUNNER BET. JOIST
@ 16" O.C.

1/8" TYP.

6"-18GA. CONT.
RUNNER

STL. BM.,
SEE PLAN

B.O. BM. = SEE PLAN

1/4" CAP P

3/16" TYP.

T.S.4"x4"x1/4" STUB
COL. @ 4'-0" O.C.

3/16" TYP.

CURVED L6"x6"x1/4" CONT.
T.B.E = SEE PLAN

3/16" 3-8" TYP.

CURVED T.S. BM.,
SEE PLAN

STL. BM.,
SEE PLAN

2'-3 1/2"

STOREFRONT, SEE
ARCH.

DRAWING TITLE:	DRAWN BY: SAAD H. TANTAWI
	(M.S.CE, B.S.CE, E.I.)
SECTION DETAIL	
	SCALE: N.T.S

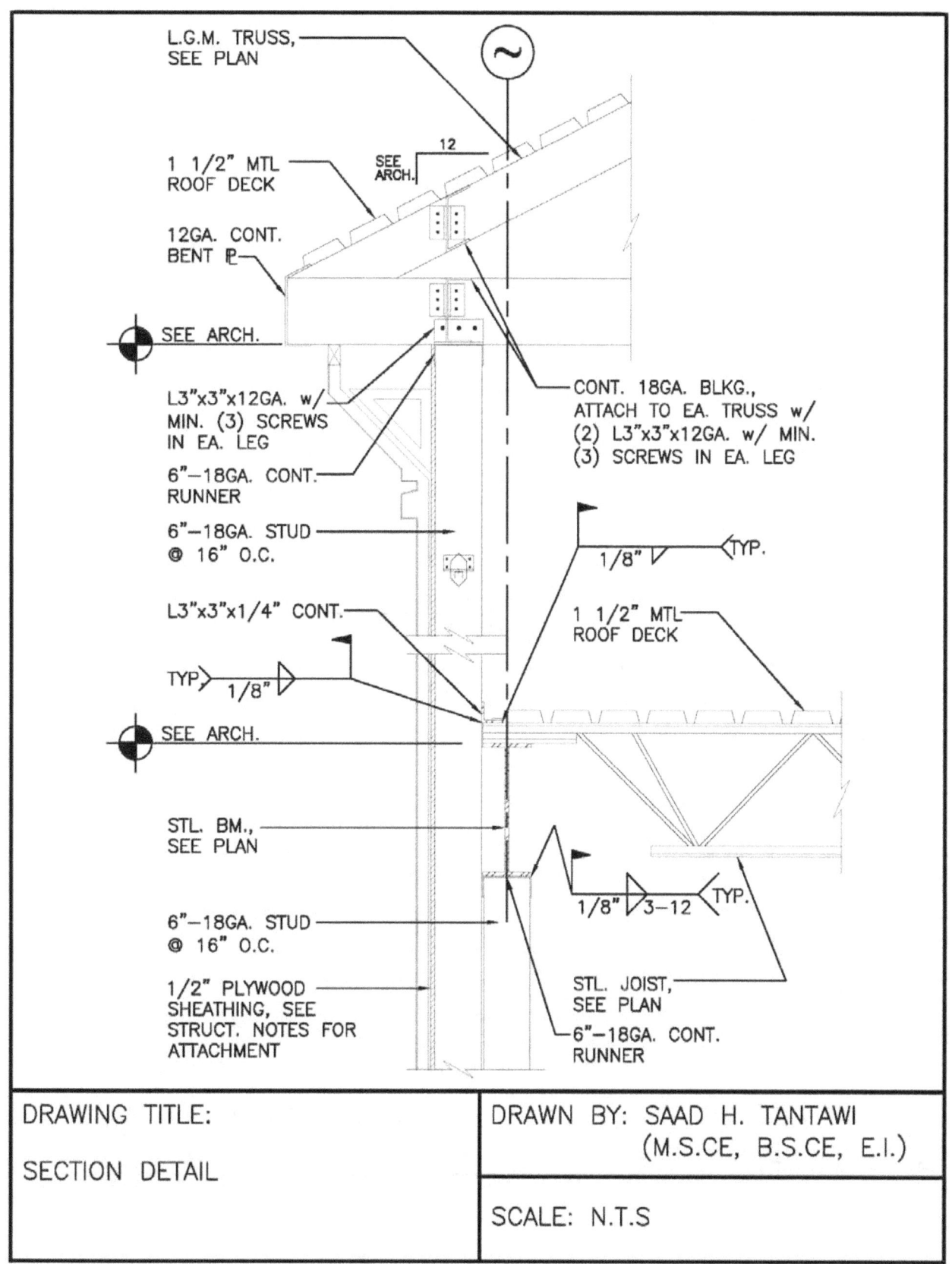

L.G.M. TRUSS,
SEE PLAN

1 1/2" MTL
ROOF DECK

SEE
ARCH.

12

12GA. CONT.
BENT ℙ

SEE ARCH.

L3"x3"x12GA. w/
MIN. (3) SCREWS
IN EA. LEG

6"−18GA. CONT.
RUNNER

6"−18GA. STUD
@ 16" O.C.

L3"x3"x1/4" CONT.

TYP   1/8"

SEE ARCH.

CONT. 18GA. BLKG.,
ATTACH TO EA. TRUSS w/
(2) L3"x3"x12GA. w/ MIN.
(3) SCREWS IN EA. LEG

1/8"   TYP.

1 1/2" MTL
ROOF DECK

STL. BM.,
SEE PLAN

6"−18GA. STUD
@ 16" O.C.

1/2" PLYWOOD
SHEATHING, SEE
STRUCT. NOTES FOR
ATTACHMENT

1/8"   3−12   TYP.

STL. JOIST,
SEE PLAN

6"−18GA. CONT.
RUNNER

DRAWING TITLE:	DRAWN BY: SAAD H. TANTAWI
	(M.S.CE, B.S.CE, E.I.)
SECTION DETAIL	
	SCALE: N.T.S

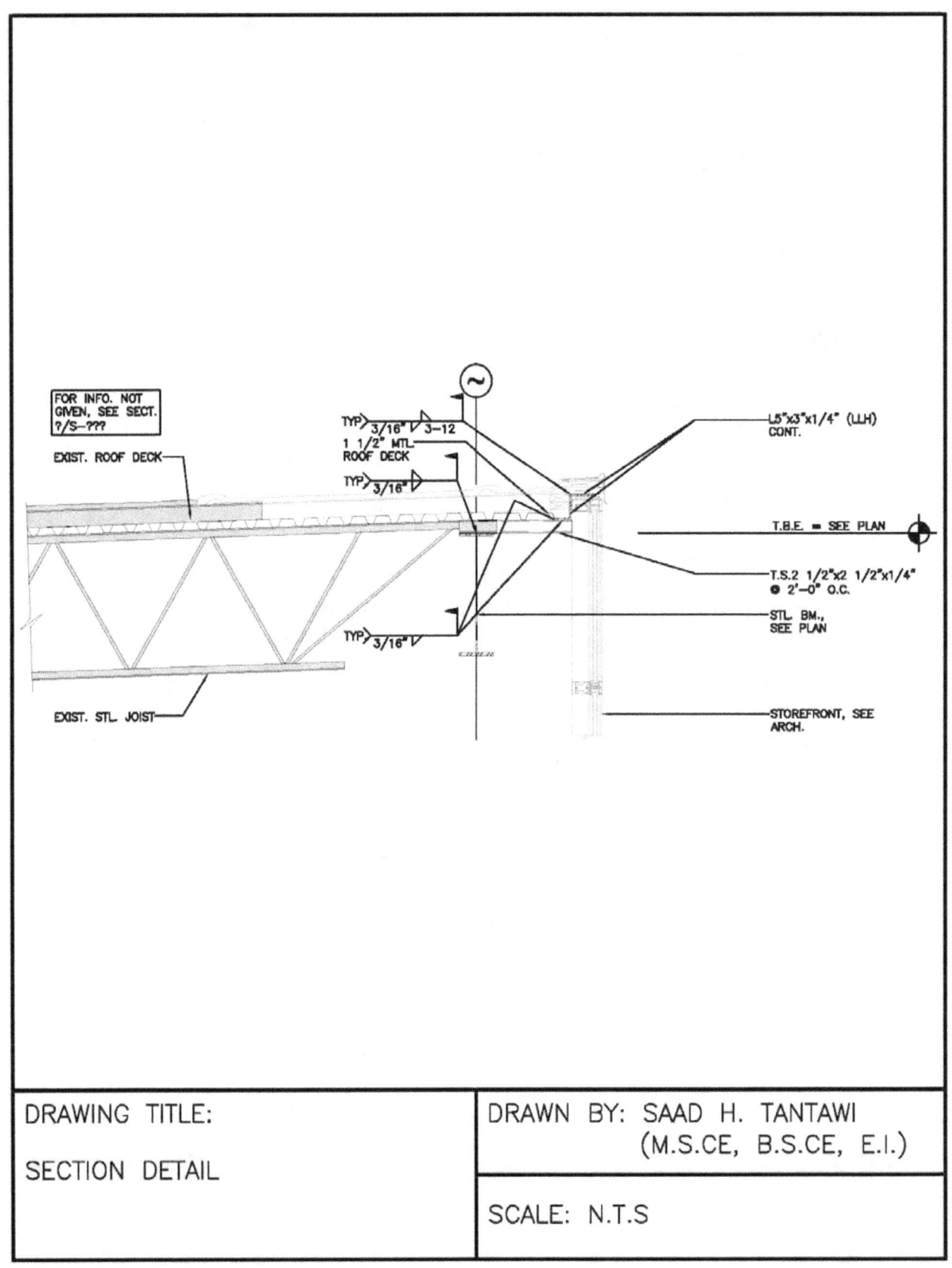

FOR INFO. NOT
GIVEN, SEE SECT.
?/S-???

EXIST. ROOF DECK

TYP 3/16" 3-12
1 1/2" MTL
ROOF DECK

TYP 3/16"

TYP 3/16"

EXIST. STL JOIST

L5"x3"x1/4" (LLH)
CONT.

T.B.E. = SEE PLAN

T.S.2 1/2"x2 1/2"x1/4"
@ 2'-0" O.C.

STL BM.,
SEE PLAN

STOREFRONT, SEE
ARCH.

DRAWING TITLE:	DRAWN BY: SAAD H. TANTAWI
	(M.S.CE, B.S.CE, E.I.)
SECTION DETAIL	
	SCALE: N.T.S

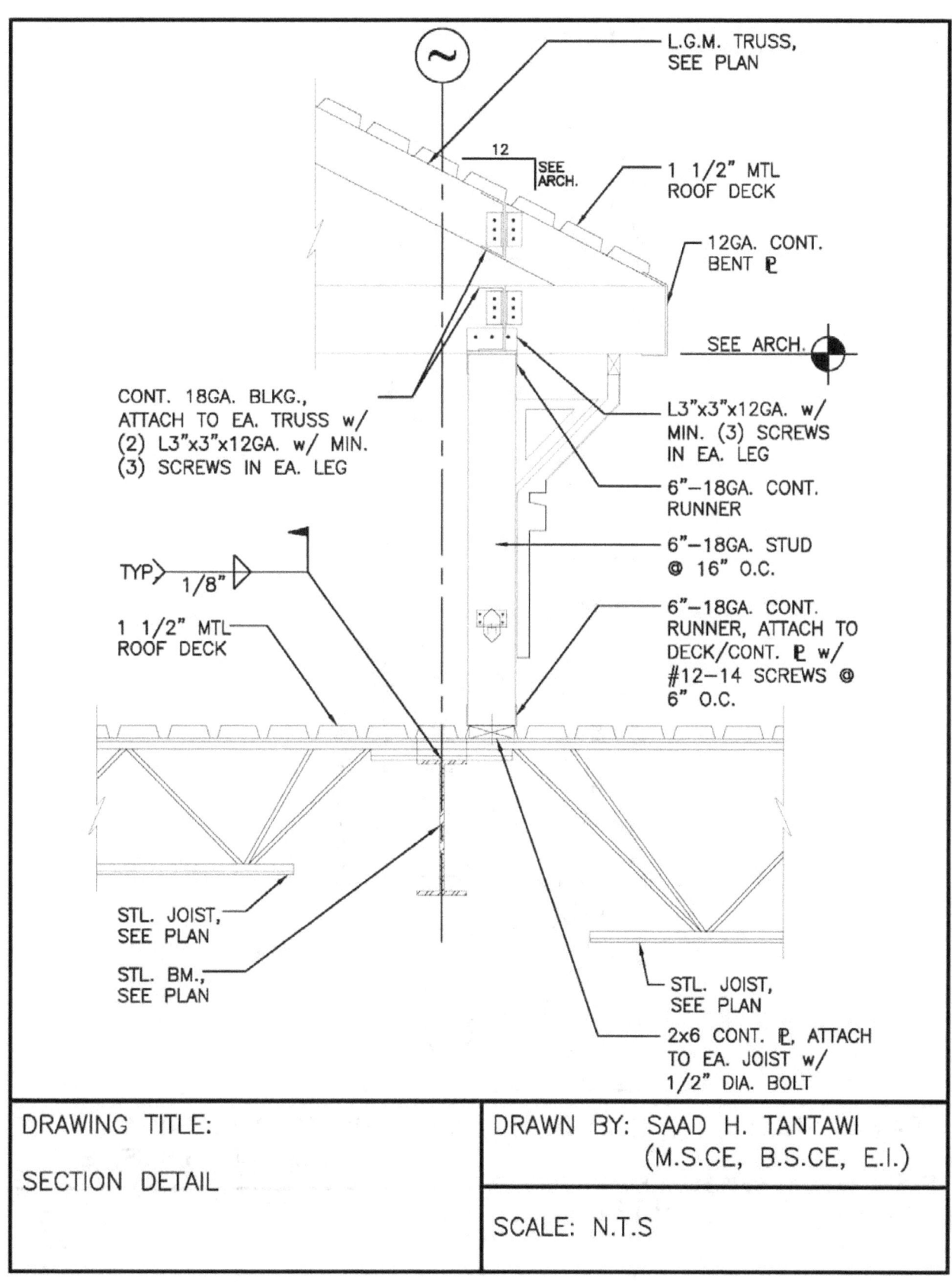

L.G.M. TRUSS,
SEE PLAN

12
SEE
ARCH.

1 1/2" MTL
ROOF DECK

12GA. CONT.
BENT ₧

SEE ARCH.

CONT. 18GA. BLKG.,
ATTACH TO EA. TRUSS w/
(2) L3"x3"x12GA. w/ MIN.
(3) SCREWS IN EA. LEG

L3"x3"x12GA. w/
MIN. (3) SCREWS
IN EA. LEG

6"-18GA. CONT.
RUNNER

6"-18GA. STUD
@ 16" O.C.

6"-18GA. CONT.
RUNNER, ATTACH TO
DECK/CONT. ₧ w/
#12-14 SCREWS @
6" O.C.

TYP> 1/8"

1 1/2" MTL
ROOF DECK

STL. JOIST,
SEE PLAN

STL. BM.,
SEE PLAN

STL. JOIST,
SEE PLAN

2x6 CONT. ₧, ATTACH
TO EA. JOIST w/
1/2" DIA. BOLT

DRAWING TITLE:	DRAWN BY: SAAD H. TANTAWI
	(M.S.CE, B.S.CE, E.I.)
SECTION DETAIL	
	SCALE: N.T.S

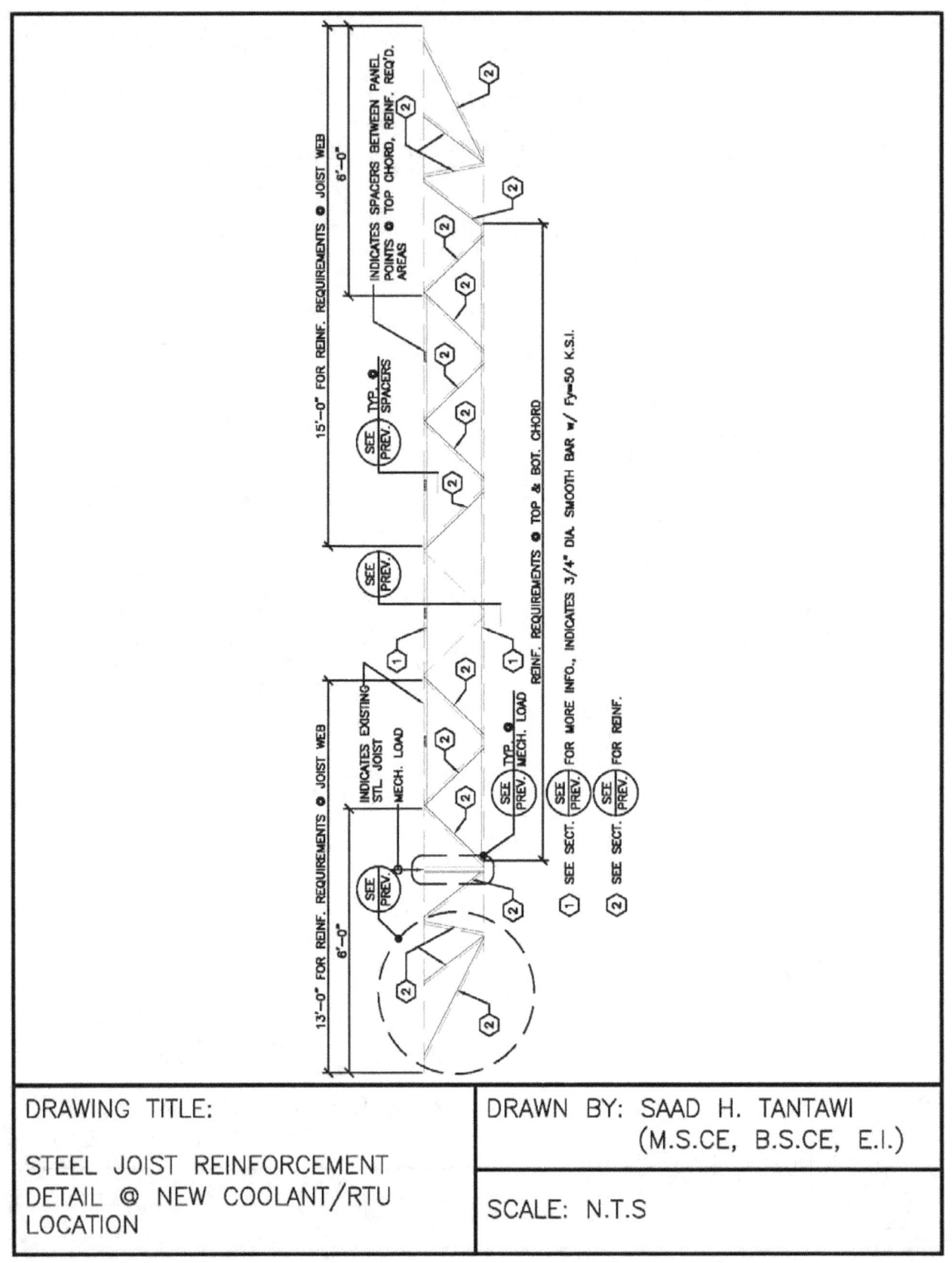

DRAWING TITLE:

STEEL JOIST REINFORCEMENT
DETAIL @ NEW COOLANT/RTU
LOCATION

DRAWN BY: SAAD H. TANTAWI
(M.S.CE, B.S.CE, E.I.)

SCALE: N.T.S

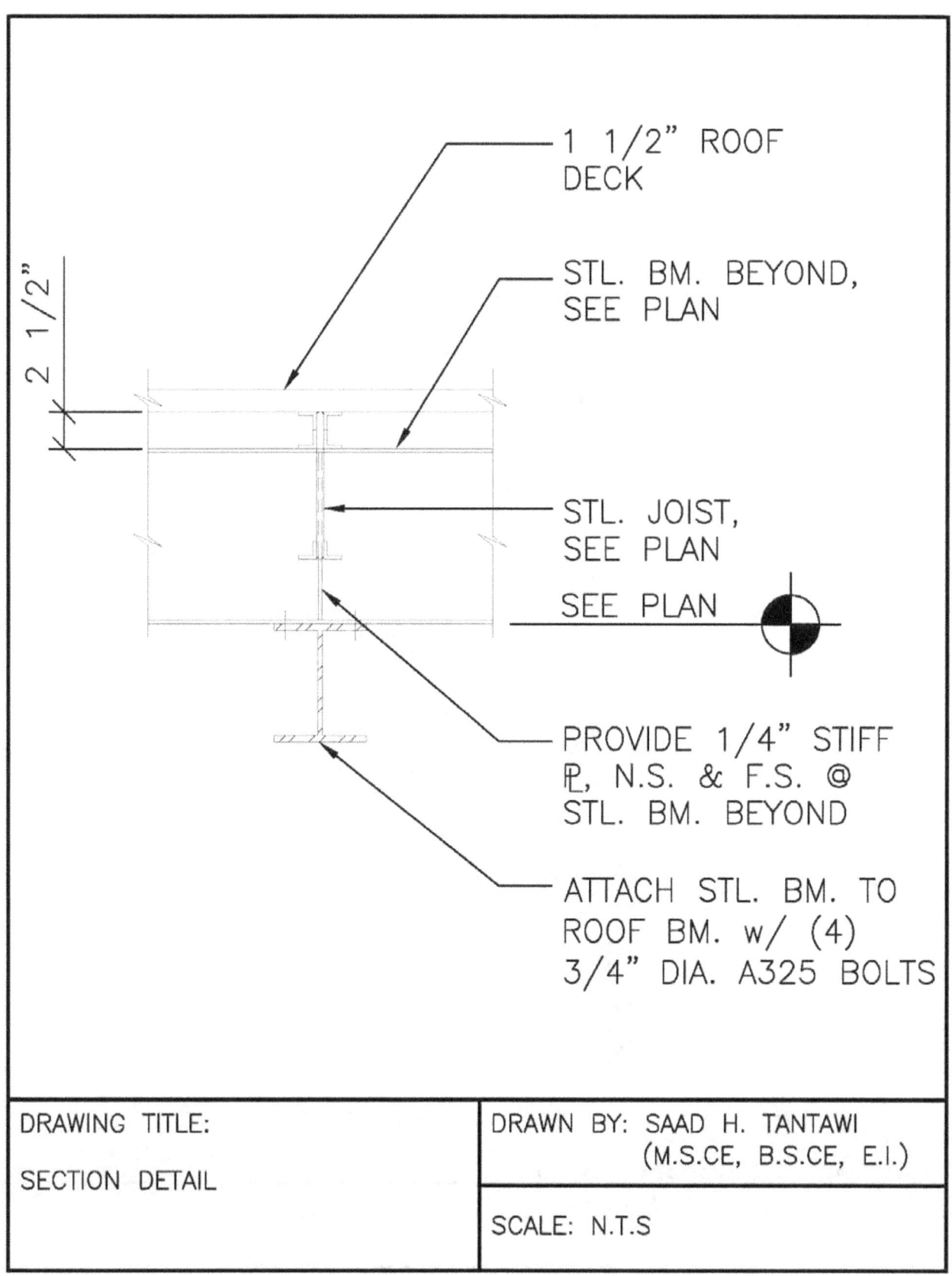

1 1/2" ROOF DECK

STL. BM. BEYOND, SEE PLAN

STL. JOIST, SEE PLAN

SEE PLAN

PROVIDE 1/4" STIFF P̶L, N.S. & F.S. @ STL. BM. BEYOND

ATTACH STL. BM. TO ROOF BM. w/ (4) 3/4" DIA. A325 BOLTS

2 1/2"

DRAWING TITLE: SECTION DETAIL	DRAWN BY: SAAD H. TANTAWI (M.S.CE, B.S.CE, E.I.)
	SCALE: N.T.S

Saad Hasan Tantawi (M.S.CE, B.S.CE, E.I., A.M.ASCE)

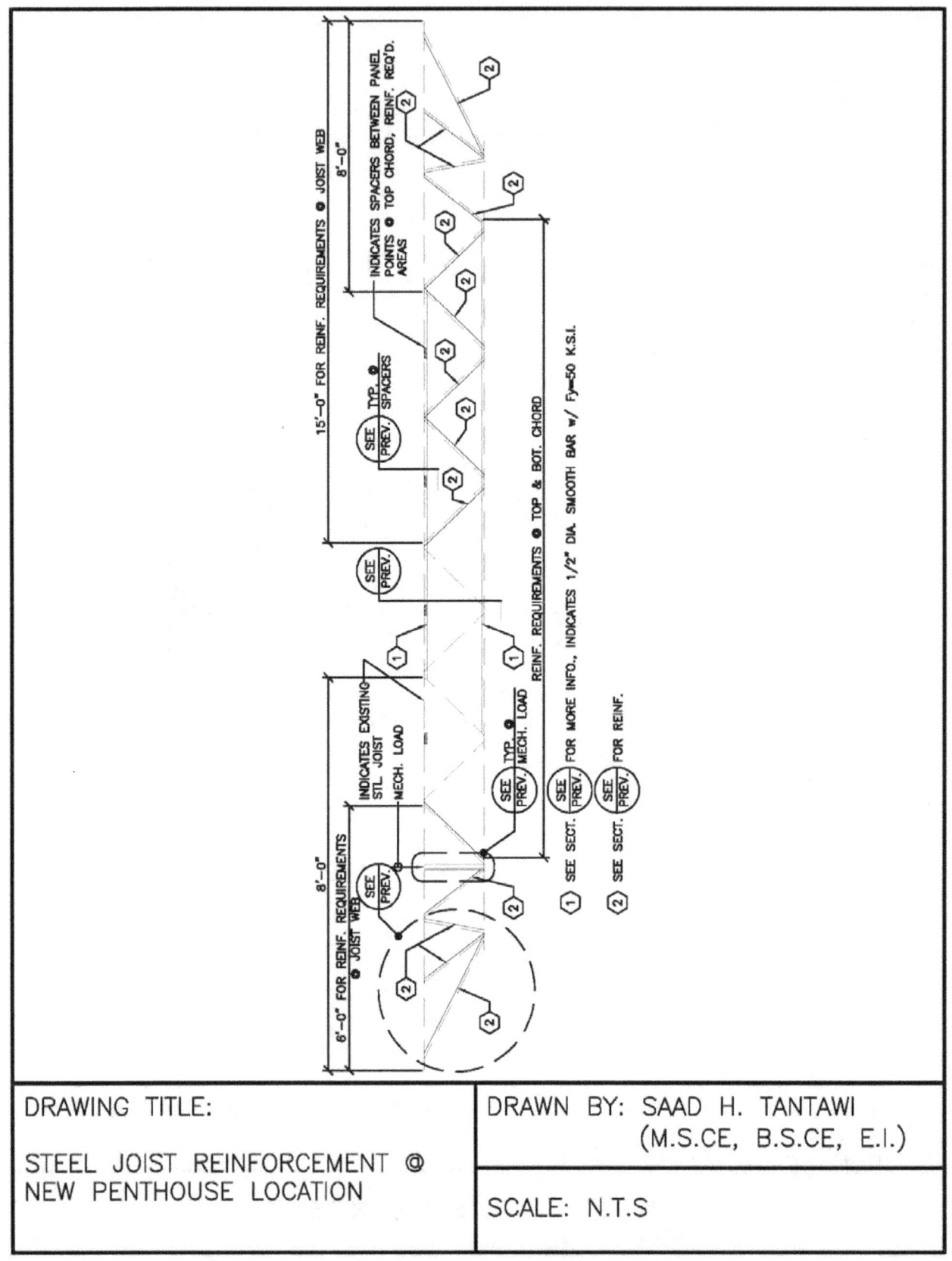

DRAWING TITLE:

STEEL JOIST REINFORCEMENT @
NEW PENTHOUSE LOCATION

DRAWN BY: SAAD H. TANTAWI
(M.S.CE, B.S.CE, E.I.)

SCALE: N.T.S

246

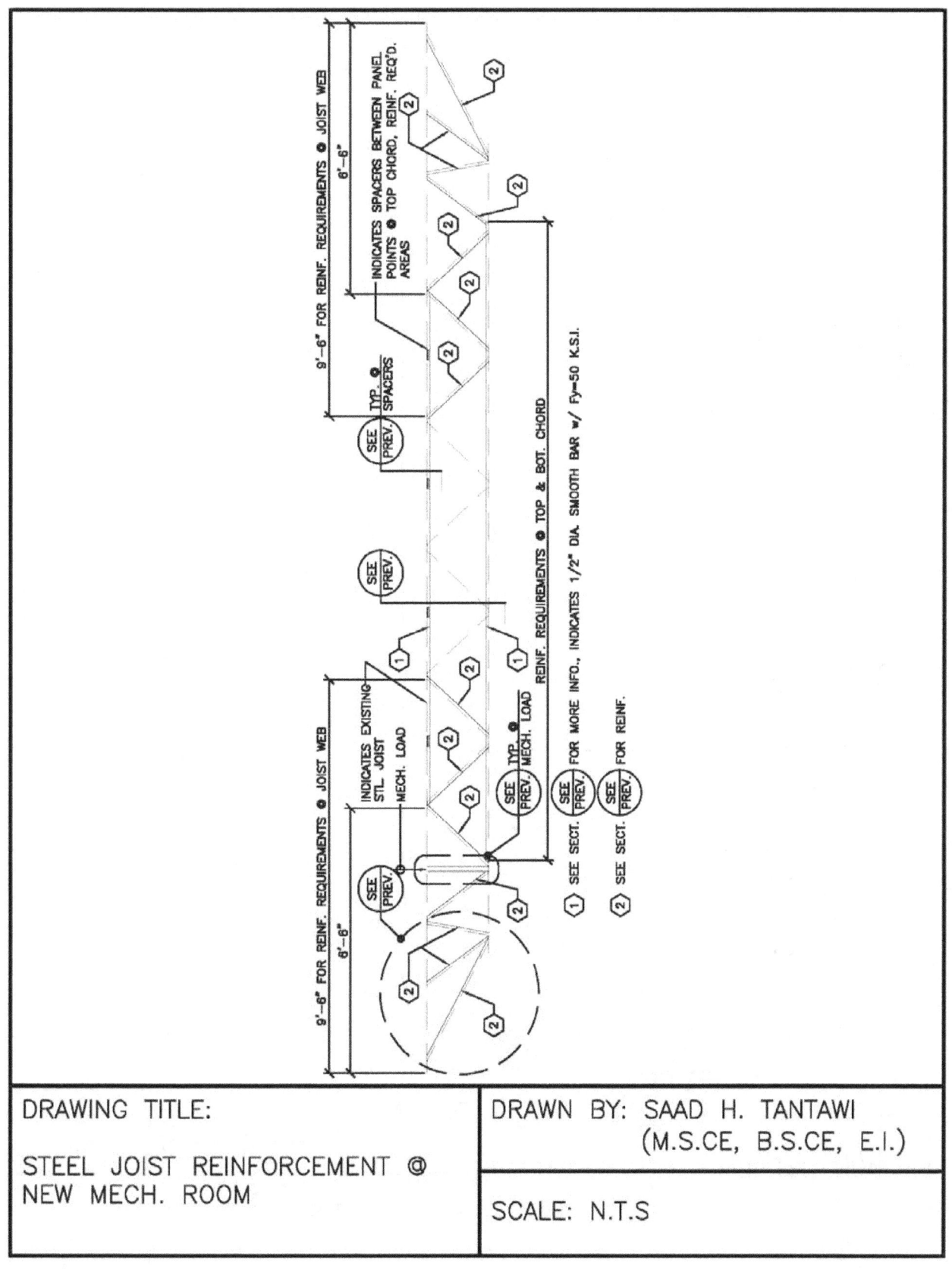

DRAWING TITLE:	DRAWN BY: SAAD H. TANTAWI
	(M.S.CE, B.S.CE, E.I.)
STEEL JOIST REINFORCEMENT @ NEW MECH. ROOM	SCALE: N.T.S

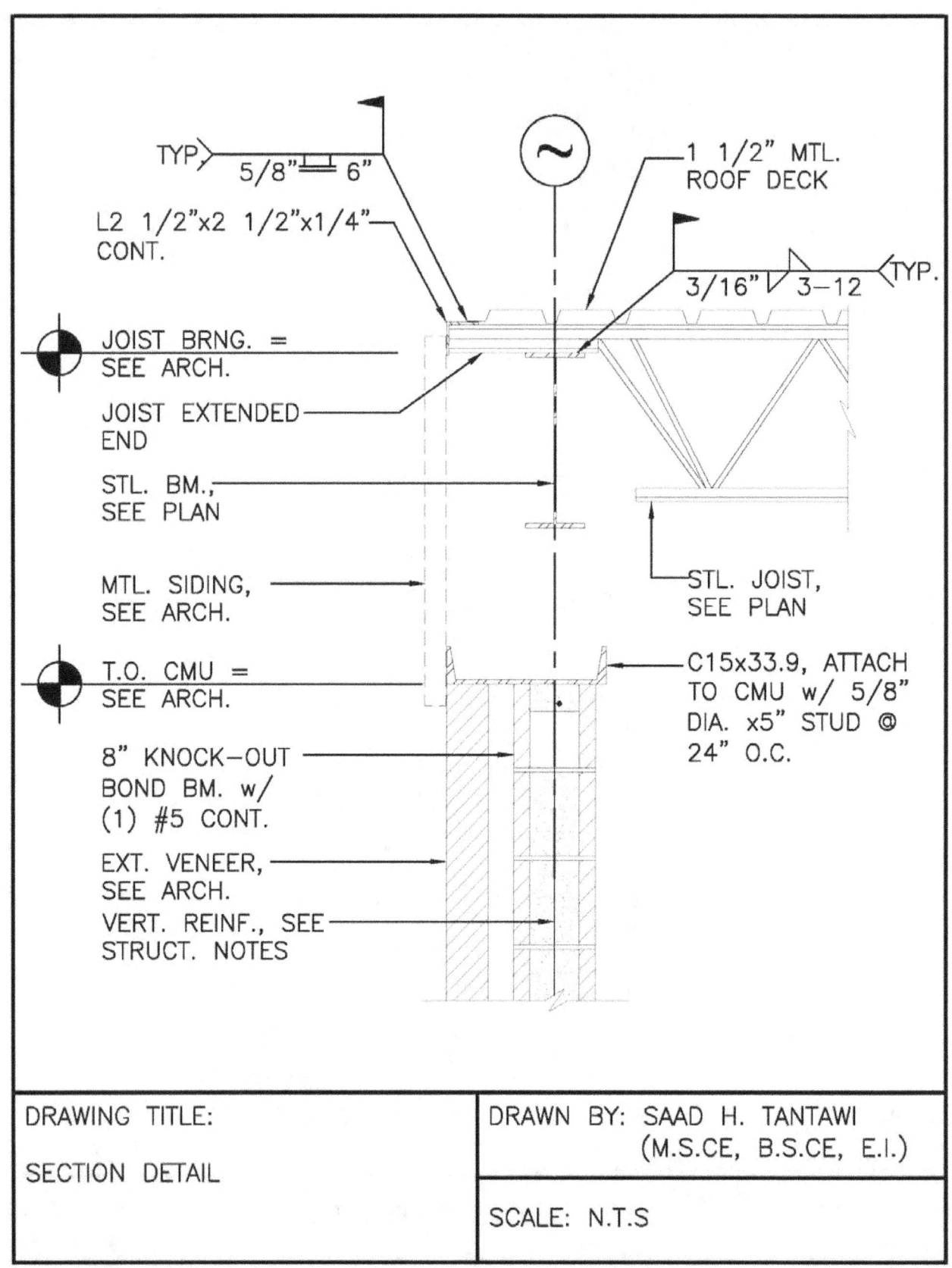

TYP. 5/8" 6"

L2 1/2"x2 1/2"x1/4" CONT.

1 1/2" MTL. ROOF DECK

3/16" 3-12 TYP.

JOIST BRNG. = SEE ARCH.

JOIST EXTENDED END

STL. BM., SEE PLAN

MTL. SIDING, SEE ARCH.

STL. JOIST, SEE PLAN

T.O. CMU = SEE ARCH.

C15x33.9, ATTACH TO CMU w/ 5/8" DIA. x5" STUD @ 24" O.C.

8" KNOCK-OUT BOND BM. w/ (1) #5 CONT.

EXT. VENEER, SEE ARCH.

VERT. REINF., SEE STRUCT. NOTES

DRAWING TITLE:	DRAWN BY: SAAD H. TANTAWI
SECTION DETAIL	(M.S.CE, B.S.CE, E.I.)
	SCALE: N.T.S

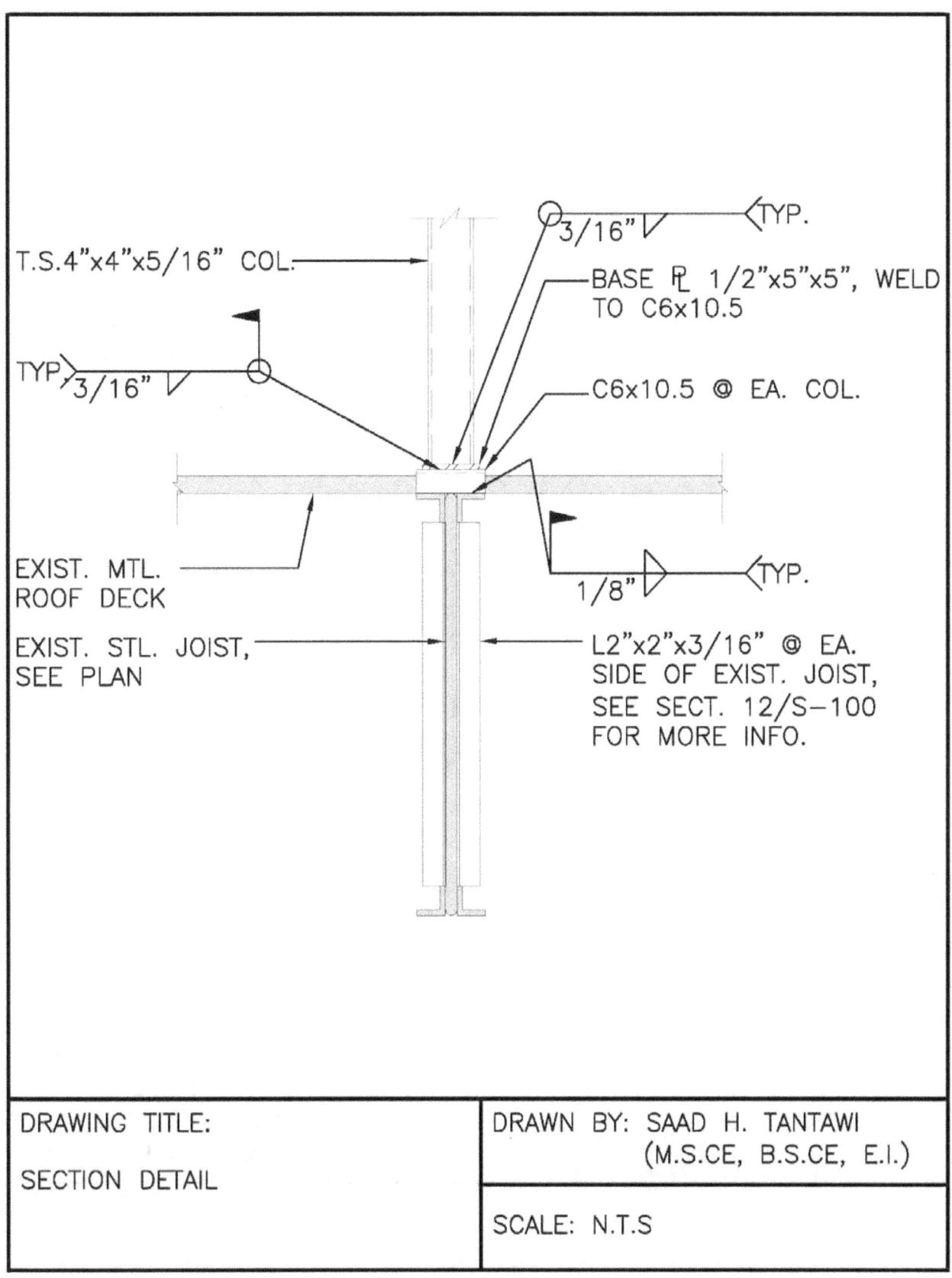

T.S.4"x4"x5/16" COL.

3/16" TYP.

BASE PL 1/2"x5"x5", WELD TO C6x10.5

C6x10.5 @ EA. COL.

TYP. 3/16"

EXIST. MTL. ROOF DECK

EXIST. STL. JOIST, SEE PLAN

1/8" TYP.

L2"x2"x3/16" @ EA. SIDE OF EXIST. JOIST, SEE SECT. 12/S-100 FOR MORE INFO.

DRAWING TITLE:	DRAWN BY: SAAD H. TANTAWI
	(M.S.CE, B.S.CE, E.I.)
SECTION DETAIL	
	SCALE: N.T.S

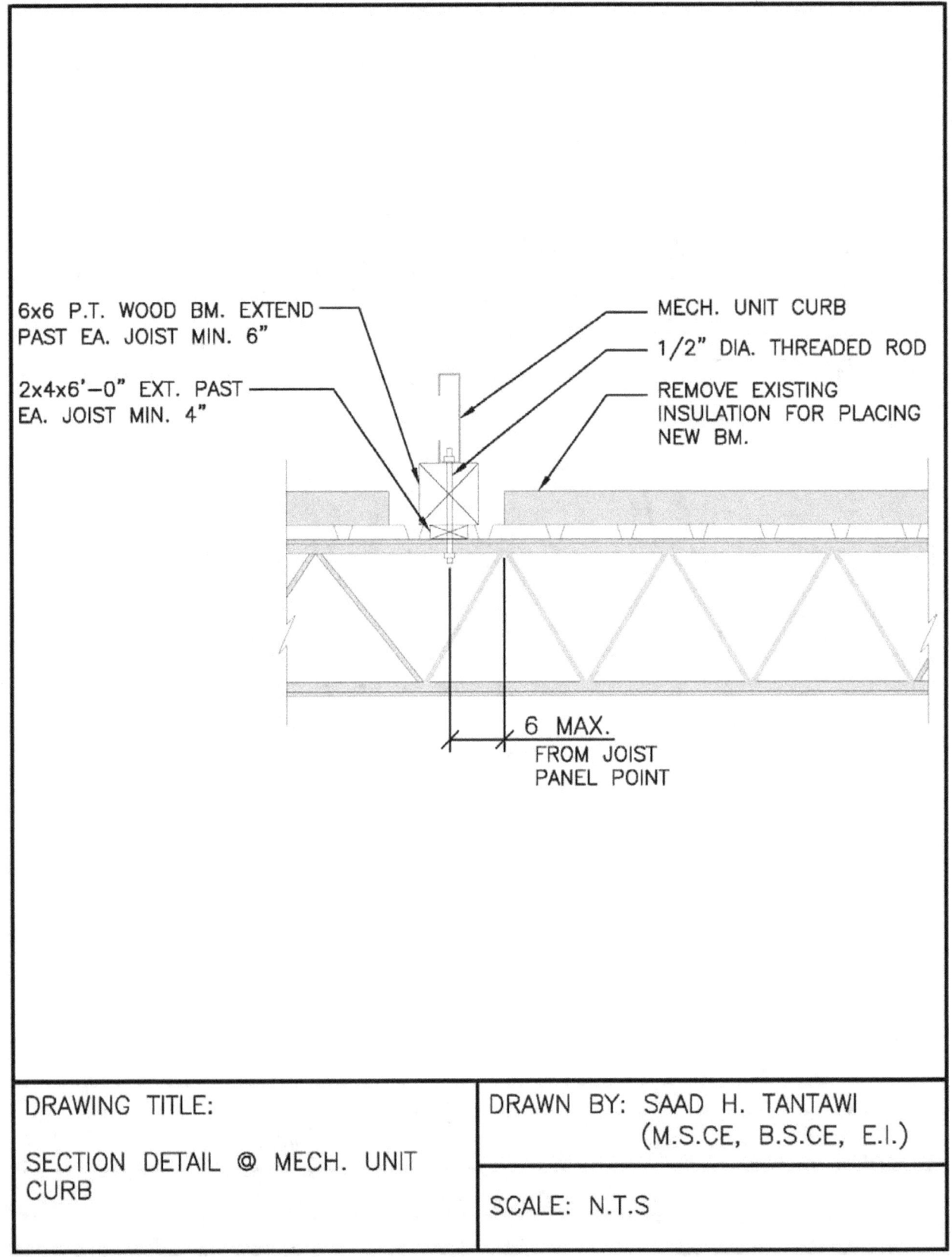

DRAWING TITLE: SECTION DETAIL @ MECH. UNIT CURB	DRAWN BY: SAAD H. TANTAWI (M.S.CE, B.S.CE, E.I.)
	SCALE: N.T.S

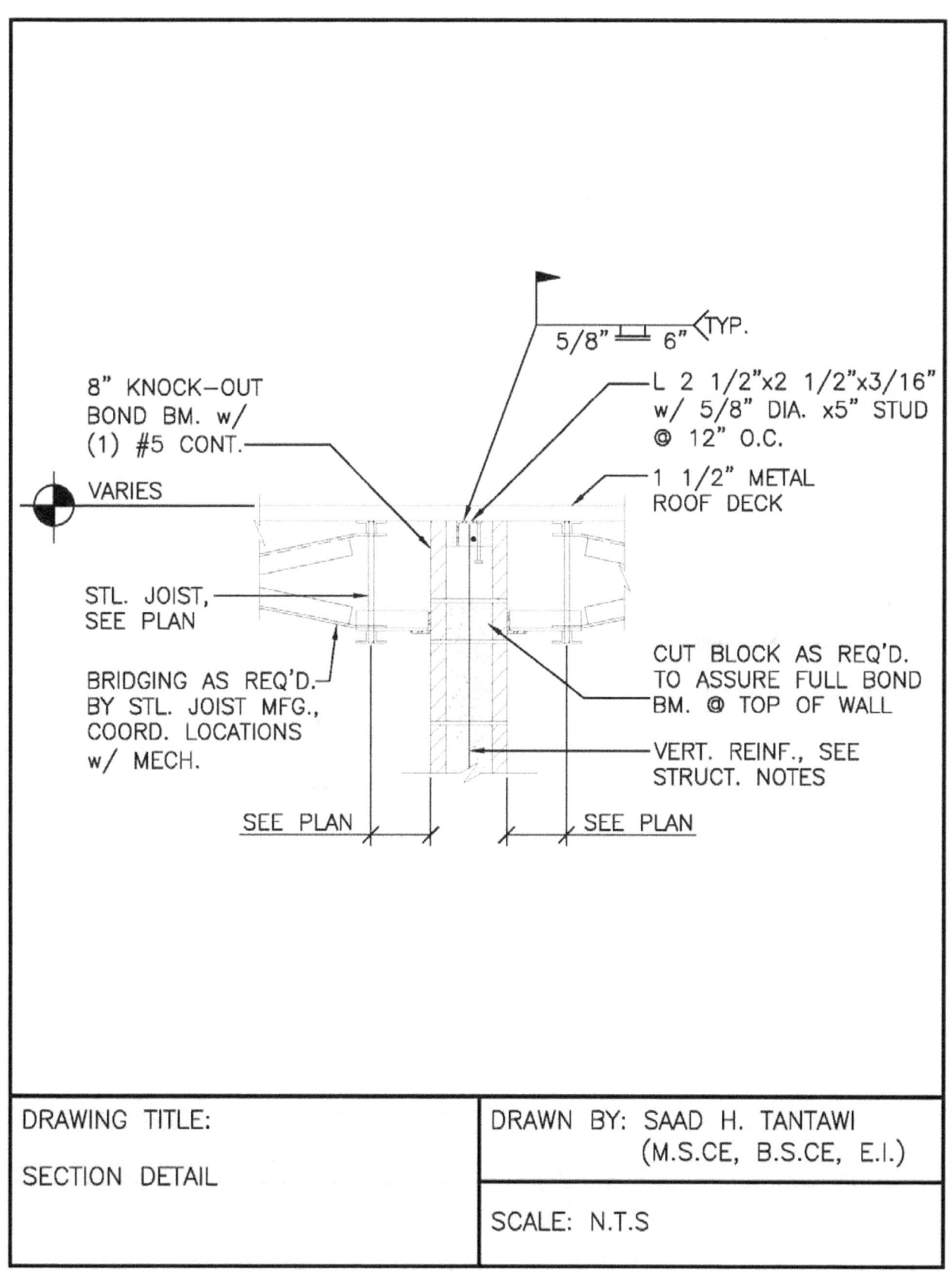

8" KNOCK-OUT
BOND BM. w/
(1) #5 CONT.

VARIES

STL. JOIST,
SEE PLAN

BRIDGING AS REQ'D.
BY STL. JOIST MFG.,
COORD. LOCATIONS
w/ MECH.

5/8" 6" TYP.

L 2 1/2"x2 1/2"x3/16"
w/ 5/8" DIA. x5" STUD
@ 12" O.C.

1 1/2" METAL
ROOF DECK

CUT BLOCK AS REQ'D.
TO ASSURE FULL BOND
BM. @ TOP OF WALL

VERT. REINF., SEE
STRUCT. NOTES

SEE PLAN   SEE PLAN

DRAWING TITLE:	DRAWN BY: SAAD H. TANTAWI
	(M.S.CE, B.S.CE, E.I.)
SECTION DETAIL	
	SCALE: N.T.S

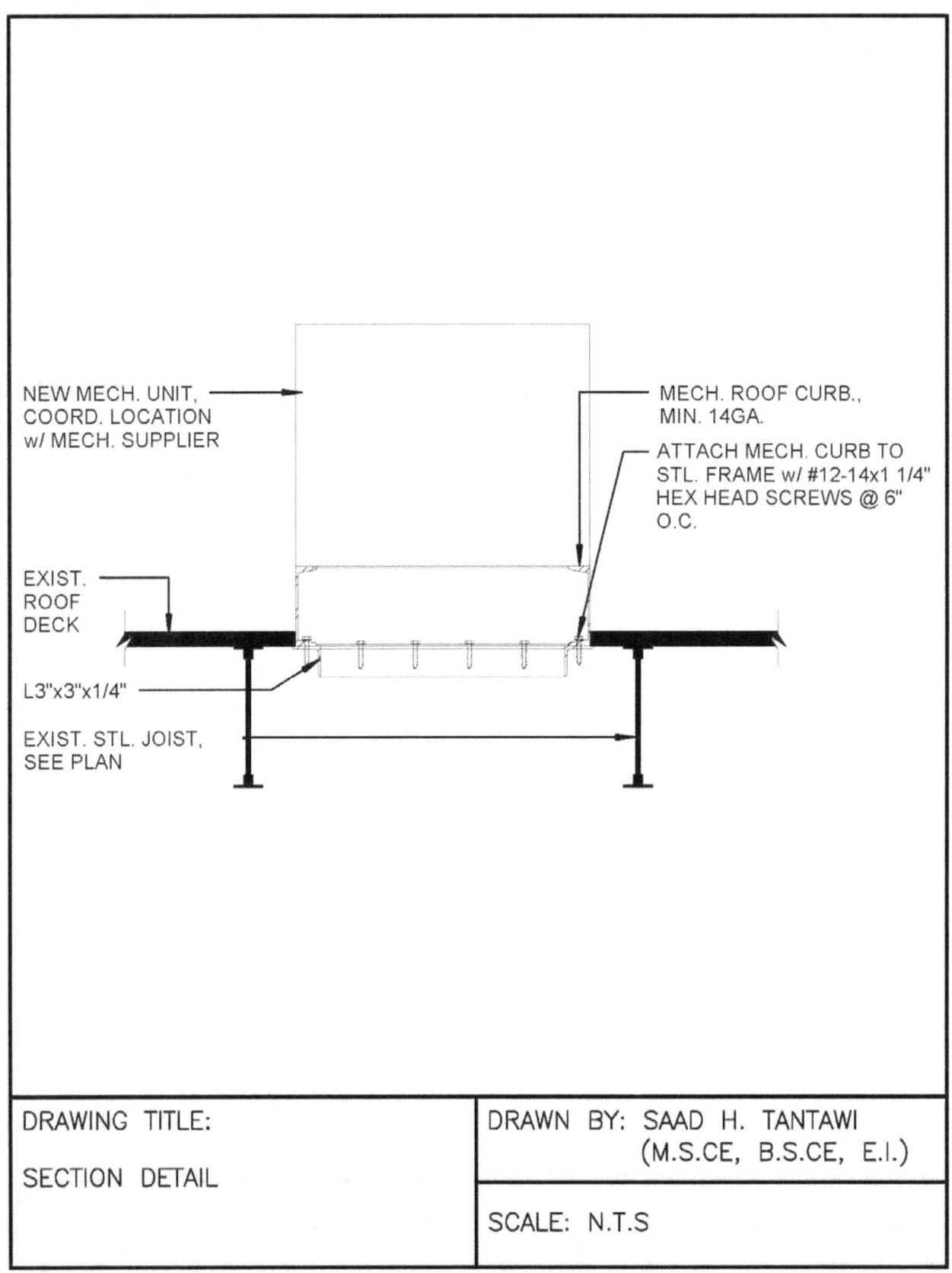

NEW MECH. UNIT,
COORD. LOCATION
w/ MECH. SUPPLIER

MECH. ROOF CURB.,
MIN. 14GA.

ATTACH MECH. CURB TO
STL. FRAME w/ #12-14x1 1/4"
HEX HEAD SCREWS @ 6"
O.C.

EXIST.
ROOF
DECK

L3"x3"x1/4"

EXIST. STL. JOIST,
SEE PLAN

DRAWING TITLE: SECTION DETAIL	DRAWN BY: SAAD H. TANTAWI (M.S.CE, B.S.CE, E.I.)
	SCALE: N.T.S

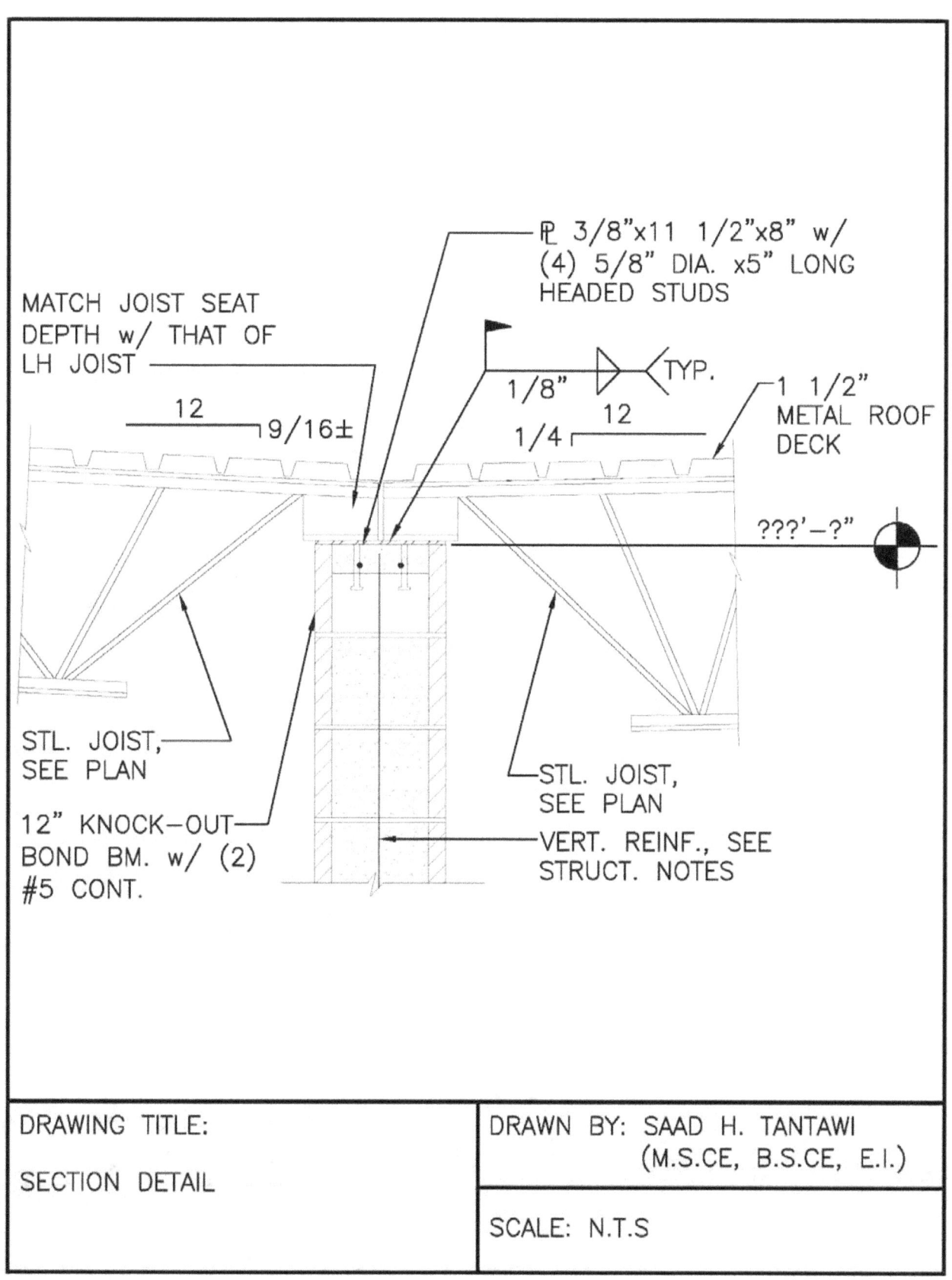

MATCH JOIST SEAT
DEPTH w/ THAT OF
LH JOIST

$\underline{12}$

9/16±

℄ 3/8"x11 1/2"x8" w/
(4) 5/8" DIA. x5" LONG
HEADED STUDS

1/8"

TYP.

1/4 $\underline{12}$

1 1/2"
METAL ROOF
DECK

???'−?"

STL. JOIST,
SEE PLAN

12" KNOCK−OUT
BOND BM. w/ (2)
#5 CONT.

STL. JOIST,
SEE PLAN

VERT. REINF., SEE
STRUCT. NOTES

DRAWING TITLE: SECTION DETAIL	DRAWN BY: SAAD H. TANTAWI (M.S.CE, B.S.CE, E.I.)
	SCALE: N.T.S

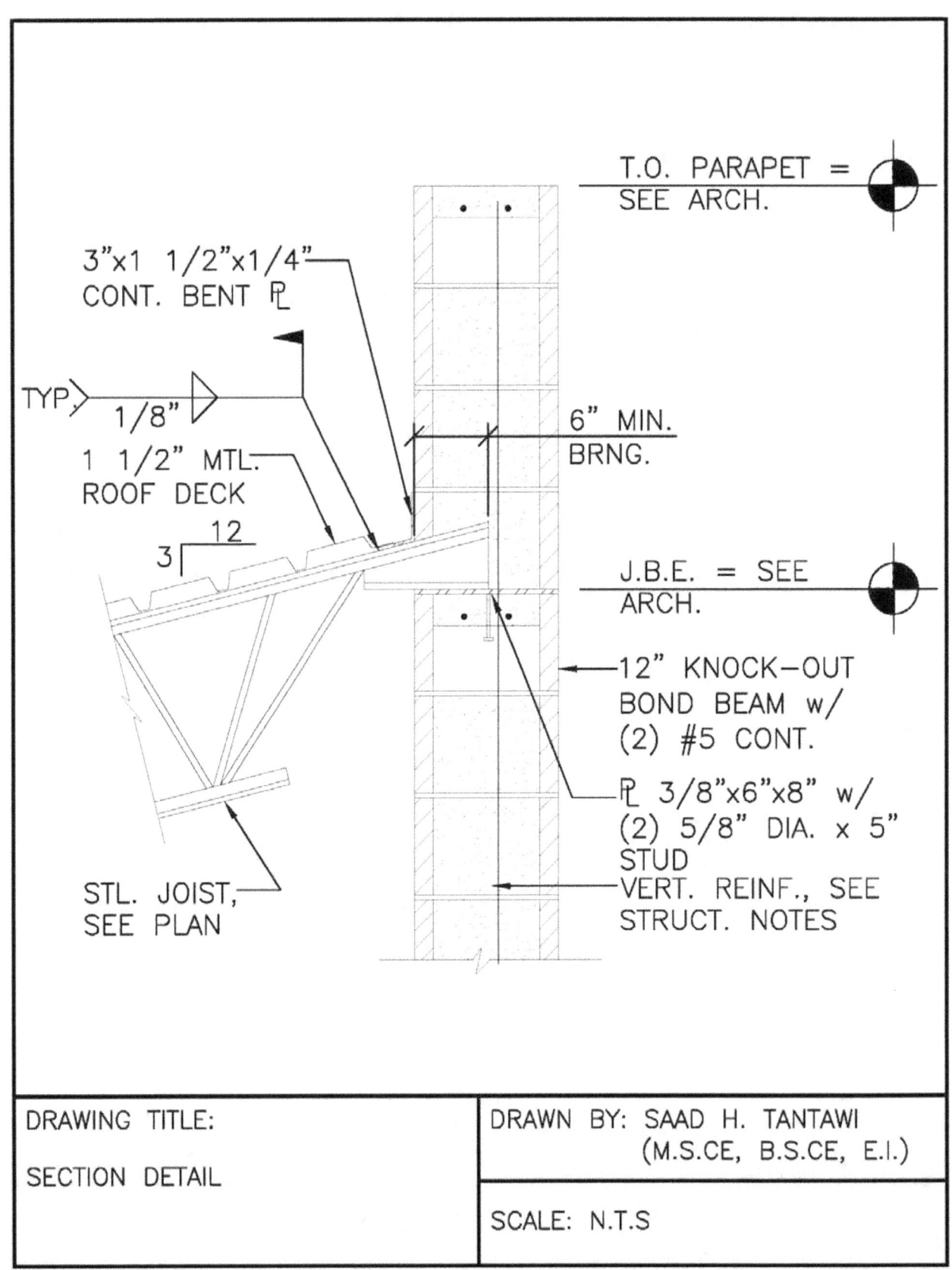

3"x1 1/2"x1/4" CONT. BENT ℄

TYP

1/8"

1 1/2" MTL. ROOF DECK

3 | 12

STL. JOIST, SEE PLAN

T.O. PARAPET = SEE ARCH.

6" MIN. BRNG.

J.B.E. = SEE ARCH.

12" KNOCK−OUT BOND BEAM w/ (2) #5 CONT.

℄ 3/8"x6"x8" w/ (2) 5/8" DIA. x 5" STUD

VERT. REINF., SEE STRUCT. NOTES

DRAWING TITLE: SECTION DETAIL	DRAWN BY: SAAD H. TANTAWI (M.S.CE, B.S.CE, E.I.)
	SCALE: N.T.S

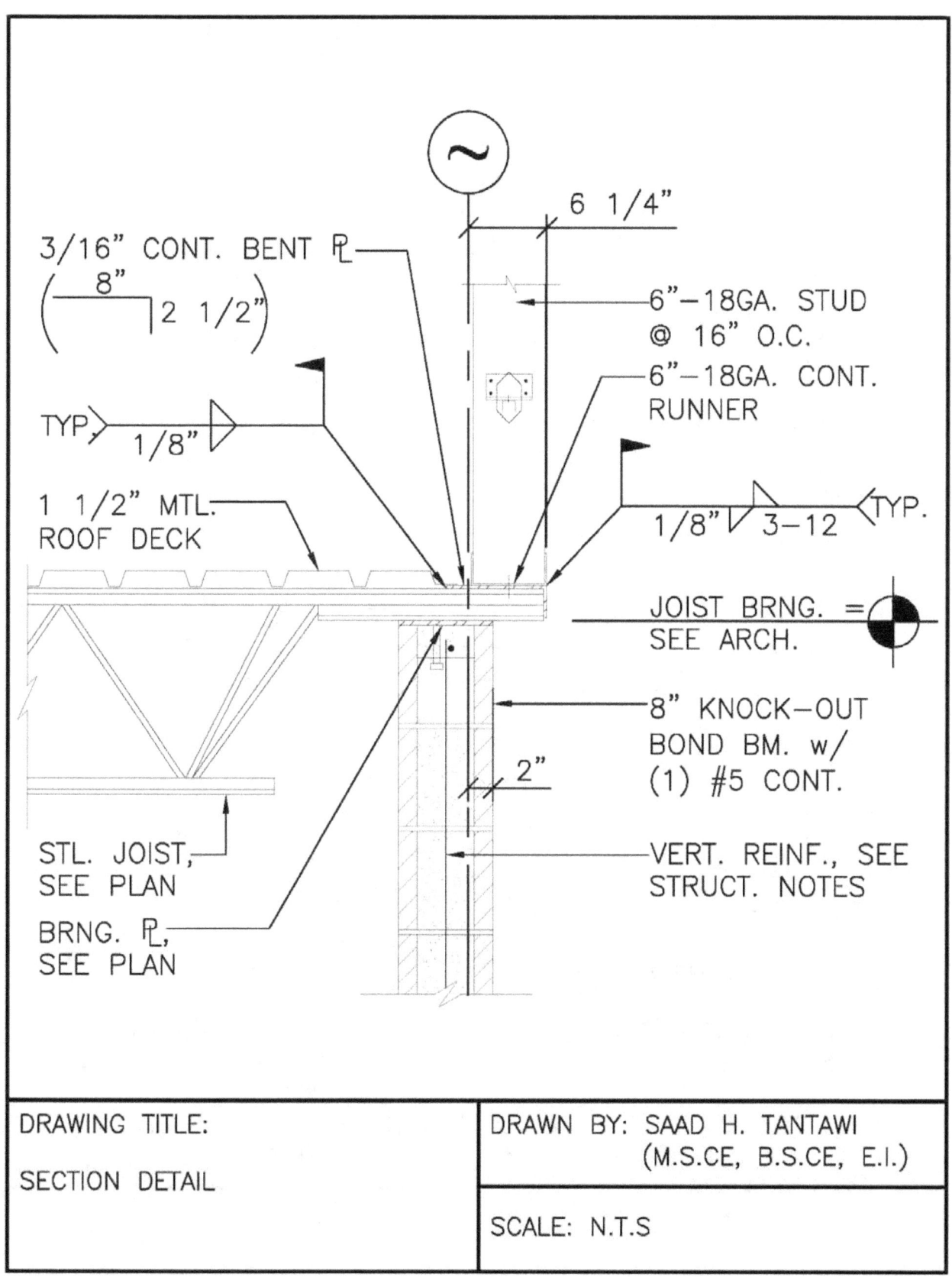

3/16" CONT. BENT ℙ

(8" ⌐ 2 1/2")

TYP. ◁ 1/8" ◁

1 1/2" MTL. ROOF DECK

STL. JOIST, SEE PLAN

BRNG. ℙ, SEE PLAN

6 1/4"

6"-18GA. STUD @ 16" O.C.

6"-18GA. CONT. RUNNER

1/8" ◁ 3-12 ◁ TYP.

JOIST BRNG. = SEE ARCH.

8" KNOCK-OUT BOND BM. w/ (1) #5 CONT.

2"

VERT. REINF., SEE STRUCT. NOTES

DRAWING TITLE:	DRAWN BY: SAAD H. TANTAWI
SECTION DETAIL	(M.S.CE, B.S.CE, E.I.)
	SCALE: N.T.S

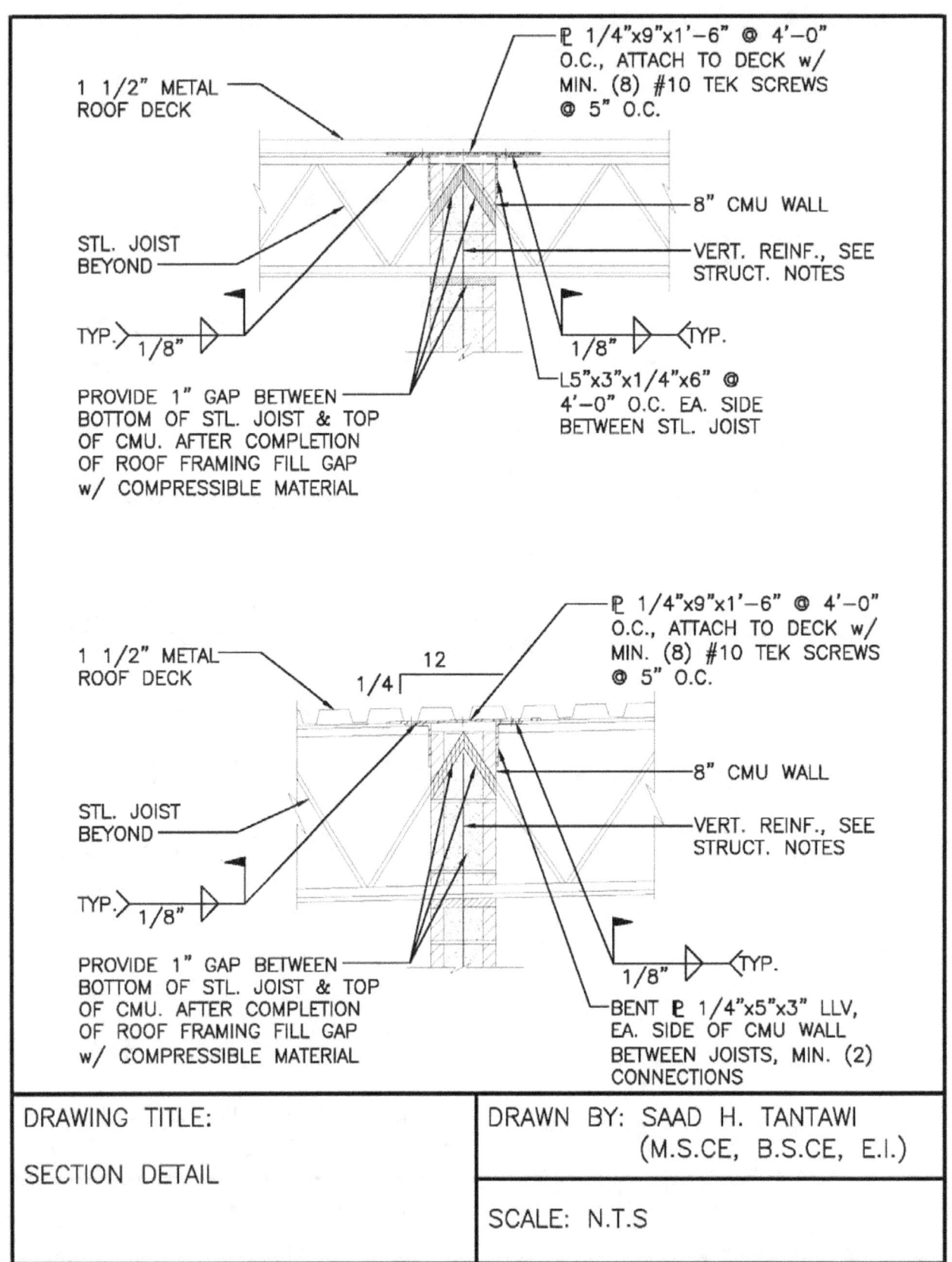

1 1/2" METAL ROOF DECK

ℙ 1/4"x9"x1'-6" @ 4'-0" O.C., ATTACH TO DECK w/ MIN. (8) #10 TEK SCREWS @ 5" O.C.

8" CMU WALL

VERT. REINF., SEE STRUCT. NOTES

STL. JOIST BEYOND

TYP. 1/8"

L5"x3"x1/4"x6" @ 4'-0" O.C. EA. SIDE BETWEEN STL. JOIST

1/8" TYP.

PROVIDE 1" GAP BETWEEN BOTTOM OF STL. JOIST & TOP OF CMU. AFTER COMPLETION OF ROOF FRAMING FILL GAP w/ COMPRESSIBLE MATERIAL

1 1/2" METAL ROOF DECK

12
1/4

ℙ 1/4"x9"x1'-6" @ 4'-0" O.C., ATTACH TO DECK w/ MIN. (8) #10 TEK SCREWS @ 5" O.C.

8" CMU WALL

VERT. REINF., SEE STRUCT. NOTES

STL. JOIST BEYOND

TYP. 1/8"

1/8" TYP.

PROVIDE 1" GAP BETWEEN BOTTOM OF STL. JOIST & TOP OF CMU. AFTER COMPLETION OF ROOF FRAMING FILL GAP w/ COMPRESSIBLE MATERIAL

BENT ℙ 1/4"x5"x3" LLV, EA. SIDE OF CMU WALL BETWEEN JOISTS, MIN. (2) CONNECTIONS

DRAWING TITLE: SECTION DETAIL	DRAWN BY: SAAD H. TANTAWI (M.S.CE, B.S.CE, E.I.)
	SCALE: N.T.S

256

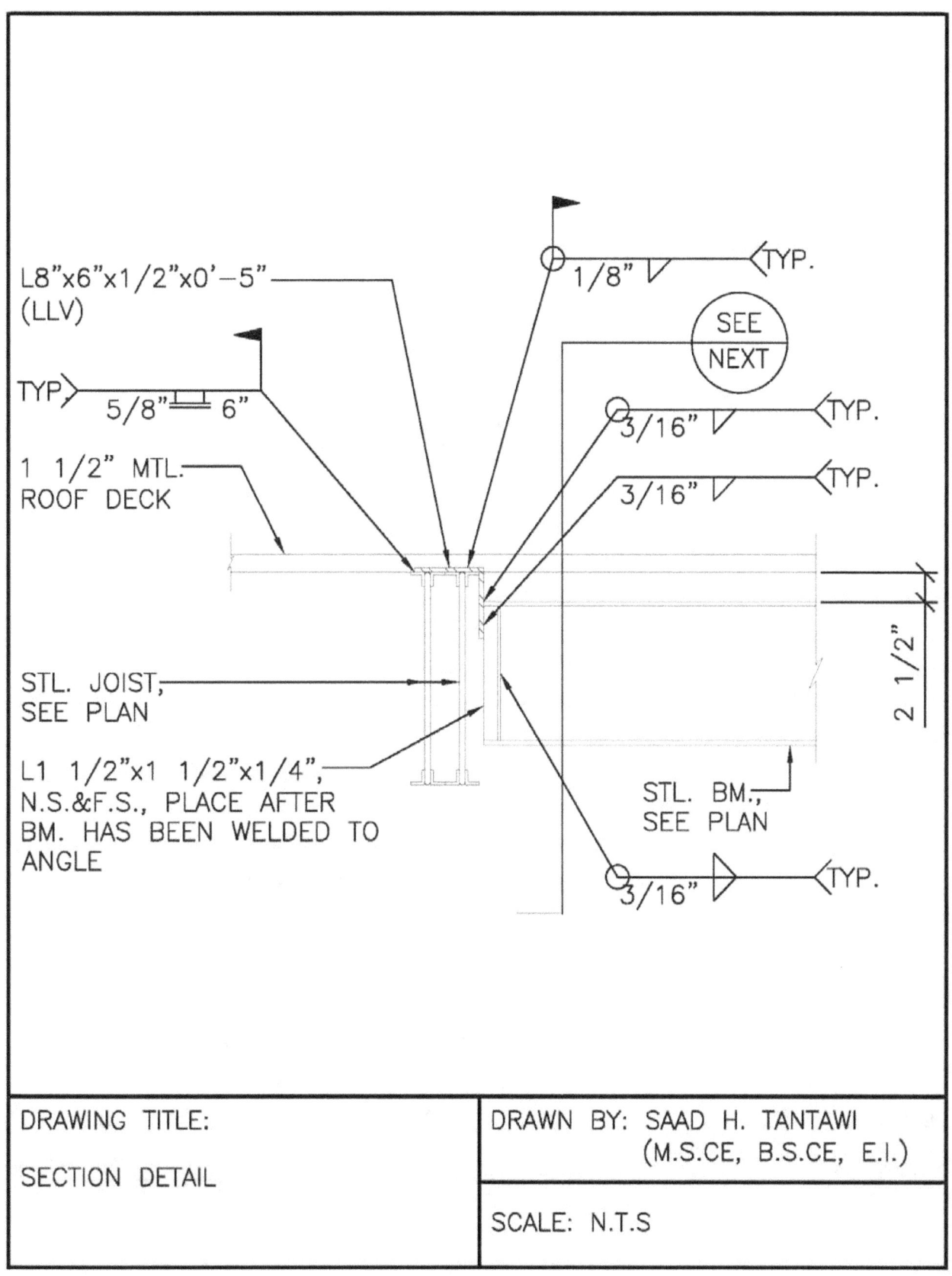

L8"x6"x1/2"x0'−5" (LLV)

TYP

5/8" 6"

1 1/2" MTL. ROOF DECK

STL. JOIST, SEE PLAN

L1 1/2"x1 1/2"x1/4", N.S.&F.S., PLACE AFTER BM. HAS BEEN WELDED TO ANGLE

1/8" TYP.

SEE NEXT

3/16" TYP.

3/16" TYP.

2 1/2"

STL. BM., SEE PLAN

3/16" TYP.

DRAWING TITLE:	DRAWN BY: SAAD H. TANTAWI (M.S.CE, B.S.CE, E.I.)
SECTION DETAIL	
	SCALE: N.T.S

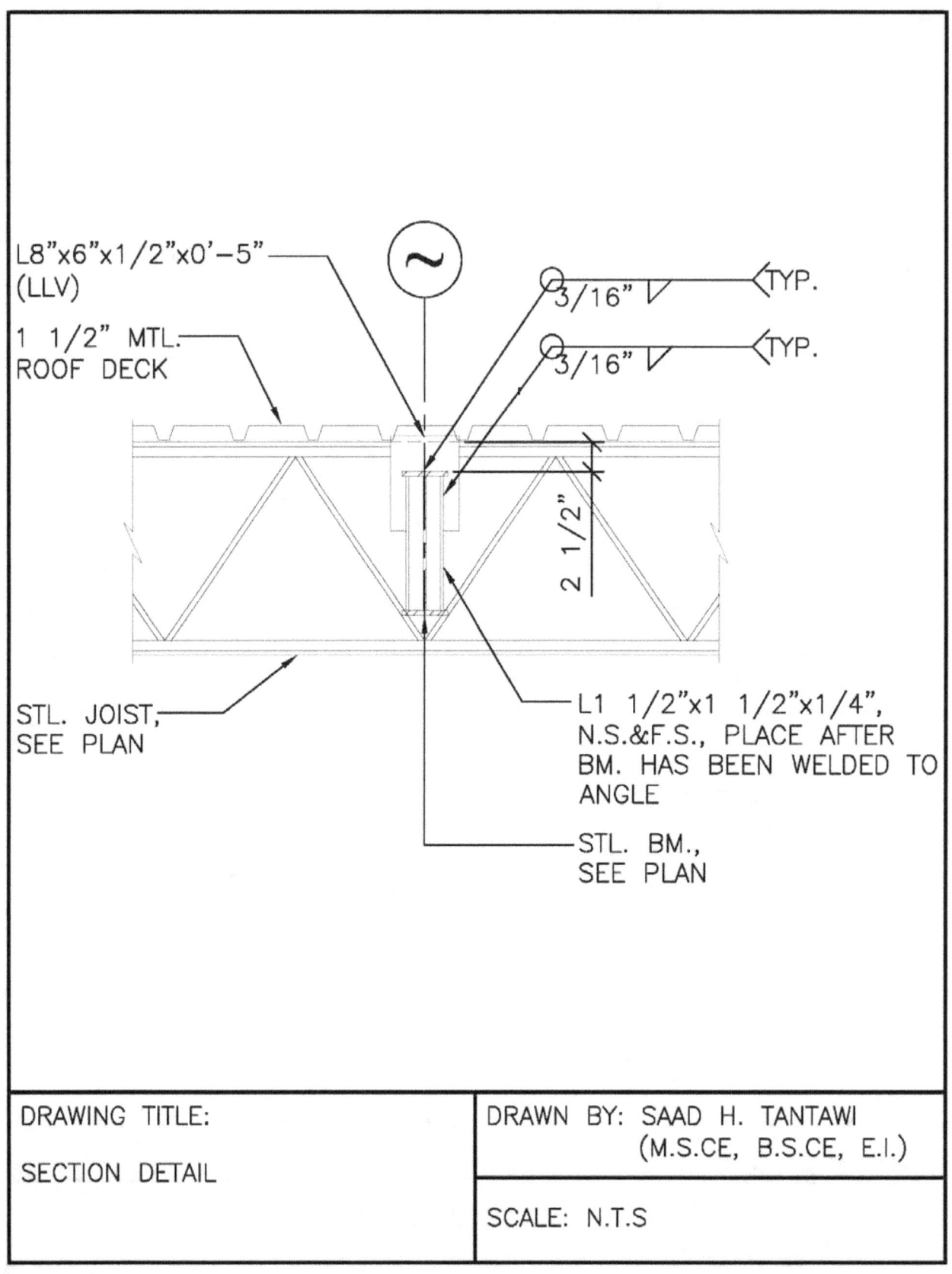

L8"x6"x1/2"x0'–5" (LLV)

1 1/2" MTL. ROOF DECK

3/16" TYP.

3/16" TYP.

2 1/2"

STL. JOIST, SEE PLAN

L1 1/2"x1 1/2"x1/4", N.S.&F.S., PLACE AFTER BM. HAS BEEN WELDED TO ANGLE

STL. BM., SEE PLAN

DRAWING TITLE:	DRAWN BY: SAAD H. TANTAWI
	(M.S.CE, B.S.CE, E.I.)
SECTION DETAIL	
	SCALE: N.T.S

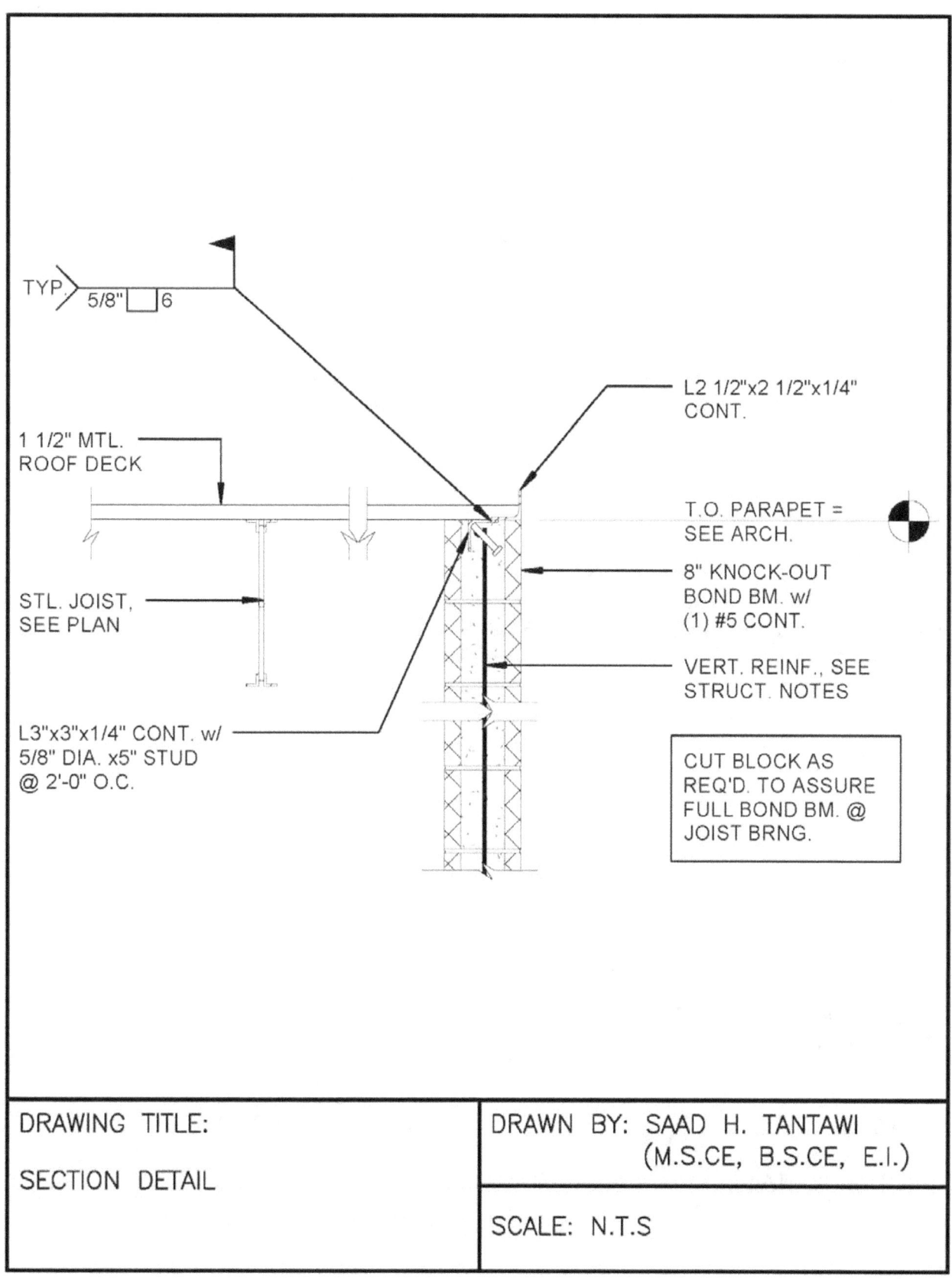

TYP.

5/8" 6

L2 1/2"x2 1/2"x1/4"
CONT.

1 1/2" MTL.
ROOF DECK

T.O. PARAPET =
SEE ARCH.

8" KNOCK-OUT
BOND BM. w/
(1) #5 CONT.

STL. JOIST,
SEE PLAN

VERT. REINF., SEE
STRUCT. NOTES

L3"x3"x1/4" CONT. w/
5/8" DIA. x5" STUD
@ 2'-0" O.C.

CUT BLOCK AS
REQ'D. TO ASSURE
FULL BOND BM. @
JOIST BRNG.

DRAWING TITLE:	DRAWN BY: SAAD H. TANTAWI
	(M.S.CE, B.S.CE, E.I.)
SECTION DETAIL	
	SCALE: N.T.S

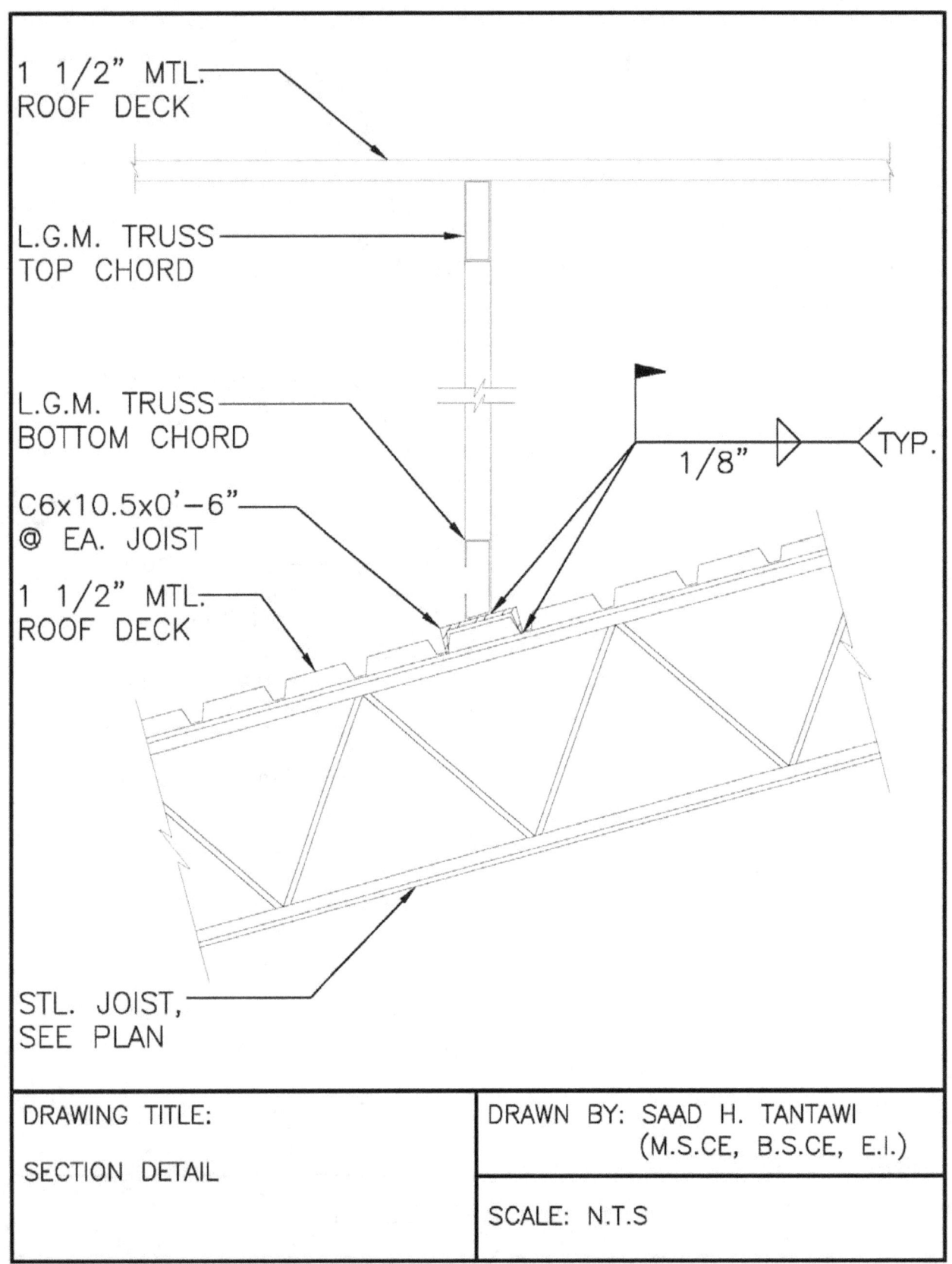

1 1/2" MTL.
ROOF DECK

L.G.M. TRUSS
TOP CHORD

L.G.M. TRUSS
BOTTOM CHORD

C6x10.5x0'-6"
@ EA. JOIST

1 1/2" MTL.
ROOF DECK

STL. JOIST,
SEE PLAN

1/8"   TYP.

DRAWING TITLE:	DRAWN BY: SAAD H. TANTAWI (M.S.CE, B.S.CE, E.I.)
SECTION DETAIL	
	SCALE: N.T.S

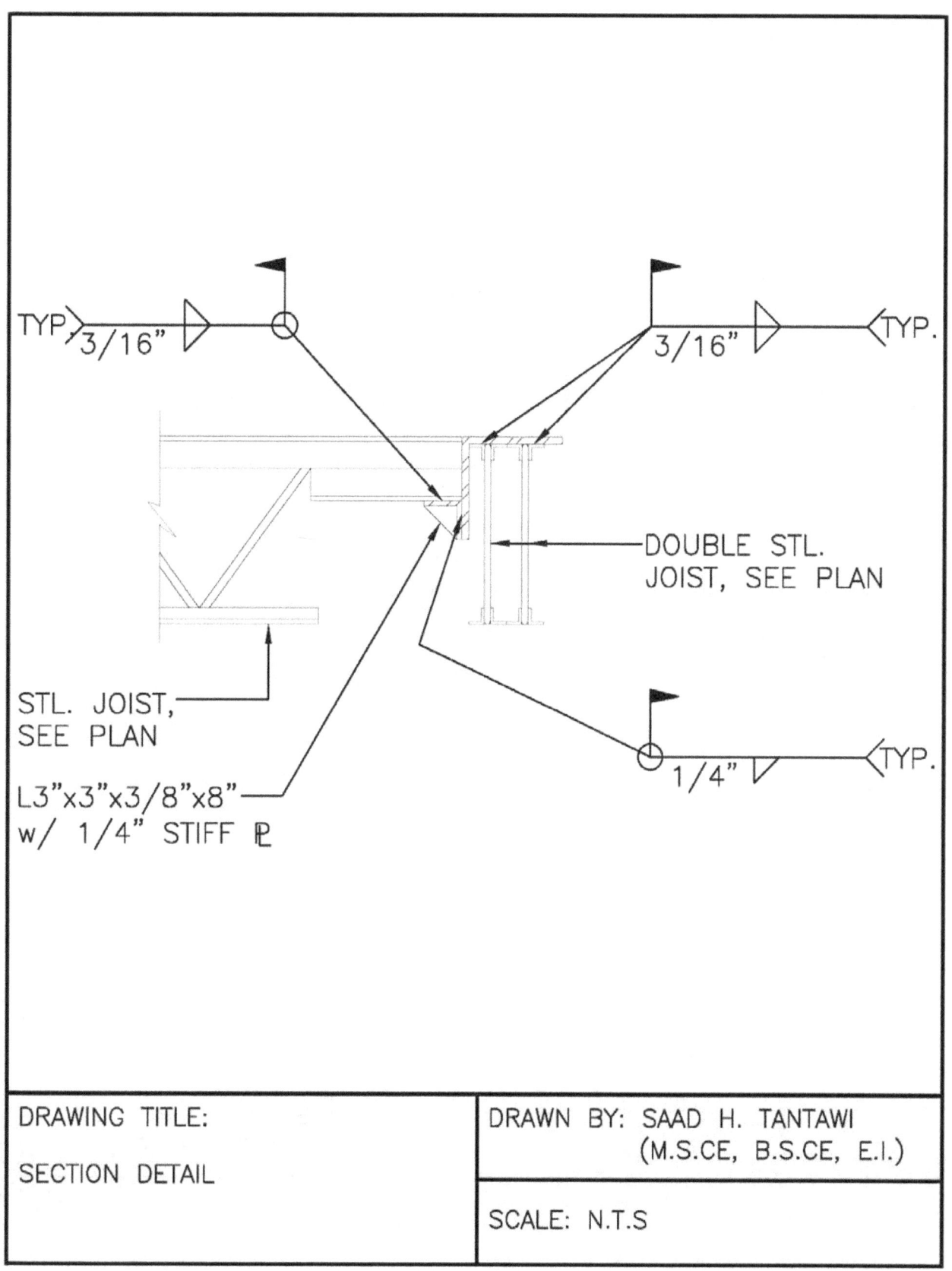

TYP. 3/16"

3/16" TYP.

DOUBLE STL.
JOIST, SEE PLAN

STL. JOIST,
SEE PLAN

L3"x3"x3/8"x8"
w/ 1/4" STIFF ℙ

1/4" TYP.

DRAWING TITLE:	DRAWN BY: SAAD H. TANTAWI
	(M.S.CE, B.S.CE, E.I.)
SECTION DETAIL	
	SCALE: N.T.S

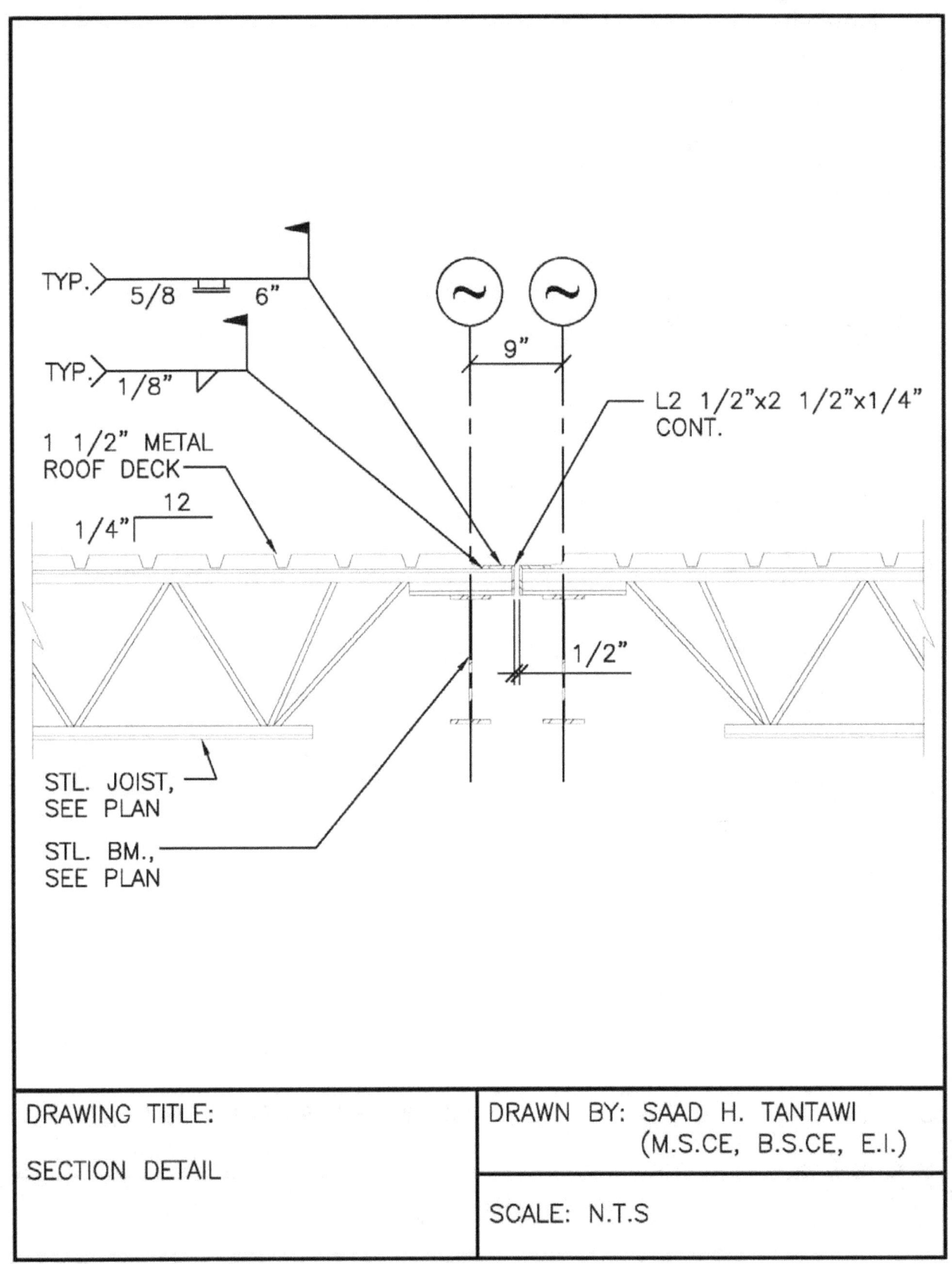

TYP. 5/8 6"

TYP. 1/8"

1 1/2" METAL
ROOF DECK

1/4" 12

9"

L2 1/2"x2 1/2"x1/4"
CONT.

1/2"

STL. JOIST,
SEE PLAN

STL. BM.,
SEE PLAN

DRAWING TITLE:	DRAWN BY: SAAD H. TANTAWI
	(M.S.CE, B.S.CE, E.I.)
SECTION DETAIL	
	SCALE: N.T.S

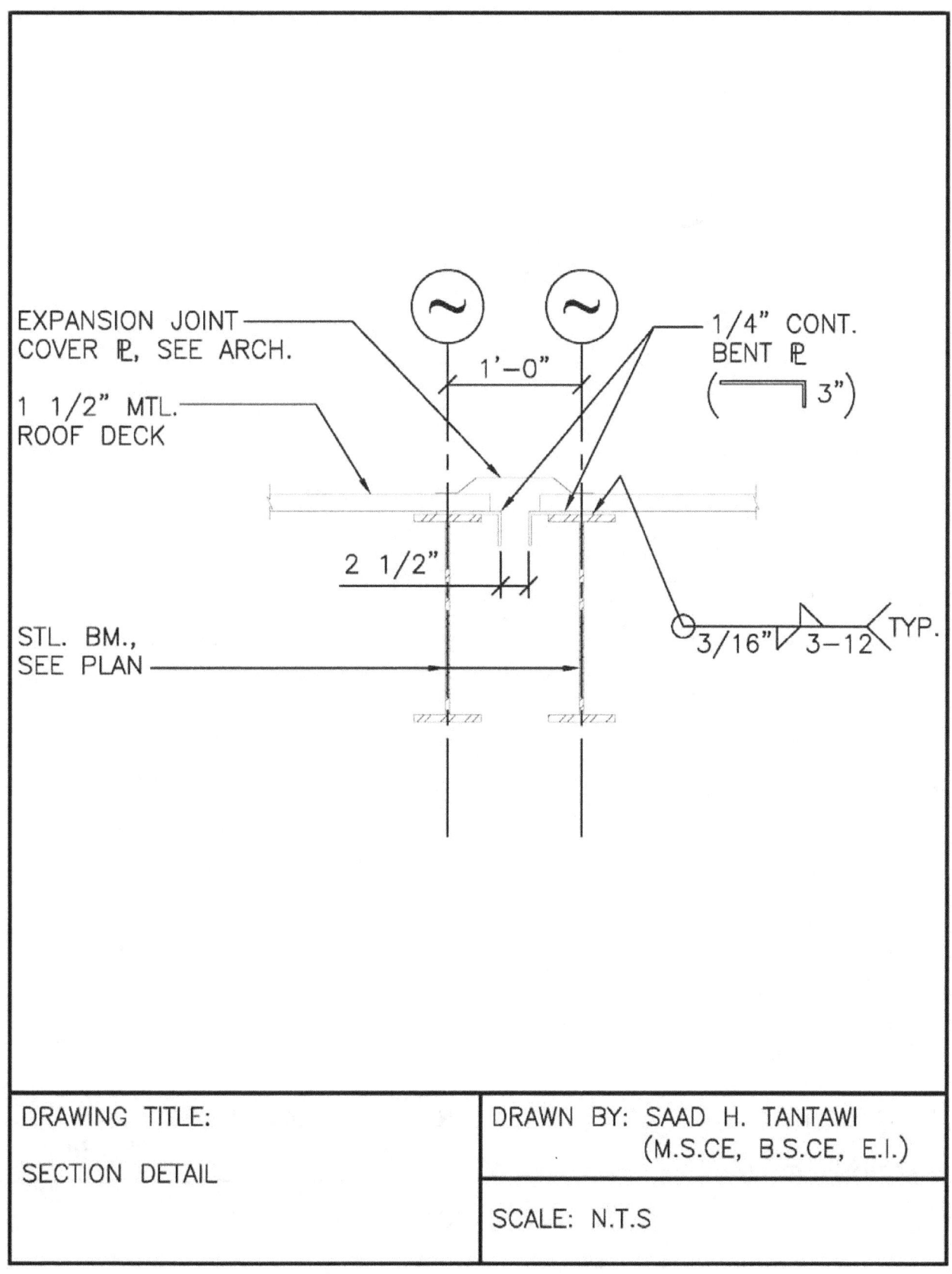

EXPANSION JOINT
COVER ₧, SEE ARCH.

1 1/2" MTL.
ROOF DECK

1'-0"

1/4" CONT.
BENT ₧

(⌐ 3")

2 1/2"

STL. BM.,
SEE PLAN

3/16" 3-12 TYP.

DRAWING TITLE:	DRAWN BY: SAAD H. TANTAWI
	(M.S.CE, B.S.CE, E.I.)
SECTION DETAIL	
	SCALE: N.T.S

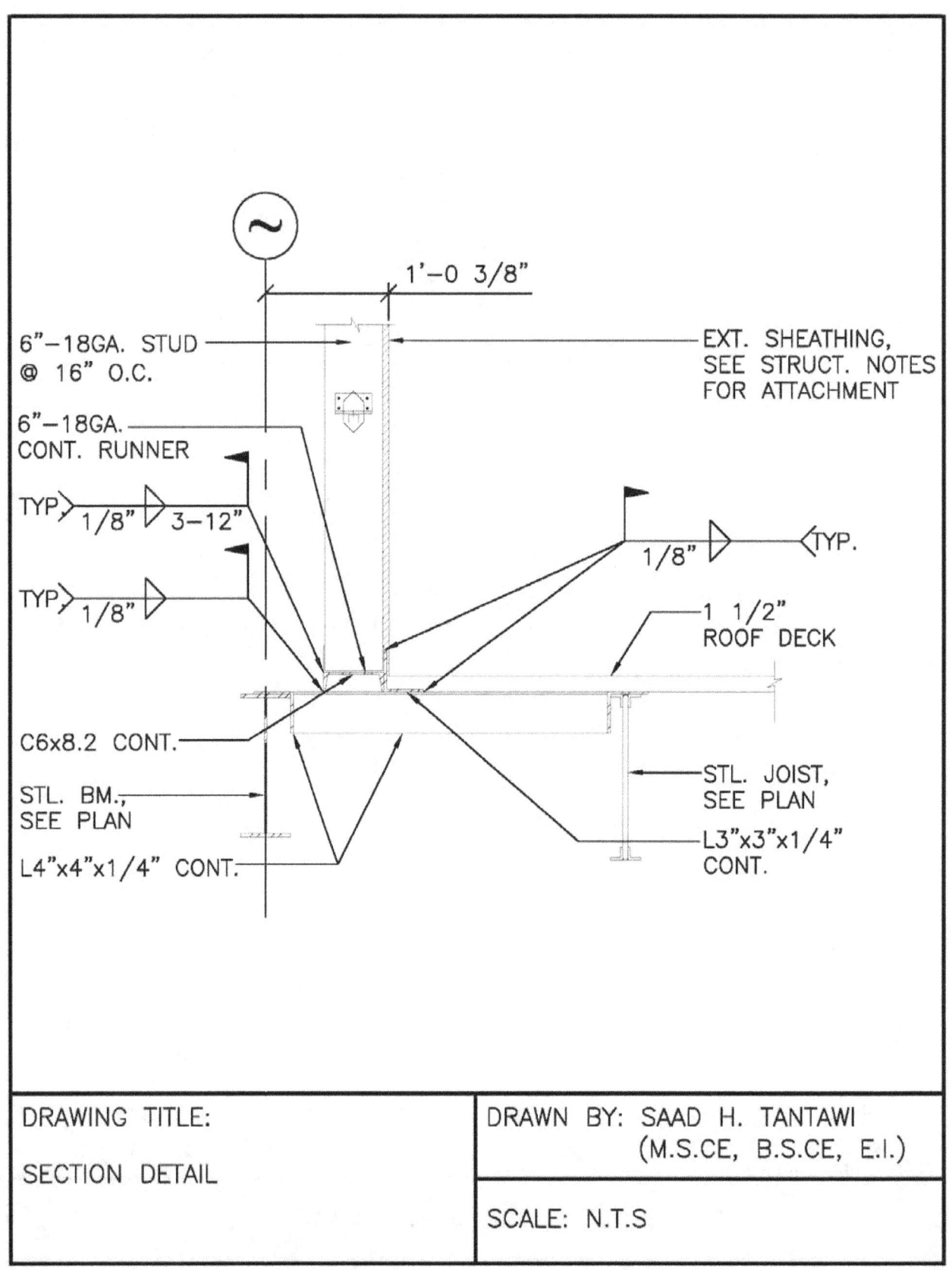

DRAWING TITLE:	DRAWN BY: SAAD H. TANTAWI
SECTION DETAIL	(M.S.CE, B.S.CE, E.I.)
	SCALE: N.T.S

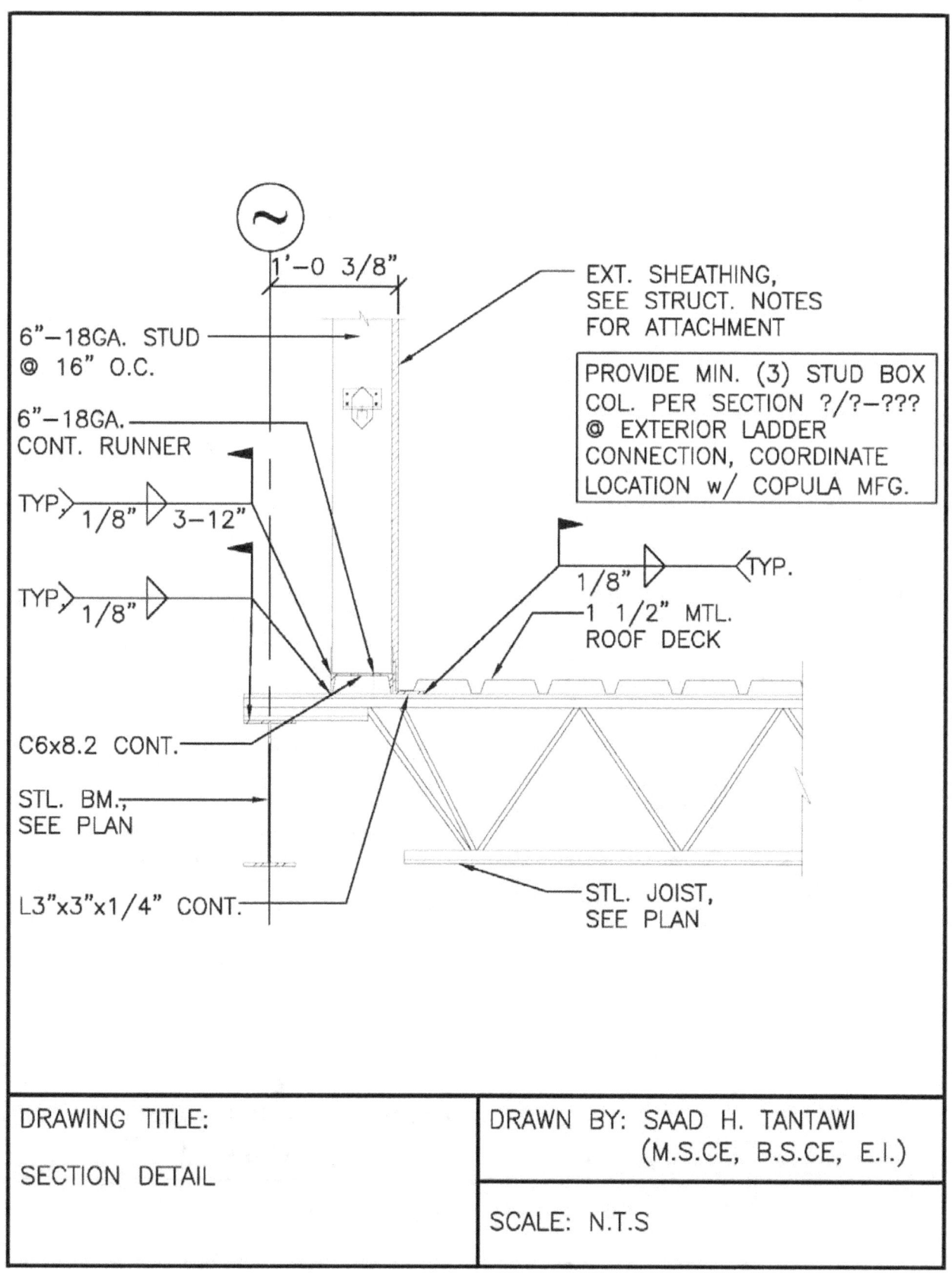

6"–18GA. STUD
@ 16" O.C.

6"–18GA.
CONT. RUNNER

TYP. 1/8" 3–12"

TYP. 1/8"

C6x8.2 CONT.

STL. BM.,
SEE PLAN

L3"x3"x1/4" CONT.

1'–0 3/8"

EXT. SHEATHING,
SEE STRUCT. NOTES
FOR ATTACHMENT

PROVIDE MIN. (3) STUD BOX
COL. PER SECTION ?/?–???
@ EXTERIOR LADDER
CONNECTION, COORDINATE
LOCATION w/ COPULA MFG.

1/8" TYP.

1 1/2" MTL.
ROOF DECK

STL. JOIST,
SEE PLAN

DRAWING TITLE: SECTION DETAIL	DRAWN BY: SAAD H. TANTAWI (M.S.CE, B.S.CE, E.I.)
	SCALE: N.T.S

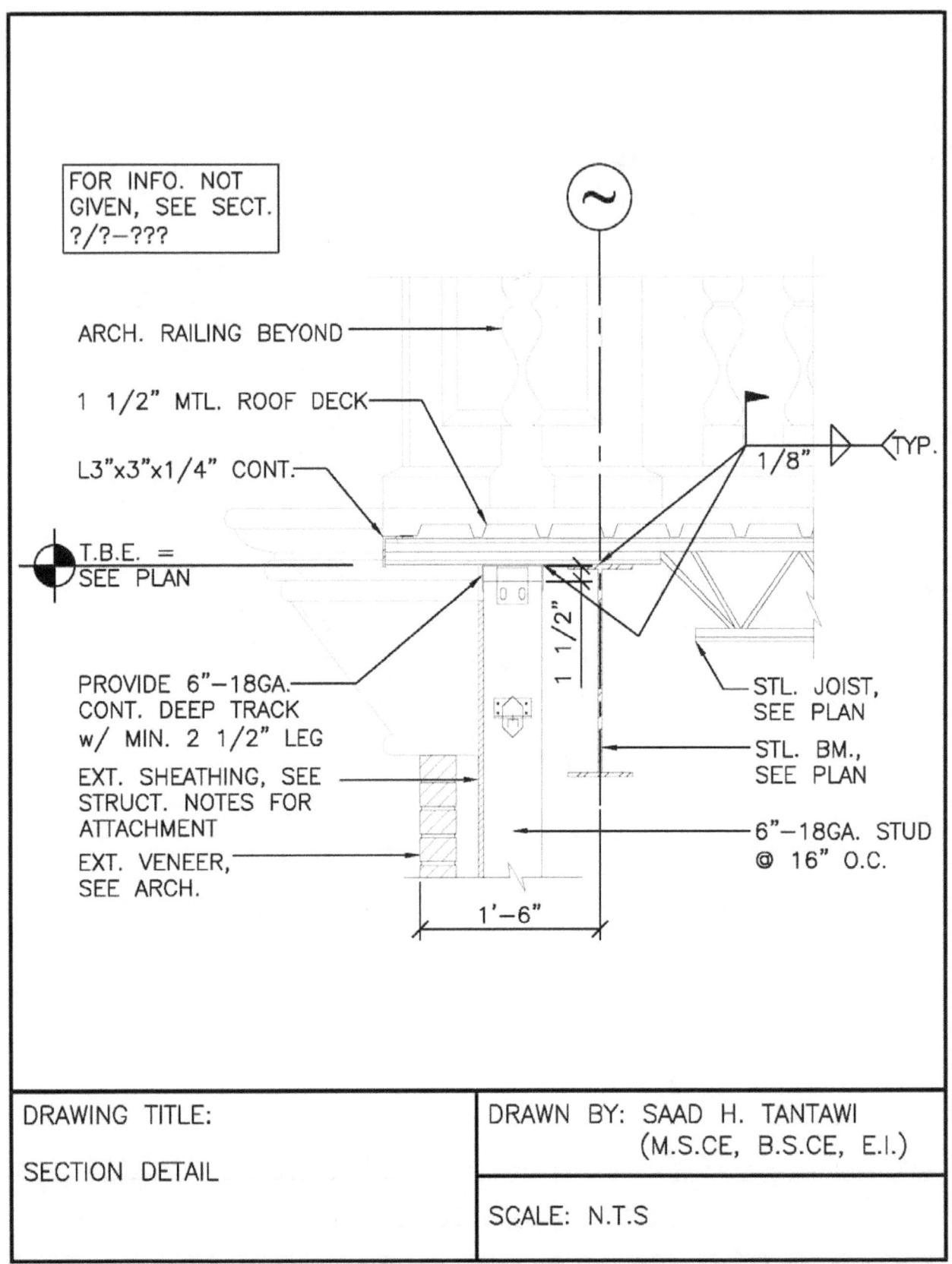

FOR INFO. NOT
GIVEN, SEE SECT.
?/?-???

ARCH. RAILING BEYOND

1 1/2" MTL. ROOF DECK

L3"x3"x1/4" CONT.

T.B.E. =
SEE PLAN

PROVIDE 6"-18GA.
CONT. DEEP TRACK
w/ MIN. 2 1/2" LEG

EXT. SHEATHING, SEE
STRUCT. NOTES FOR
ATTACHMENT

EXT. VENEER,
SEE ARCH.

1/8"

TYP.

1 1/2"

STL. JOIST,
SEE PLAN

STL. BM.,
SEE PLAN

6"-18GA. STUD
@ 16" O.C.

1'-6"

DRAWING TITLE:	DRAWN BY: SAAD H. TANTAWI
	(M.S.CE, B.S.CE, E.I.)
SECTION DETAIL	
	SCALE: N.T.S

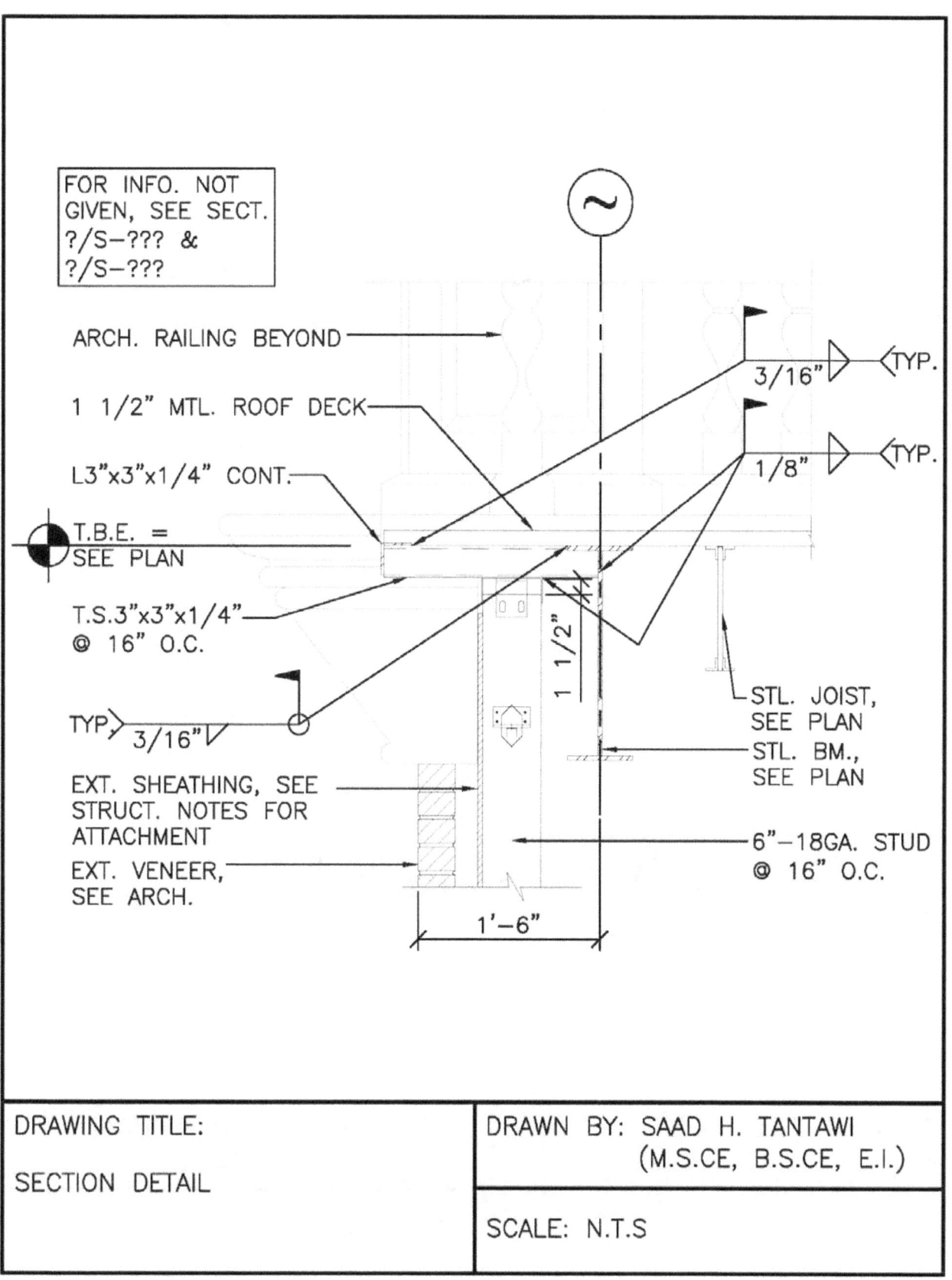

FOR INFO. NOT
GIVEN, SEE SECT.
?/S-??? &
?/S-???

ARCH. RAILING BEYOND

1 1/2" MTL. ROOF DECK

L3"x3"x1/4" CONT.

T.B.E. =
SEE PLAN

T.S.3"x3"x1/4"
@ 16" O.C.

TYP 3/16"

EXT. SHEATHING, SEE
STRUCT. NOTES FOR
ATTACHMENT

EXT. VENEER,
SEE ARCH.

3/16" TYP.

1/8" TYP.

1 1/2"

STL. JOIST,
SEE PLAN

STL. BM.,
SEE PLAN

6"-18GA. STUD
@ 16" O.C.

1'-6"

DRAWING TITLE:

SECTION DETAIL

DRAWN BY: SAAD H. TANTAWI
(M.S.CE, B.S.CE, E.I.)

SCALE: N.T.S

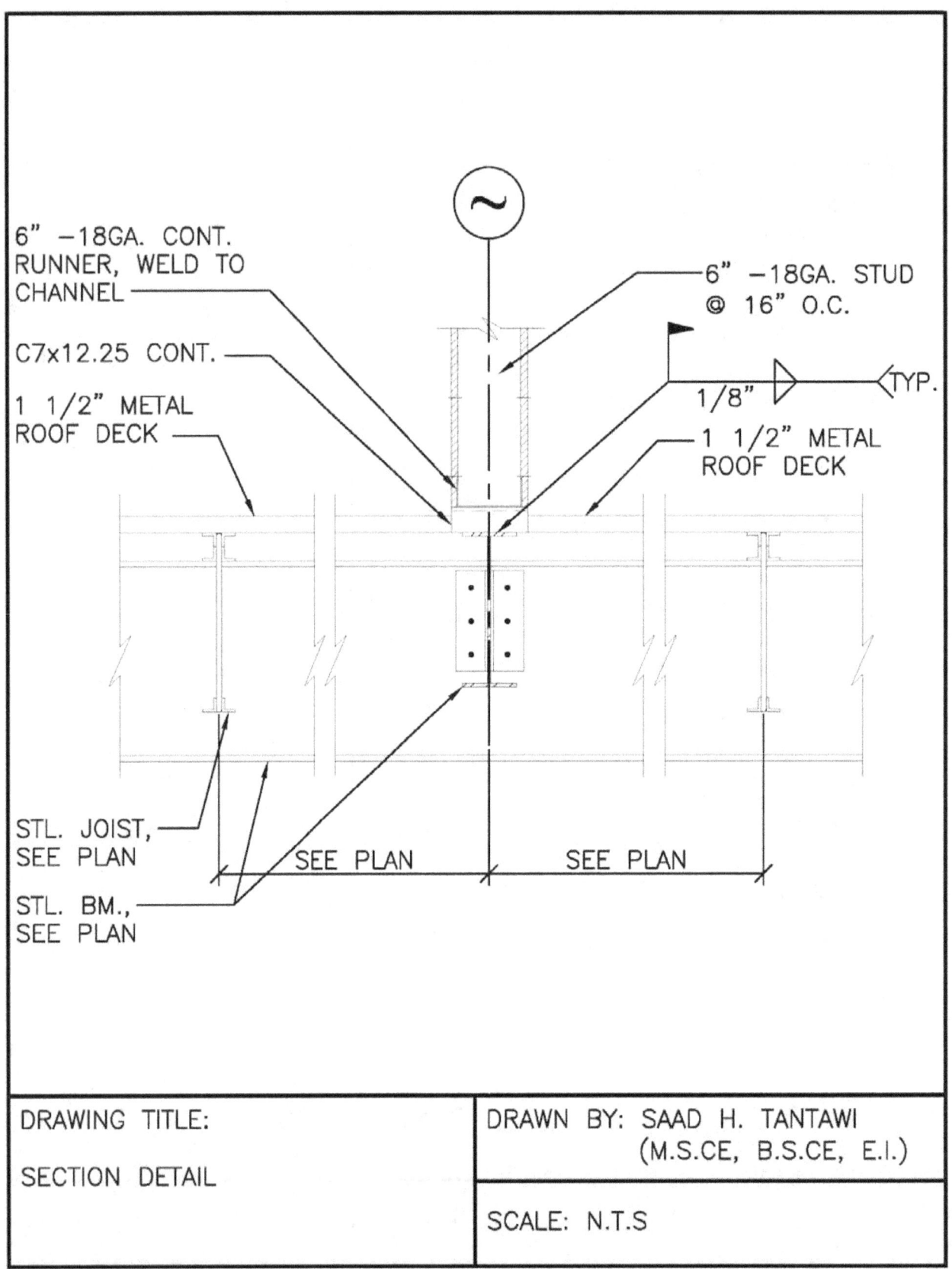

6" —18GA. CONT.
RUNNER, WELD TO
CHANNEL

C7x12.25 CONT.

1 1/2" METAL
ROOF DECK

6" —18GA. STUD
@ 16" O.C.

1/8"   TYP.

1 1/2" METAL
ROOF DECK

STL. JOIST,
SEE PLAN

STL. BM.,
SEE PLAN

SEE PLAN

SEE PLAN

DRAWING TITLE:	DRAWN BY: SAAD H. TANTAWI
	(M.S.CE, B.S.CE, E.I.)
SECTION DETAIL	
	SCALE: N.T.S

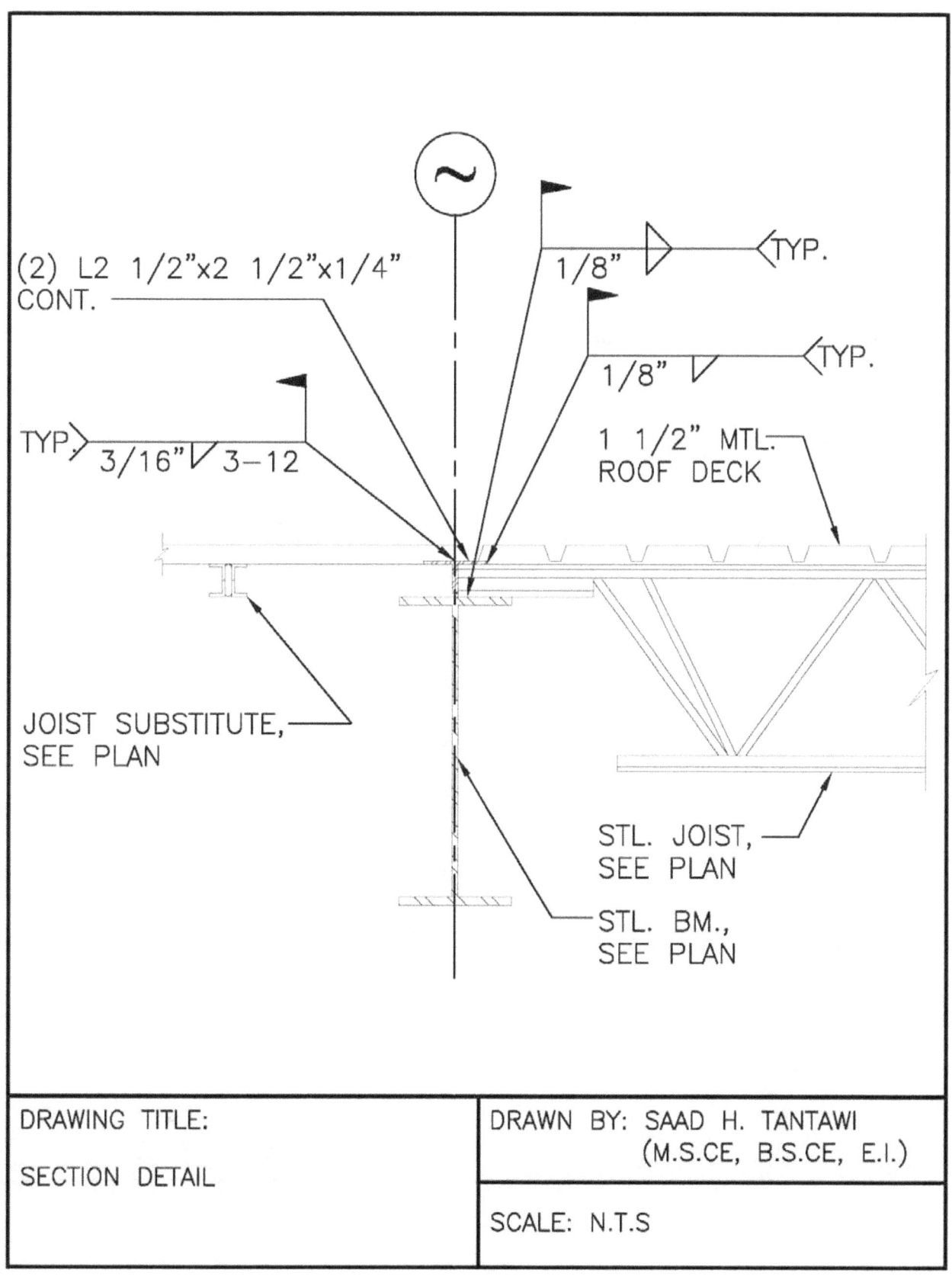

(2) L2 1/2"x2 1/2"x1/4" CONT.

1/8"   TYP.

1/8"   TYP.

TYP. 3/16" 3-12

1 1/2" MTL. ROOF DECK

JOIST SUBSTITUTE, SEE PLAN

STL. JOIST, SEE PLAN

STL. BM., SEE PLAN

DRAWING TITLE: SECTION DETAIL	DRAWN BY: SAAD H. TANTAWI (M.S.CE, B.S.CE, E.I.)
	SCALE: N.T.S

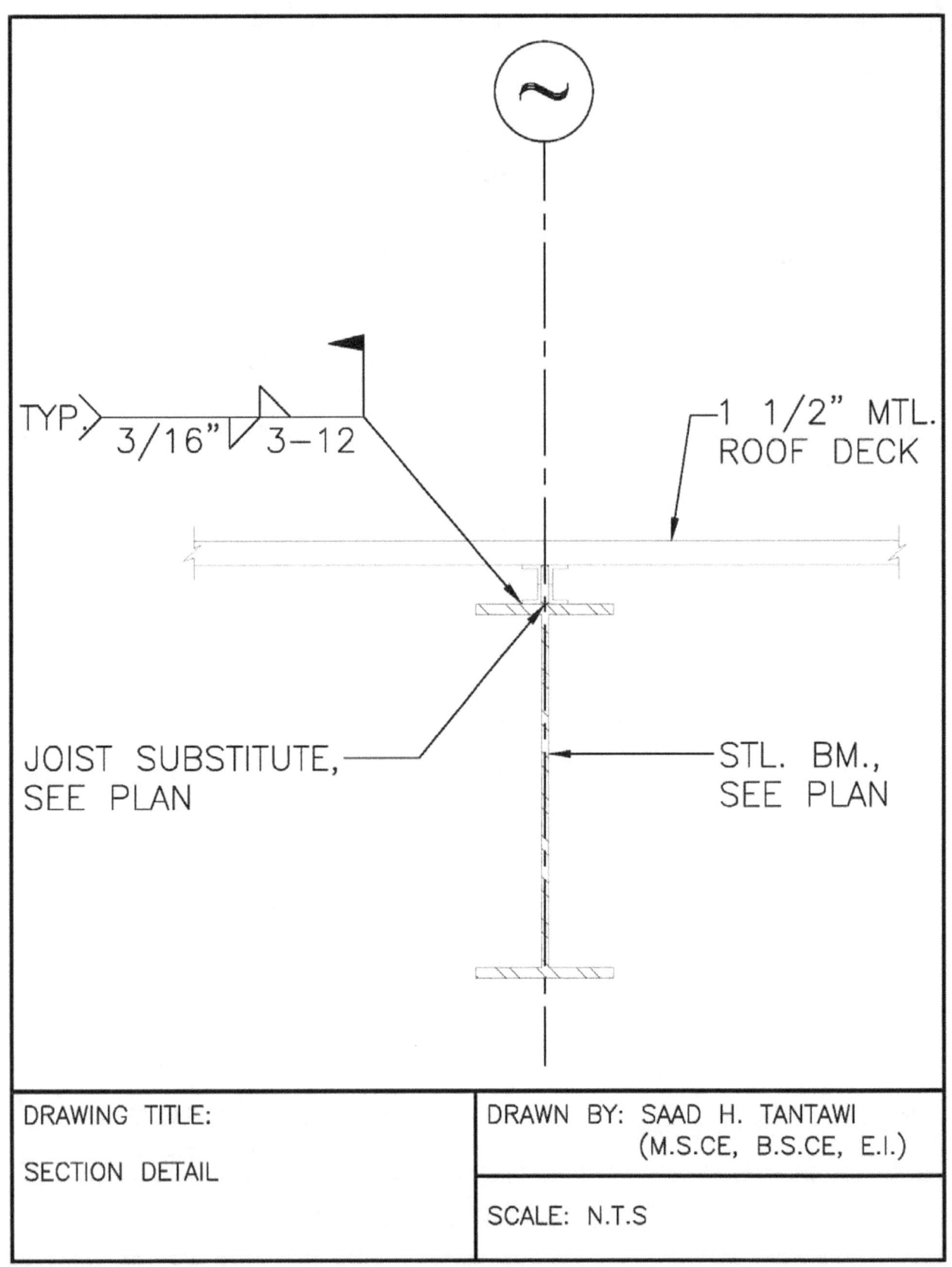

TYP.
3/16" 3-12

1 1/2" MTL.
ROOF DECK

JOIST SUBSTITUTE,
SEE PLAN

STL. BM.,
SEE PLAN

DRAWING TITLE:

SECTION DETAIL

DRAWN BY: SAAD H. TANTAWI
(M.S.CE, B.S.CE, E.I.)

SCALE: N.T.S

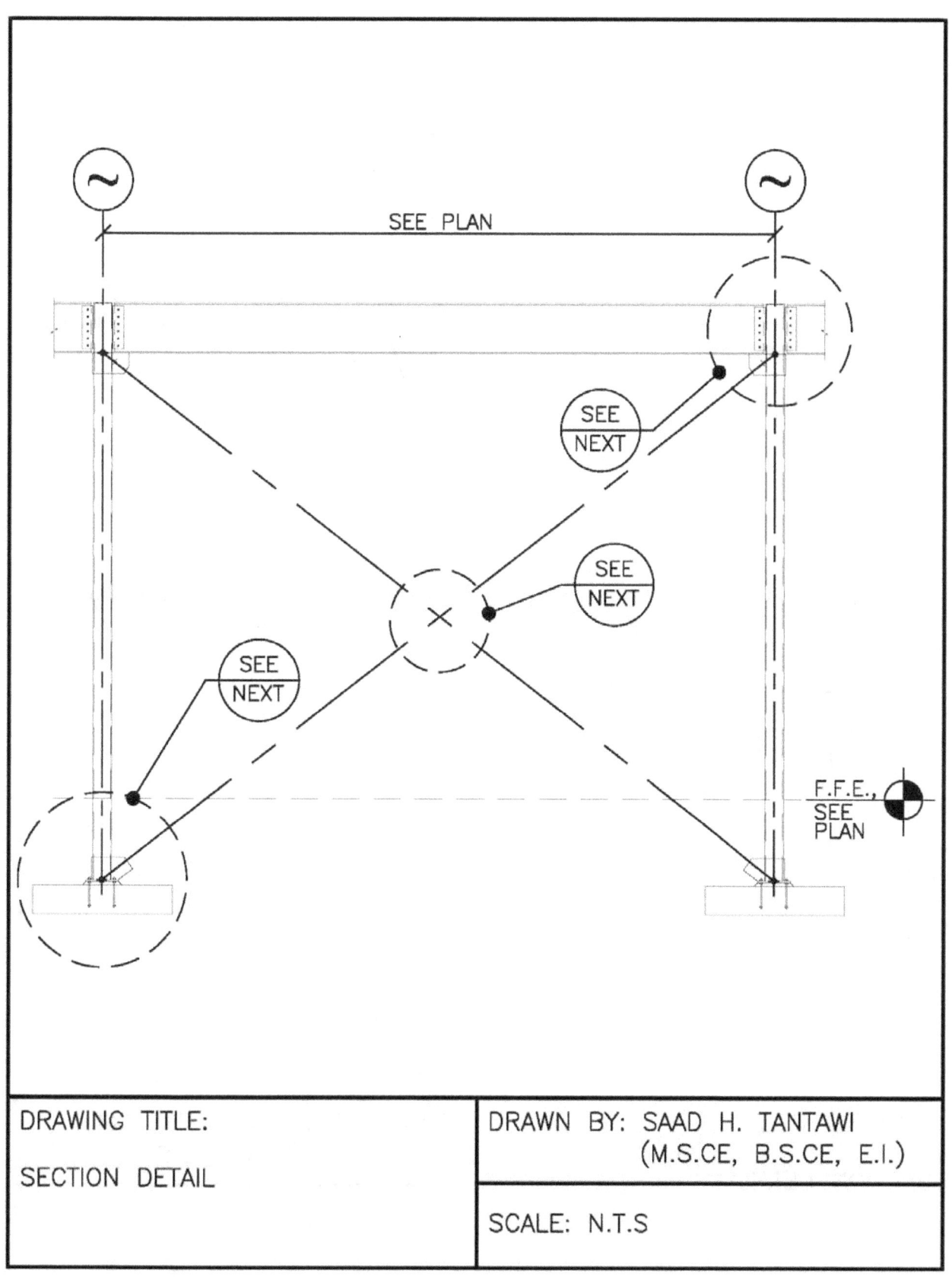

SEE PLAN

SEE
NEXT

SEE
NEXT

SEE
NEXT

F.F.E.,
SEE
PLAN

DRAWING TITLE:	DRAWN BY: SAAD H. TANTAWI
	(M.S.CE, B.S.CE, E.I.)
SECTION DETAIL	
	SCALE: N.T.S

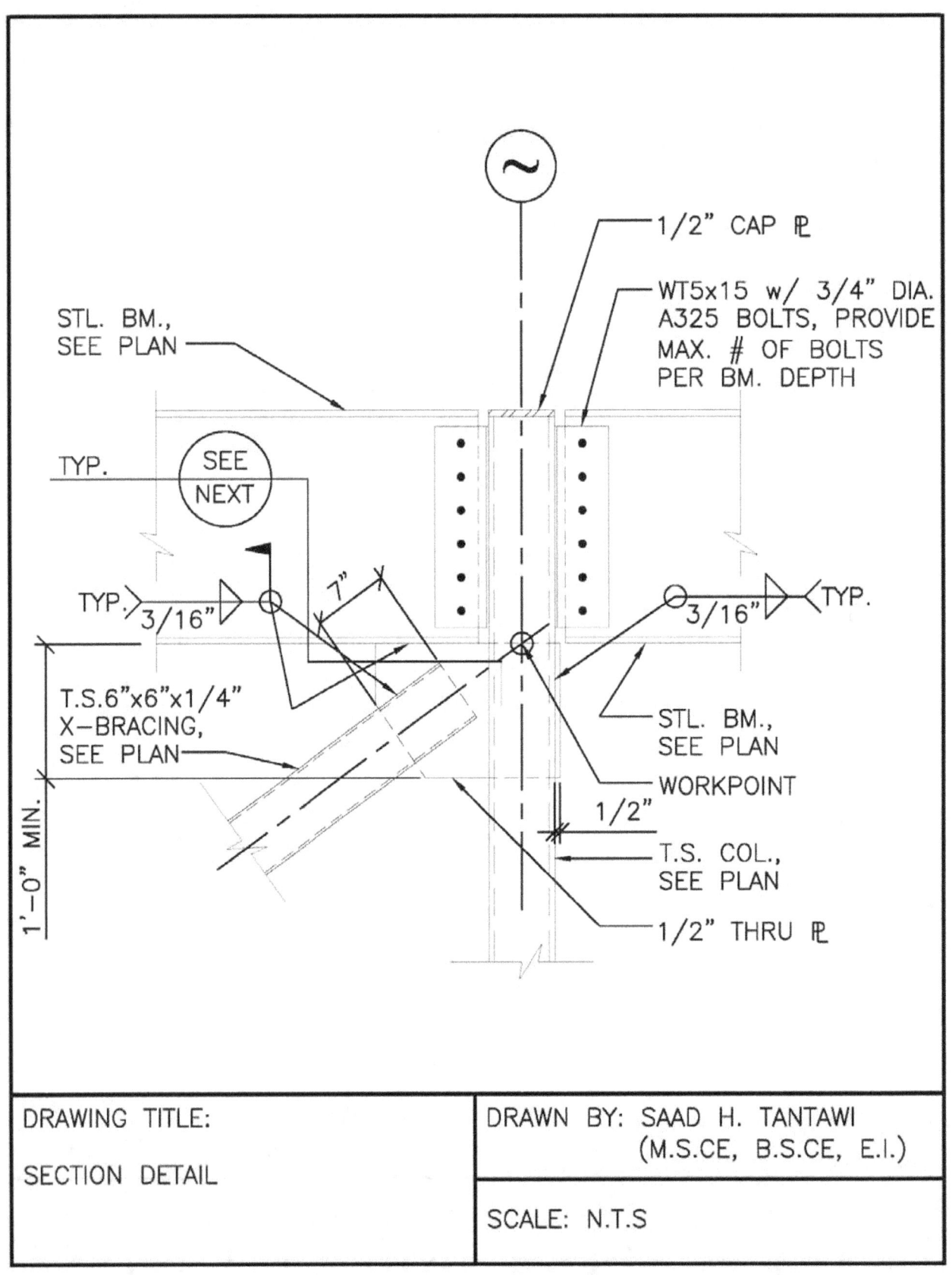

1/2" CAP ℄

WT5x15 w/ 3/4" DIA.
A325 BOLTS, PROVIDE
MAX. # OF BOLTS
PER BM. DEPTH

STL. BM.,
SEE PLAN

TYP.

SEE
NEXT

TYP.
3/16"

7"

3/16"

TYP.

T.S.6"x6"x1/4"
X-BRACING,
SEE PLAN

STL. BM.,
SEE PLAN

WORKPOINT

1/2"

1'-0" MIN.

T.S. COL.,
SEE PLAN

1/2" THRU ℄

DRAWING TITLE:	DRAWN BY: SAAD H. TANTAWI
	(M.S.CE, B.S.CE, E.I.)
SECTION DETAIL	
	SCALE: N.T.S

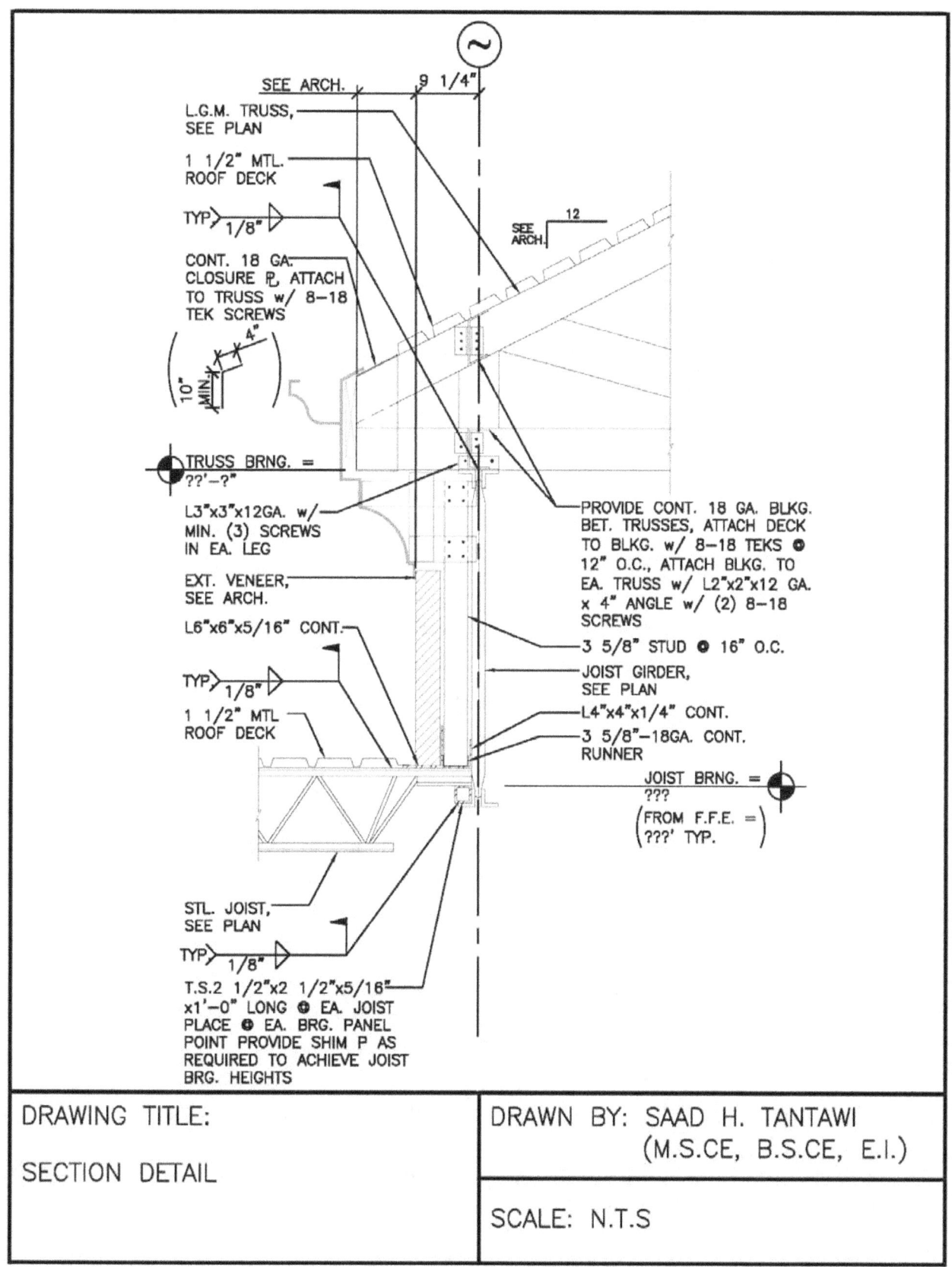

SEE ARCH.

9 1/4"

L.G.M. TRUSS,
SEE PLAN

1 1/2" MTL.
ROOF DECK

TYP 1/8"

SEE
ARCH. 12

CONT. 18 GA.
CLOSURE ℗, ATTACH
TO TRUSS w/ 8-18
TEK SCREWS

4"

10" MIN.

TRUSS BRNG. =
??'-?"

L3"x3"x12GA. w/
MIN. (3) SCREWS
IN EA. LEG

EXT. VENEER,
SEE ARCH.

L6"x6"x5/16" CONT.

TYP 1/8"

1 1/2" MTL
ROOF DECK

PROVIDE CONT. 18 GA. BLKG.
BET. TRUSSES, ATTACH DECK
TO BLKG. w/ 8-18 TEKS ⊕
12" O.C., ATTACH BLKG. TO
EA. TRUSS w/ L2"x2"x12 GA.
x 4" ANGLE w/ (2) 8-18
SCREWS

3 5/8" STUD ⊕ 16" O.C.

JOIST GIRDER,
SEE PLAN

L4"x4"x1/4" CONT.

3 5/8"-18GA. CONT.
RUNNER

JOIST BRNG. =
???

(FROM F.F.E. =
???' TYP.)

STL. JOIST,
SEE PLAN

TYP 1/8"

T.S.2 1/2"x2 1/2"x5/16"
x1'-0" LONG ⊕ EA. JOIST
PLACE ⊕ EA. BRG. PANEL
POINT PROVIDE SHIM ℗ AS
REQUIRED TO ACHIEVE JOIST
BRG. HEIGHTS

DRAWING TITLE:	DRAWN BY: SAAD H. TANTAWI
	(M.S.CE, B.S.CE, E.I.)
SECTION DETAIL	
	SCALE: N.T.S

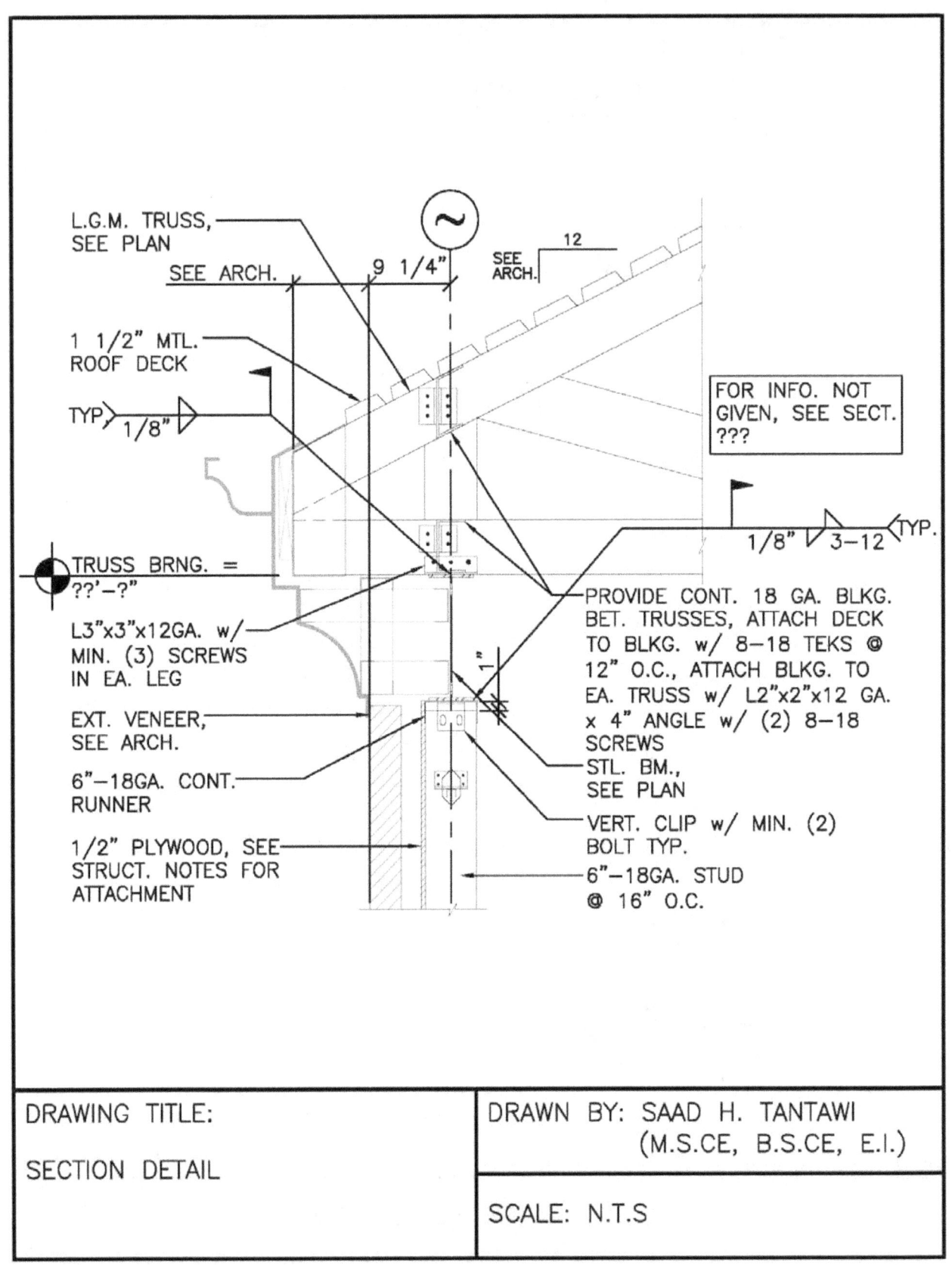

L.G.M. TRUSS,
SEE PLAN

SEE ARCH.

9 1/4"

SEE ARCH.

12

FOR INFO. NOT
GIVEN, SEE SECT.
???

1 1/2" MTL.
ROOF DECK

TYP 1/8"

1/8" 3-12 TYP.

TRUSS BRNG. =
??'-?"

PROVIDE CONT. 18 GA. BLKG.
BET. TRUSSES, ATTACH DECK
TO BLKG. w/ 8-18 TEKS @
12" O.C., ATTACH BLKG. TO
EA. TRUSS w/ L2"x2"x12 GA.
x 4" ANGLE w/ (2) 8-18
SCREWS

L3"x3"x12GA. w/
MIN. (3) SCREWS
IN EA. LEG

EXT. VENEER,
SEE ARCH.

STL. BM.,
SEE PLAN

6"-18GA. CONT.
RUNNER

VERT. CLIP w/ MIN. (2)
BOLT TYP.

1/2" PLYWOOD, SEE
STRUCT. NOTES FOR
ATTACHMENT

1"

6"-18GA. STUD
@ 16" O.C.

DRAWING TITLE: SECTION DETAIL	DRAWN BY: SAAD H. TANTAWI (M.S.CE, B.S.CE, E.I.)
	SCALE: N.T.S

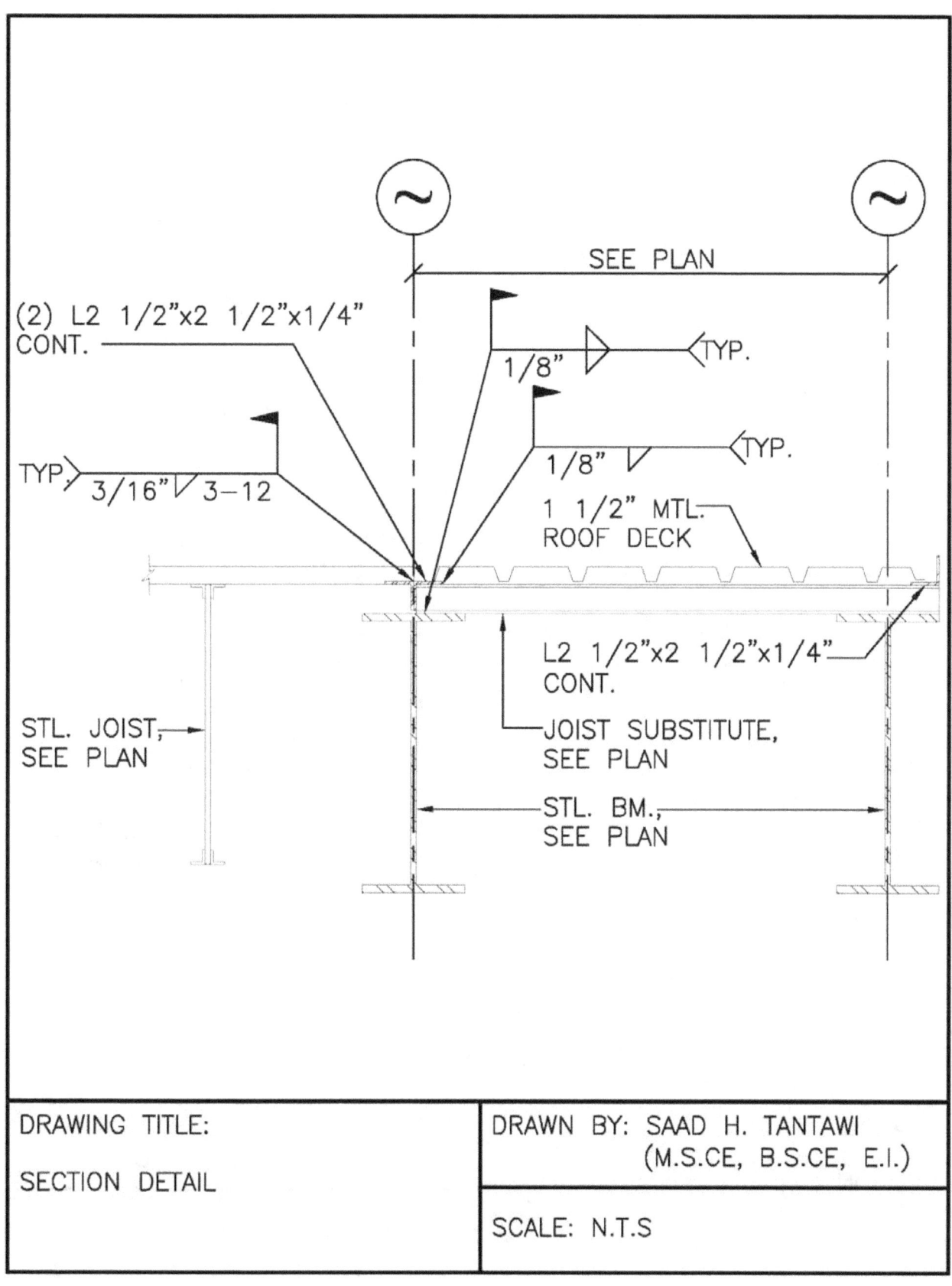

DRAWING TITLE:	DRAWN BY: SAAD H. TANTAWI
SECTION DETAIL	(M.S.CE, B.S.CE, E.I.)
	SCALE: N.T.S

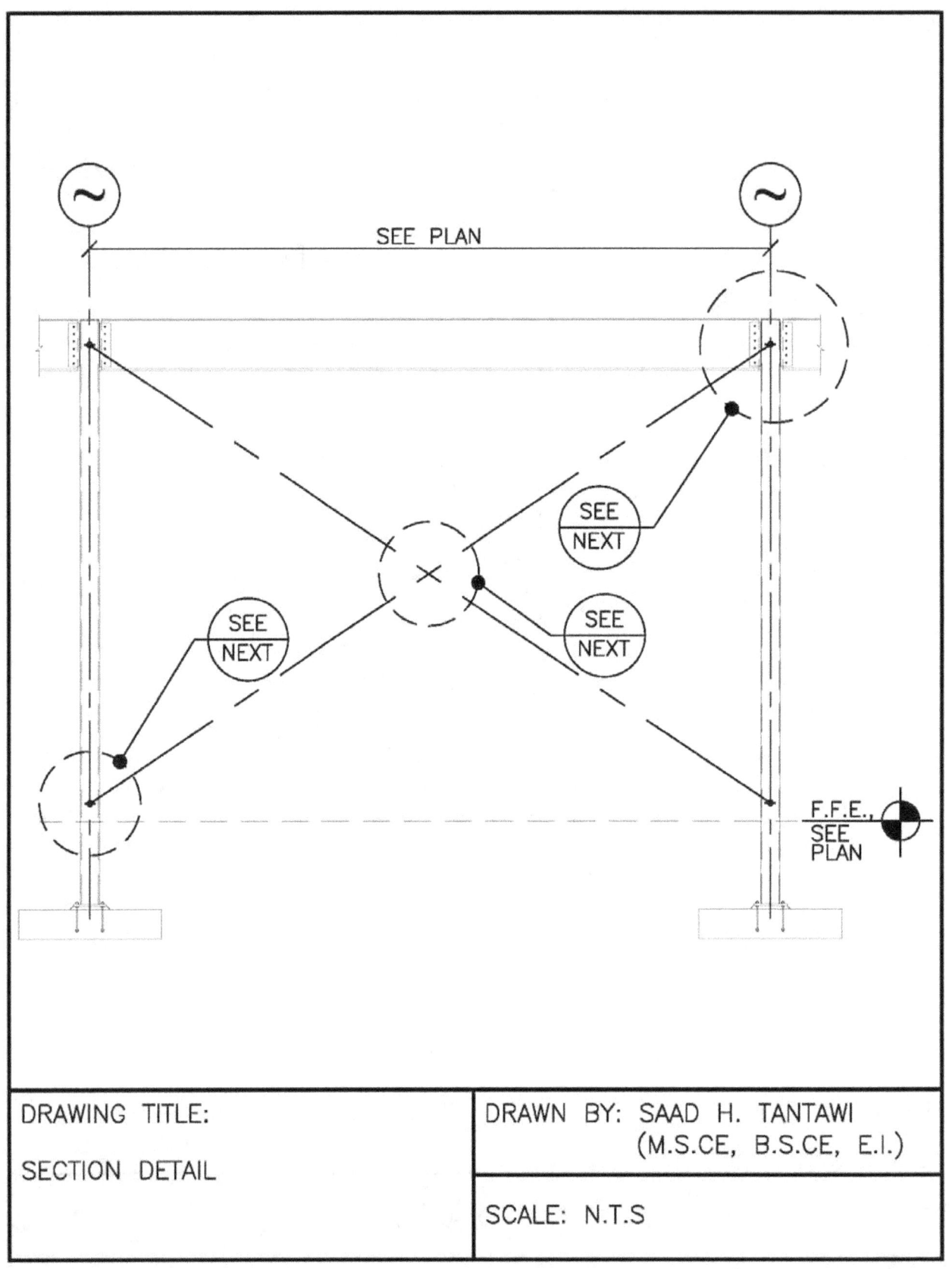

SEE PLAN

SEE
NEXT

SEE
NEXT

SEE
NEXT

F.F.E.,
SEE
PLAN

DRAWING TITLE: SECTION DETAIL	DRAWN BY: SAAD H. TANTAWI (M.S.CE, B.S.CE, E.I.)
	SCALE: N.T.S

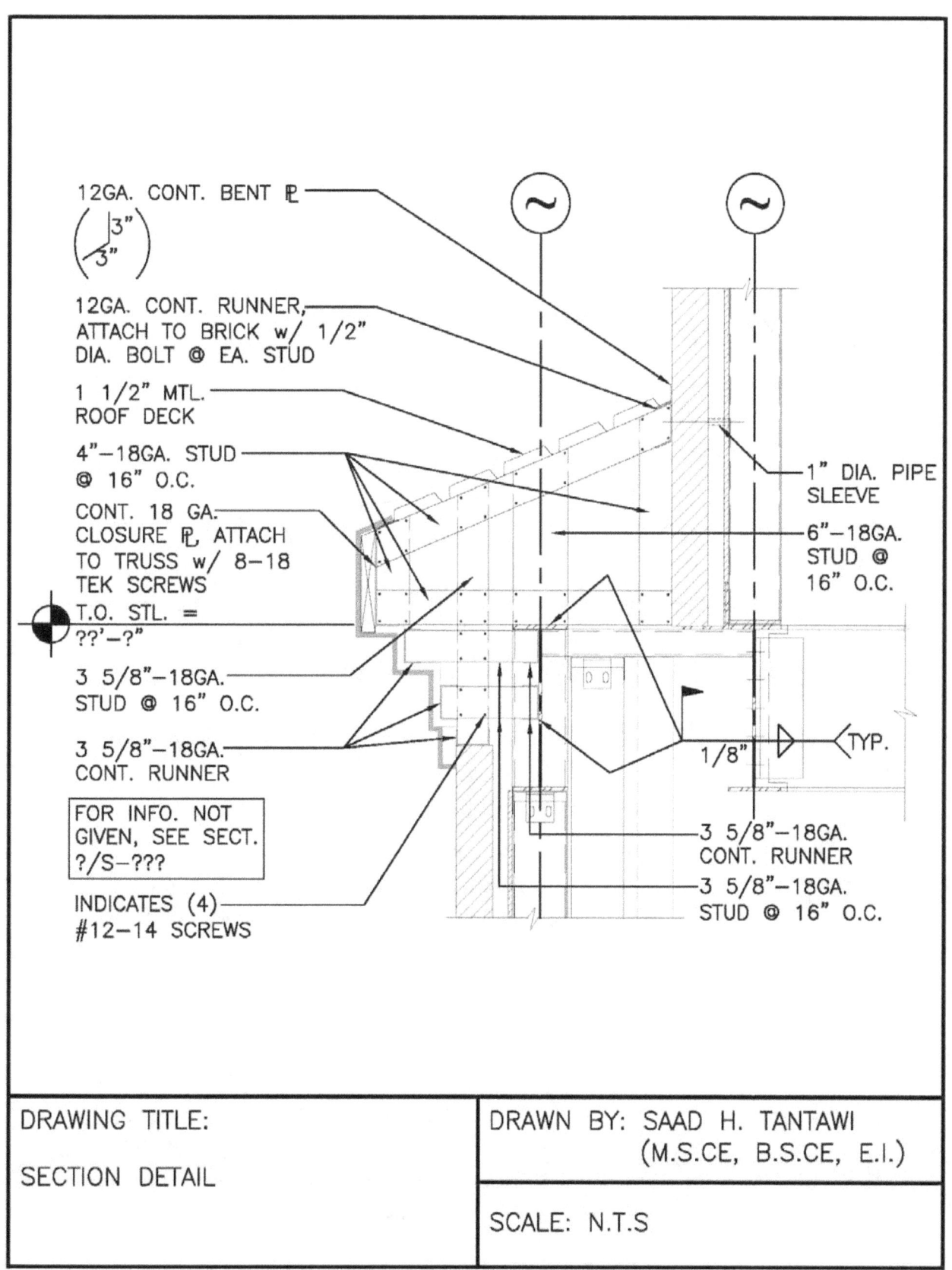

12GA. CONT. BENT ℄

3"
3"

12GA. CONT. RUNNER,
ATTACH TO BRICK w/ 1/2"
DIA. BOLT @ EA. STUD

1 1/2" MTL.
ROOF DECK

4"–18GA. STUD
@ 16" O.C.

CONT. 18 GA.
CLOSURE ℄, ATTACH
TO TRUSS w/ 8–18
TEK SCREWS

T.O. STL. =
??'–?"

3 5/8"–18GA.
STUD @ 16" O.C.

3 5/8"–18GA.
CONT. RUNNER

FOR INFO. NOT
GIVEN, SEE SECT.
?/S–???

INDICATES (4)
#12–14 SCREWS

1" DIA. PIPE
SLEEVE

6"–18GA.
STUD @
16" O.C.

1/8"   TYP.

3 5/8"–18GA.
CONT. RUNNER

3 5/8"–18GA.
STUD @ 16" O.C.

DRAWING TITLE:	DRAWN BY: SAAD H. TANTAWI
	(M.S.CE, B.S.CE, E.I.)
SECTION DETAIL	
	SCALE: N.T.S

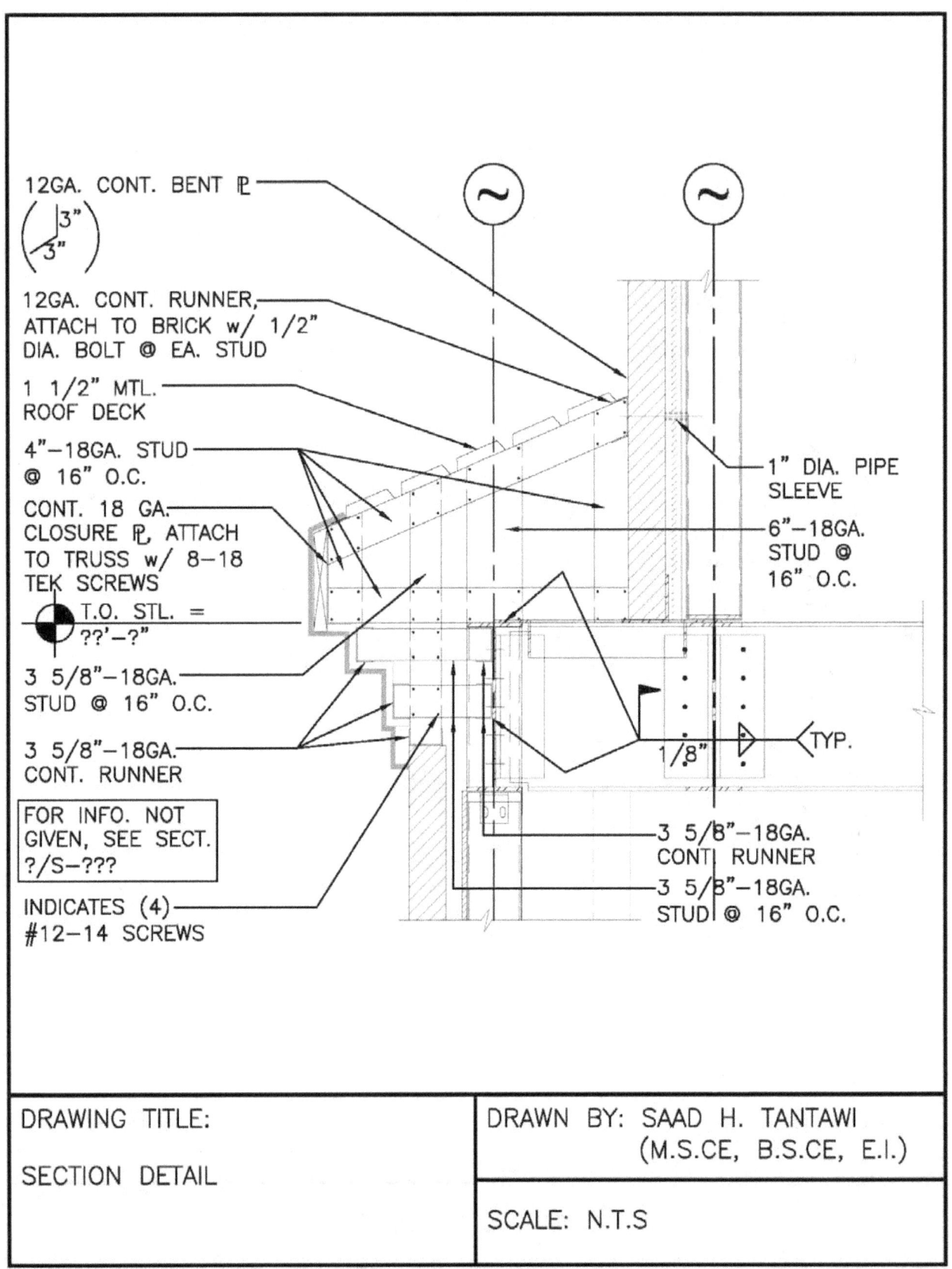

12GA. CONT. BENT ℙ

12GA. CONT. RUNNER, ATTACH TO BRICK w/ 1/2" DIA. BOLT @ EA. STUD

1 1/2" MTL. ROOF DECK

4"−18GA. STUD @ 16" O.C.

CONT. 18 GA. CLOSURE ℙ, ATTACH TO TRUSS w/ 8−18 TEK SCREWS

T.O. STL. = ??'−?"

3 5/8"−18GA. STUD @ 16" O.C.

3 5/8"−18GA. CONT. RUNNER

FOR INFO. NOT GIVEN, SEE SECT. ?/S−???

INDICATES (4) #12−14 SCREWS

1" DIA. PIPE SLEEVE

6"−18GA. STUD @ 16" O.C.

TYP.

1/8"

3 5/8"−18GA. CONT. RUNNER

3 5/8"−18GA. STUD @ 16" O.C.

DRAWING TITLE:	DRAWN BY: SAAD H. TANTAWI (M.S.CE, B.S.CE, E.I.)
SECTION DETAIL	
	SCALE: N.T.S

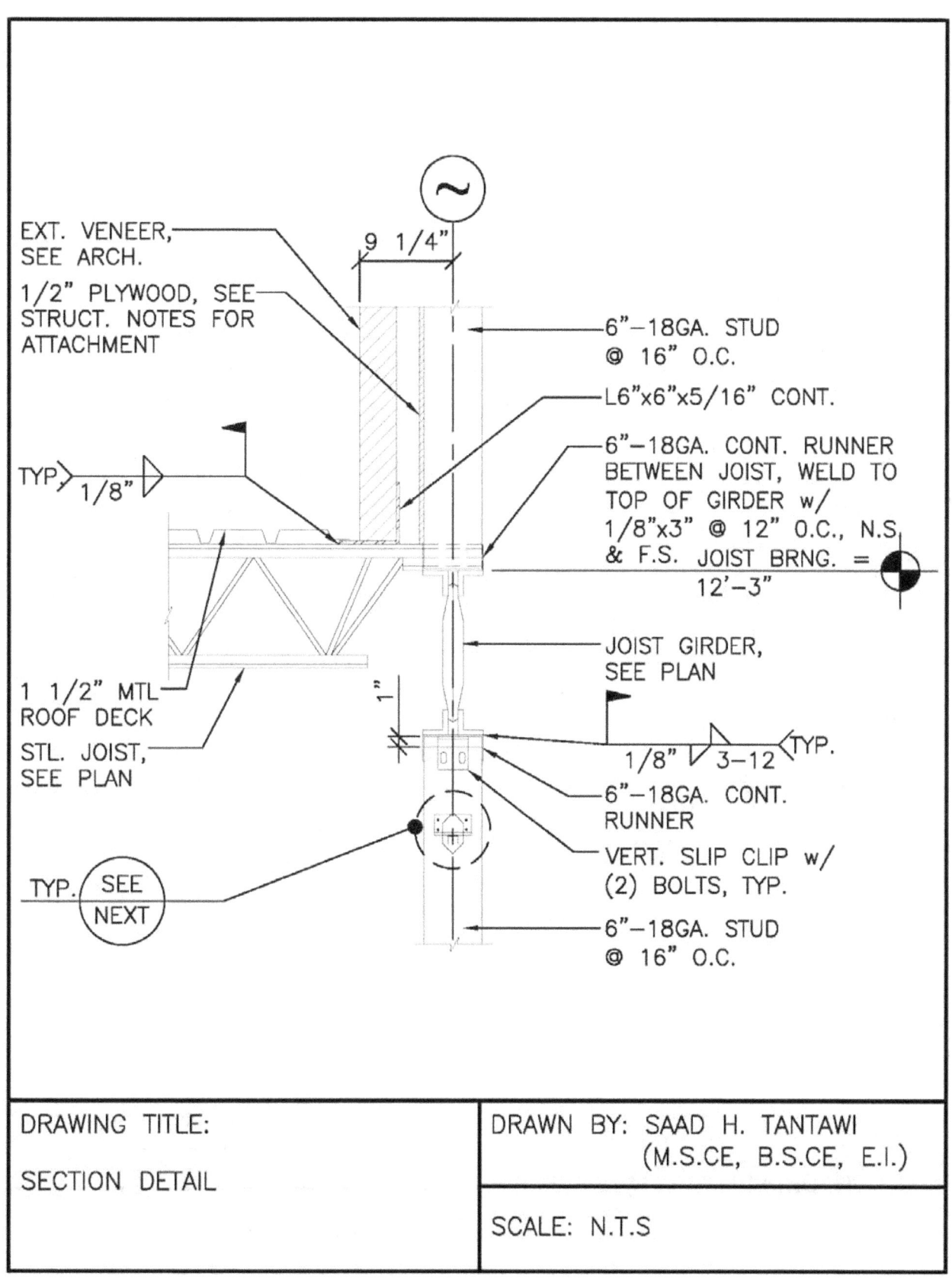

EXT. VENEER,
SEE ARCH.

1/2" PLYWOOD, SEE
STRUCT. NOTES FOR
ATTACHMENT

9 1/4"

6"-18GA. STUD
@ 16" O.C.

L6"x6"x5/16" CONT.

6"-18GA. CONT. RUNNER
BETWEEN JOIST, WELD TO
TOP OF GIRDER w/
1/8"x3" @ 12" O.C., N.S.
& F.S. JOIST BRNG. =
12'-3"

TYP. 1/8"

JOIST GIRDER,
SEE PLAN

1 1/2" MTL
ROOF DECK

STL. JOIST,
SEE PLAN

1"

1/8" 3-12 TYP.

6"-18GA. CONT.
RUNNER

VERT. SLIP CLIP w/
(2) BOLTS, TYP.

TYP. SEE
NEXT

6"-18GA. STUD
@ 16" O.C.

DRAWING TITLE:	DRAWN BY: SAAD H. TANTAWI
	(M.S.CE, B.S.CE, E.I.)
SECTION DETAIL	
	SCALE: N.T.S

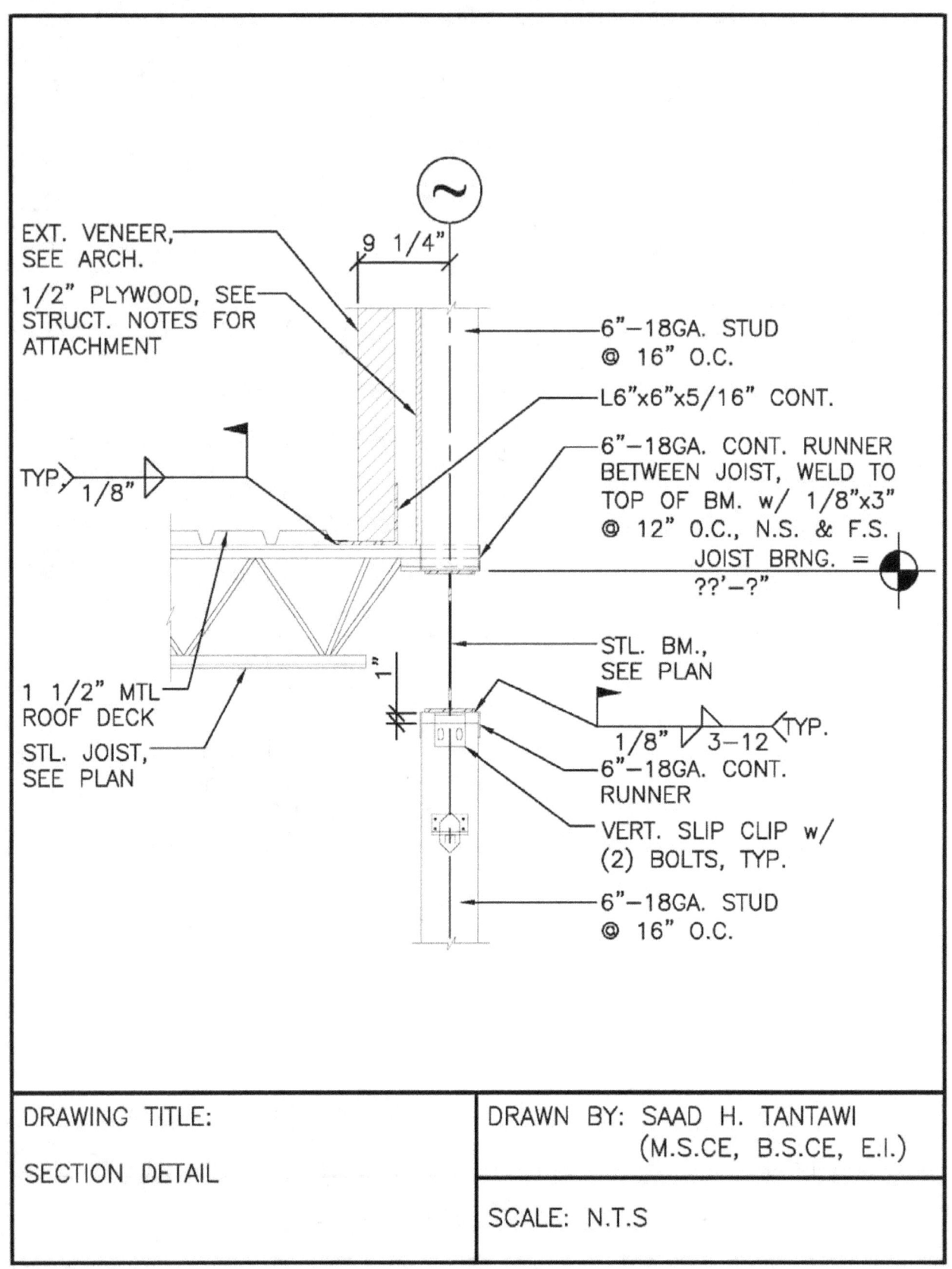

EXT. VENEER,
SEE ARCH.

1/2" PLYWOOD, SEE
STRUCT. NOTES FOR
ATTACHMENT

9 1/4"

6"-18GA. STUD
@ 16" O.C.

L6"x6"x5/16" CONT.

6"-18GA. CONT. RUNNER
BETWEEN JOIST, WELD TO
TOP OF BM. w/ 1/8"x3"
@ 12" O.C., N.S. & F.S.

JOIST BRNG. =
??'-?"

TYP.
1/8"

STL. BM.,
SEE PLAN

1 1/2" MTL
ROOF DECK

1"

1/8"   3-12   TYP.

6"-18GA. CONT.
RUNNER

STL. JOIST,
SEE PLAN

VERT. SLIP CLIP w/
(2) BOLTS, TYP.

6"-18GA. STUD
@ 16" O.C.

DRAWING TITLE:	DRAWN BY: SAAD H. TANTAWI (M.S.CE, B.S.CE, E.I.)
SECTION DETAIL	
	SCALE: N.T.S

280

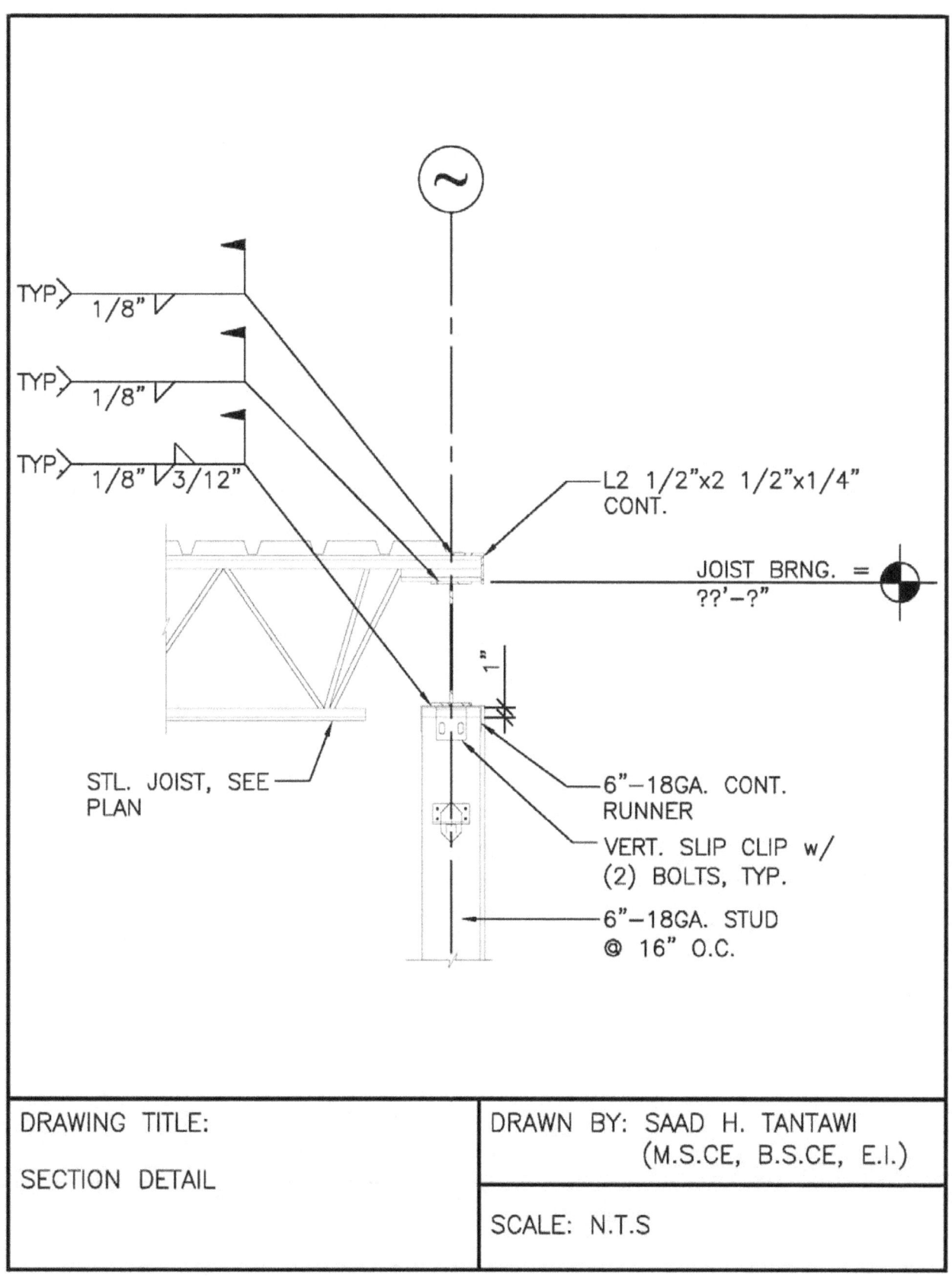

TYP〉 1/8" ∨

TYP〉 1/8" ∨

TYP〉 1/8" ∨ 3/12"

L2 1/2"x2 1/2"x1/4"
CONT.

JOIST BRNG. =
??'-?"

1"

STL. JOIST, SEE
PLAN

6"-18GA. CONT.
RUNNER

VERT. SLIP CLIP w/
(2) BOLTS, TYP.

6"-18GA. STUD
@ 16" O.C.

DRAWING TITLE:	DRAWN BY: SAAD H. TANTAWI
	(M.S.CE, B.S.CE, E.I.)
SECTION DETAIL	
	SCALE: N.T.S

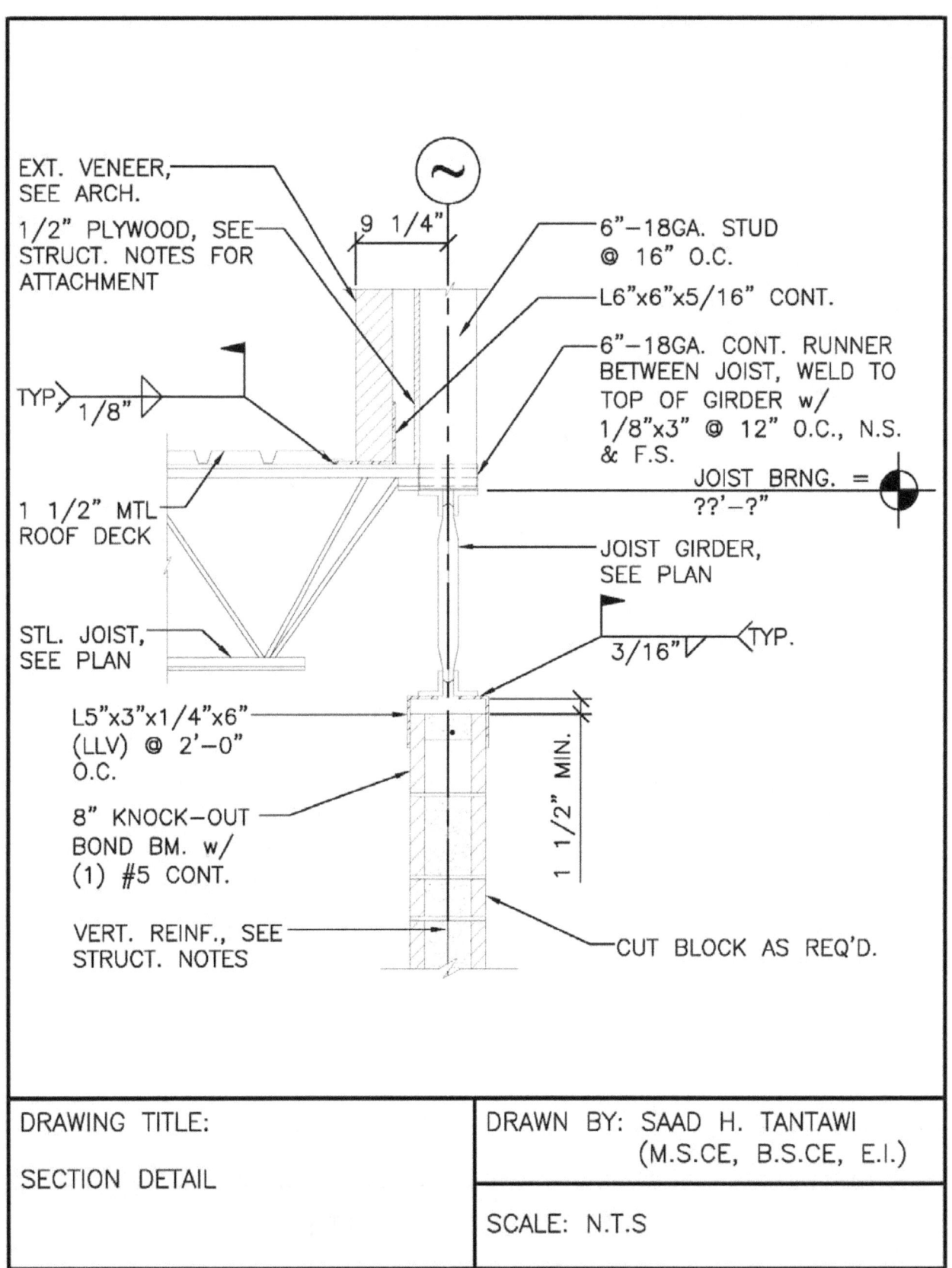

EXT. VENEER,
SEE ARCH.

1/2" PLYWOOD, SEE
STRUCT. NOTES FOR
ATTACHMENT

9 1/4"

6"−18GA. STUD
@ 16" O.C.

L6"x6"x5/16" CONT.

6"−18GA. CONT. RUNNER
BETWEEN JOIST, WELD TO
TOP OF GIRDER w/
1/8"x3" @ 12" O.C., N.S.
& F.S.

TYP 1/8"

$\dfrac{\text{JOIST BRNG.}}{??'-?"} =$

1 1/2" MTL
ROOF DECK

JOIST GIRDER,
SEE PLAN

STL. JOIST,
SEE PLAN

3/16" TYP.

L5"x3"x1/4"x6"
(LLV) @ 2'−0"
O.C.

1 1/2" MIN.

8" KNOCK−OUT
BOND BM. w/
(1) #5 CONT.

VERT. REINF., SEE
STRUCT. NOTES

CUT BLOCK AS REQ'D.

DRAWING TITLE:	DRAWN BY: SAAD H. TANTAWI
	(M.S.CE, B.S.CE, E.I.)
SECTION DETAIL	
	SCALE: N.T.S

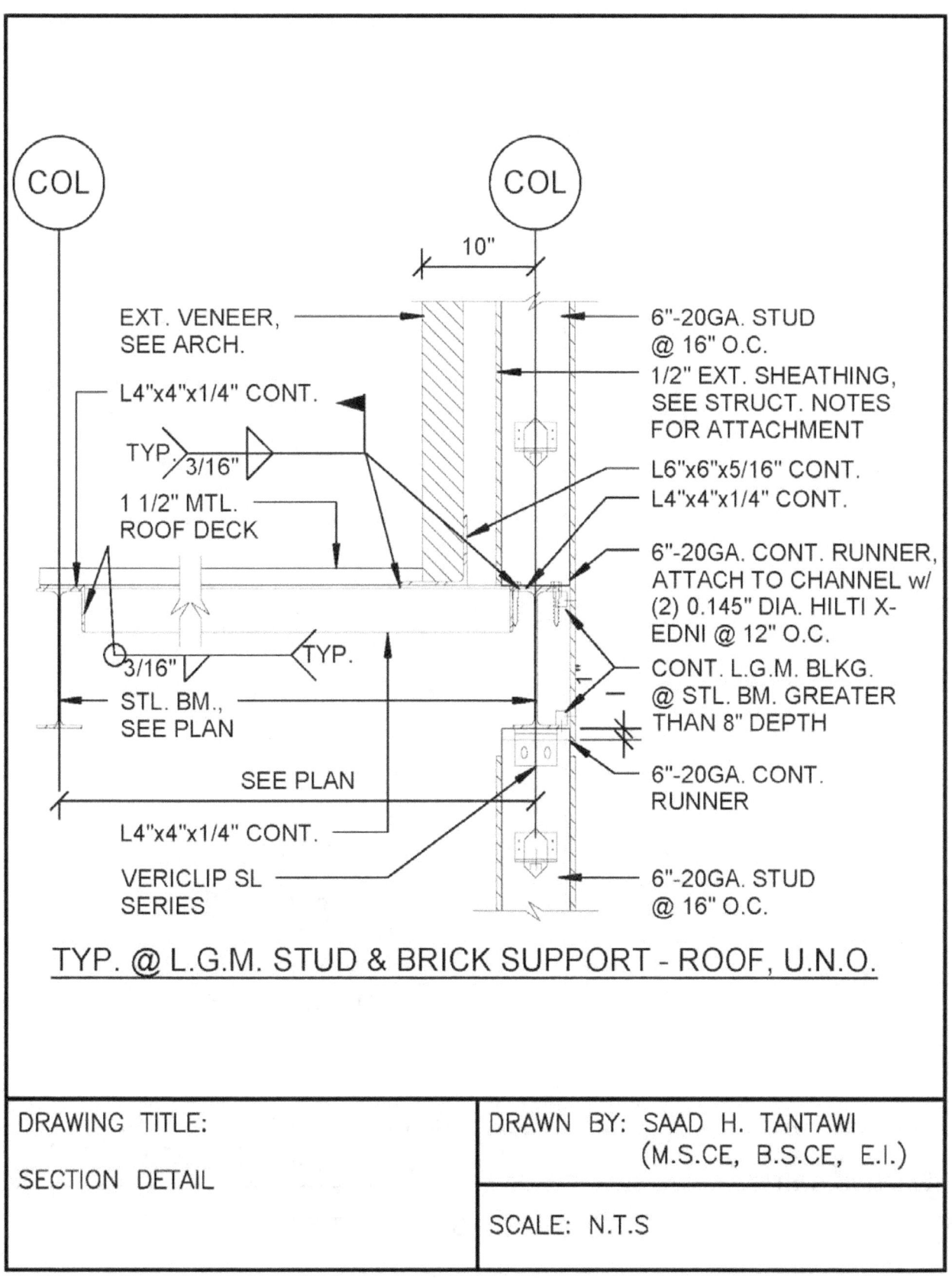

TYP. @ L.G.M. STUD & BRICK SUPPORT - ROOF, U.N.O.

DRAWING TITLE:	DRAWN BY: SAAD H. TANTAWI
SECTION DETAIL	(M.S.CE, B.S.CE, E.I.)
	SCALE: N.T.S

Text labels in diagram:

COL   COL

10"

EXT. VENEER, SEE ARCH.

6"-20GA. STUD @ 16" O.C.

1/2" EXT. SHEATHING, SEE STRUCT. NOTES FOR ATTACHMENT

L4"x4"x1/4" CONT.

TYP. 3/16"

L6"x6"x5/16" CONT.

L4"x4"x1/4" CONT.

1 1/2" MTL. ROOF DECK

6"-20GA. CONT. RUNNER, ATTACH TO CHANNEL w/ (2) 0.145" DIA. HILTI X-EDNI @ 12" O.C.

3/16"   TYP.

CONT. L.G.M. BLKG. @ STL. BM. GREATER THAN 8" DEPTH

STL. BM., SEE PLAN

SEE PLAN

6"-20GA. CONT. RUNNER

L4"x4"x1/4" CONT.

VERICLIP SL SERIES

6"-20GA. STUD @ 16" O.C.

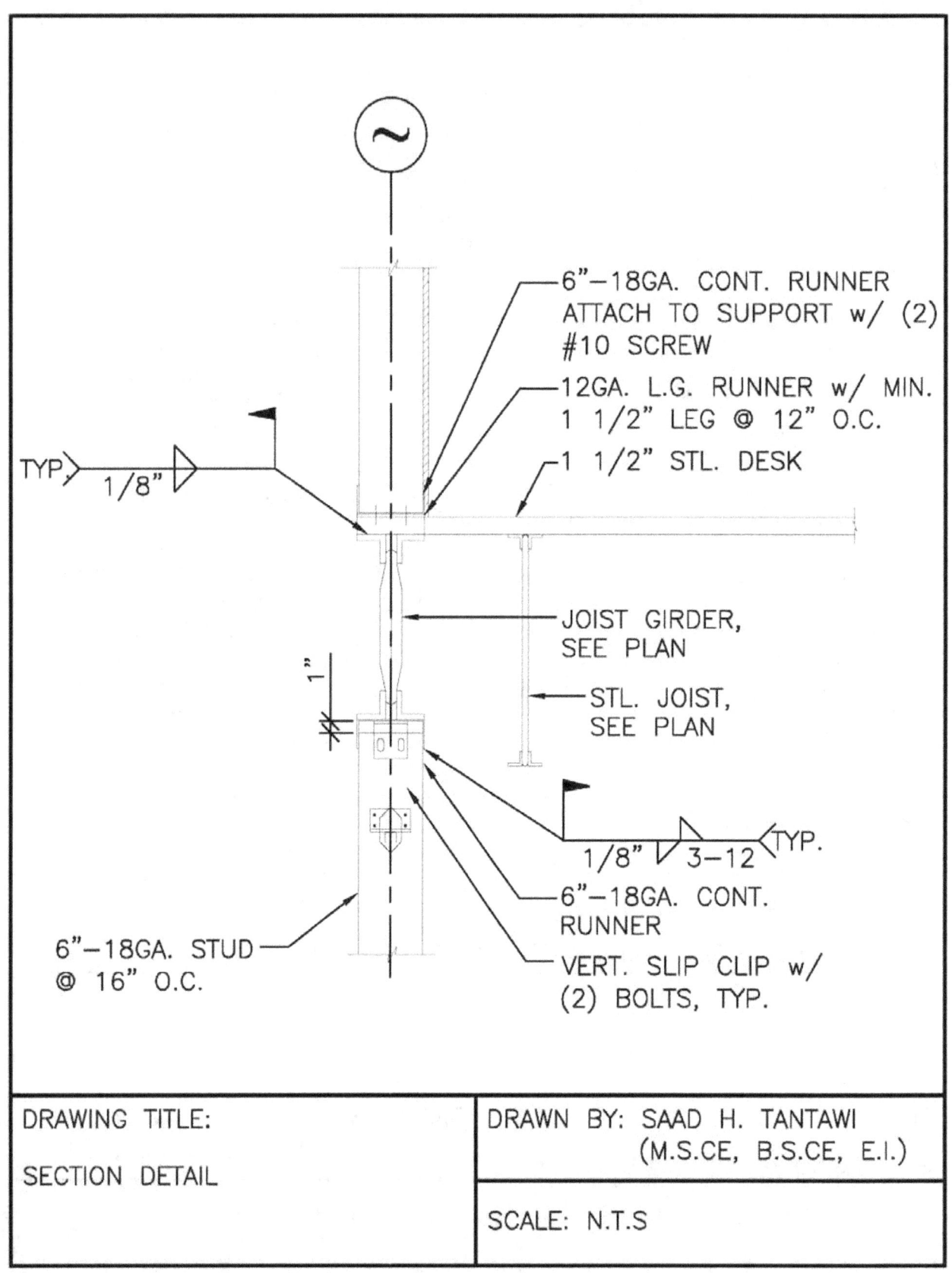

6"–18GA. CONT. RUNNER ATTACH TO SUPPORT w/ (2) #10 SCREW

12GA. L.G. RUNNER w/ MIN. 1 1/2" LEG @ 12" O.C.

1 1/2" STL. DESK

TYP⟩ 1/8"

JOIST GIRDER, SEE PLAN

STL. JOIST, SEE PLAN

1"

1/8" 3–12 ⟨TYP.

6"–18GA. CONT. RUNNER

6"–18GA. STUD @ 16" O.C.

VERT. SLIP CLIP w/ (2) BOLTS, TYP.

DRAWING TITLE: SECTION DETAIL	DRAWN BY: SAAD H. TANTAWI (M.S.CE, B.S.CE, E.I.)
	SCALE: N.T.S

284

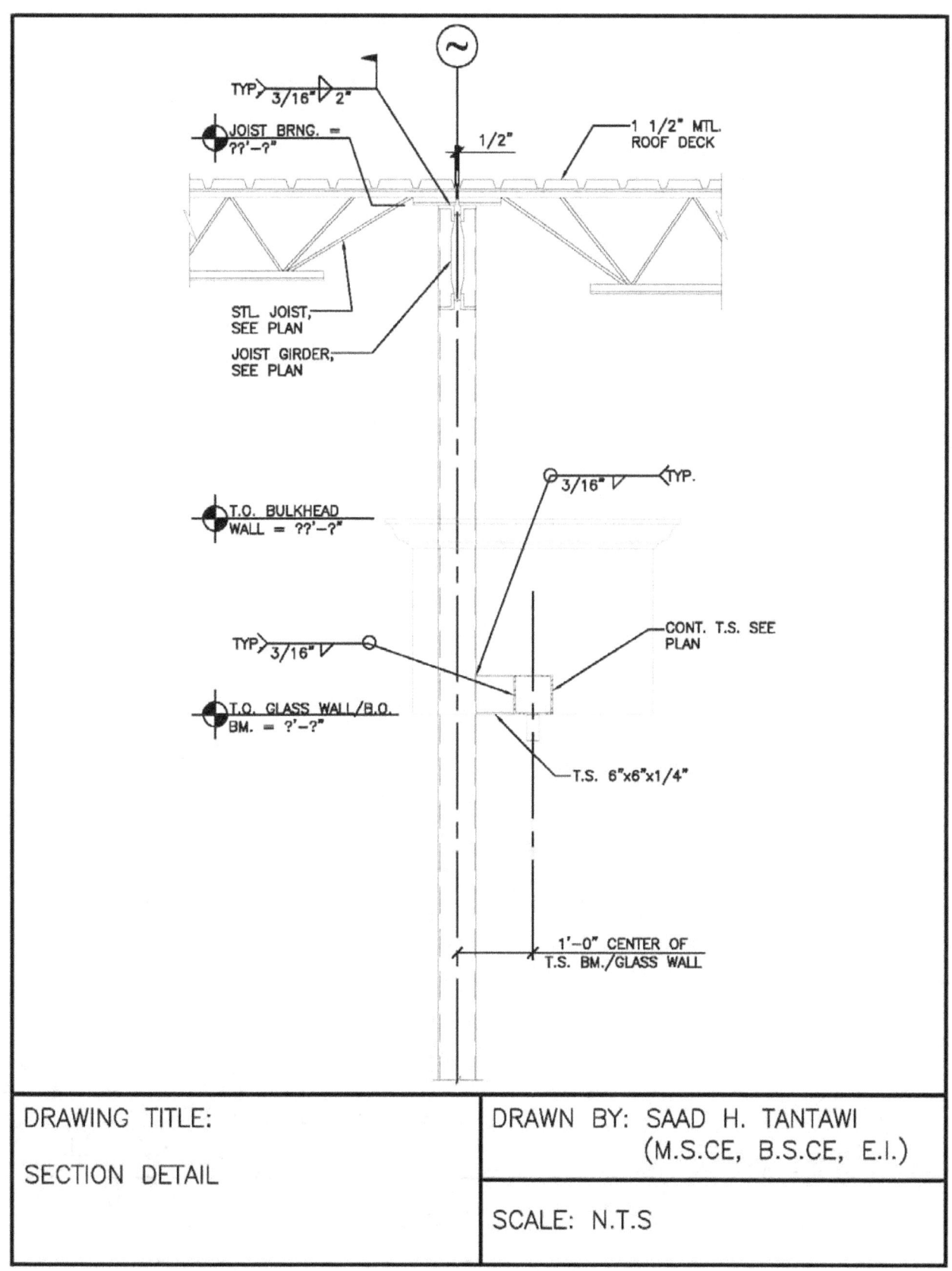

DRAWING TITLE:	DRAWN BY: SAAD H. TANTAWI
	(M.S.CE, B.S.CE, E.I.)
SECTION DETAIL	SCALE: N.T.S

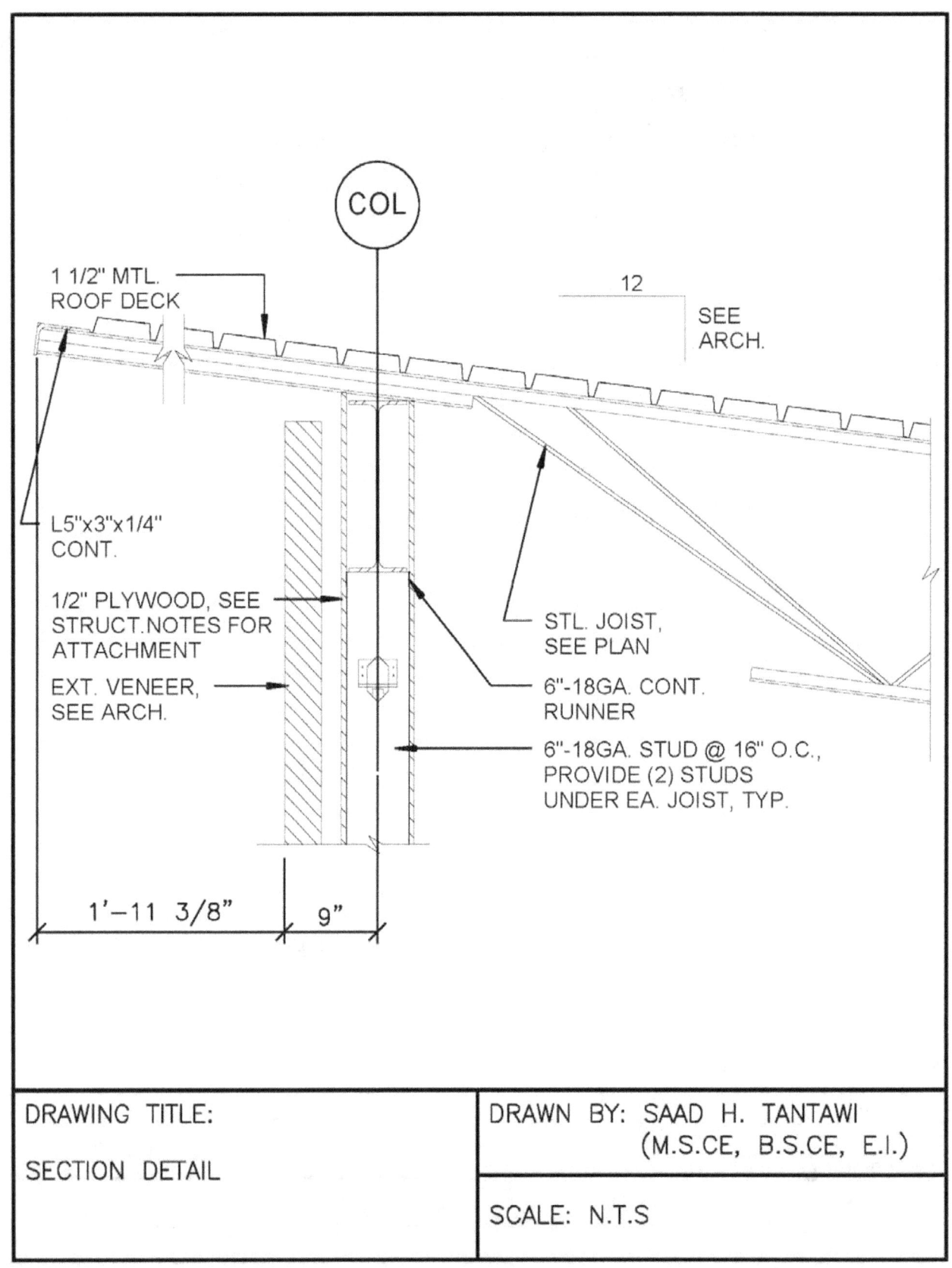

1 1/2" MTL.
ROOF DECK

COL

12

SEE
ARCH.

L5"x3"x1/4"
CONT.

1/2" PLYWOOD, SEE
STRUCT.NOTES FOR
ATTACHMENT

EXT. VENEER,
SEE ARCH.

STL. JOIST,
SEE PLAN

6"-18GA. CONT.
RUNNER

6"-18GA. STUD @ 16" O.C.,
PROVIDE (2) STUDS
UNDER EA. JOIST, TYP.

1'-11 3/8"   9"

DRAWING TITLE:

SECTION DETAIL

DRAWN BY: SAAD H. TANTAWI
(M.S.CE, B.S.CE, E.I.)

SCALE: N.T.S

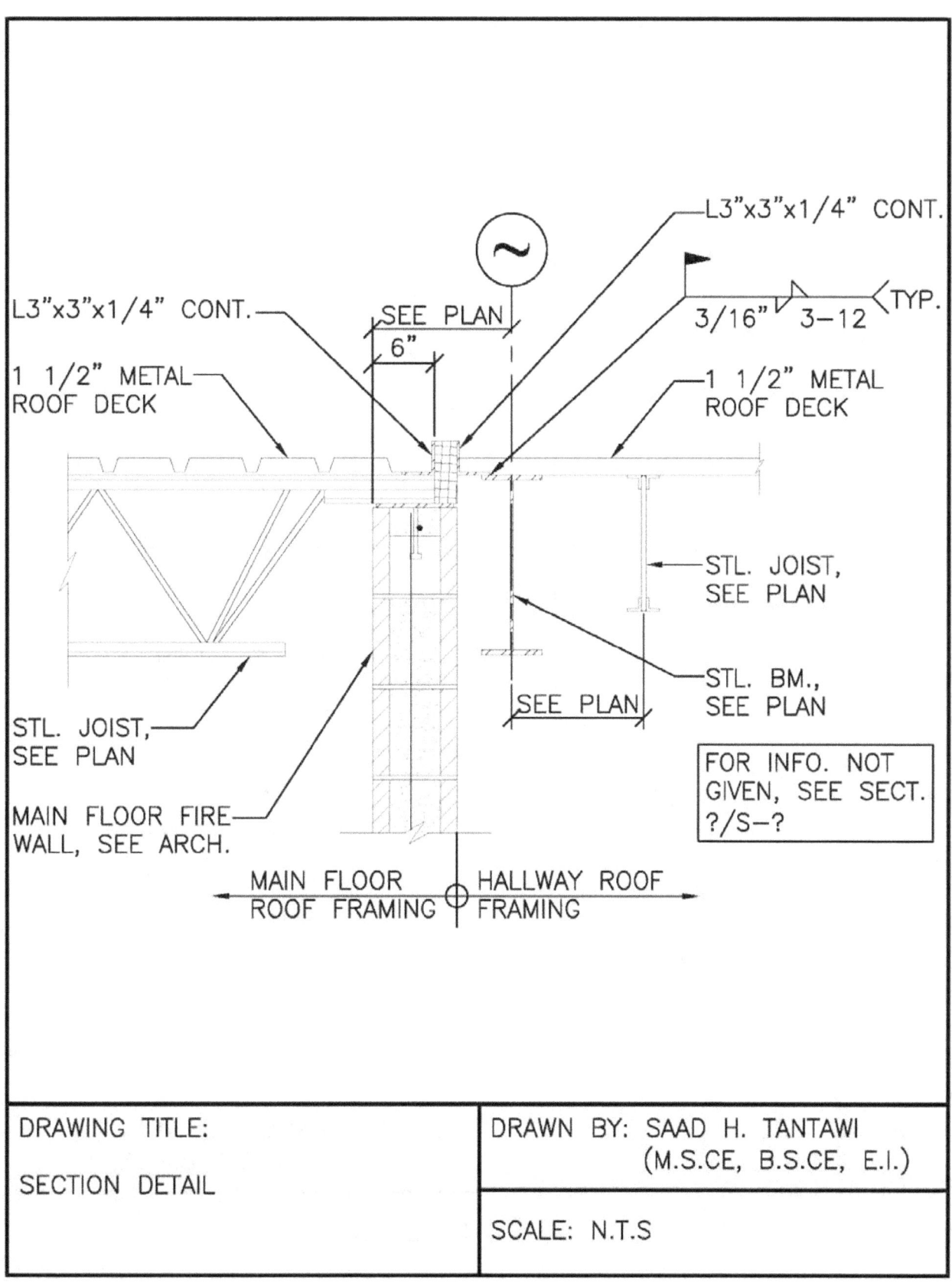

L3"x3"x1/4" CONT.

SEE PLAN

6"

1 1/2" METAL
ROOF DECK

L3"x3"x1/4" CONT.

1 1/2" METAL
ROOF DECK

3/16"   3-12   TYP.

STL. JOIST,
SEE PLAN

STL. BM.,
SEE PLAN

SEE PLAN

STL. JOIST,
SEE PLAN

MAIN FLOOR FIRE
WALL, SEE ARCH.

FOR INFO. NOT
GIVEN, SEE SECT.
?/S-?

MAIN FLOOR
ROOF FRAMING

HALLWAY ROOF
FRAMING

DRAWING TITLE:	DRAWN BY: SAAD H. TANTAWI
SECTION DETAIL	(M.S.CE, B.S.CE, E.I.)
	SCALE: N.T.S

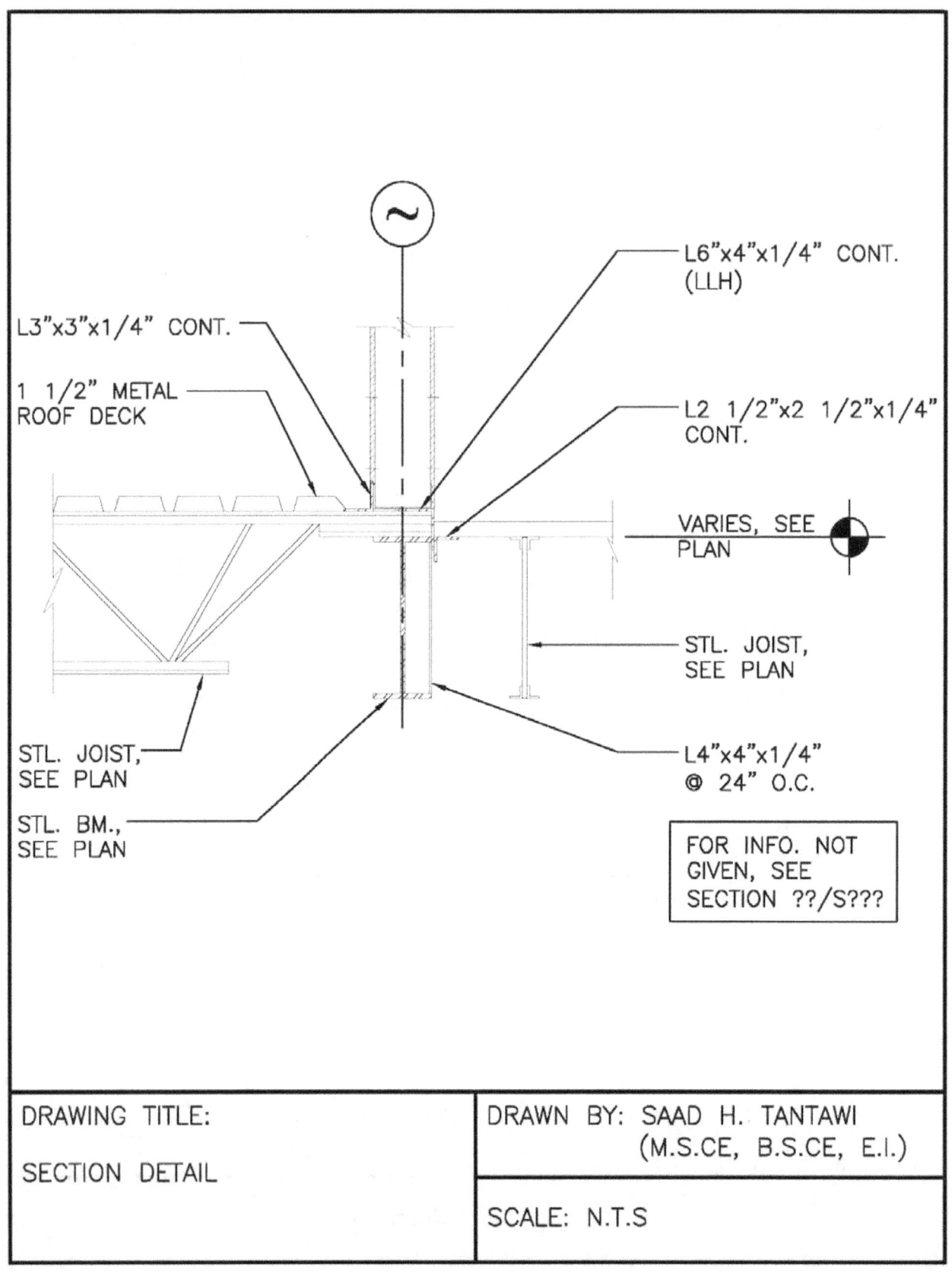

L3"x3"x1/4" CONT.

1 1/2" METAL
ROOF DECK

L6"x4"x1/4" CONT.
(LLH)

L2 1/2"x2 1/2"x1/4"
CONT.

VARIES, SEE
PLAN

STL. JOIST,
SEE PLAN

STL. JOIST,
SEE PLAN

STL. BM.,
SEE PLAN

L4"x4"x1/4"
@ 24" O.C.

FOR INFO. NOT
GIVEN, SEE
SECTION ??/S???

DRAWING TITLE:	DRAWN BY: SAAD H. TANTAWI
SECTION DETAIL	(M.S.CE, B.S.CE, E.I.)
	SCALE: N.T.S

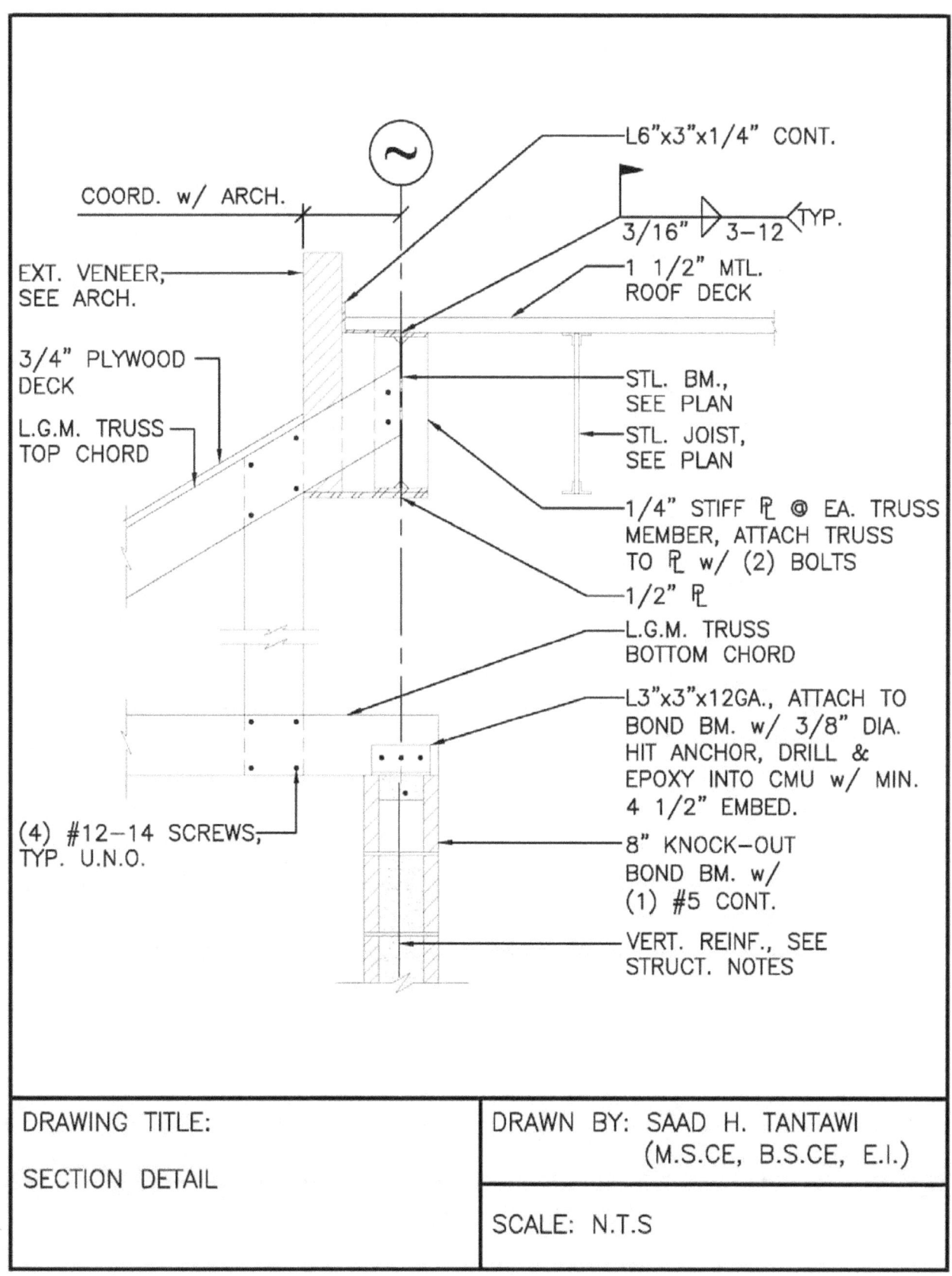

COORD. w/ ARCH.

EXT. VENEER,
SEE ARCH.

3/4" PLYWOOD
DECK

L.G.M. TRUSS
TOP CHORD

(4) #12-14 SCREWS,
TYP. U.N.O.

L6"x3"x1/4" CONT.

3/16" 3-12 TYP.

1 1/2" MTL.
ROOF DECK

STL. BM.,
SEE PLAN

STL. JOIST,
SEE PLAN

1/4" STIFF ℟ @ EA. TRUSS
MEMBER, ATTACH TRUSS
TO ℟ w/ (2) BOLTS

1/2" ℟

L.G.M. TRUSS
BOTTOM CHORD

L3"x3"x12GA., ATTACH TO
BOND BM. w/ 3/8" DIA.
HIT ANCHOR, DRILL &
EPOXY INTO CMU w/ MIN.
4 1/2" EMBED.

8" KNOCK-OUT
BOND BM. w/
(1) #5 CONT.

VERT. REINF., SEE
STRUCT. NOTES

DRAWING TITLE:	DRAWN BY: SAAD H. TANTAWI
	(M.S.CE, B.S.CE, E.I.)
SECTION DETAIL	
	SCALE: N.T.S

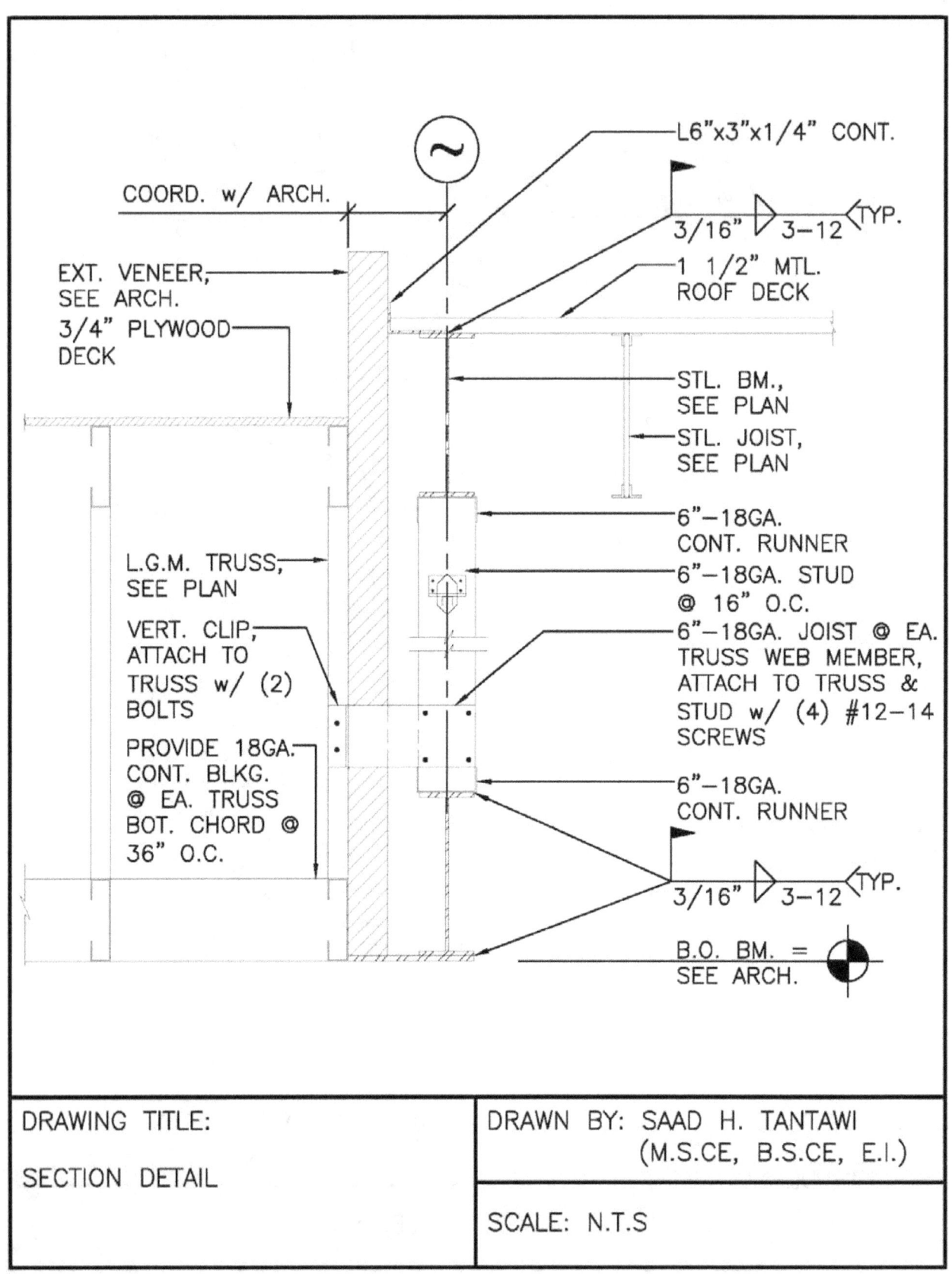

COORD. w/ ARCH.

L6"x3"x1/4" CONT.

3/16" 3-12 TYP.

EXT. VENEER,
SEE ARCH.

1 1/2" MTL.
ROOF DECK

3/4" PLYWOOD
DECK

STL. BM.,
SEE PLAN

STL. JOIST,
SEE PLAN

6"-18GA.
CONT. RUNNER

L.G.M. TRUSS,
SEE PLAN

6"-18GA. STUD
@ 16" O.C.

VERT. CLIP,
ATTACH TO
TRUSS w/ (2)
BOLTS

6"-18GA. JOIST @ EA.
TRUSS WEB MEMBER,
ATTACH TO TRUSS &
STUD w/ (4) #12-14
SCREWS

PROVIDE 18GA.
CONT. BLKG.
@ EA. TRUSS
BOT. CHORD @
36" O.C.

6"-18GA.
CONT. RUNNER

3/16" 3-12 TYP.

B.O. BM. =
SEE ARCH.

DRAWING TITLE:	DRAWN BY: SAAD H. TANTAWI
	(M.S.CE, B.S.CE, E.I.)
SECTION DETAIL	
	SCALE: N.T.S

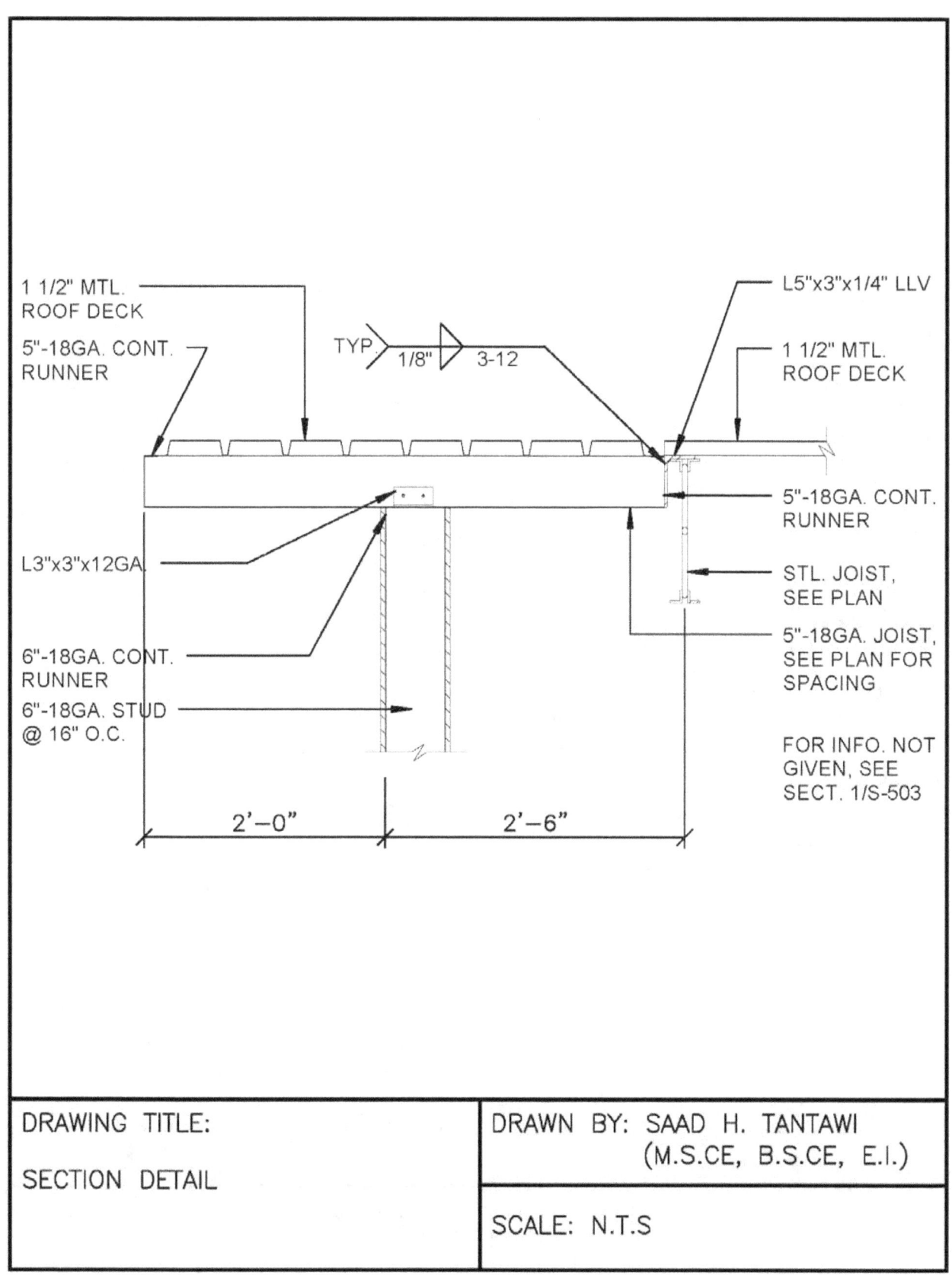

1 1/2" MTL. ROOF DECK

5"-18GA. CONT. RUNNER

TYP. 1/8" 3-12

L5"x3"x1/4" LLV

1 1/2" MTL. ROOF DECK

L3"x3"x12GA.

6"-18GA. CONT. RUNNER

6"-18GA. STUD @ 16" O.C.

5"-18GA. CONT. RUNNER

STL. JOIST, SEE PLAN

5"-18GA. JOIST, SEE PLAN FOR SPACING

FOR INFO. NOT GIVEN, SEE SECT. 1/S-503

2'-0"

2'-6"

DRAWING TITLE:	DRAWN BY: SAAD H. TANTAWI
	(M.S.CE, B.S.CE, E.I.)
SECTION DETAIL	
	SCALE: N.T.S

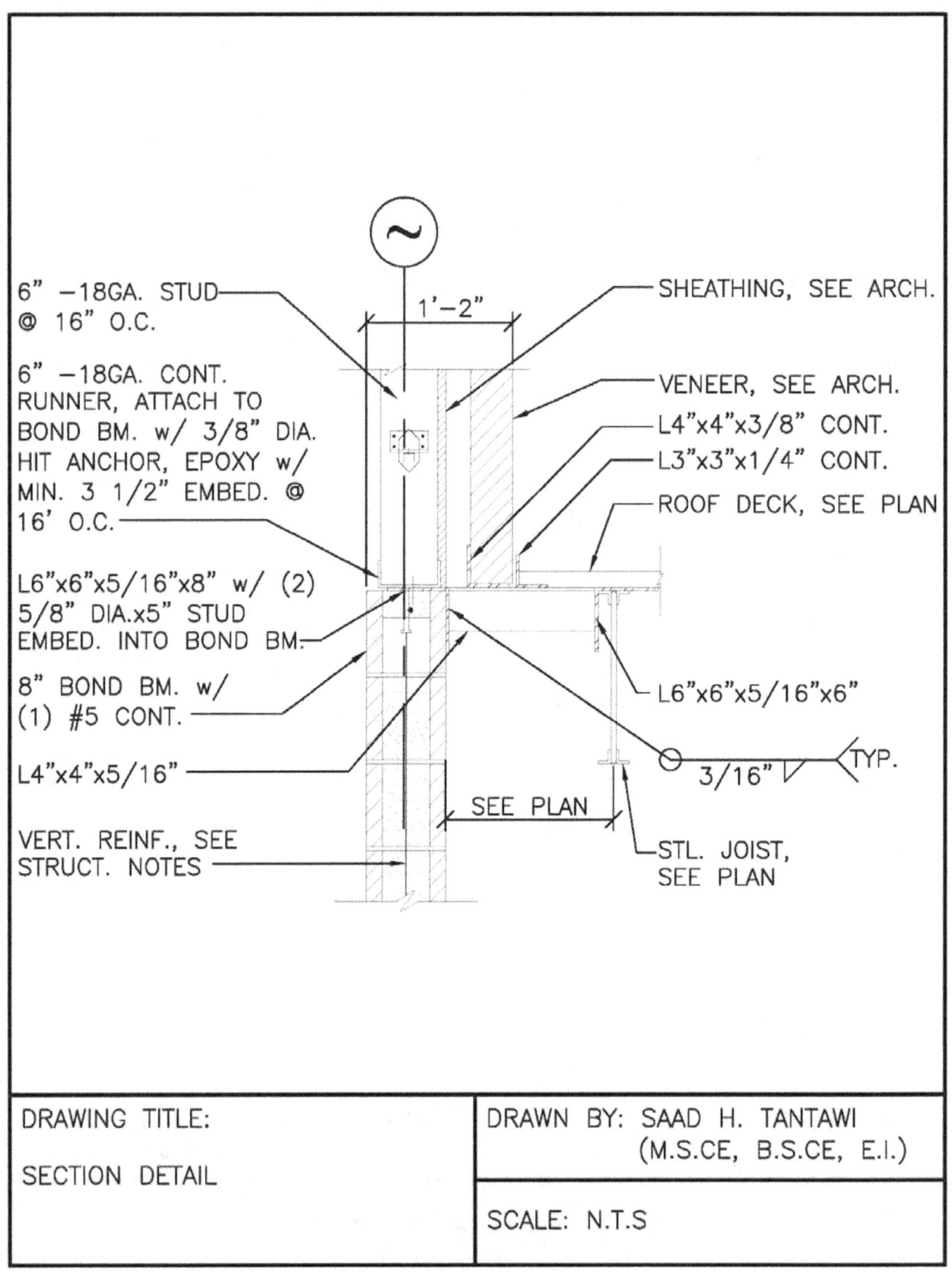

6" —18GA. STUD
@ 16" O.C.

6" —18GA. CONT.
RUNNER, ATTACH TO
BOND BM. w/ 3/8" DIA.
HIT ANCHOR, EPOXY w/
MIN. 3 1/2" EMBED. @
16' O.C.

L6"x6"x5/16"x8" w/ (2)
5/8" DIA.x5" STUD
EMBED. INTO BOND BM.

8" BOND BM. w/
(1) #5 CONT.

L4"x4"x5/16"

VERT. REINF., SEE
STRUCT. NOTES

1'—2"

SHEATHING, SEE ARCH.

VENEER, SEE ARCH.

L4"x4"x3/8" CONT.

L3"x3"x1/4" CONT.

ROOF DECK, SEE PLAN

L6"x6"x5/16"x6"

3/16"   TYP.

SEE PLAN

STL. JOIST,
SEE PLAN

DRAWING TITLE:	DRAWN BY: SAAD H. TANTAWI
	(M.S.CE, B.S.CE, E.I.)
SECTION DETAIL	
	SCALE: N.T.S

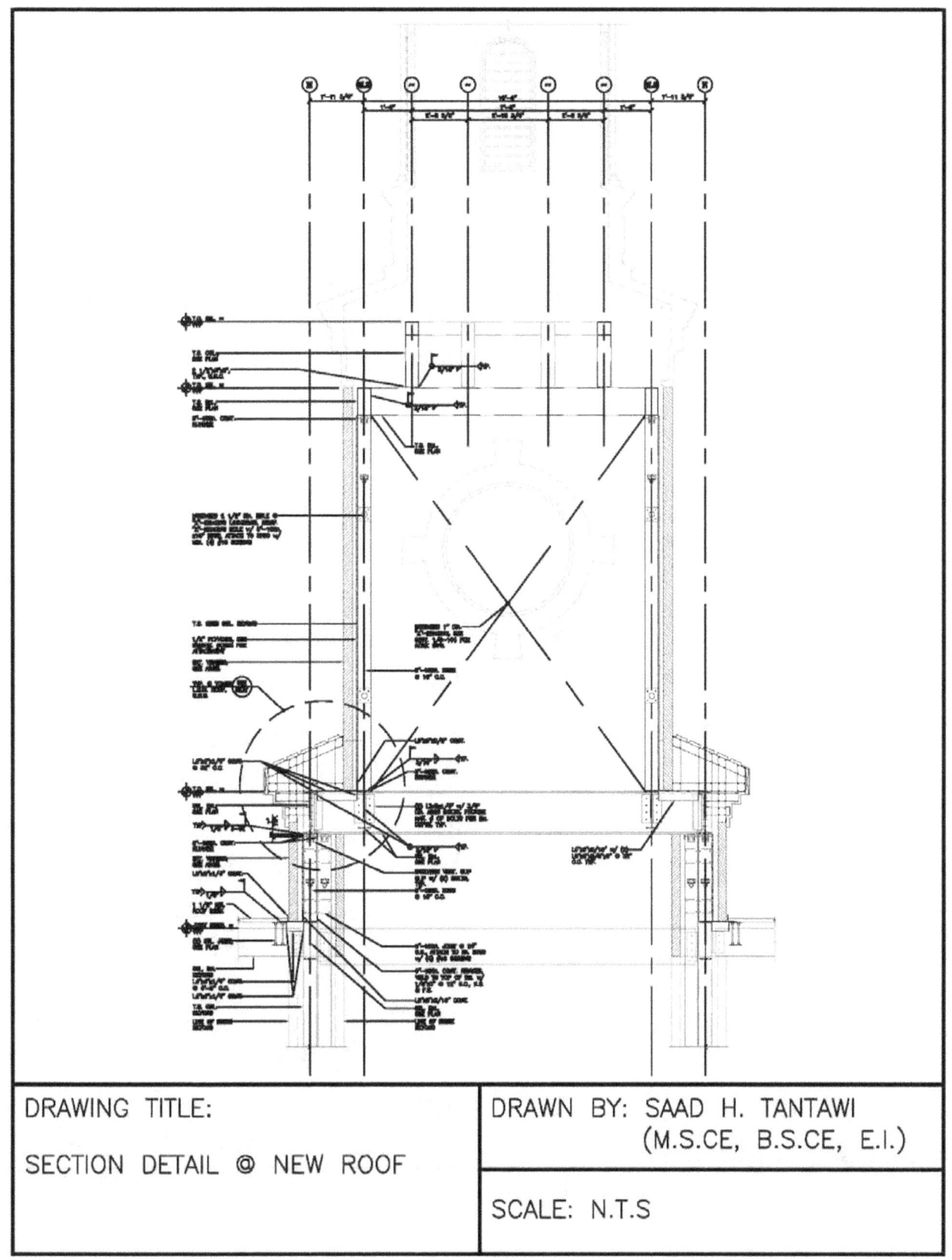

DRAWING TITLE:	DRAWN BY: SAAD H. TANTAWI
	(M.S.CE, B.S.CE, E.I.)
SECTION DETAIL @ NEW ROOF	
	SCALE: N.T.S

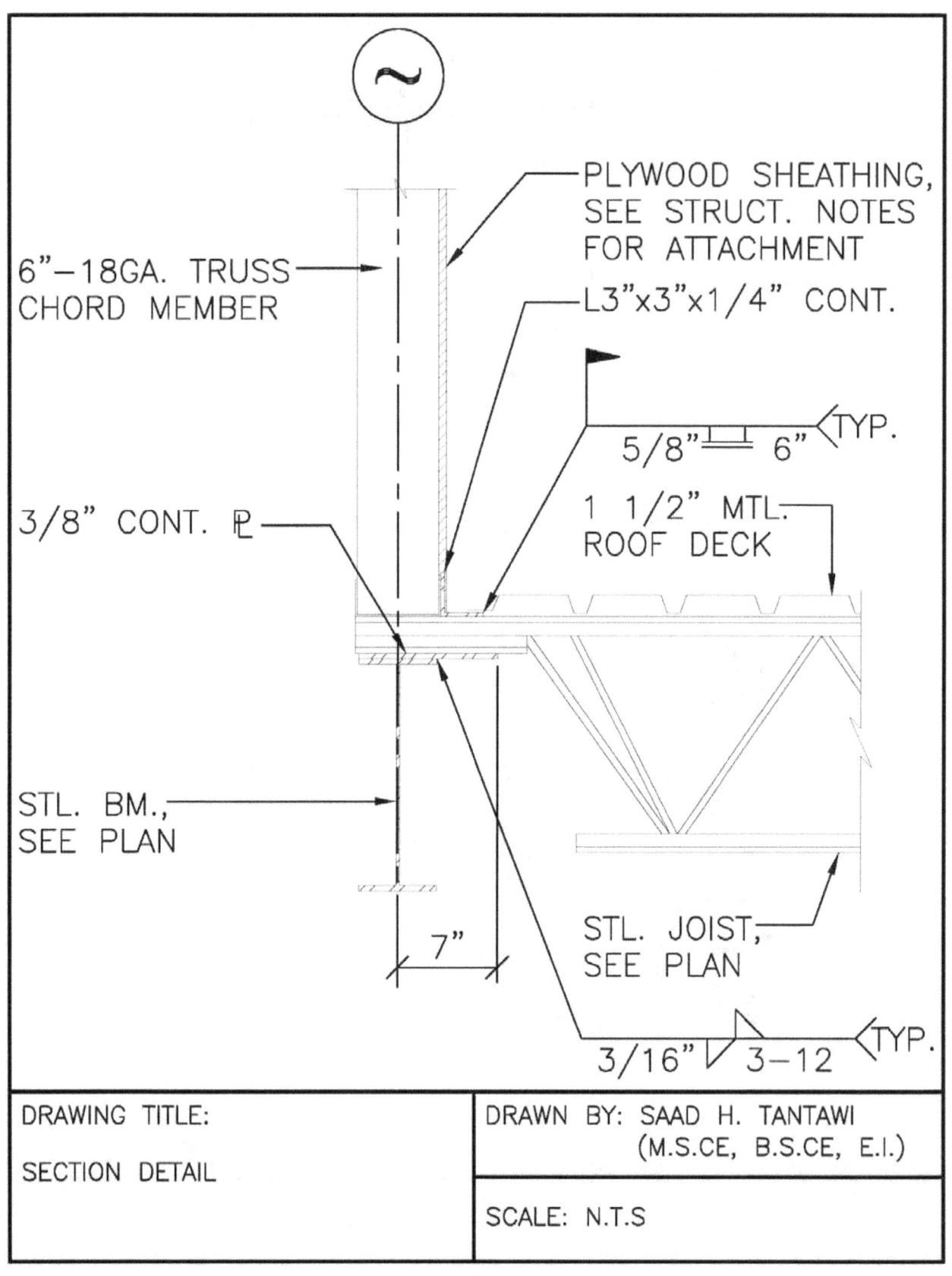

6"-18GA. TRUSS CHORD MEMBER

PLYWOOD SHEATHING, SEE STRUCT. NOTES FOR ATTACHMENT

L3"x3"x1/4" CONT.

5/8" 6" TYP.

1 1/2" MTL. ROOF DECK

3/8" CONT. ℟

STL. BM., SEE PLAN

7"

STL. JOIST, SEE PLAN

3/16" 3-12 TYP.

DRAWING TITLE:	DRAWN BY: SAAD H. TANTAWI (M.S.CE, B.S.CE, E.I.)
SECTION DETAIL	
	SCALE: N.T.S

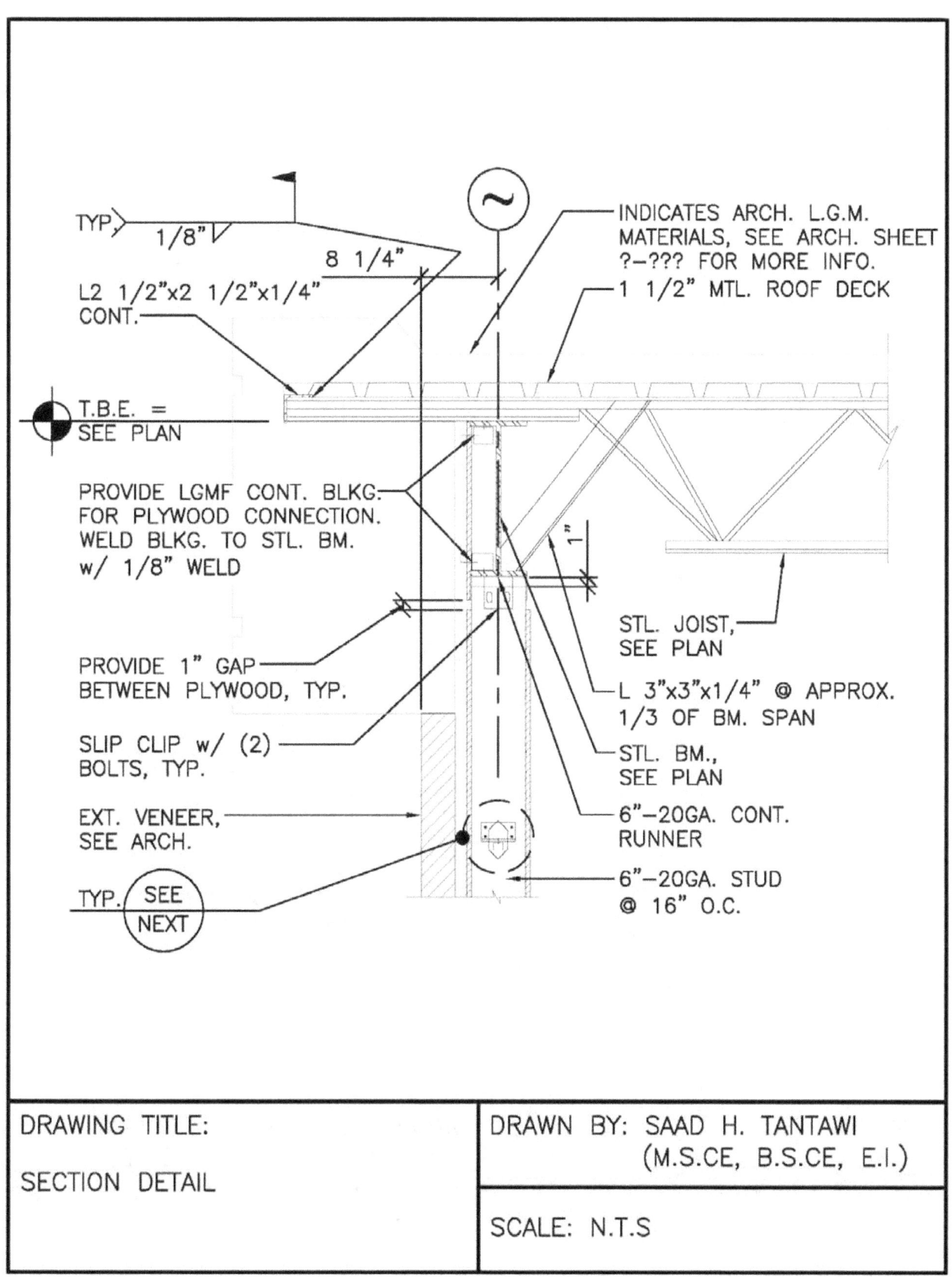

TYP. 1/8"

8 1/4"

L2 1/2"x2 1/2"x1/4" CONT.

INDICATES ARCH. L.G.M. MATERIALS, SEE ARCH. SHEET ?-??? FOR MORE INFO.

1 1/2" MTL. ROOF DECK

T.B.E. = SEE PLAN

PROVIDE LGMF CONT. BLKG. FOR PLYWOOD CONNECTION. WELD BLKG. TO STL. BM. w/ 1/8" WELD

1"

STL. JOIST, SEE PLAN

PROVIDE 1" GAP BETWEEN PLYWOOD, TYP.

L 3"x3"x1/4" @ APPROX. 1/3 OF BM. SPAN

SLIP CLIP w/ (2) BOLTS, TYP.

STL. BM., SEE PLAN

EXT. VENEER, SEE ARCH.

6"-20GA. CONT. RUNNER

TYP. SEE NEXT

6"-20GA. STUD @ 16" O.C.

DRAWING TITLE:	DRAWN BY: SAAD H. TANTAWI
	(M.S.CE, B.S.CE, E.I.)
SECTION DETAIL	
	SCALE: N.T.S

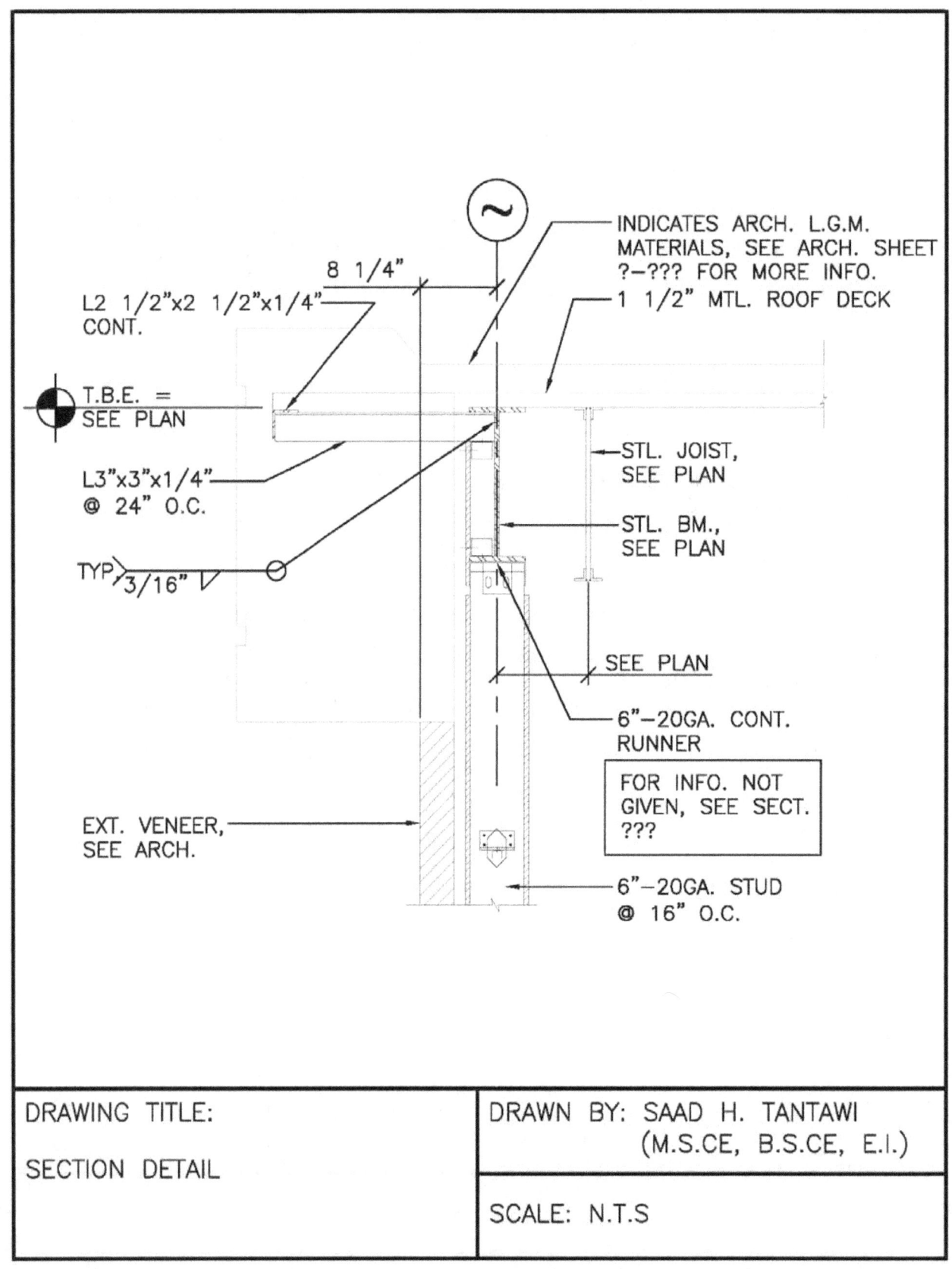

L2 1/2"x2 1/2"x1/4" CONT.

8 1/4"

INDICATES ARCH. L.G.M. MATERIALS, SEE ARCH. SHEET ?-??? FOR MORE INFO.

1 1/2" MTL. ROOF DECK

T.B.E. = SEE PLAN

STL. JOIST, SEE PLAN

L3"x3"x1/4" @ 24" O.C.

STL. BM., SEE PLAN

TYP 3/16"

SEE PLAN

6"-20GA. CONT. RUNNER

FOR INFO. NOT GIVEN, SEE SECT. ???

EXT. VENEER, SEE ARCH.

6"-20GA. STUD @ 16" O.C.

DRAWING TITLE: SECTION DETAIL	DRAWN BY: SAAD H. TANTAWI (M.S.CE, B.S.CE, E.I.)
	SCALE: N.T.S

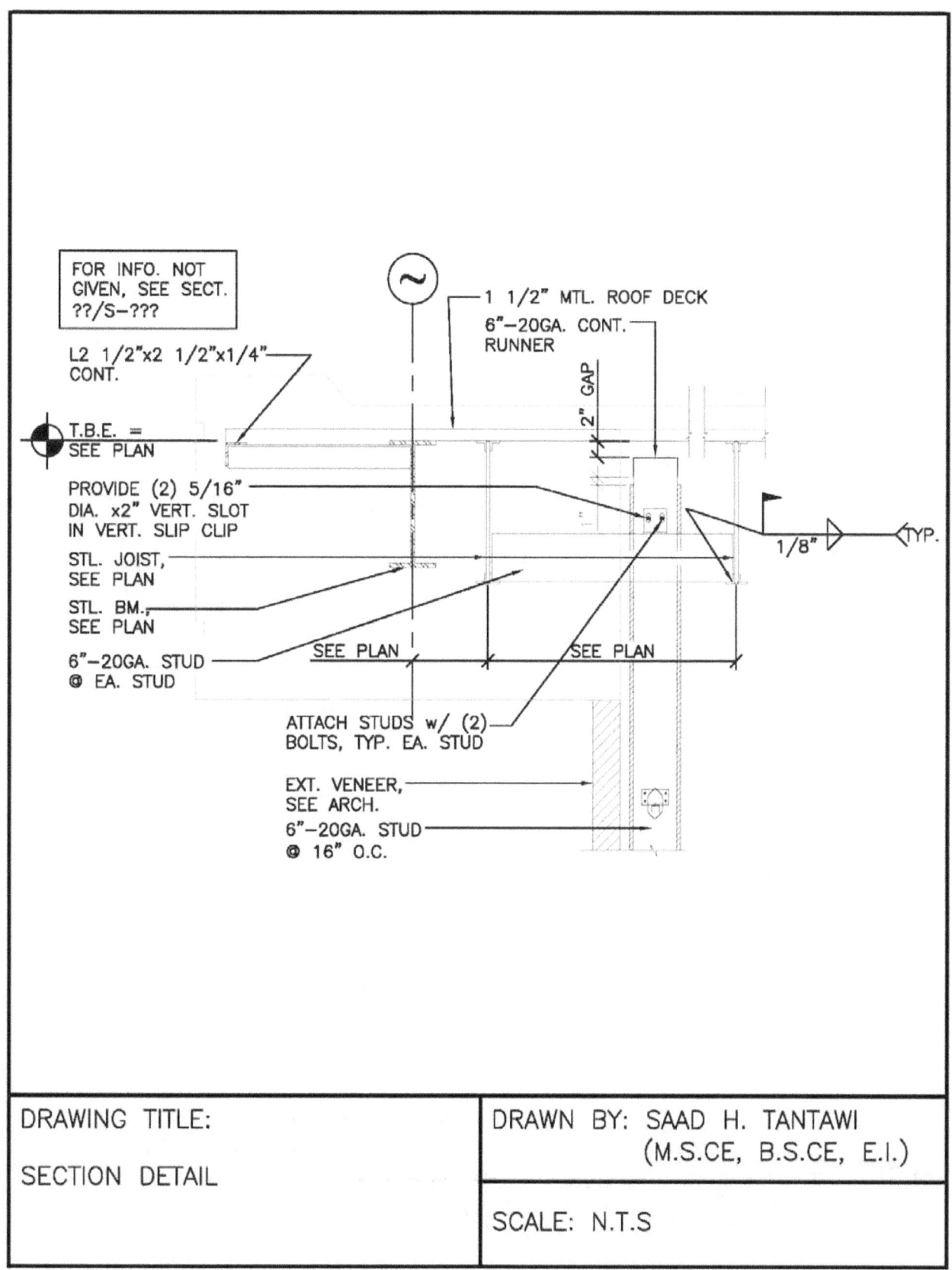

FOR INFO. NOT
GIVEN, SEE SECT.
??/S-???

L2 1/2"x2 1/2"x1/4"
CONT.

1 1/2" MTL. ROOF DECK
6"-20GA. CONT.
RUNNER

2" GAP

T.B.E. =
SEE PLAN

PROVIDE (2) 5/16"
DIA. x2" VERT. SLOT
IN VERT. SLIP CLIP

STL. JOIST,
SEE PLAN

STL. BM.,
SEE PLAN

6"-20GA. STUD
@ EA. STUD

SEE PLAN

SEE PLAN

1/8"

TYP.

ATTACH STUDS w/ (2)
BOLTS, TYP. EA. STUD

EXT. VENEER,
SEE ARCH.

6"-20GA. STUD
@ 16" O.C.

DRAWING TITLE:	DRAWN BY: SAAD H. TANTAWI
	(M.S.CE, B.S.CE, E.I.)
SECTION DETAIL	
	SCALE: N.T.S

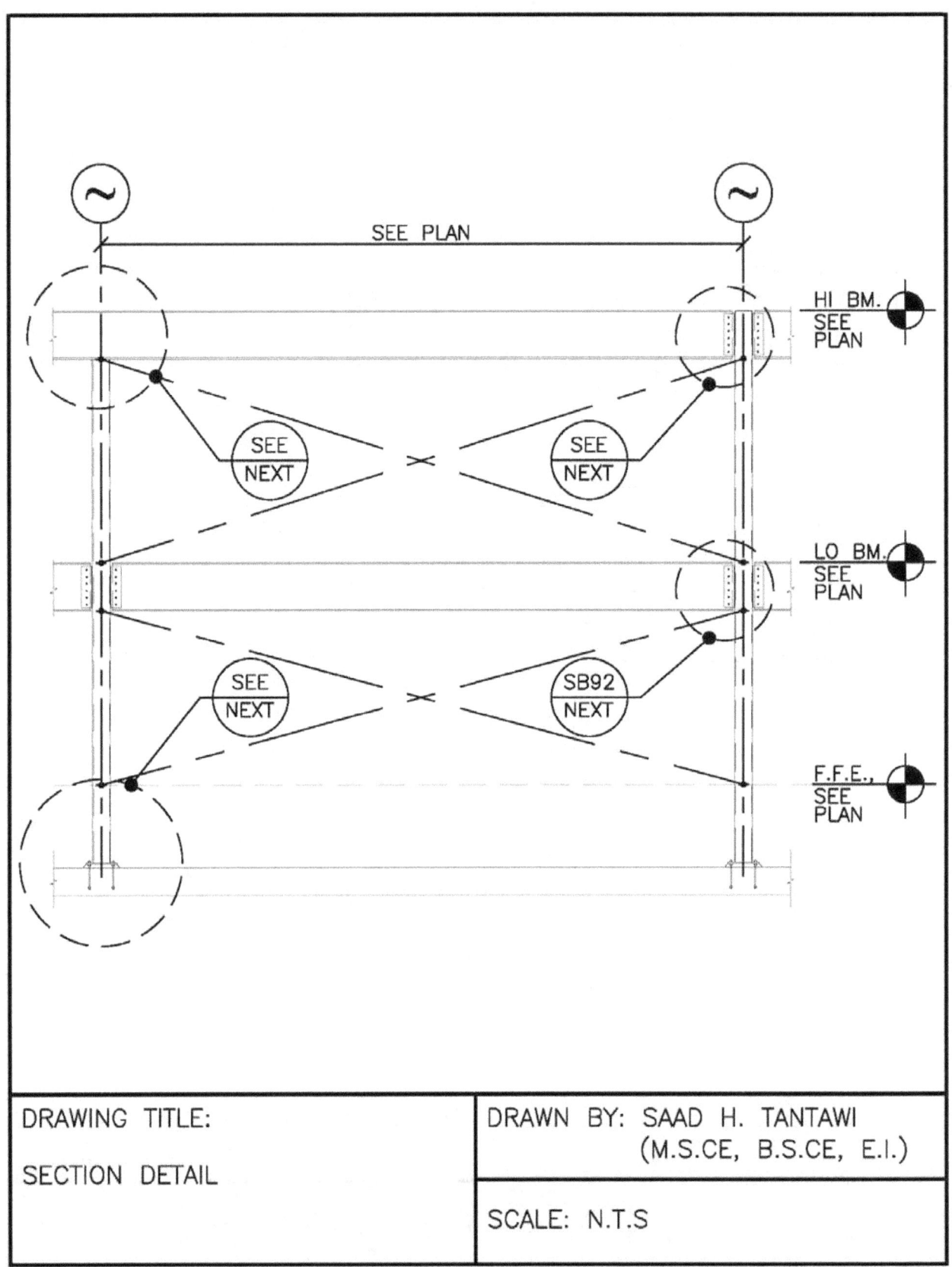

SEE PLAN

HI BM.
SEE
PLAN

SEE
NEXT

SEE
NEXT

LO BM.
SEE
PLAN

SEE
NEXT

SB92
NEXT

F.F.E.,
SEE
PLAN

DRAWING TITLE: SECTION DETAIL	DRAWN BY: SAAD H. TANTAWI (M.S.CE, B.S.CE, E.I.)
	SCALE: N.T.S

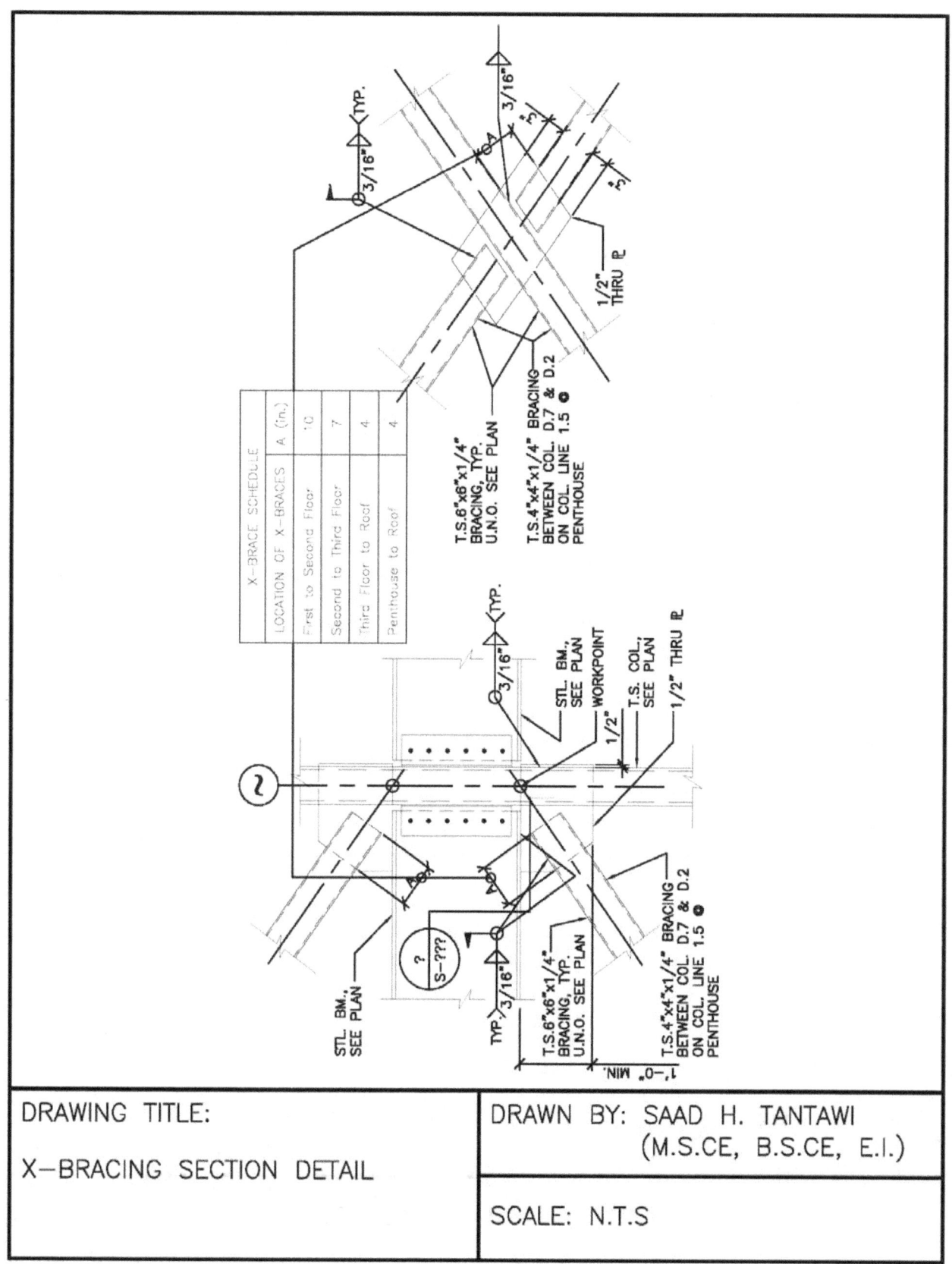

DRAWING TITLE:	DRAWN BY: SAAD H. TANTAWI
X—BRACING SECTION DETAIL	(M.S.CE, B.S.CE, E.I.)
	SCALE: N.T.S

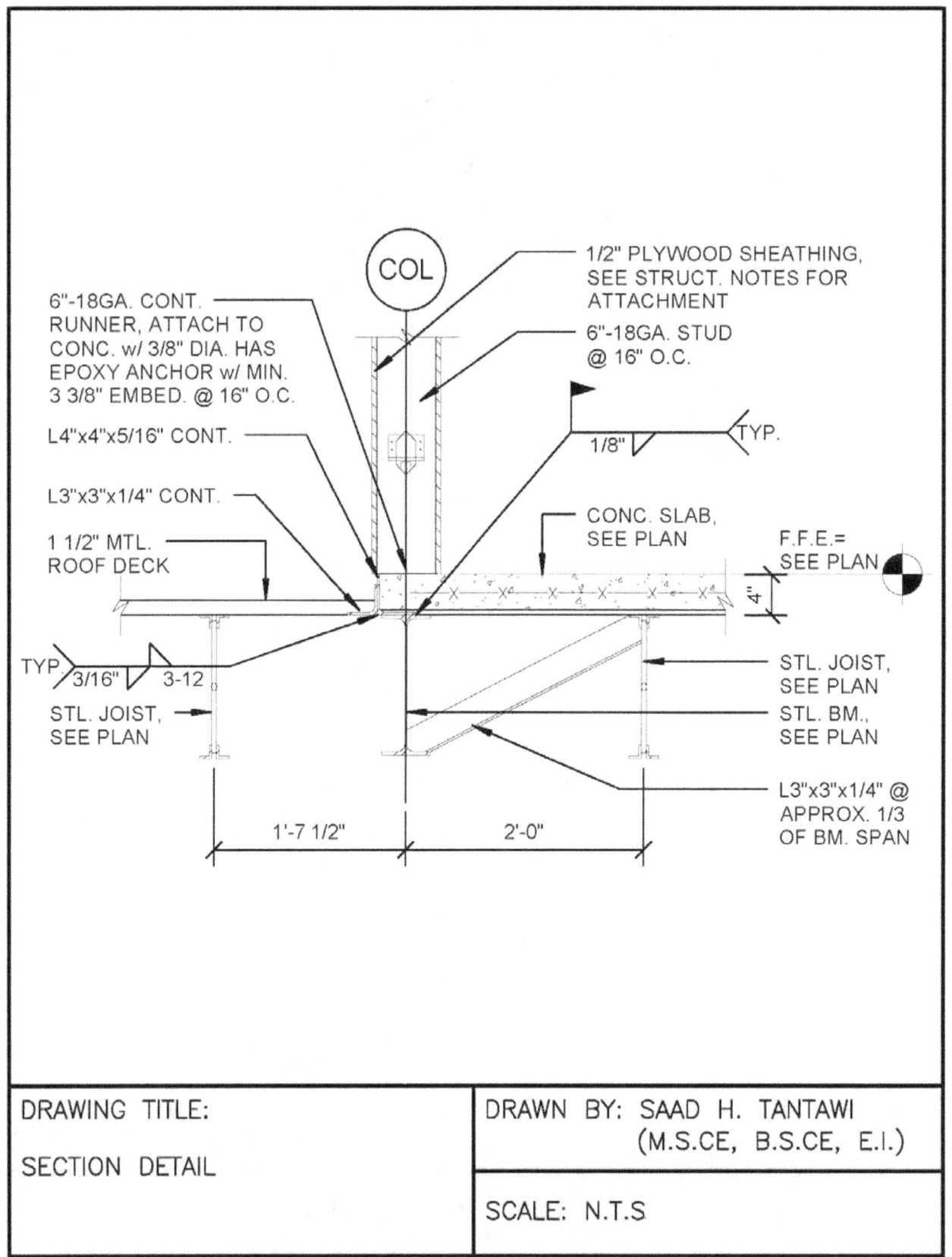

6"-18GA. CONT.
RUNNER, ATTACH TO
CONC. w/ 3/8" DIA. HAS
EPOXY ANCHOR w/ MIN.
3 3/8" EMBED. @ 16" O.C.

L4"x4"x5/16" CONT.

L3"x3"x1/4" CONT.

1 1/2" MTL.
ROOF DECK

COL

1/2" PLYWOOD SHEATHING,
SEE STRUCT. NOTES FOR
ATTACHMENT

6"-18GA. STUD
@ 16" O.C.

1/8"

TYP.

CONC. SLAB,
SEE PLAN

F.F.E.=
SEE PLAN

4"

TYP.
3/16"   3-12

STL. JOIST,
SEE PLAN

STL. JOIST,
SEE PLAN

STL. BM.,
SEE PLAN

L3"x3"x1/4" @
APPROX. 1/3
OF BM. SPAN

1'-7 1/2"   2'-0"

DRAWING TITLE:	DRAWN BY: SAAD H. TANTAWI
	(M.S.CE, B.S.CE, E.I.)
SECTION DETAIL	
	SCALE: N.T.S

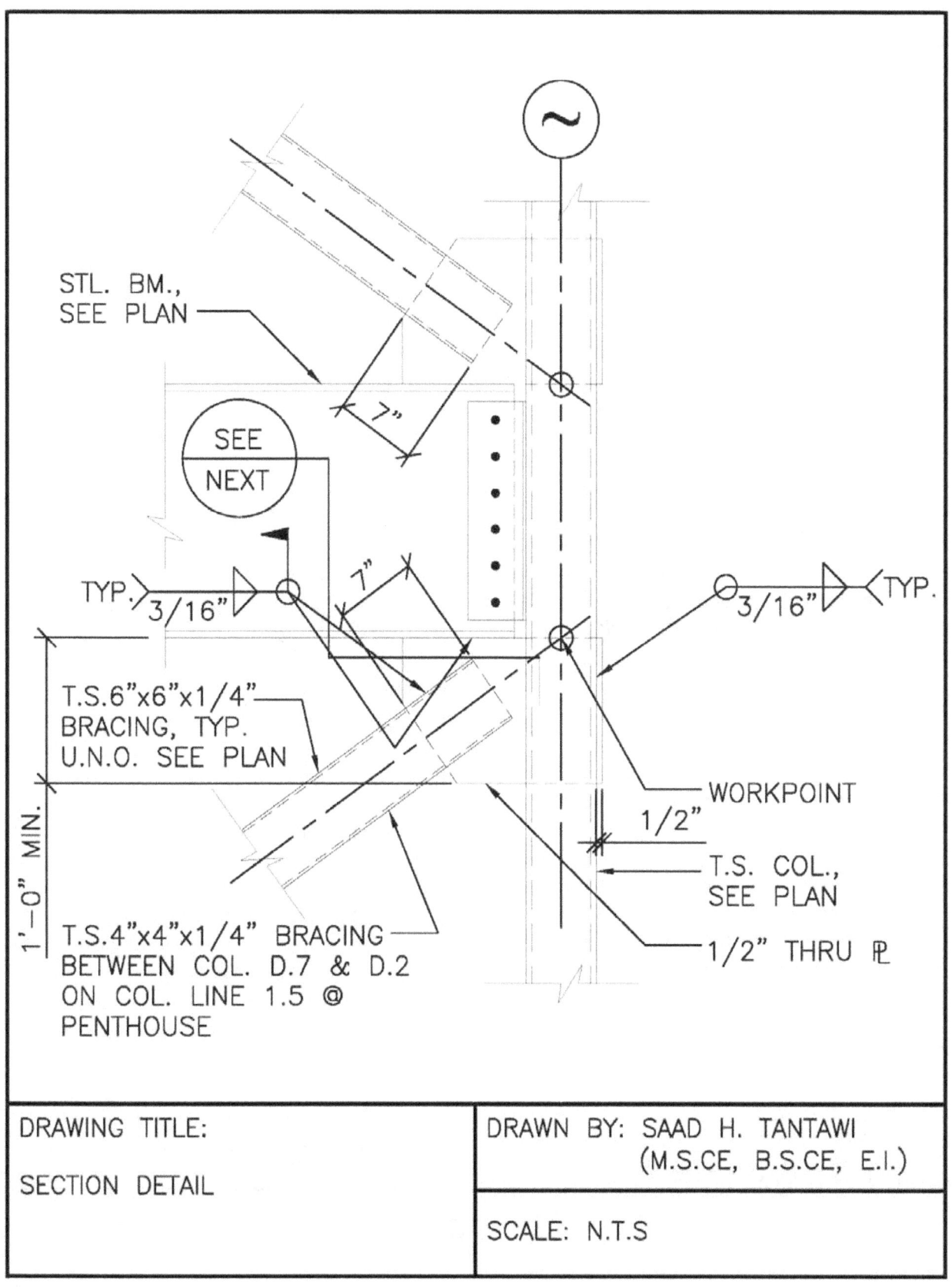

DRAWING TITLE:	DRAWN BY: SAAD H. TANTAWI
	(M.S.CE, B.S.CE, E.I.)
SECTION DETAIL	SCALE: N.T.S

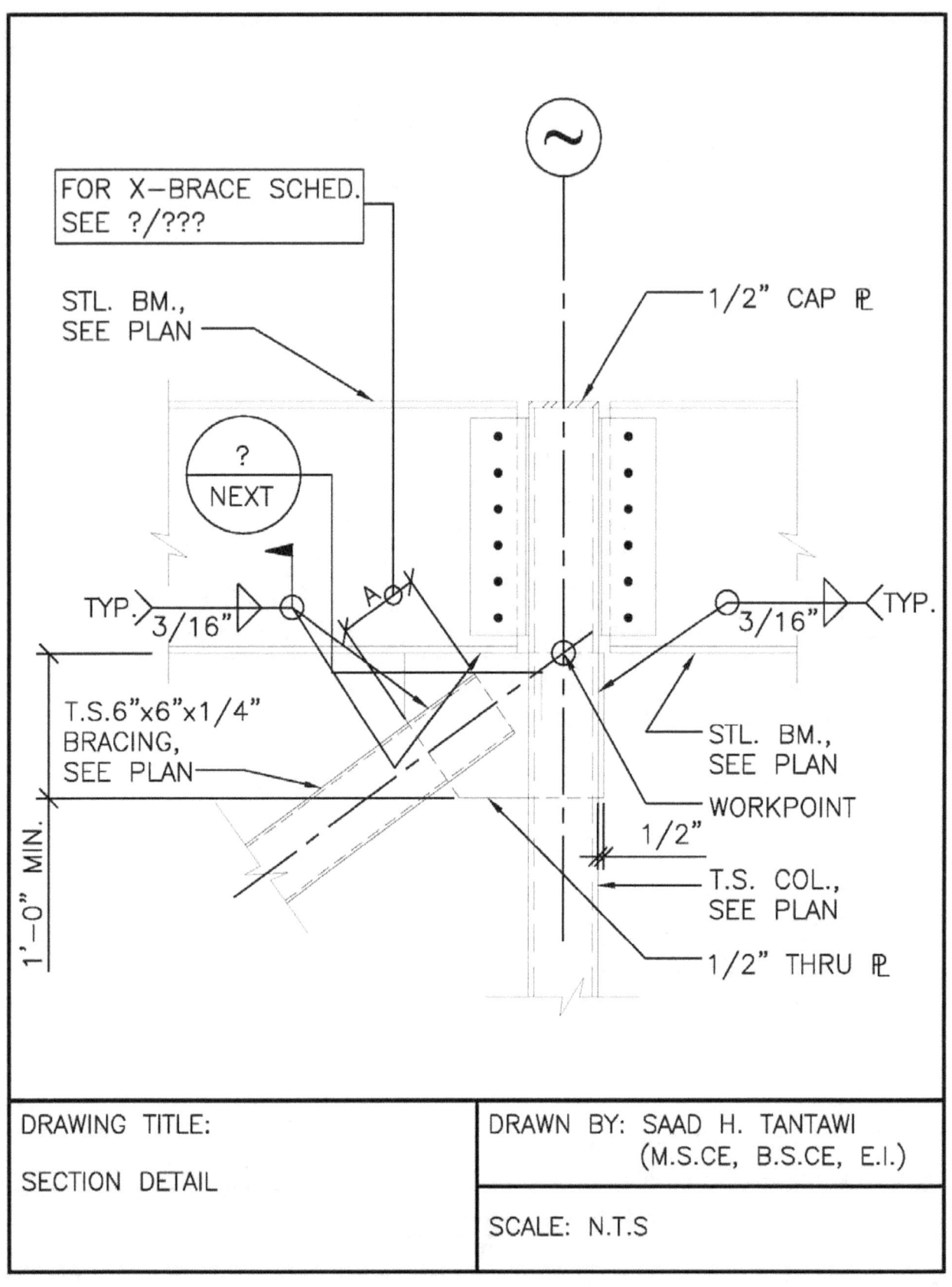

FOR X-BRACE SCHED.
SEE ?/???

STL. BM.,
SEE PLAN

1/2" CAP ℙ

?
NEXT

TYP.
3/16"

A

3/16"
TYP.

T.S.6"x6"x1/4"
BRACING,
SEE PLAN

STL. BM.,
SEE PLAN

WORKPOINT

1/2"

1'-0" MIN.

T.S. COL.,
SEE PLAN

1/2" THRU ℙ

DRAWING TITLE:	DRAWN BY: SAAD H. TANTAWI (M.S.CE, B.S.CE, E.I.)
SECTION DETAIL	
	SCALE: N.T.S

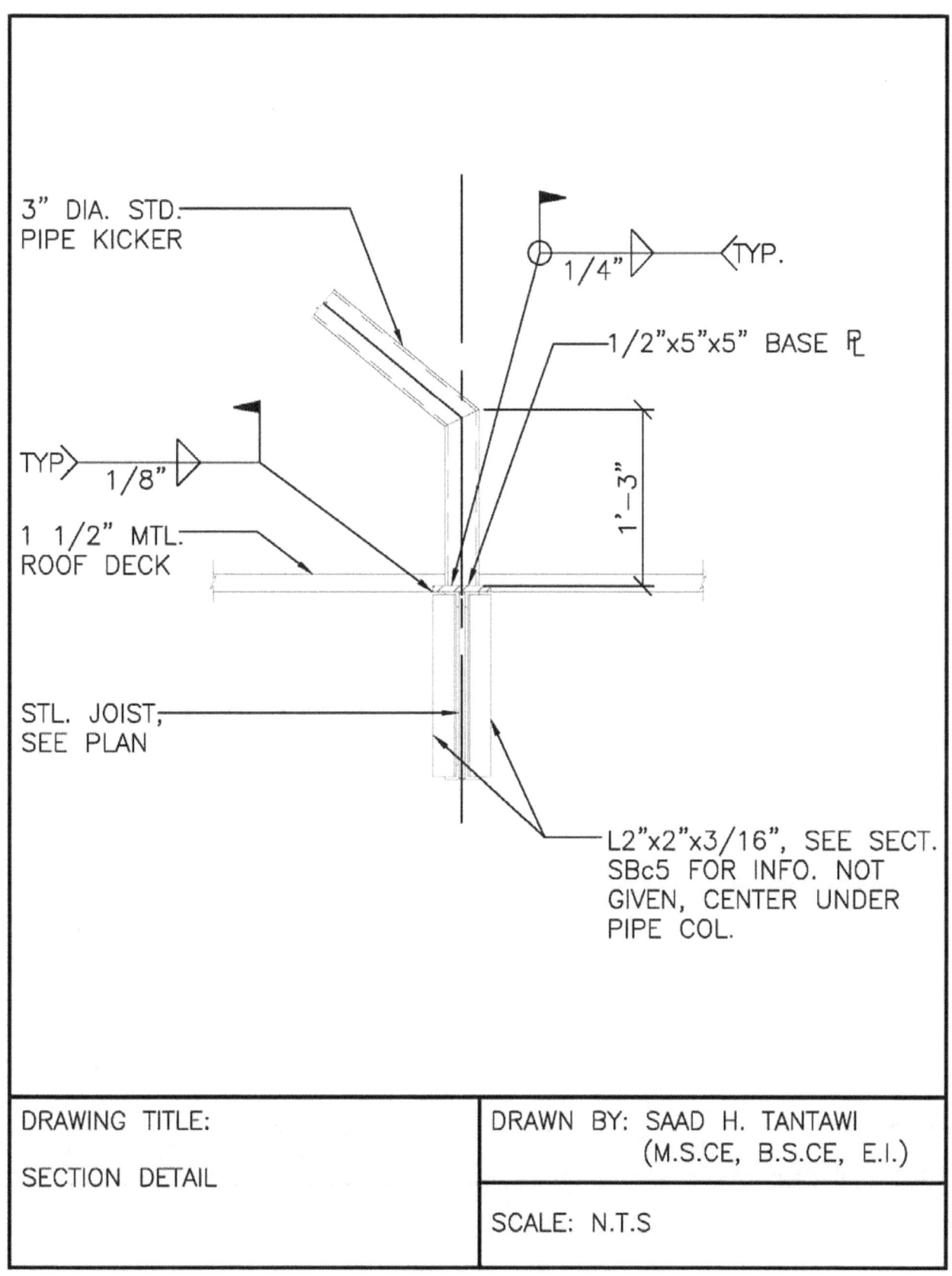

3" DIA. STD.
PIPE KICKER

1/4" TYP.

1/2"x5"x5" BASE ℞

1'-3"

TYP 1/8"

1 1/2" MTL.
ROOF DECK

STL. JOIST,
SEE PLAN

L2"x2"x3/16", SEE SECT.
SBc5 FOR INFO. NOT
GIVEN, CENTER UNDER
PIPE COL.

DRAWING TITLE:	DRAWN BY: SAAD H. TANTAWI
	(M.S.CE, B.S.CE, E.I.)
SECTION DETAIL	
	SCALE: N.T.S

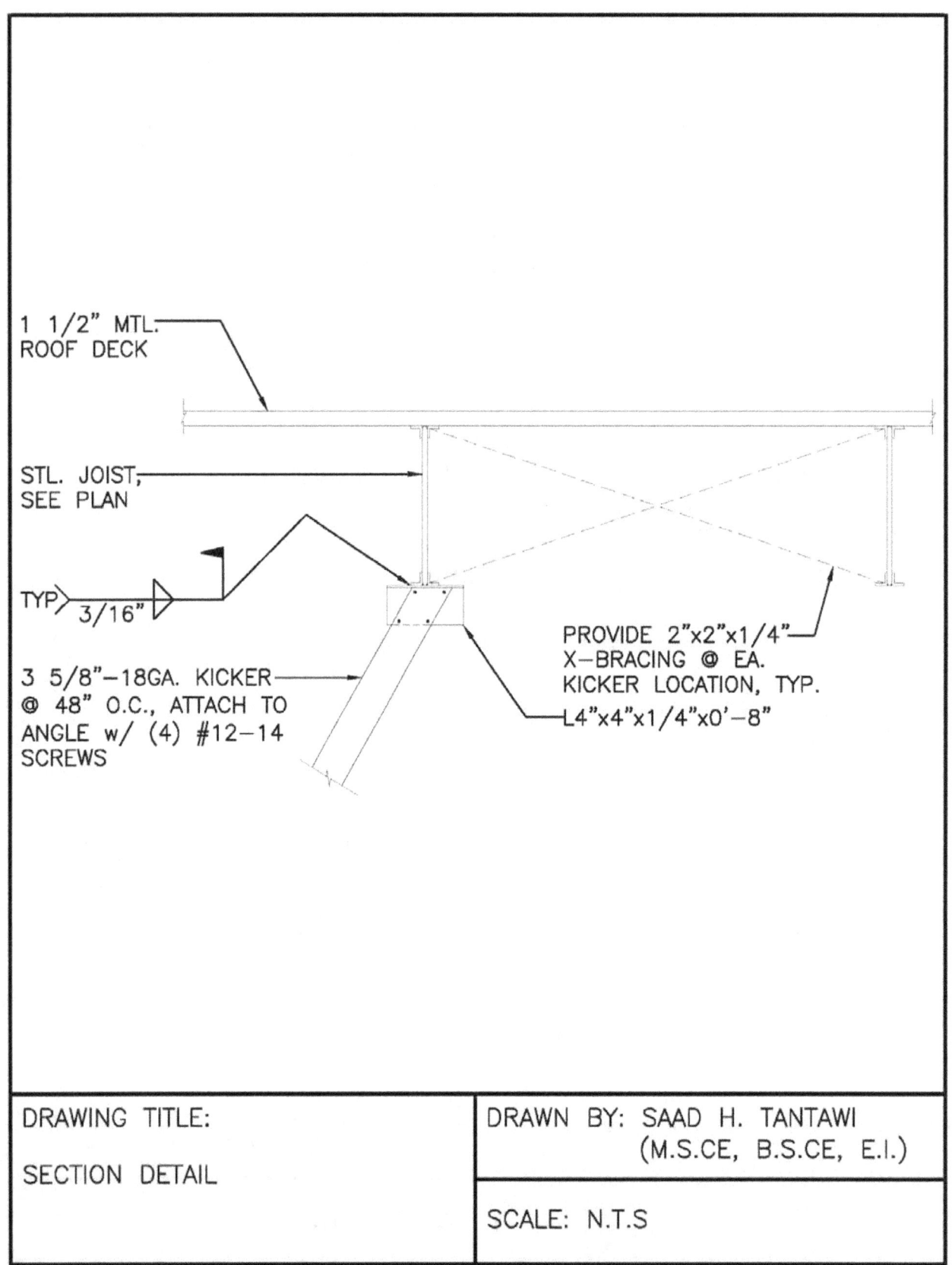

1 1/2" MTL. ROOF DECK

STL. JOIST, SEE PLAN

TYP 3/16"

3 5/8"–18GA. KICKER @ 48" O.C., ATTACH TO ANGLE w/ (4) #12–14 SCREWS

PROVIDE 2"x2"x1/4" X–BRACING @ EA. KICKER LOCATION, TYP.

L4"x4"x1/4"x0'–8"

DRAWING TITLE: SECTION DETAIL	DRAWN BY: SAAD H. TANTAWI (M.S.CE, B.S.CE, E.I.)
	SCALE: N.T.S

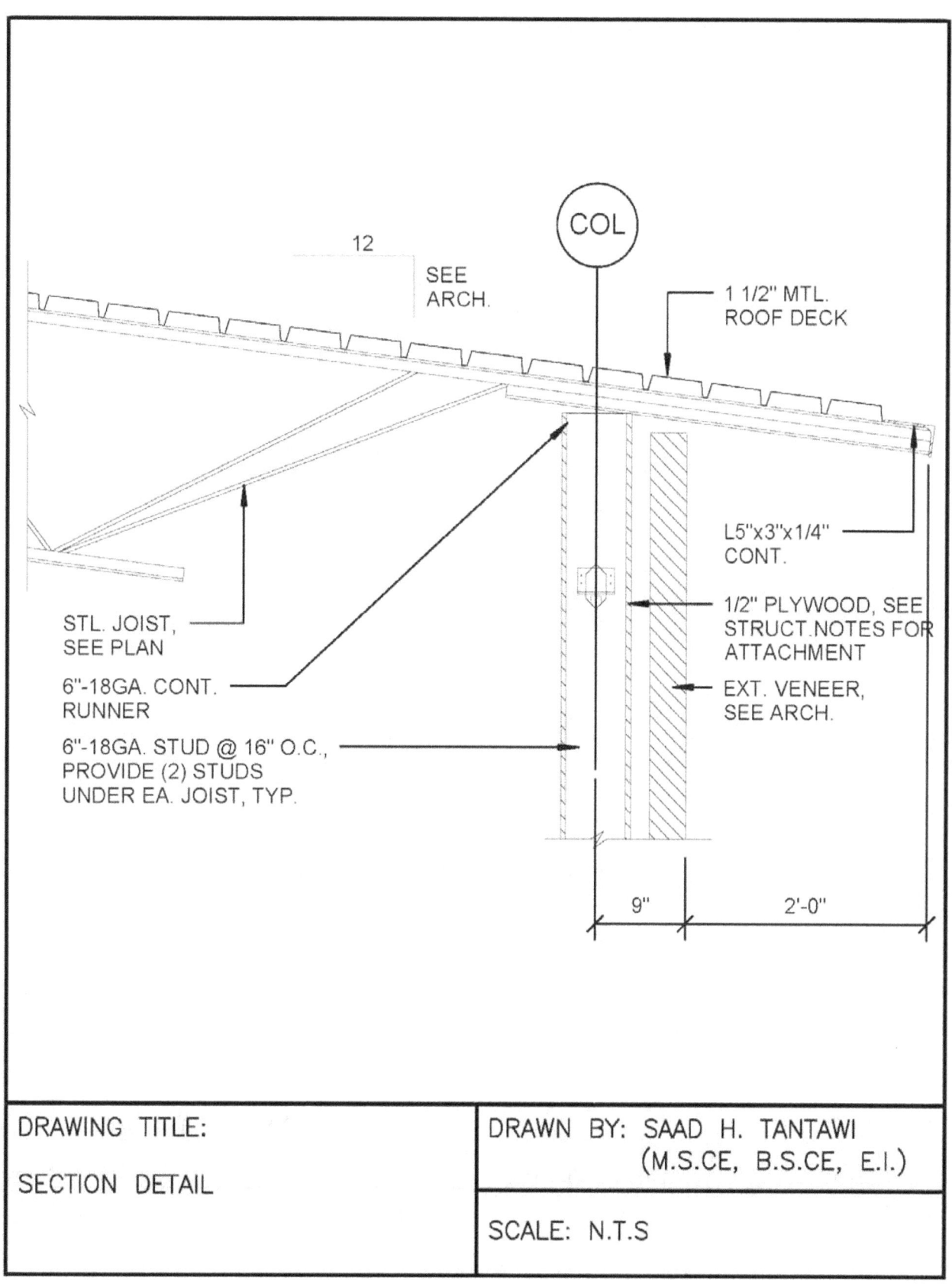

DRAWING TITLE:	DRAWN BY: SAAD H. TANTAWI
	(M.S.CE, B.S.CE, E.I.)
SECTION DETAIL	
	SCALE: N.T.S

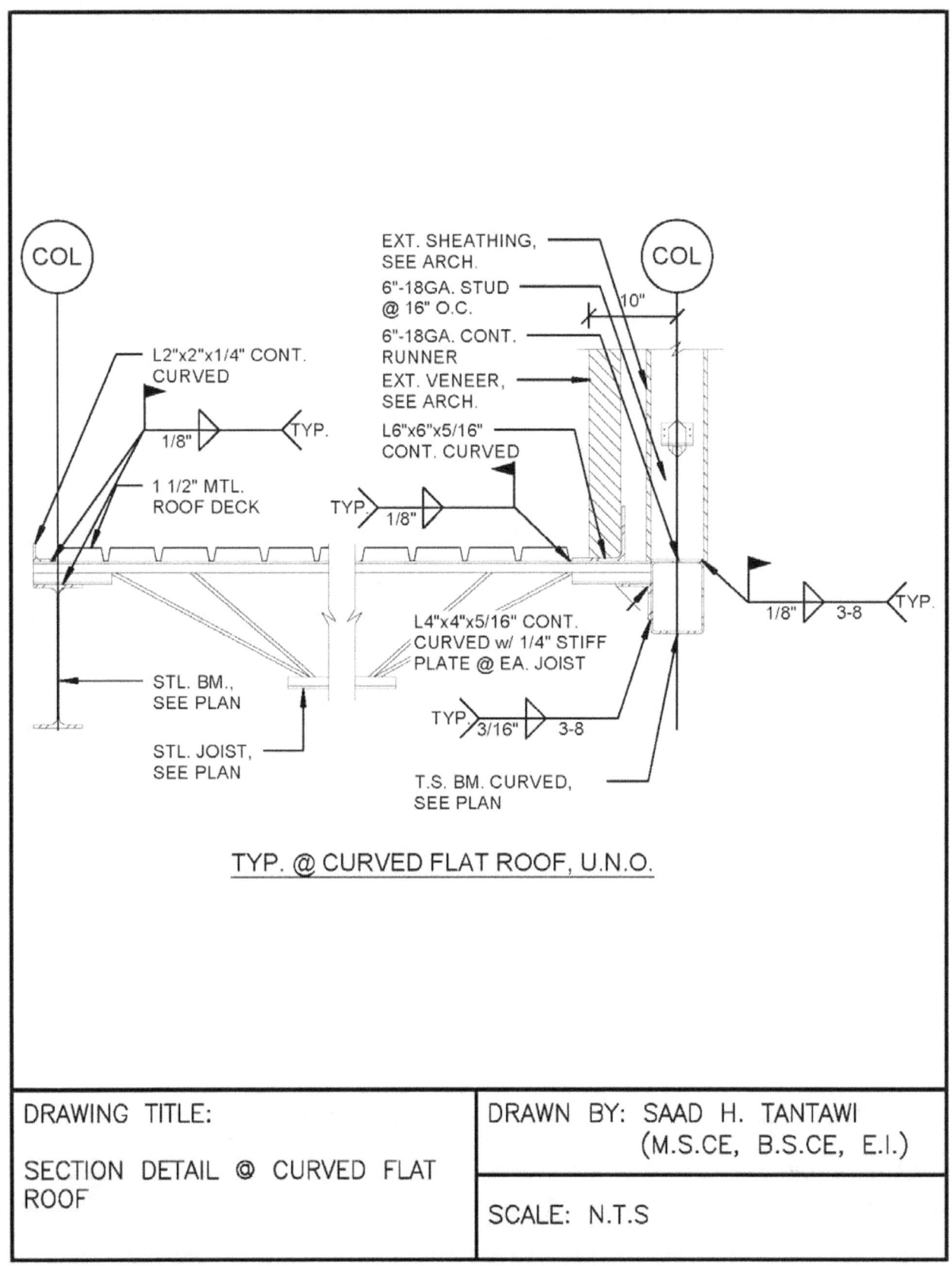

TYP. @ CURVED FLAT ROOF, U.N.O.

DRAWING TITLE:	DRAWN BY: SAAD H. TANTAWI
SECTION DETAIL @ CURVED FLAT ROOF	(M.S.CE, B.S.CE, E.I.)
	SCALE: N.T.S

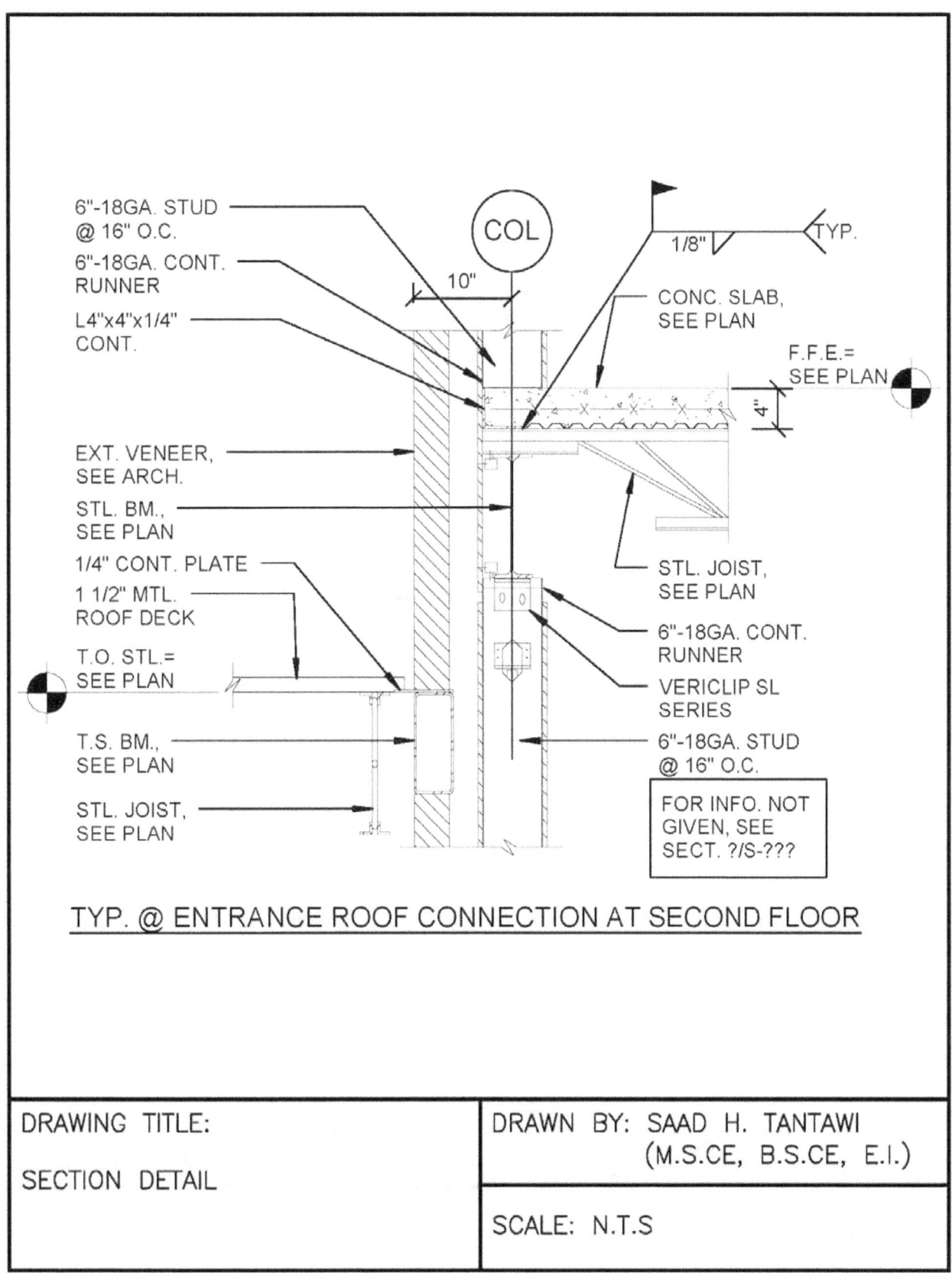

6"-18GA. STUD
@ 16" O.C.

6"-18GA. CONT.
RUNNER

L4"x4"x1/4"
CONT.

EXT. VENEER,
SEE ARCH.

STL. BM.,
SEE PLAN

1/4" CONT. PLATE

1 1/2" MTL.
ROOF DECK

T.O. STL.=
SEE PLAN

T.S. BM.,
SEE PLAN

STL. JOIST,
SEE PLAN

COL

10"

1/8"   TYP.

CONC. SLAB,
SEE PLAN

F.F.E.=
SEE PLAN

4"

STL. JOIST,
SEE PLAN

6"-18GA. CONT.
RUNNER

VERICLIP SL
SERIES

6"-18GA. STUD
@ 16" O.C.

FOR INFO. NOT
GIVEN, SEE
SECT. ?/S-???

TYP. @ ENTRANCE ROOF CONNECTION AT SECOND FLOOR

DRAWING TITLE:	DRAWN BY: SAAD H. TANTAWI
	(M.S.CE, B.S.CE, E.I.)
SECTION DETAIL	
	SCALE: N.T.S

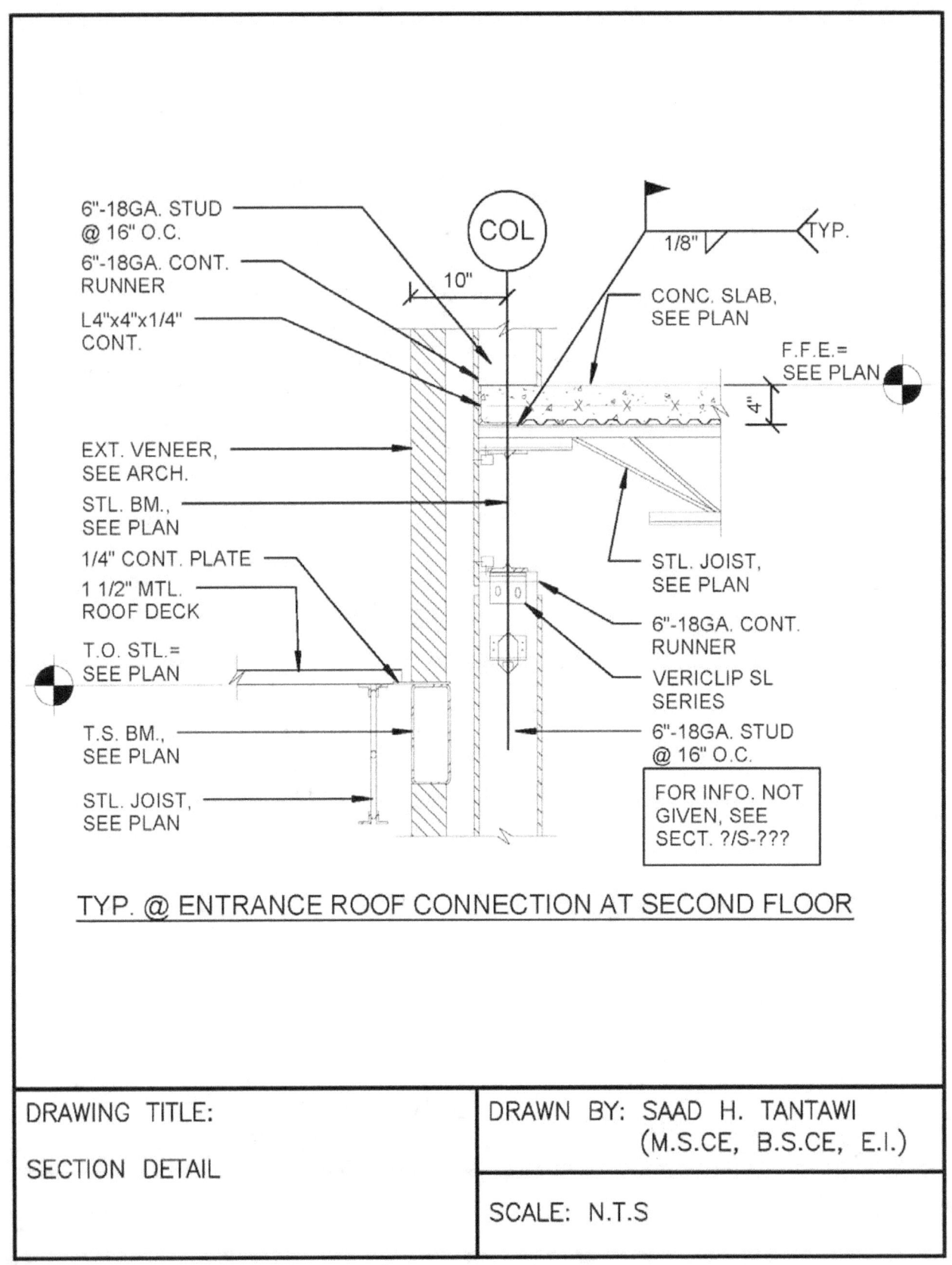

6"-18GA. STUD
@ 16" O.C.

6"-18GA. CONT.
RUNNER

L4"x4"x1/4"
CONT.

COL

10"

CONC. SLAB,
SEE PLAN

1/8"   TYP.

F.F.E.=
SEE PLAN

4"

EXT. VENEER,
SEE ARCH.

STL. BM.,
SEE PLAN

STL. JOIST,
SEE PLAN

1/4" CONT. PLATE

1 1/2" MTL.
ROOF DECK

6"-18GA. CONT.
RUNNER

T.O. STL.=
SEE PLAN

VERICLIP SL
SERIES

T.S. BM.,
SEE PLAN

6"-18GA. STUD
@ 16" O.C.

STL. JOIST,
SEE PLAN

FOR INFO. NOT
GIVEN, SEE
SECT. ?/S-???

TYP. @ ENTRANCE ROOF CONNECTION AT SECOND FLOOR

DRAWING TITLE:	DRAWN BY: SAAD H. TANTAWI
	(M.S.CE, B.S.CE, E.I.)
SECTION DETAIL	
	SCALE: N.T.S

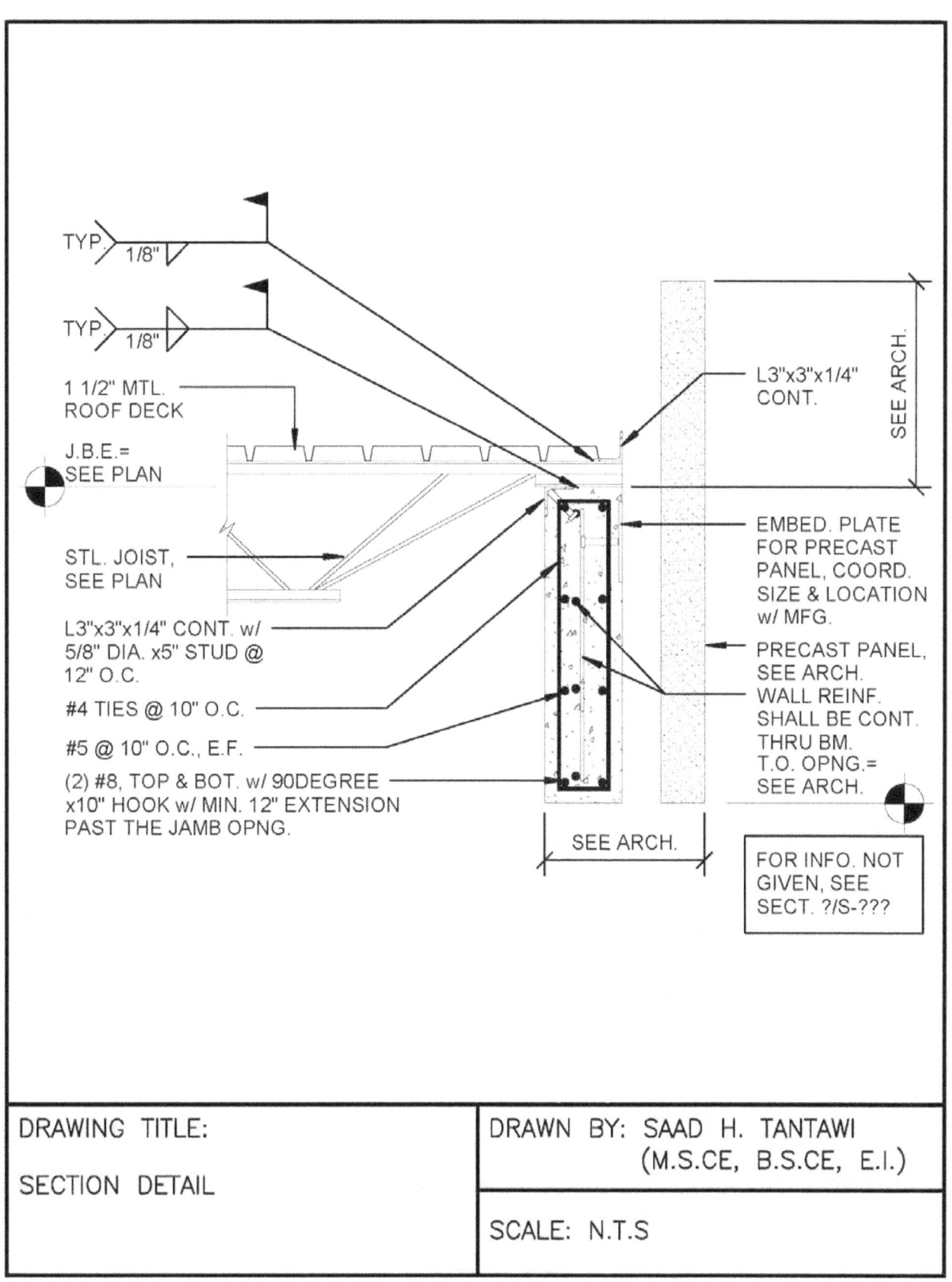

TYP. 1/8"

TYP. 1/8"

1 1/2" MTL.
ROOF DECK

J.B.E.=
SEE PLAN

STL. JOIST,
SEE PLAN

L3"x3"x1/4" CONT. w/
5/8" DIA. x5" STUD @
12" O.C.

#4 TIES @ 10" O.C.

#5 @ 10" O.C., E.F.

(2) #8, TOP & BOT. w/ 90DEGREE
x10" HOOK w/ MIN. 12" EXTENSION
PAST THE JAMB OPNG.

L3"x3"x1/4"
CONT.

SEE ARCH.

EMBED. PLATE
FOR PRECAST
PANEL, COORD.
SIZE & LOCATION
w/ MFG.

PRECAST PANEL,
SEE ARCH.
WALL REINF.
SHALL BE CONT.
THRU BM.
T.O. OPNG.=
SEE ARCH.

SEE ARCH.

FOR INFO. NOT
GIVEN, SEE
SECT. ?/S-???

DRAWING TITLE:	DRAWN BY: SAAD H. TANTAWI
	(M.S.CE, B.S.CE, E.I.)
SECTION DETAIL	
	SCALE: N.T.S

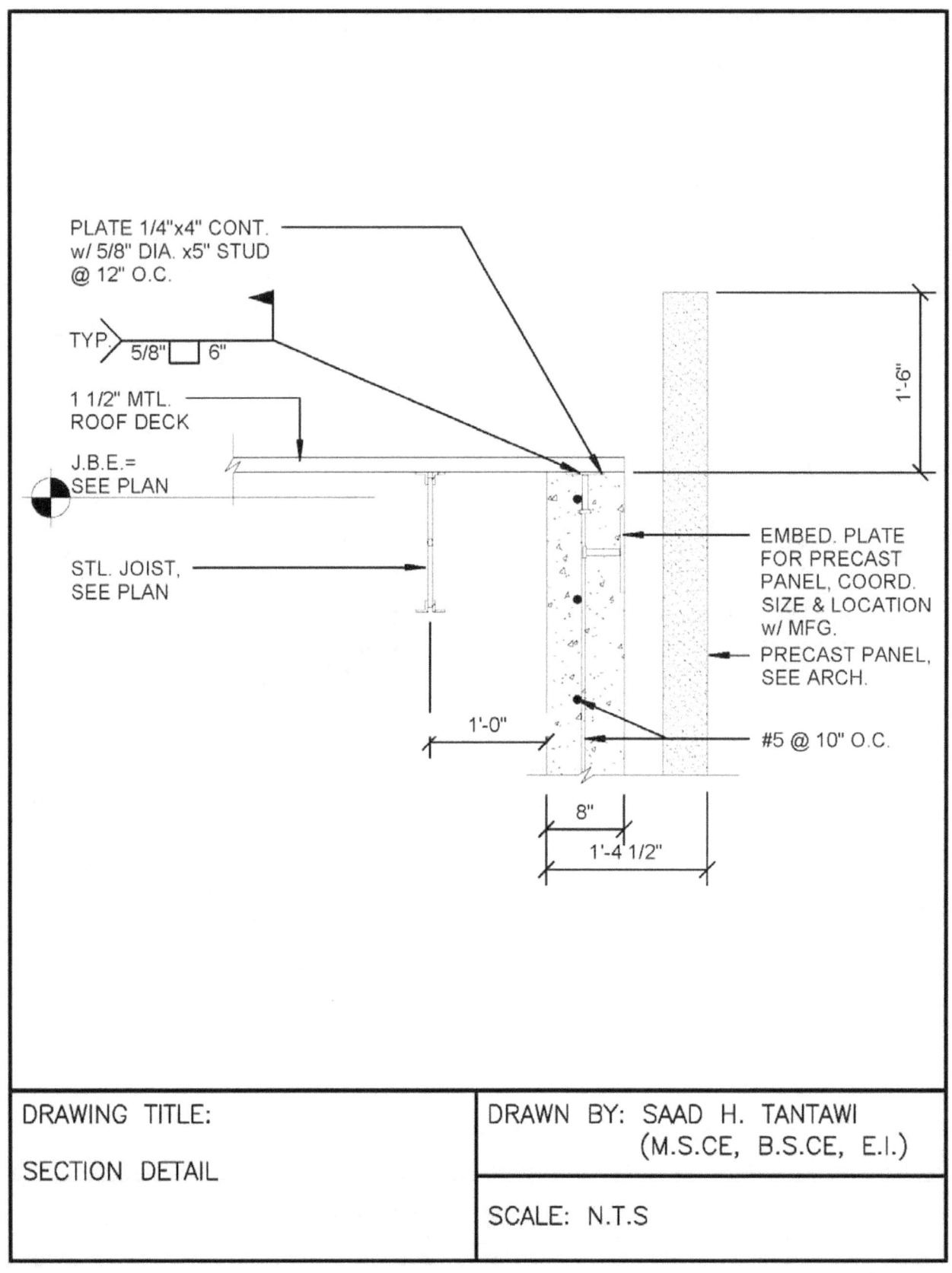

PLATE 1/4"x4" CONT. w/ 5/8" DIA. x5" STUD @ 12" O.C.

TYP. 5/8" 6"

1 1/2" MTL. ROOF DECK

J.B.E.= SEE PLAN

STL. JOIST, SEE PLAN

1'-6"

EMBED. PLATE FOR PRECAST PANEL, COORD. SIZE & LOCATION w/ MFG.

PRECAST PANEL, SEE ARCH.

#5 @ 10" O.C.

1'-0"

8"

1'-4 1/2"

DRAWING TITLE: SECTION DETAIL	DRAWN BY: SAAD H. TANTAWI (M.S.CE, B.S.CE, E.I.)
	SCALE: N.T.S

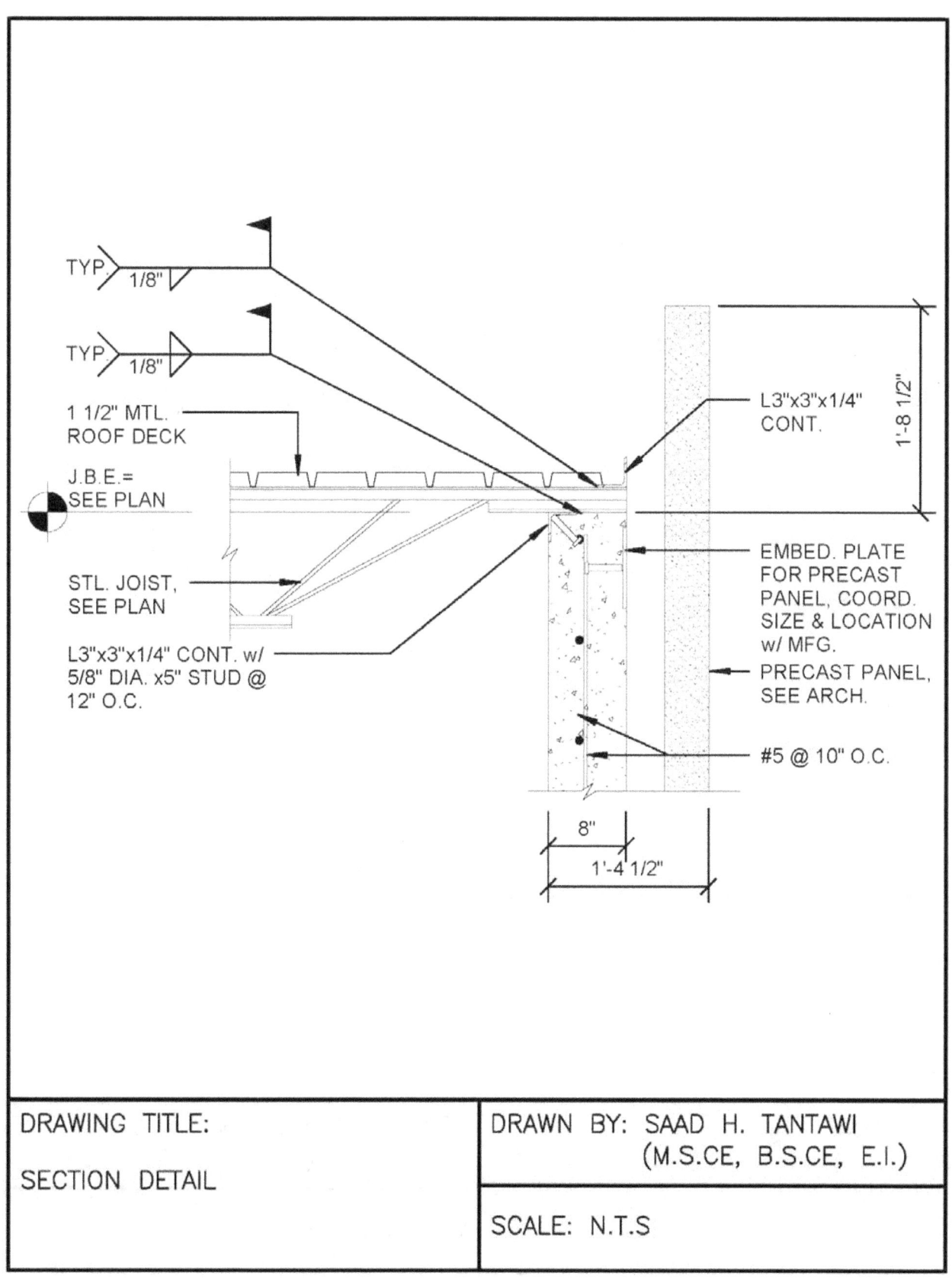

TYP. 1/8"

TYP. 1/8"

1 1/2" MTL.
ROOF DECK

J.B.E.=
SEE PLAN

STL. JOIST,
SEE PLAN

L3"x3"x1/4" CONT. w/
5/8" DIA. x5" STUD @
12" O.C.

L3"x3"x1/4"
CONT.

1'-8 1/2"

EMBED. PLATE
FOR PRECAST
PANEL, COORD.
SIZE & LOCATION
w/ MFG.

PRECAST PANEL,
SEE ARCH.

#5 @ 10" O.C.

8"

1'-4 1/2"

DRAWING TITLE:	DRAWN BY: SAAD H. TANTAWI
	(M.S.CE, B.S.CE, E.I.)
SECTION DETAIL	
	SCALE: N.T.S

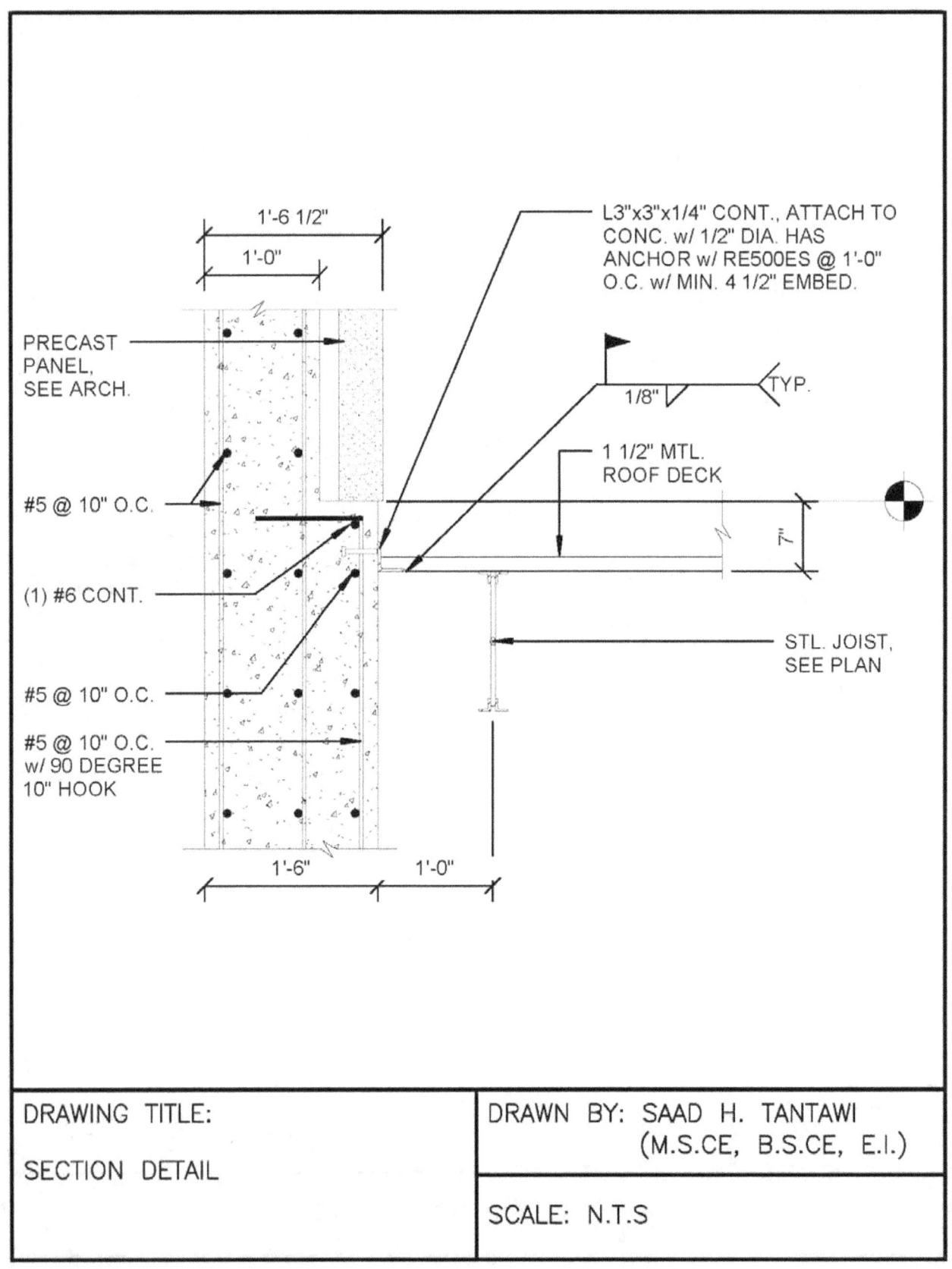

DRAWING TITLE:	DRAWN BY: SAAD H. TANTAWI
SECTION DETAIL	(M.S.CE, B.S.CE, E.I.)
	SCALE: N.T.S

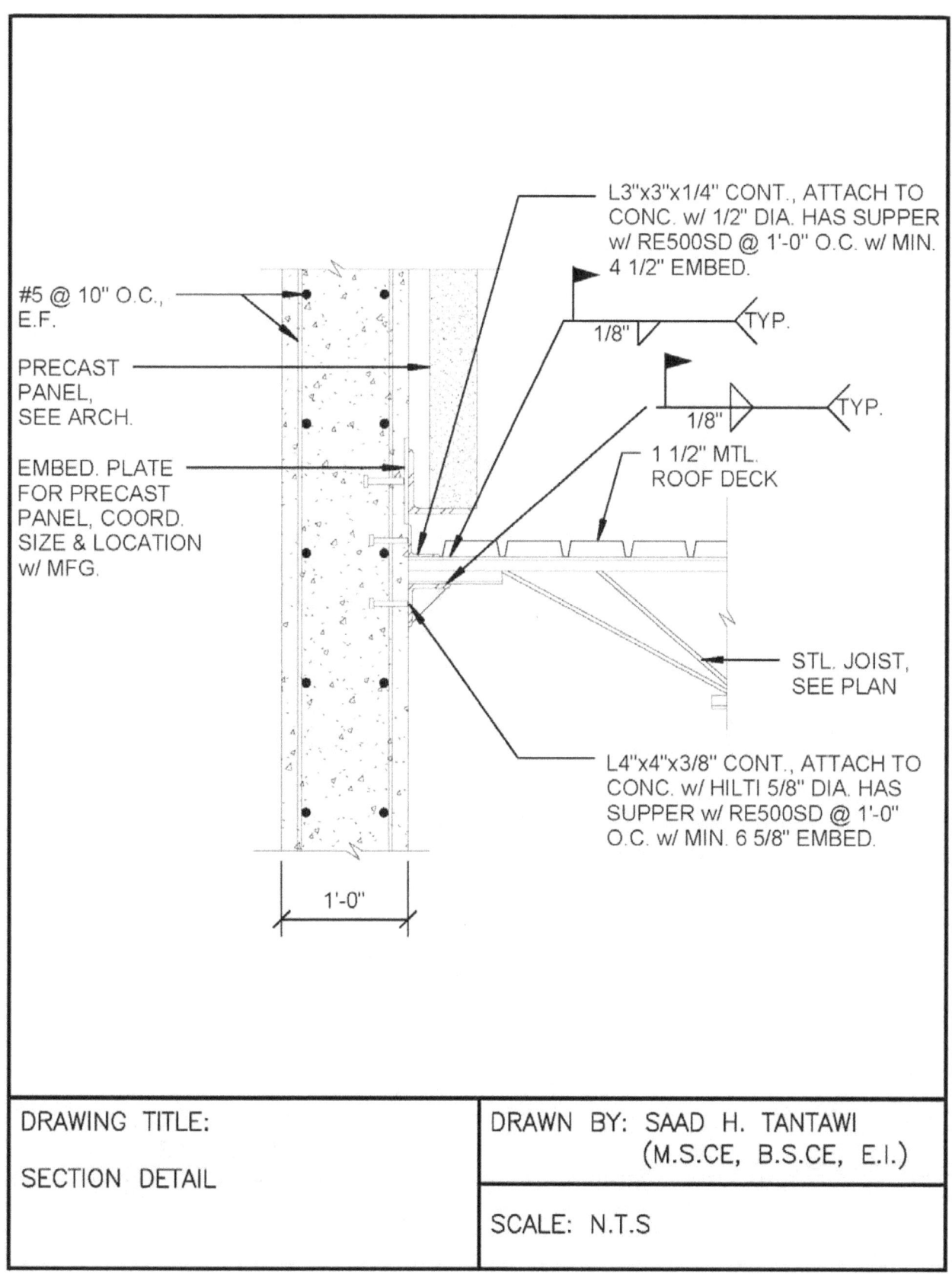

#5 @ 10" O.C.,
E.F.

PRECAST
PANEL,
SEE ARCH.

EMBED. PLATE
FOR PRECAST
PANEL, COORD.
SIZE & LOCATION
w/ MFG.

L3"x3"x1/4" CONT., ATTACH TO
CONC. w/ 1/2" DIA. HAS SUPPER
w/ RE500SD @ 1'-0" O.C. w/ MIN.
4 1/2" EMBED.

1/8"   TYP.

1/8"   TYP.

1 1/2" MTL.
ROOF DECK

STL. JOIST,
SEE PLAN

L4"x4"x3/8" CONT., ATTACH TO
CONC. w/ HILTI 5/8" DIA. HAS
SUPPER w/ RE500SD @ 1'-0"
O.C. w/ MIN. 6 5/8" EMBED.

1'-0"

DRAWING TITLE:	DRAWN BY: SAAD H. TANTAWI
	(M.S.CE, B.S.CE, E.I.)
SECTION DETAIL	
	SCALE: N.T.S

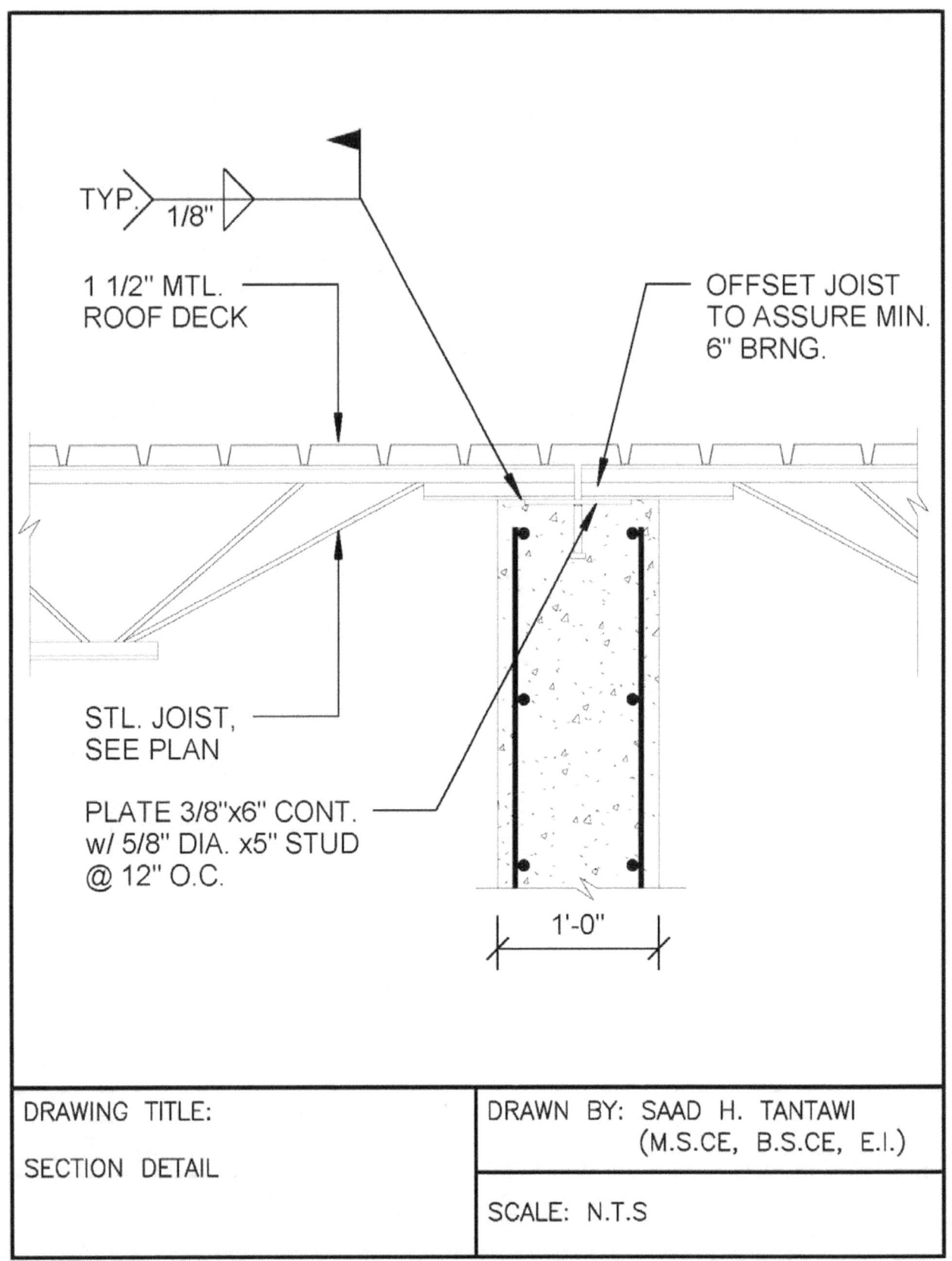

TYP. 1/8"

1 1/2" MTL.
ROOF DECK

OFFSET JOIST
TO ASSURE MIN.
6" BRNG.

STL. JOIST,
SEE PLAN

PLATE 3/8"x6" CONT.
w/ 5/8" DIA. x5" STUD
@ 12" O.C.

1'-0"

DRAWING TITLE:	DRAWN BY: SAAD H. TANTAWI
	(M.S.CE, B.S.CE, E.I.)
SECTION DETAIL	
	SCALE: N.T.S

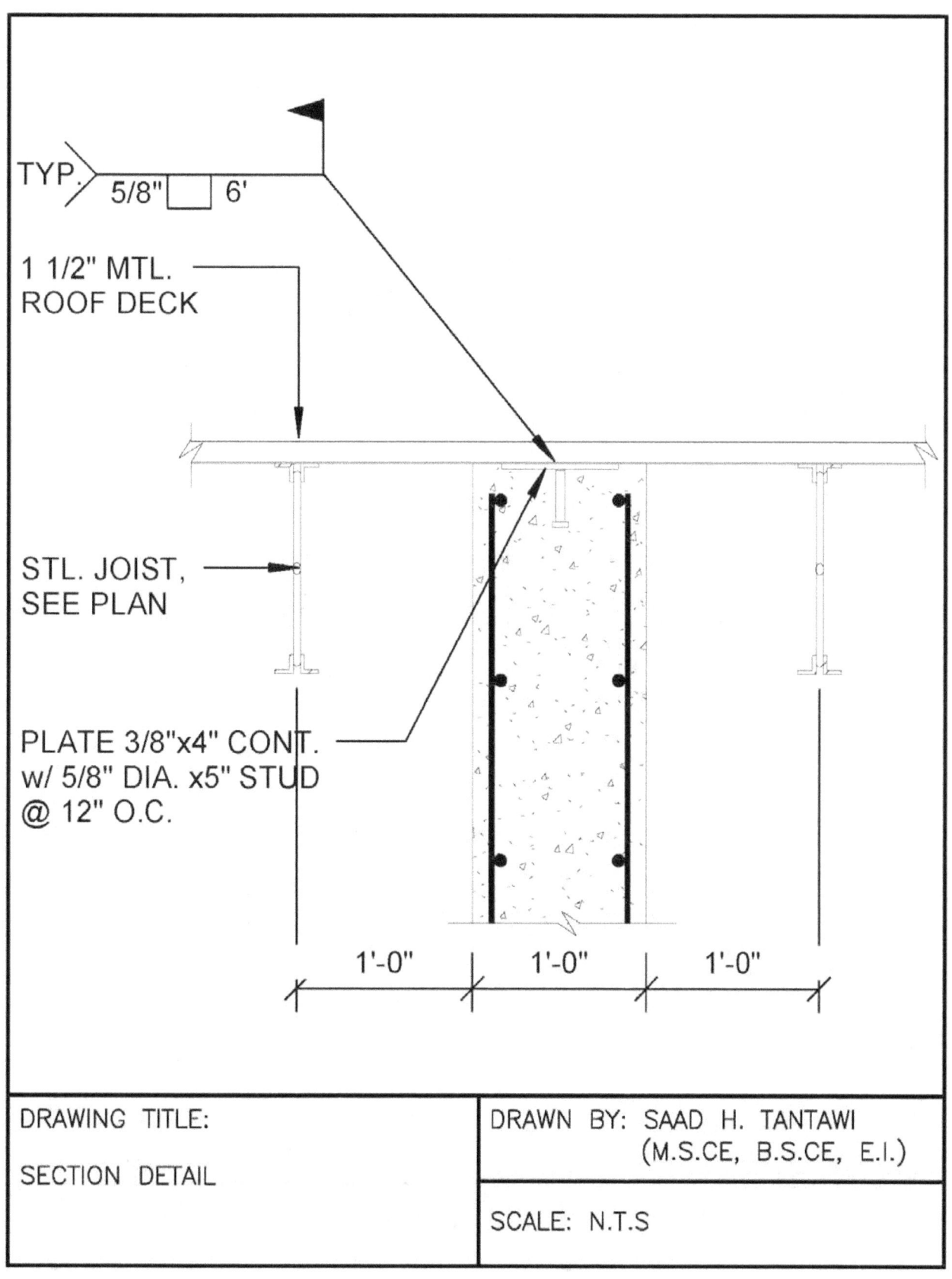

TYP. 5/8" 6'

1 1/2" MTL.
ROOF DECK

STL. JOIST,
SEE PLAN

PLATE 3/8"x4" CONT.
w/ 5/8" DIA. x5" STUD
@ 12" O.C.

1'-0"   1'-0"   1'-0"

DRAWING TITLE:	DRAWN BY: SAAD H. TANTAWI
	(M.S.CE, B.S.CE, E.I.)
SECTION DETAIL	
	SCALE: N.T.S

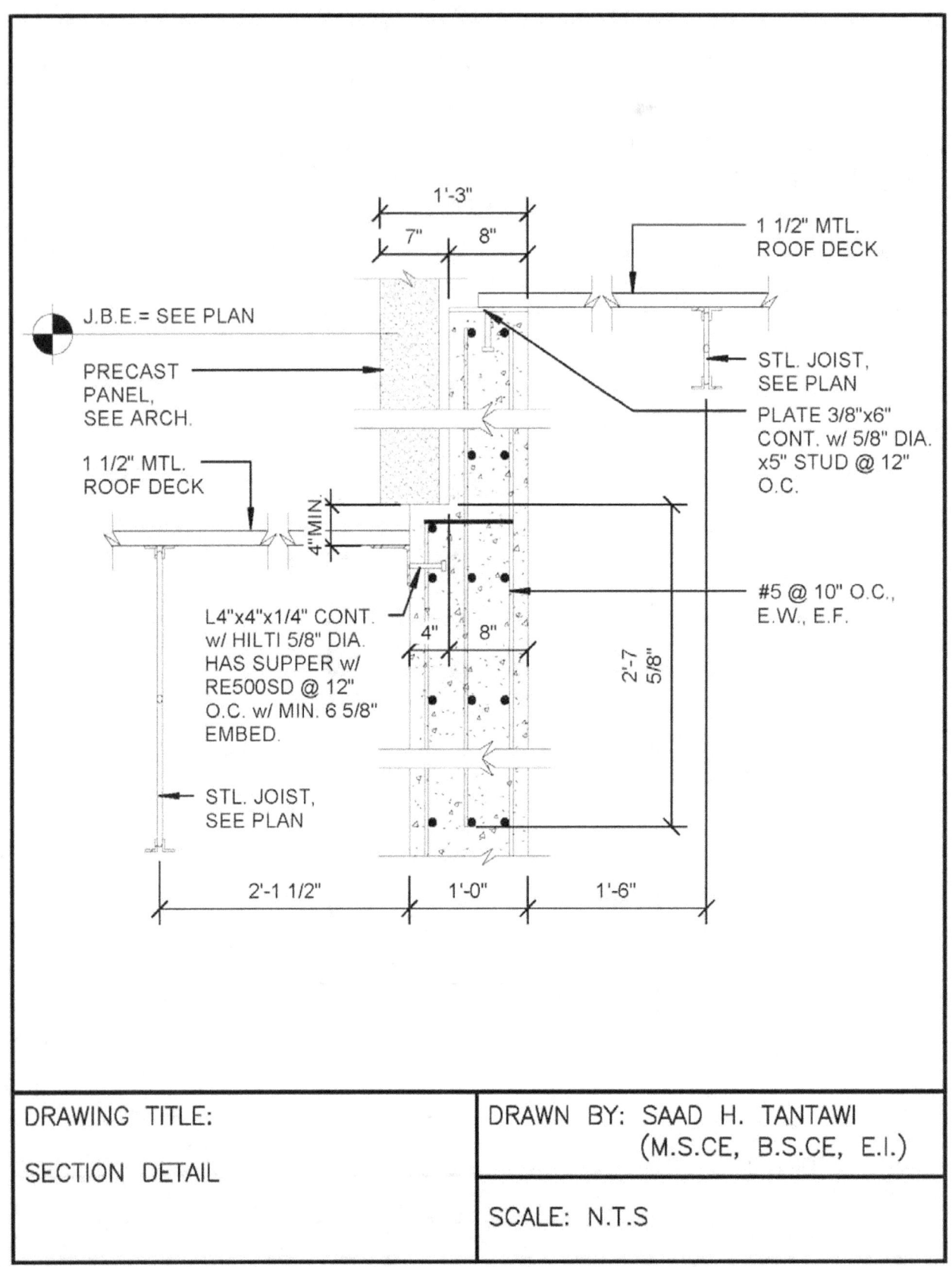

DRAWING TITLE:	DRAWN BY: SAAD H. TANTAWI
SECTION DETAIL	(M.S.CE, B.S.CE, E.I.)
	SCALE: N.T.S

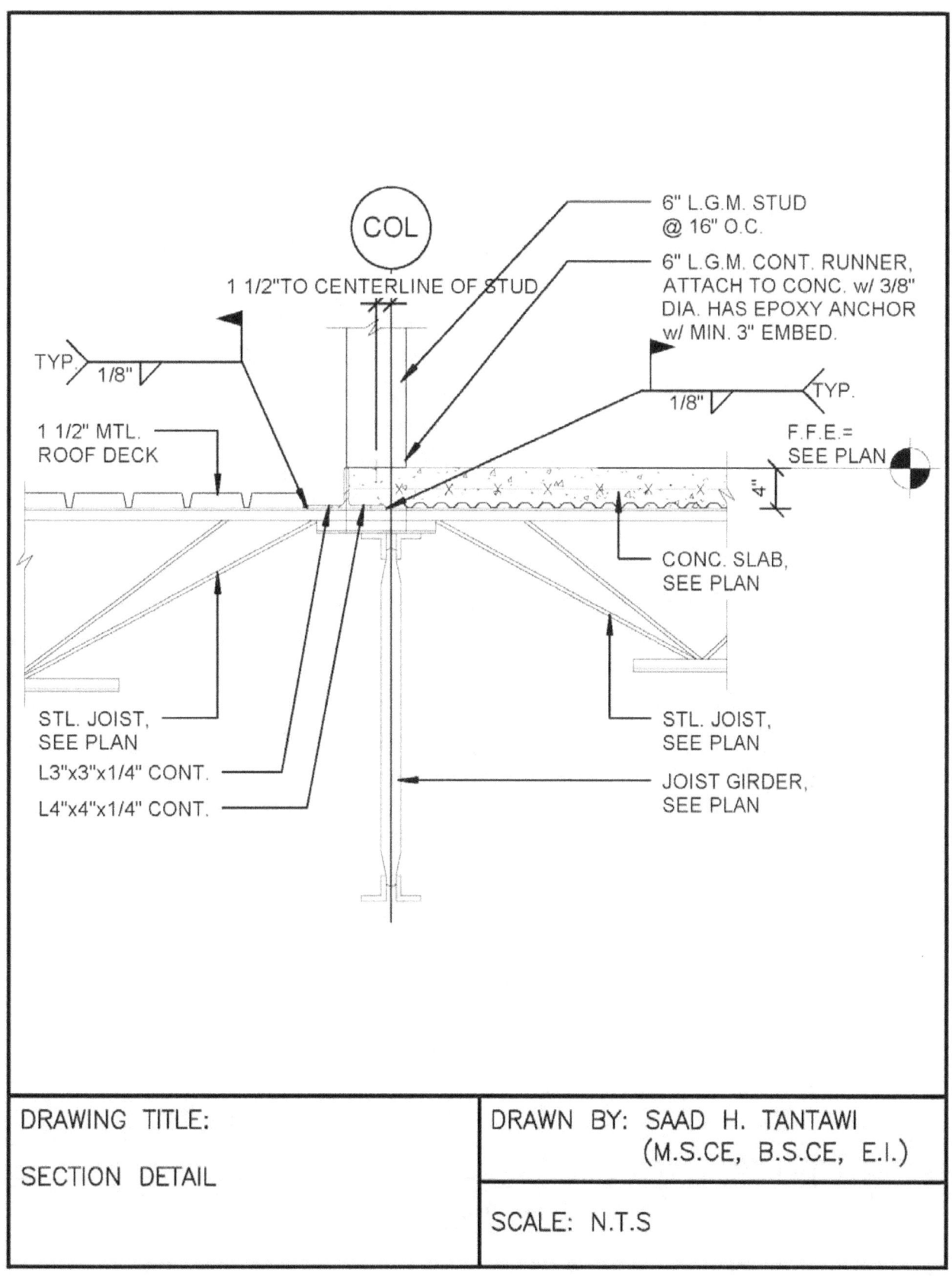

COL

1 1/2"TO CENTERLINE OF STUD

6" L.G.M. STUD
@ 16" O.C.

6" L.G.M. CONT. RUNNER,
ATTACH TO CONC. w/ 3/8"
DIA. HAS EPOXY ANCHOR
w/ MIN. 3" EMBED.

TYP.
1/8"

TYP.
1/8"

1 1/2" MTL.
ROOF DECK

F.F.E.=
SEE PLAN

4"

CONC. SLAB,
SEE PLAN

STL. JOIST,
SEE PLAN

L3"x3"x1/4" CONT.

L4"x4"x1/4" CONT.

STL. JOIST,
SEE PLAN

JOIST GIRDER,
SEE PLAN

DRAWING TITLE:	DRAWN BY: SAAD H. TANTAWI
	(M.S.CE, B.S.CE, E.I.)
SECTION DETAIL	
	SCALE: N.T.S

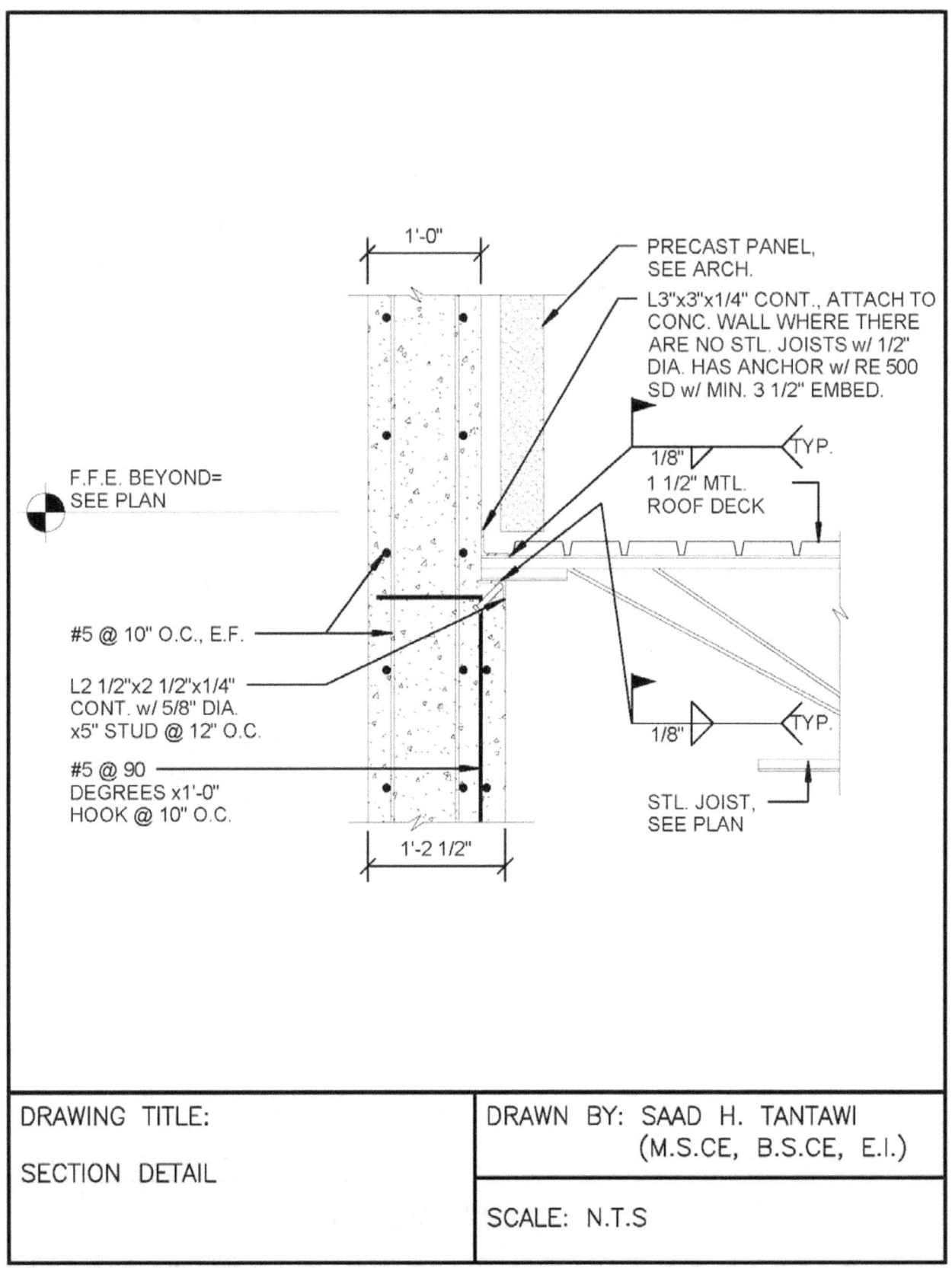

1'-0"

PRECAST PANEL,
SEE ARCH.

L3"x3"x1/4" CONT., ATTACH TO
CONC. WALL WHERE THERE
ARE NO STL. JOISTS w/ 1/2"
DIA. HAS ANCHOR w/ RE 500
SD w/ MIN. 3 1/2" EMBED.

TYP.

1/8"

1 1/2" MTL.
ROOF DECK

F.F.E. BEYOND=
SEE PLAN

#5 @ 10" O.C., E.F.

L2 1/2"x2 1/2"x1/4"
CONT. w/ 5/8" DIA.
x5" STUD @ 12" O.C.

#5 @ 90
DEGREES x1'-0"
HOOK @ 10" O.C.

1/8"

TYP.

STL. JOIST,
SEE PLAN

1'-2 1/2"

DRAWING TITLE:	DRAWN BY: SAAD H. TANTAWI
	(M.S.CE, B.S.CE, E.I.)
SECTION DETAIL	
	SCALE: N.T.S

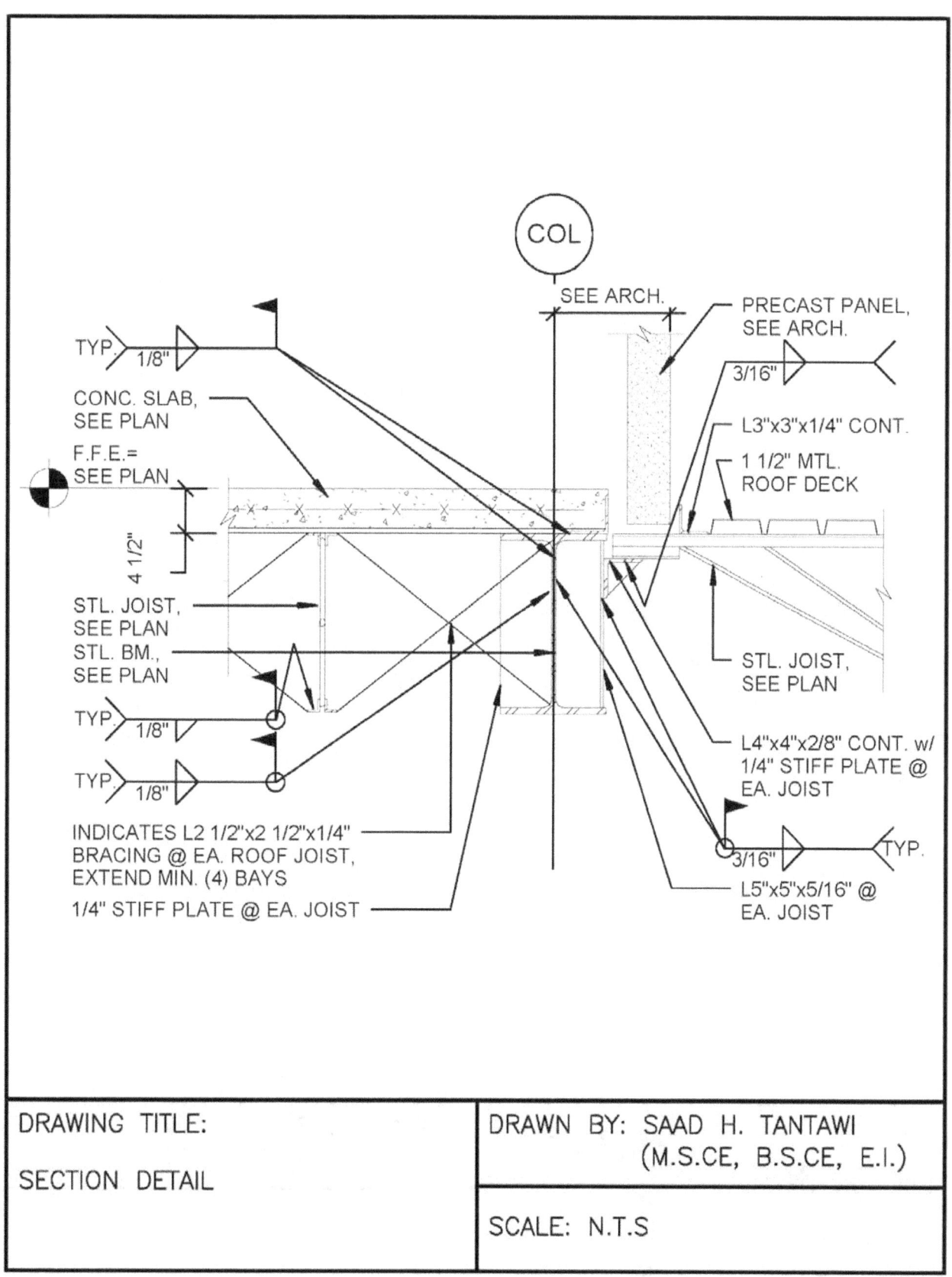

COL

SEE ARCH.

PRECAST PANEL,
SEE ARCH.

3/16"

TYP. 1/8"

CONC. SLAB,
SEE PLAN

F.F.E.=
SEE PLAN

4 1/2"

L3"x3"x1/4" CONT.

1 1/2" MTL.
ROOF DECK

STL. JOIST,
SEE PLAN

STL. JOIST,
SEE PLAN

STL. BM.,
SEE PLAN

TYP. 1/8"

TYP. 1/8"

L4"x4"x2/8" CONT. w/
1/4" STIFF PLATE @
EA. JOIST

INDICATES L2 1/2"x2 1/2"x1/4"
BRACING @ EA. ROOF JOIST,
EXTEND MIN. (4) BAYS

1/4" STIFF PLATE @ EA. JOIST

3/16"

TYP.

L5"x5"x5/16" @
EA. JOIST

DRAWING TITLE:	DRAWN BY: SAAD H. TANTAWI
	(M.S.CE, B.S.CE, E.I.)
SECTION DETAIL	
	SCALE: N.T.S

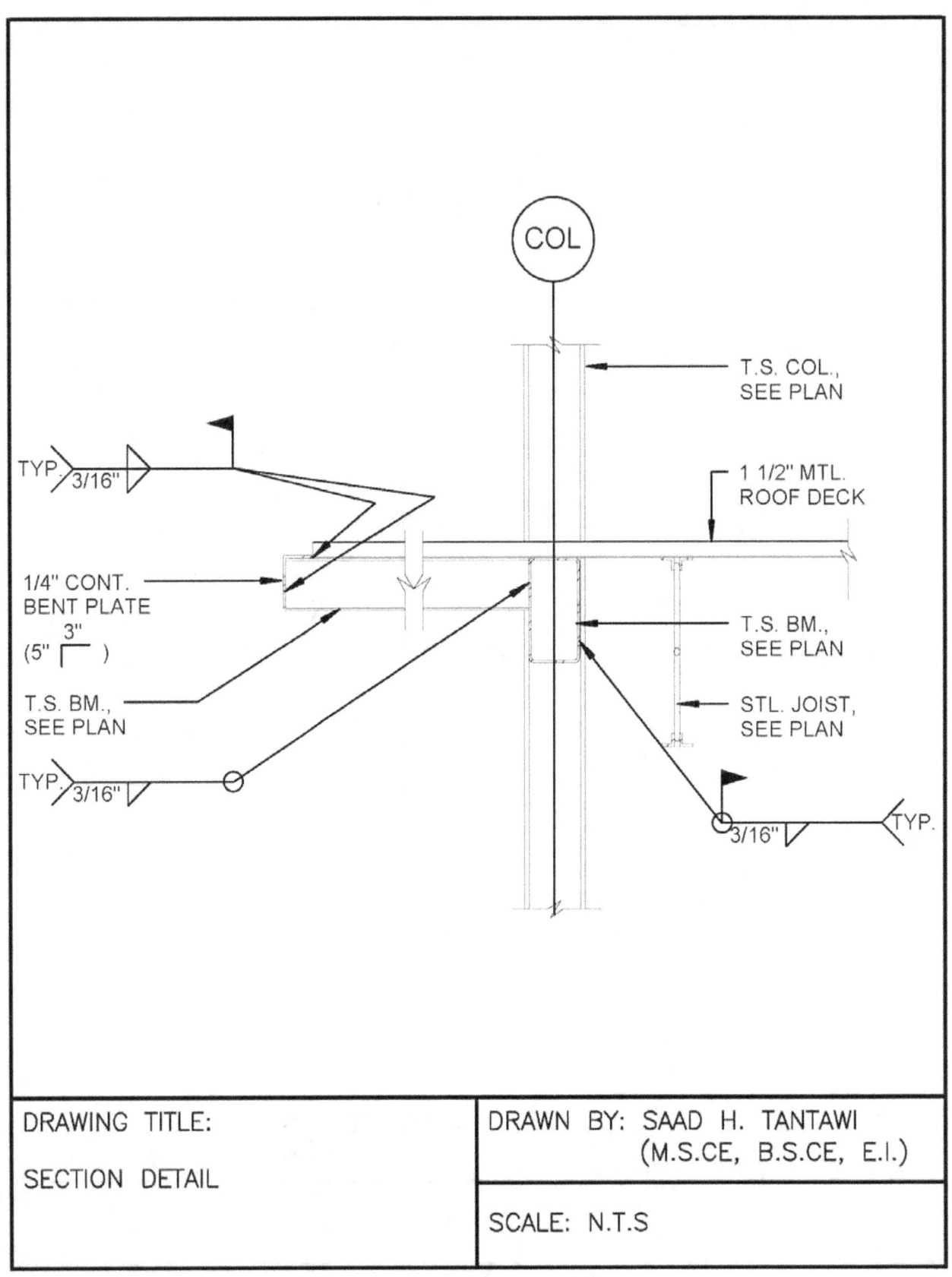

TYP. 3/16"

1/4" CONT.
BENT PLATE
3"
(5" ⌐)

T.S. BM.,
SEE PLAN

TYP. 3/16"

COL

T.S. COL.,
SEE PLAN

1 1/2" MTL.
ROOF DECK

T.S. BM.,
SEE PLAN

STL. JOIST,
SEE PLAN

3/16" TYP.

DRAWING TITLE:	DRAWN BY: SAAD H. TANTAWI
SECTION DETAIL	(M.S.CE, B.S.CE, E.I.)
	SCALE: N.T.S

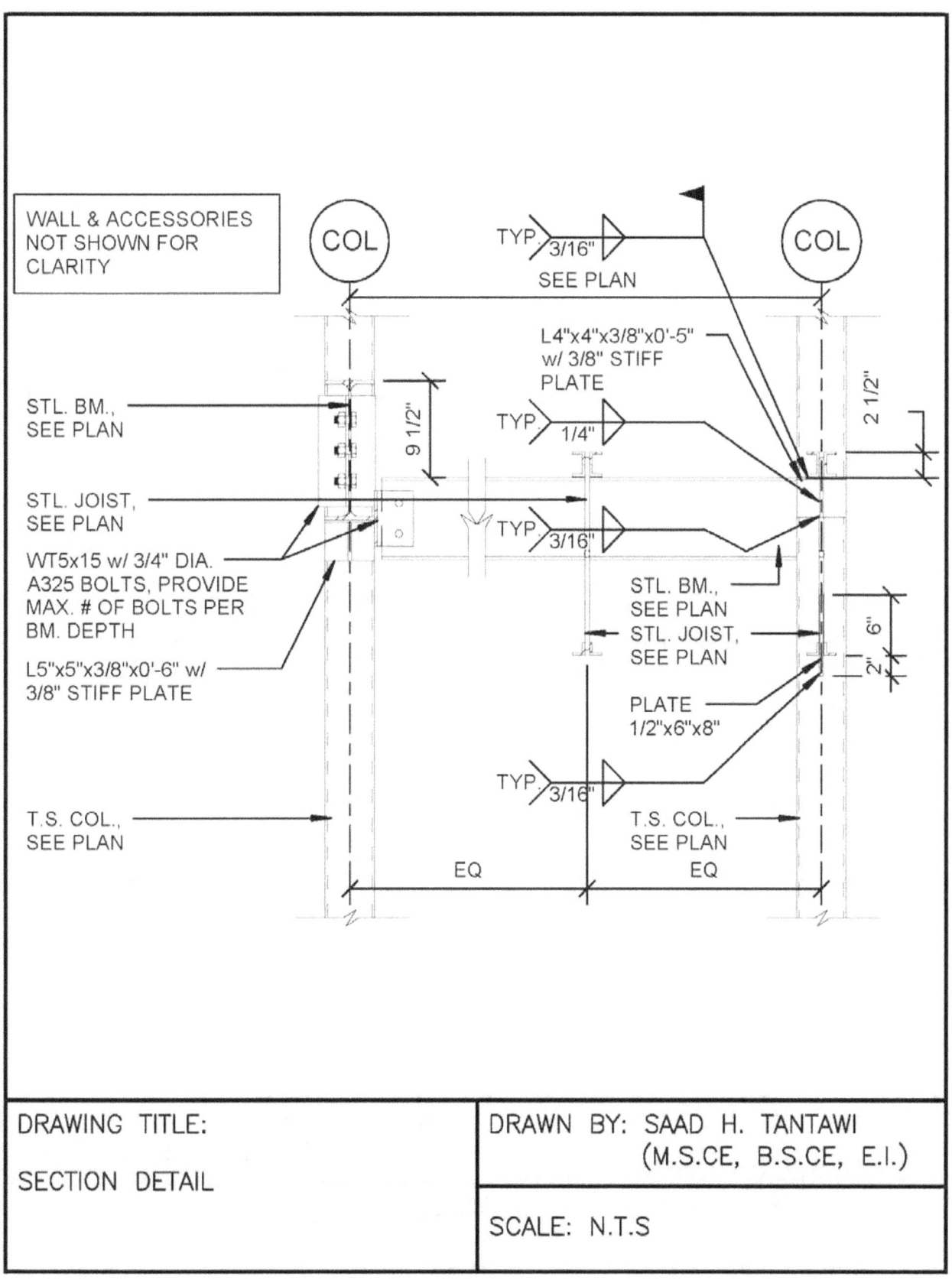

WALL & ACCESSORIES
NOT SHOWN FOR
CLARITY

COL

COL

TYP $\underset{3/16"}{\triangleright}$

SEE PLAN

L4"x4"x3/8"x0'-5"
w/ 3/8" STIFF
PLATE

2 1/2"

STL. BM.,
SEE PLAN

9 1/2"

TYP $\underset{1/4"}{\triangleright}$

STL. JOIST,
SEE PLAN

TYP $\underset{3/16"}{\triangleright}$

WT5x15 w/ 3/4" DIA.
A325 BOLTS, PROVIDE
MAX. # OF BOLTS PER
BM. DEPTH

STL. BM.,
SEE PLAN

STL. JOIST,
SEE PLAN

6"

L5"x5"x3/8"x0'-6" w/
3/8" STIFF PLATE

PLATE
1/2"x6"x8"

2"

TYP $\underset{3/16"}{\triangleright}$

T.S. COL.,
SEE PLAN

T.S. COL.,
SEE PLAN

EQ

EQ

DRAWING TITLE:	DRAWN BY: SAAD H. TANTAWI
	(M.S.CE, B.S.CE, E.I.)
SECTION DETAIL	
	SCALE: N.T.S

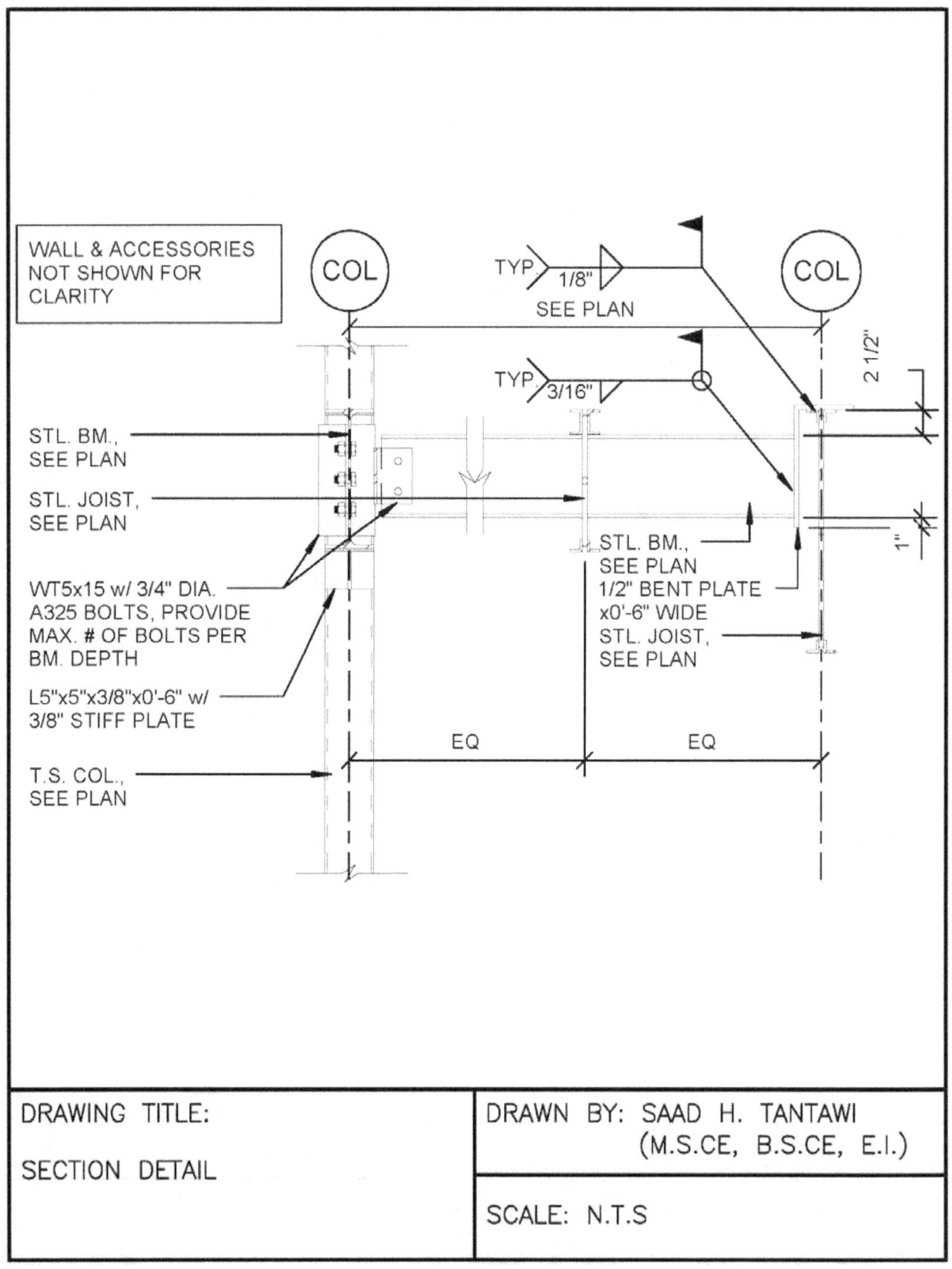

WALL & ACCESSORIES NOT SHOWN FOR CLARITY

COL

TYP. 1/8"
SEE PLAN

COL

TYP. 3/16"

2 1/2"

STL. BM., SEE PLAN

STL. JOIST, SEE PLAN

STL. BM., SEE PLAN
1/2" BENT PLATE x0'-6" WIDE

WT5x15 w/ 3/4" DIA. A325 BOLTS, PROVIDE MAX. # OF BOLTS PER BM. DEPTH

STL. JOIST, SEE PLAN

1"

L5"x5"x3/8"x0'-6" w/ 3/8" STIFF PLATE

T.S. COL., SEE PLAN

EQ

EQ

DRAWING TITLE:	DRAWN BY: SAAD H. TANTAWI
	(M.S.CE, B.S.CE, E.I.)
SECTION DETAIL	
	SCALE: N.T.S

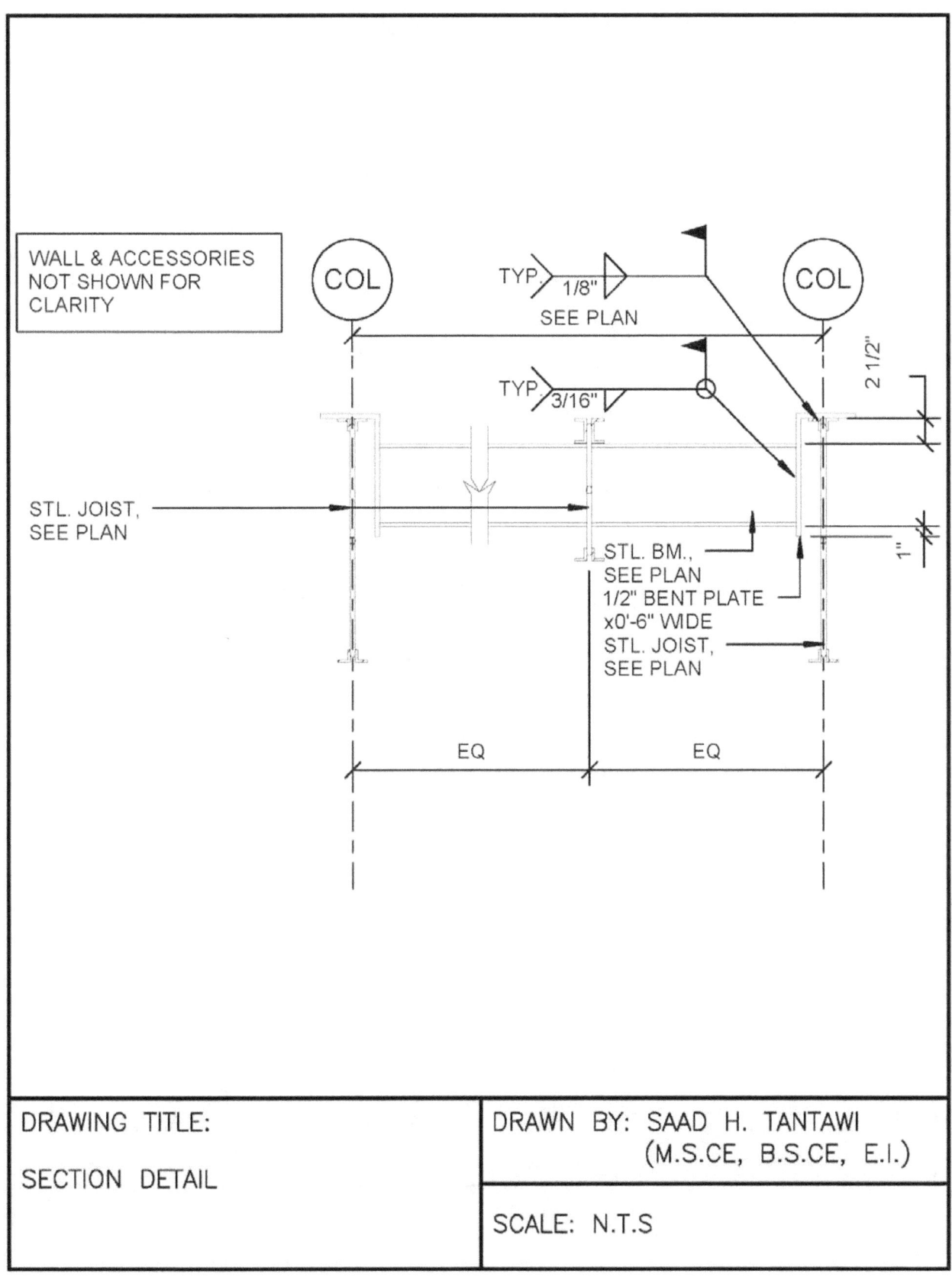

Saad Hasan Tantawi (M.S.CE, B.S.CE, E.I., A.M.ASCE)

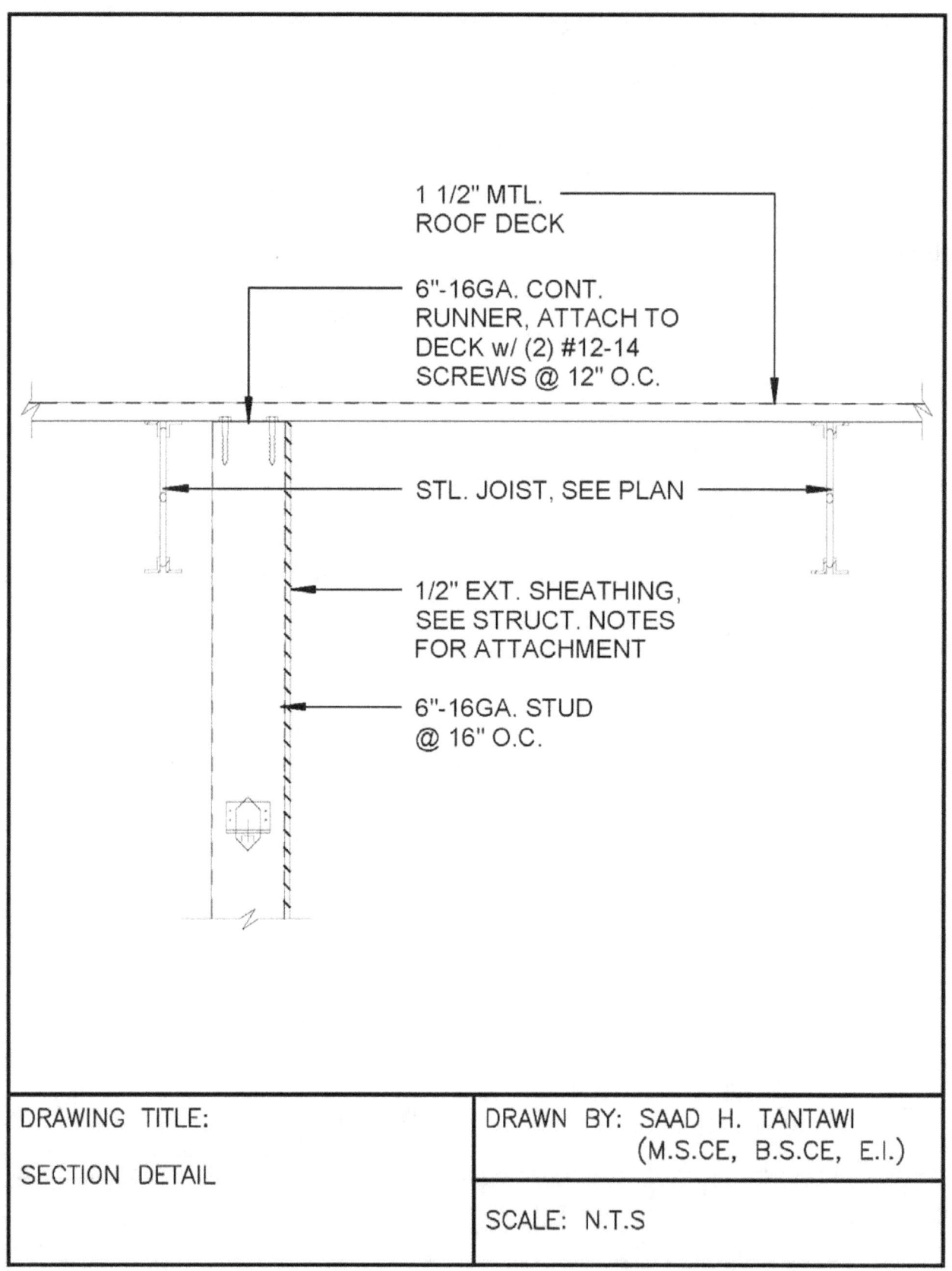

1 1/2" MTL.
ROOF DECK

6"-16GA. CONT.
RUNNER, ATTACH TO
DECK w/ (2) #12-14
SCREWS @ 12" O.C.

STL. JOIST, SEE PLAN

1/2" EXT. SHEATHING,
SEE STRUCT. NOTES
FOR ATTACHMENT

6"-16GA. STUD
@ 16" O.C.

DRAWING TITLE:	DRAWN BY: SAAD H. TANTAWI
	(M.S.CE, B.S.CE, E.I.)
SECTION DETAIL	
	SCALE: N.T.S

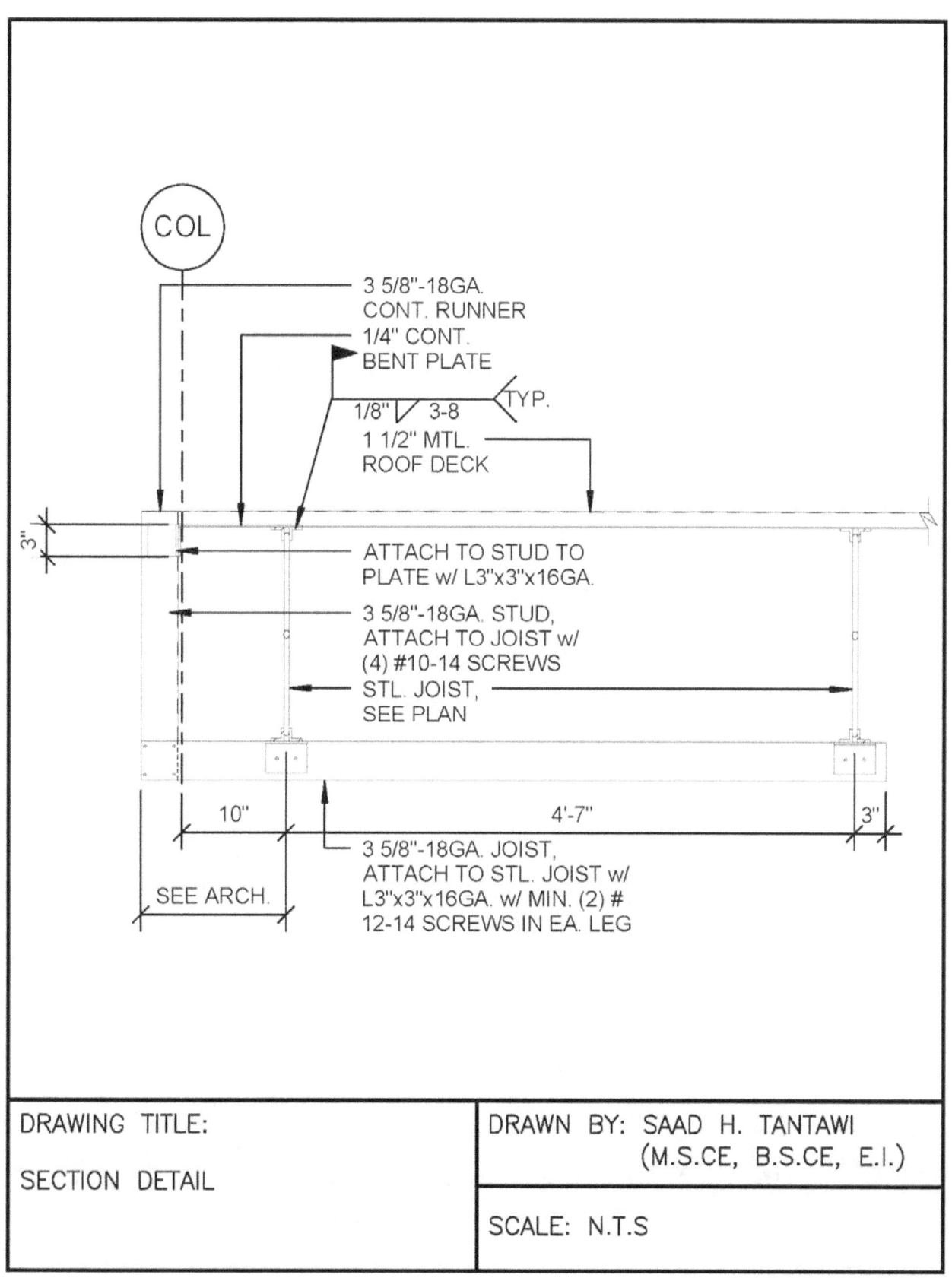

COL

3 5/8"-18GA.
CONT. RUNNER
1/4" CONT.
BENT PLATE

1/8" ◁ 3-8   TYP.

1 1/2" MTL.
ROOF DECK

ATTACH TO STUD TO
PLATE w/ L3"x3"x16GA.

3 5/8"-18GA. STUD,
ATTACH TO JOIST w/
(4) #10-14 SCREWS

STL. JOIST,
SEE PLAN

3"

10"   4'-7"   3"

SEE ARCH.

3 5/8"-18GA. JOIST,
ATTACH TO STL. JOIST w/
L3"x3"x16GA. w/ MIN. (2) #
12-14 SCREWS IN EA. LEG

DRAWING TITLE:

SECTION DETAIL

DRAWN BY: SAAD H. TANTAWI
(M.S.CE, B.S.CE, E.I.)

SCALE: N.T.S

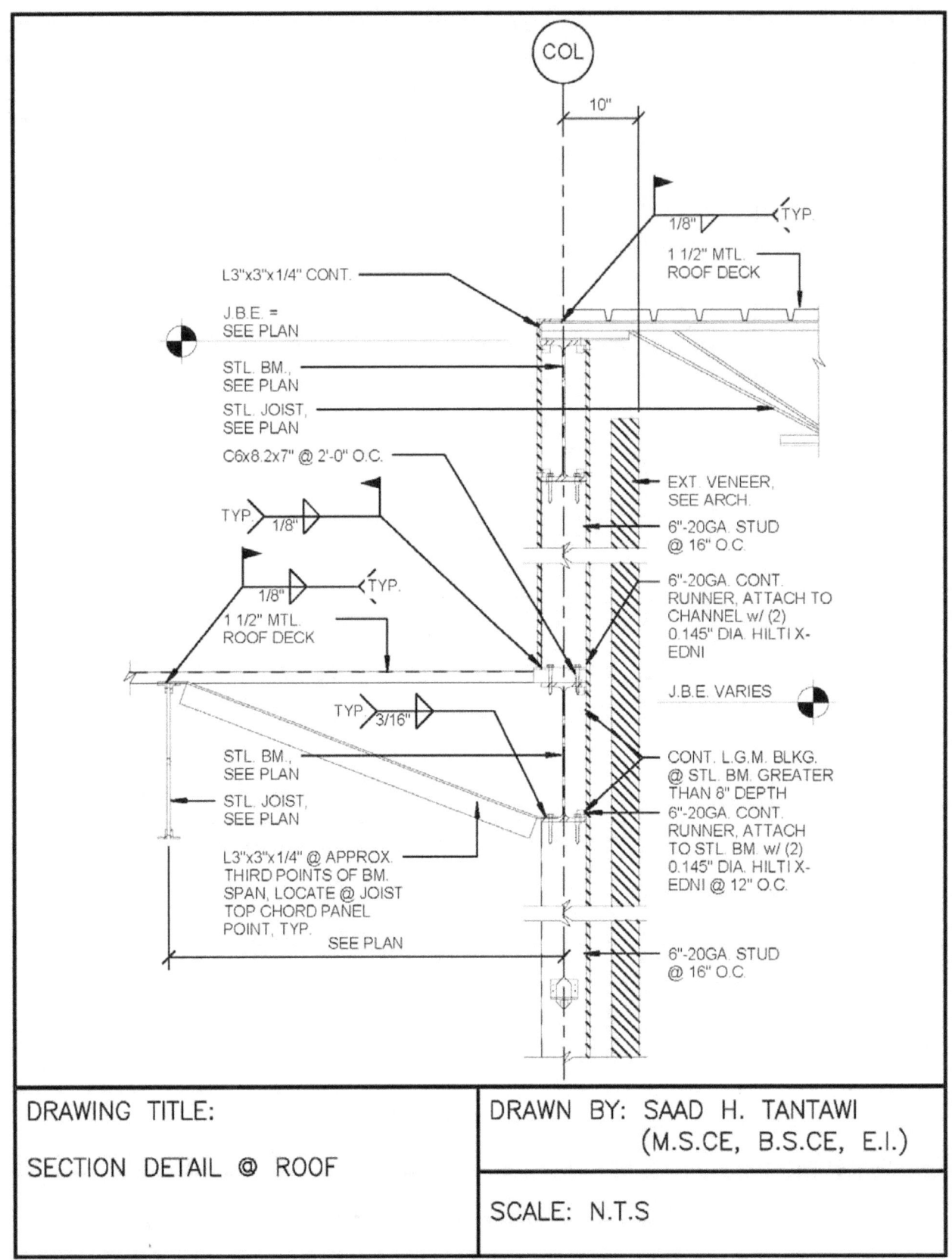

COL

10"

1/8" TYP.

1 1/2" MTL.
ROOF DECK

L3"x3"x1/4" CONT.

J.B.E. =
SEE PLAN

STL. BM.,
SEE PLAN

STL. JOIST,
SEE PLAN

C6x8.2x7" @ 2'-0" O.C.

EXT. VENEER,
SEE ARCH.

6"-20GA. STUD
@ 16" O.C.

6"-20GA. CONT.
RUNNER, ATTACH TO
CHANNEL w/ (2)
0.145" DIA. HILTI X-
EDNI

TYP 1/8"

1/8" TYP.

1 1/2" MTL.
ROOF DECK

J.B.E. VARIES

TYP 3/16"

STL. BM.,
SEE PLAN

STL. JOIST,
SEE PLAN

L3"x3"x1/4" @ APPROX.
THIRD POINTS OF BM.
SPAN, LOCATE @ JOIST
TOP CHORD PANEL
POINT, TYP.
SEE PLAN

CONT. L.G.M. BLKG.
@ STL. BM. GREATER
THAN 8" DEPTH

6"-20GA. CONT.
RUNNER, ATTACH
TO STL. BM. w/ (2)
0.145" DIA. HILTI X-
EDNI @ 12" O.C.

6"-20GA. STUD
@ 16" O.C.

DRAWING TITLE:	DRAWN BY: SAAD H. TANTAWI
SECTION DETAIL @ ROOF	(M.S.CE, B.S.CE, E.I.)
	SCALE: N.T.S

GLOSSARY

A/C- An abbreviation for air conditioner or air conditioning.

A/C Condenser- The outside fan unit of the Air Conditioning system. It removes the heat from the freon gas and "turns" the gas back into a liquid and pumps the liquid back to the coil in the furnace.

A/C Disconnect- The main electrical ON-OFF switch near the A/C Condenser.

Aerator- The round screened screw-on tip of a sink spout. It mixes water and air for a smooth flow.

Aggregate- A mixture of sand and stone and a major component of concrete.

Air space - The area between insulation facing and interior of exterior wall coverings. Normally a 1" air gap.

Anchor bolts- Bolts to secure a wooden sill plate to concrete , or masonry floor or wall.

Bat - A half-brick.

Batt - A section of fiber-glass or rock-wool insulation measuring 15 or 23 inches wide by four to eight feet long and various thickness'. Sometimes "faced" (meaning to have a paper covering on one side) or "unfaced" (without paper).

Batten- Narrow strips of wood used to cover joints or as decorative vertical members over plywood or wide boards.

Bay window- Any window space projecting outward from the walls of a building, either square or polygonal in plan.

Beam- A structural member transversely supporting a load. A structural member carrying building loads (weight) from one support to another. Sometimes called a "girder".

Bearing partition- A partition that supports any vertical load in addition to its own weight.

Bearing point- A point where a bearing or structural weight is concentrated and transferred to the foundation

Bearing wall- A wall that supports any vertical load in addition to its own weight.

Bearing header- (a) A beam placed perpendicular to joists and to which joists are nailed in framing for a chimney, stairway, or other opening. (b) A wood lintel. (c) The horizontal structural member over an opening (for example over a door or window).

Bedrock- A subsurface layer of earth that is suitable to support a structure.

Blankets- Fiber-glass or rock-wool insulation that comes in long rolls 15 or 23 inches wide.

Blocked (door blocking)- Wood shims used between the door frame and the vertical structural wall framing members.

Blocked (rafters)- Short "2 by 4's" used to keep rafters from twisting, and installed at the ends and at mid-span.

Blocking- Small wood pieces to brace framing members or to provide a nailing base for gypsum board or paneling.

Block out- To install a box or barrier within a foundation wall to prevent the concrete from entering an area. For example, foundation walls are sometimes "blocked" in order for mechanical pipes to pass through the wall, to install a crawl space door, and to depress the concrete at a garage door.

Butt hinge- The most common type. One leaf attaches to the door's edge, the other to its jamb.

Butt joint- The junction where the ends of two timbers meet, and also where sheets of drywall meet on the 4 foot edge. To place materials end-to-end or end-to-edge without overlapping

Caisson- A 10" or 12" diameter hole drilled into the earth and embedded into bedrock 3 - 4 feet. The structural support for a type of foundation wall, porch, patio, monopost, or other structure. Two or more "sticks" of reinforcing bars (rebar) are inserted into and run the full length of the hole and concrete is poured into the caisson hole

Cantilever- An overhang. Where one floor extends beyond and over a foundation wall. For example at a fireplace location or bay window cantilever. Normally, not extending over 2 feet.

Cantilevered void- Foundation void material used in unusually expansive soils conditions. This void is "trapezoid" shaped and has vertical sides of 6" and 4" respectively.

Cap- The upper member of a column, pilaster, door cornice, molding, or fireplace.

Cap flashing- The portion of the flashing attached to a vertical surface to prevent water from migrating behind the base flashing.

Capital- The principal part of a loan, i.e. the original amount borrowed.

Capital and interest- A repayment loan and the most conventional form of home loan. The borrower pays an amount each month to cover the amount borrowed (or capital or principal) *plus* the interest charged on capital.

Capped rate- The mortgage interest rate will not exceed a specified value during a certain period of time, but it will fluctuate up and down below that level.

Casement- Frames of wood or metal enclosing part (or all) of a window sash. May be opened by means of hinges affixed to the vertical edges.

Casement Window- A window with hinges on one of the vertical sides and swings open like a normal door

Casing- Wood trim molding installed around a door or window opening.

Caulking- (1) A flexible material used to seal a gap between two surfaces e.g. between pieces of siding or the corners in tub walls. (2) To fill a joint with mastic or asphalt plastic cement to prevent leaks.

Collar- Preformed flange placed over a vent pipe to seal the roofing above the vent pipe opening. Also called a vent sleeve.

Collar beam- Nominal 1- or 2-inch-thick members connecting opposite roof rafters. They serve to stiffen the roof structure.

Column- A vertical structural compression member which supports loads.

Combustion air- The duct work installed to bring fresh, outside air to the furnace and/or hot water heater. Normally 2 separate supplies of air are brought in: One high and One low.

Combustion chamber- The part of a boiler, furnace or woodstove where the burn occurs; normally lined with firebrick or molded or sprayed insulation.

Compression web- A member of a truss system which connects the bottom and top chords and which provides downward support.

Compressor- A mechanical device that pressurizes a gas in order to turn it into a liquid, thereby allowing heat to be removed or added. A compressor is the main component of conventional heat pumps and air conditioners. In an air conditioning system, the compressor normally sits outside and has a large fan (to remove heat).

Concrete- The mixture of Portland cement, sand, gravel, and water. Used to make garage and basement floors, sidewalks, patios, foundation walls, etc. It is commonly reinforced with steel rods (rebar) or wire screening (mesh).

Concrete block - A hollow concrete 'brick' often 8" x 8" x 16" in size.

Concrete board - A panel made out of concrete and fiberglass usually used as a tile backing material.

Condensation- Beads or drops of water (and frequently frost in extremely cold weather) that accumulate on the inside of the exterior covering of a building. Use of louvers or attic ventilators will reduce moisture condensation in attics. A vapor barrier under the gypsum lath or dry wall on exposed walls will reduce condensation.

Condensing unit - The outdoor component of a cooling system. It includes a compressor and condensing coil designed to give off heat.

Control joint- Tooled, straight grooves made on concrete floors to "control" where the concrete should crack

Convection- Currents created by heating air, which then rises and pulls cooler air behind it. Also see radiation.

Dormer- An opening in a sloping roof, the framing of which projects out to form a vertical wall suitable for windows or other openings.

Double glass- Window or door in which two panes of glass are used with a sealed air space between. Also known as Insulating Glass.

Double hung window- A window with two vertically sliding sashes, both of which can move up and down.

Down payment- The difference between the sales price and the mortgage amount. A downpayment is usually paid at closing.

Downspout- A pipe, usually of metal, for carrying rainwater down from the roof's horizontal gutters.

Drain tile- A perforated, corrugated plastic pipe laid at the bottom of the foundation wall and used to drain excess water away from the foundation. It prevents ground water from seeping through the foundation wall. Sometimes called perimeter drain.

Draw- The amount of progress billings on a contract that is currently available to a contractor under a contract with a fixed payment schedule.

Drip- (a) A member of a cornice or other horizontal exterior finish course that has a projection beyond the other parts for throwing off water.(b) A groove in the underside of a sill or drip cap to cause water to drop off on the outer edge instead of drawing back and running down the face of the building.

Drip cap- A molding or metal flashing placed on the exterior topside of a door or window frame to cause water to drip beyond the outside of the frame.

Dry in- To install the black roofing felt (tar paper) on the roof.

Drywall (or Gypsum Wallboard (GWB), Sheet rock or Plasterboard)- Wall board or gypsum- A manufactured panel made out of gypsum plaster and encased in a thin cardboard. Usually 1/2" thick and 4' x 8' or 4' x 12' in size. The panels are nailed or screwed onto the framing and the joints are taped and covered with a 'joint compound'. 'Green board' type drywall has a greater resistance to moisture than regular (white) plasterboard and is used in bathrooms and other "wet areas".

Ducts- The heating system. Usually round or rectangular metal pipes installed for distributing warm (or cold) air from the furnace to rooms in the home. Also a tunnel made of galvanized metal or rigid fiberglass, which carries air from the heater or ventilation opening to the rooms in a building.

Expansion joint- Fibrous material (@1/2" thick) installed in and around a concrete slab to permit it to move up and down (seasonally) along the non-moving foundation wall.

Expansive soils- Earth that swells and contracts depending on the amount of water that is present. ("Betonite" is an expansive soil).

Exposed aggregate finish- A method of finishing concrete which washes the cement/sand mixture off the top layer of the aggregate - usually gravel. Often used in driveways, patios and other exterior surfaces.

Frost line- The depth of frost penetration in soil and/or the depth at which the earth will freeze and swell. This depth varies in different parts of the country.

Furring strips- Strips of wood, often 1 X 2 and used to shim out and provide a level fastening surface for a wall or ceiling.

Gusset- A flat wood, plywood, or similar type member used to provide a connection at the intersection of wood members. Most commonly used at joints of wood trusses. They are fastened by nails, screws, bolts, or adhesives.

Gutter- A shallow channel or conduit of metal or wood set below and along the (fascia) eaves of a house to catch and carry off rainwater from the roof.

Gyp board- Drywall. Wall board or gypsum- A panel (normally 4' X 8', 10', 12', or 16')made with a core of Gypsum (chalk-like) rock, which covers interior walls and ceilings.

Gypsum plaster- Gypsum formulated to be used with the addition of sand and water for base-coat plaster.

H Clip- Small metal clips formed like an "H" that fits at the joints of two plywood (or wafer board) sheets to stiffen the joint. Normally used on the roof sheeting.

Hardware- All of the "metal" fittings that go into the home when it is near completion. For example, door knobs, towel bars, handrail brackets, closet rods, house numbers, door closers, etc. The Interior Trim Carpenter installs the "hardware".

Haunch- An extension, knee like protrusion of the foundation wall that a concrete porch or patio will rest upon for support.

Hazard insurance - Protection against damage caused by fire, windstorms, or other common hazards. Many lenders require borrowers to carry it in an amount at least equal to the mortgage.

Header- (a) A beam placed perpendicular to joists and to which joists are nailed inframing for a chimney, stairway, or other opening. (b) A wood lintel. (c) The horizontal structural member over an opening (for example over a door or window).

Hearth- The fireproof area directly in front of a fireplace. The inner or outer floor of a fireplace, usually made of brick, tile, or stone.

Heating load- The amount of heating required to keep a building at a specified temperature during the winter, usually 65° F, regardless of outside temperature.

Heat meter- An electrical municipal inspection of the electric meter breaker panel box.

Heat pump- A mechanical device which uses compression and decompression of gas to heat and/or cool a house.

Heat Rough- Work performed by the Heating Contractor after the stairs and interior walls are built. This includes installing all duct work and flue pipes. Sometimes, the furnace and fireplaces are installed at this stage of construction.

Heat Trim- Work done by the Heating Contractor to get the home ready for the municipal Final Heat Inspection. This includes venting the hot water heater, installing all vent grills, registers, air conditioning services, turning on the furnace, installing thermostats, venting ranges and hoods, and all other heat related work.

Heel cut- A notch cut in the end of a rafter to permit it to fit flat on a wall and on the top, doubled, exterior wall plate.

Highlights- A light spot, area, or streak on a painted surface.

Hip- A roof with four sloping sides. The external angle formed by the meeting of two sloping sides of a roof.

Hip roof- A roof that rises by inclined planes from all four sides of a building.

Home run (electrical)- The electrical cable that carries power from the main circuit breaker panel to the first electrical box, plug, or switch in the circuit.

Honey combs- The appearance concrete makes when rocks in the concrete are visible and where there are void areas in the foundation wall, especially around concrete foundation windows.

Hose bib- An exterior water faucet (sill cock).

Hot wire- The wire that carries electrical energy to a receptacle or other device—in contrast to a neutral, which carries electricity away again. Normally the black wire. Also see ground.

Humidifier- An appliance normally attached to the furnace, or portable unit device designed to increase the humidity within a room or a house by means of the discharge of water vapor.

Hurricane clip- Metal straps that are nailed and secure the roof rafters and trusses to the top horizontal wall plate. Sometimes called a Teco clip.

I-beam- A steel beam with a cross section resembling the letter **I**. It is used for long spans as basement beams or over wide wall openings, such as a double garage door, when wall and roof loads bear down on the opening.

I-joist- Manufactured structural building component resembling the letter "I". Used as floor joists and rafters. I-joists include two key parts: **flanges** and **webs**. The **flange** of the I joist may be made of laminated veneer lumber or dimensional lumber, usually formed into a 1 ½" width. The **web** or center of the I-joist is commonly made of plywood or oriented strand board (OSB). Large holes can be cut in the web to accommodate duct work and plumbing waste lines. I-joists are available in lengths up to 60 feet long

Incandescent lamp- A lamp employing an electrically charged metal filament that glows at white heat. A typical light bulb.

Index- The interest rate or adjustment standard that determines the changes in monthly payments for an adjustable rate loan.

Infiltration- The passage of air from indoors to outdoors and vice versa; term is usually associated with drafts from cracks, seams or holes in buildings.

Inside corner- The point at which two walls form an internal angle, as in the corner of a room.

Insulating glass- Window or door in which two panes of glass are used with a sealed air space between. Also known as **Double glass**.

Insulation board, rigid- A structural building board made of coarse wood or cane fiber in ½- and 25/32-inch thickness. It can be obtained in various size sheets and densities.

Insulation- Any material high in resistance to heat transmission that, when placed in the walls, ceiling, or floors of a structure, and will reduce the rate of heat flow.

Interest - The cost paid to a lender for borrowed money.

Interior finish- Material used to cover the interior framed areas of walls and ceilings.

Laminated shingles - Shingles that have added dimensionality because of extra layers or tabs, giving a shake-like appearance. May also be called "architectural shingles" or "three-dimensional shingles."

Laminating- Bonding together two or more layers of materials.

Landing- A platform between flights of stairs or at the termination of a flight of stairs. Often used when stairs change direction. Normally no less than 3 ft. X 3 ft. square.

Lap- To cover the surface of one shingle or roll with another.

Latch- A beveled metal tongue operated by a spring-loaded knob or lever. The tongue's bevel lets you close the door and engage the locking mechanism, if any, without using a key. Contrasts with dead bolt.

Lateral (electric, gas, telephone, sewer and water)- The underground trench and related services (i.e., electric, gas, telephone, sewer and water lines) that will be buried within the trench.

Lath- A building material of narrow wood, metal, gypsum, or insulating board that is fastened to the frame of a building to act as a base for plaster, shingles, or tiles.

Lattice- An open framework of criss-crossed wood or metal strips that form regular, patterned spaces.

Ledger (for a Structural Floor)- The wooden perimeter frame lumber member that bolts onto the face of a foundation wall and supports the wood structural floor.

Ledger strip- A strip of lumber nailed along the bottom of the side of a girder on which joists rest.

Leech field- A method used to treat/dispose of sewage in rural areas not accessible to a municipal sewer system. Sewage is permitted to be filtered and eventually discharged into a section of the lot called a leech field.

Let-in brace- Nominal 1 inch-thick boards applied into notched studs diagonally. Also, an "L" shaped, long (@ 10') metal strap that are installed by the framer at the rough stage to give support to an exterior wall or wall corner.

Level- True horizontal. Also a tool used to determine level.

Level Payment Mortgage- A mortgage with identical monthly payments over the life of the loan.

Lien- An encumbrance that usually makes real or personal property the security for payment of a debt or discharge of an obligation.

Light- Space in a window sash for a single pane of glass. Also, a pane of glass.

Limit switch- A safety control that automatically shuts off a furnace if it gets too hot. Most also control blower cycles.

Lineal foot- A unit of measure for lumber equal to 1 inch thick by 12 inches wide by 12 inches long. Examples: 1" x 12" x 16' = 16 board feet, 2" x 12" x 16' = 32 board feet.

Lintel- A horizontal structural member that supports the load over an opening such as a door or window.

Load bearing wall- Includes all exterior walls and any interior wall that is aligned above a support beam or girder. Normally, any wall that has a double horizontal top plate.

Loan- The amount to be borrowed.

Loan to value ratio- The ratio of the loan amount to the property valuation and expressed as a percentage. E.g. if a borrower is seeking a loan of $200,000 on a property worth $400,000 it has a 50% loan to value rate. If the loan were $300,000, the LTV would be 75%. The higher the loan to value, the greater the lender's perceived risk. Loans above normal lending LTV ratios may require additional security.

Lookout- A short wood bracket or cantilever that supports an overhang portion of a roof.

Louver- A vented opening into the home that has a series of horizontal slats and arranged to permit ventilation but to exclude rain, snow, light, insects, or other living creatures.

Lumens- Unit of measure for total light output. The amount of light falling on a surface of one square foot.

Single hung window- A window with one vertically sliding sash or window vent.

Skylight- A more or less horizontal window located on the roof of a building.

Slab, concrete- Concrete pavement, i.e. driveways, garages, and basement floors.

Slab, door- A rectangular door without hinges or frame.

Slab on grade- A type of foundation with a concrete floor which is placed directly on the soil. The edge of the slab is usually thicker and acts as the footing for the walls.

Slag- Concrete cement that sometimes covers the vertical face of the foundation void material.

Sleeper- Usually, a wood member embedded in concrete, as in a floor, that serves to support and to fasten the subfloor or flooring.

Sleeve(s)- Pipe installed under the concrete driveway or sidewalk, and that will be used later to run sprinkler pipe or low voltage wire.

Slope- The incline angle of a roof surface, given as a ratio of the rise (in inches) to the run (in feet). See also pitch.

Slump- The "wetness" of concrete. A 3 inch slump is dryer and stiffer than a 5 inch slump.

Soffit- The area below the eaves and overhangs. The underside where the roof overhangs the walls. Usually the underside of an overhanging cornice.

Soil pipe- A large pipe that carries liquid and solid wastes to a sewer or septic tank.

Soil stack- A plumbing vent pipe that penetrates the roof.

Sole plate- The bottom, horizontal framing member of a wall that's attached to the floor sheeting and vertical wall studs.

Solid bridging- A solid member placed between adjacent floor joists near the center of the span to prevent joists or rafters from twisting.

Sonotube- Round, large cardboard tubes designed to hold wet concrete in place until it hardens.

Sound attenuation- Sound proofing a wall or subfloor, generally with fiberglass insulation.

Space heat- Heat supplied to the living space, for example, to a room or the living area of a building.

Spacing- The distance between individual members or shingles in building construction.

Span- The clear distance that a framing member carries a load without support between structural supports. The horizontal distance from eaves to eaves.

Spec home- A house built before it is sold. The builder speculates that he can sell it at a profit.

Specifications or Specs- A narrative list of materials, methods, model numbers, colors, allowances, and other details which supplement the information contained in the blue prints. Written elaboration in specific detail about construction materials and methods. Written to supplement working drawings.

Splash block- Portable concrete (or vinyl) channel generally placed beneath an exterior sill cock (water faucet) or downspout in order to receive roof drainage from downspouts and to divert it away from the building.

Square- A unit of measure-100 square feet-usually applied to roofing and siding material. Also, a situation that exists when two elements are at right angles to each other. Also a tool for checking this.

Square-tab shingles- Shingles on which tabs are all the same size and exposure.

Squeegie- Fine pea gravel used to grade a floor (normally before concrete is placed).

Stack (trusses)- To position trusses on the walls in their correct location.

Standard practices of the trade(s)- One of the more common basic and minimum construction standards. This is another way of saying that the work should be done in the way it is normally done by the average professional in the field.

Starter strip- Asphalt roofing applied at the eaves that provides protection by filling in the spaces under the cutouts and joints of the first course of shingles.

Stair carriage or stringer- Supporting member for stair treads. Usually a 2 X 12 inch plank notched to receive the treads; sometimes called a "rough horse."

Stair landing- A platform between flights of stairs or at the termination of a flight of stairs. Often used when stairs change direction. Normally no less than 3 ft. X 3 ft. square.

Stair rise- The vertical distance from stair tread to stair tread (and not to exceed 7 ½").

Static vent- A vent that does not include a fan.

STC (Sound Transmission Class)- The measure of sound stopping of ordinary noise.

Steel inspection- A municipal and/or engineers inspection of the concrete foundation wall, conducted before concrete is poured into the foundation panels. Done to insure that the rebar (reinforcing bar), rebar nets, void material, beam pocket plates, and basement window bucks are installed and wrapped with rebar and complies with the foundation plan.

Step flashing- Flashing application method used where a vertical surface meets a sloping roof plane. 6" X 6" galvanized metal bent at a 90 degree angle, and installed beneath siding and over the top of shingles. Each piece overlaps the one beneath it the entire length of the sloping roof (step by step).

Stick built- A house built without prefabricated parts. Also called conventional building.

Stile- An upright framing member in a panel door.

Stool- The flat molding fitted over the window sill between jambs and contacting the bottom rail of the lower sash. Also another name for toilet.

Stop box- Normally a cast iron pipe with a lid (@ 5" in diameter) that is placed vertically into the ground, situated near the water tap in the yard, and where a water cut-off valve to the home is located (underground). A long pole with a special end is inserted into the curb stop to turn off/on the water.

Stop Order- A formal, written notification to a contractor to discontinue some or all work on a project for reasons such as safety violations, defective materials or workmanship, or cancellation of the contract.

Stops- Moldings along the inner edges of a door or window frame. Also valves used to shut off water to a fixture.

Stop valve- A device installed in a water supply line, usually near a fixture, that permits an individual to shut off the water supply to one fixture without interrupting service to the rest of the system.

Storm sash or storm window-. An extra window usually placed outside of an existing one, as additional protection against cold weather.

Storm sewer- A sewer system designed to collect storm water and is separated from the waste water system.

Story- That part of a building between any floor or between the floor and roof.

Strike- The plate on a door frame that engages a latch or dead bolt.

String, stringer- A timber or other support for cross members in floors or ceilings. In stairs, the supporting member for stair treads. Usually a 2 X 12 inch plank notched to receive the treads

Strip flooring- Wood flooring consisting of narrow, matched strips.

Structural floor- A framed lumber floor that is installed as a basement floor *instead* of concrete. This is done on very expansive soils.

Stub, stubbed- To push through.

Stucco- Refers to an outside plaster finish made with Portland cement as its base.

Stud- A vertical wood framing member, also referred to as a wall stud, attached to the horizontal sole plate below and the top plate above. Normally 2 X 4's or 2 X 6's, 8' long (sometimes 92 5/8"). One of a series of wood or metal vertical structural members placed as supporting elements in walls and partitions.

Stud framing- A building method that distributes structural loads to each of a series of relatively lightweight studs. Contrasts with post-and-beam.

Stud shoe- A metal, structural bracket that reinforces a vertical stud. Used on an outside bearing wall where holes are drilled to accommodate a plumbing waste line.

Subfloor- The framing components of a floor to include the sill plate, floor joists, and deck sheeting over which a finish floor is to be laid.

Sump- Pit or large plastic bucket/barrel inside the home designed to collect ground water from a perimeter drain system.

Sump pump- A submersible pump in a sump pit that pumps any excess ground water to the outside of the home.

Suspended ceiling- A ceiling system supported by hanging it from the overhead structural framing.

Sway brace- Metal straps or wood blocks installed diagonally on the inside of a wall from bottom to top plate, to prevent the wall from twisting, racking, or falling over "domino" fashion.

Switch- A device that completes or disconnects an electrical circuit.

Valley- The "V" shaped area of a roof where two sloping roofs meet. Water drains off the roof at the valleys.

Valley flashing- Sheet metal that lays in the "V" area of a roof valley.

Valuation- An inspection carried out for the benefit of the mortgage lender to ascertain if a property is a good security for a loan.

Valuation fee- Th fee paid by the prospective borrower for the lender's inspection of the property. Normally paid upon loan application.

Vapor barrier- A building product installed on exterior walls and ceilings under the drywall and on the warm side of the insulation. It is used to retard the movement of water vapor into walls and prevent condensation within them. Normally, polyethylene plastic sheeting is used.

Variable rate- An interest rate that will vary over the term of the loan.

Veneer- Extremely thin sheets of wood. Also a thin slice of wood or brick or stone covering a framed wall.

Vent- A pipe or duct which allows the flow of air and gasses to the outside. Also, another word for the moving glass part of a window sash, i.e. window vent.

Weep holes- Small holes in storm window frames that allow moisture to escape.

Whole house fan- A fan designed to move air through and out of a home and normally installed in the ceiling.

Wind bracing- Metal straps or wood blocks installed diagonally on the inside of a wall from bottom to top plate, to prevent the wall from twisting, racking, or falling over "domino" fashion.

Window buck- Square or rectangular box that is installed within a concrete foundation or block wall. A window will eventually be installed in this "buck" during the siding stage of construction

Window frame- The stationary part of a window unit; window sash fits into the window frame.

Window sash- The operating or movable part of a window; the sash is made of window panes and their border.

Wire nut- A plastic device used to connect bare wires together.